Elliptic
Problem Solvers II

Academic Press Rapid Manuscript Reproduction

Proceedings of the Elliptic Problem Solvers Conference
Sponsored by the Mathematics and Mechanics Branch,
Office of Naval Research, Held in Monterey, California,
January 10–12, 1983

Elliptic Problem Solvers II

Edited by

Garrett Birkhoff
Department of Mathematics
Harvard University
Cambridge, Massachusetts

Arthur Schoenstadt
Department of Mathematics
Naval Postgraduate School
Monterey, California

1984

ACADEMIC PRESS, INC.
(Harcourt Brace Jovanovich, Publishers)
Orlando San Diego San Francisco New York London
Toronto Montreal Sydney Tokyo São Paulo

ACADEMIC PRESS, INC.
Orlando, Florida 32887

United Kingdom Edition published by
ACADEMIC PRESS, INC. (LONDON) LTD.
24/28 Oval Road, London NW1 7DX

Main entry under title:

Elliptic problem solvers II.

Includes index.
1. Differential equations, Elliptic--Numerical solutions--Congresses. I. Birkhoff, Garrett, Date II. Schoenstadt, Arthur
QA377.E533 1984 515.3'53 83-25664
ISBN 0-12-100560-7 (alk. paper)

PRINTED IN THE UNITED STATES OF AMERICA

84 85 86 87 9 8 7 6 5 4 3 2 1

Contents

*Keynote paper.

V
Advances in Modeling and Physical Applications

Contributors

Numbers in parentheses indicate the pages on which the authors' contributions begin.

O. Axelsson (219), *Department of Mathematics, Catholic University, Toernooiveld, Nijmegen, The Netherlands*

Ivo Babuška (345), *Institute for Physical Science and Technology, University of Maryland, College Park, Maryland 20742*

Randolph E. Bank (493), *Computer Science Department, University of California, San Diego, La Jolla, California 92093*

Alvin Bayliss (233), *Exxon Research and Engineering Company, Linden, New Jersey 07036*

D. Bégis (23), *INRIA, 78153 Rocquencourt, France*

Garrett Birkhoff (427), *Department of Mathematics, Harvard University, Cambridge, Massachusetts 02138*

Petter E. Bjørstad (245, 531), *Det Norske Veritas, FDIV, 1322 Høvik, Norway*

James L. Blue (521), *National Bureau of Standards, Washington, D.C. 20234*

Ronald F. Boisvert (35), *Center for Applied Mathematics, National Bureau of Standards, Washington, D.C. 20234*

John F. Brophy (427), *Mathematics Department, Purdue University, West Lafayette, Indiana 46205*

W. M. Coughran, Jr. (531), *AT&T Bell Laboratories, Murray Hill, New Jersey 07974*

Joel E. Dendy, Jr. (439), *Theoretical Division, Los Alamos National Laboratory, Los Alamos, New Mexico 87545*

Q. V. Dihn (395), *AMD/BA, 92214 St. Cloud, France*

Craig C. Douglas (453), *Department of Computer Science, Yale University, New Haven, Connecticut 06520*

Wayne R. Dyksen (467), *Department of Mathematics and Computer Science, Calvin College, Grand Rapids, Michigan 49506*

L. W. Ehrlich (257), *Milton S. Eisenhower Research Center, Johns Hopkins University Applied Physics Laboratory, Laurel, Maryland 20707*

Stanley C. Eisenstat (45), *Research Center for Scientific Computation, Department of Computer Science, Yale University, New Haven, Connecticut 06520*

Howard C. Elman (45, 271), *Research Center for Scientific Computation, Department of Computer Science, Yale University, New Haven, Connecticut 06520*

David J. Evans (285), *Department of Computer Studies, Loughborough University of Technology, Loughborough, Leicestershire, United Kingdom*

George J. Fix (379), *Department of Mathematics, Carnegie–Mellon University, Pittsburgh, Pennsylvania 15213*

Kelton Flinn (481), *Department of Applied Mathematics and Computer Science, University of Virginia, Charlottesville, Virginia 22904*

Dennis Gannon (93), *Department of Computer Sciences, Purdue University, West Lafayette, Indiana 47907*

Eugene C. Gartland, Jr. (545), *Department of Mathematics, Southern Methodist University, Dallas, Texas 75275*

R. Glowinski (395), *Université Pierre et Marie Curie, 75230 Paris Cedex 05, France, and INRIA, 78153 Rocquencourt, France*

Charles I. Goldstein (233), *Applied Mathematics Department, Brookhaven National Laboratory, Upton, New York 11973*

Joseph F. Grcar (557), *Applied Mathematics Division, Sandia National Laboratories, Livermore, California 94550*

H. S. Greenside[1] (531), *AT&T Bell Laboratories, Murray Hill, New Jersey 07974*

F. Hecht (23), *INRIA, 78153 Rocquencourt, France*

James M. Hyman (301), *Center for Nonlinear Studies, Los Alamos National Laboratory, Los Alamos, New Mexico 87545*

Kang C. Jea[2] (323), *Center for Numerical Analysis, The University of Texas at Austin, Austin, Texas 78712*

Joseph W. Jerome (507), *Mathematics Department, Northwestern University, Evanston, Illinois 60201*

Lennart Johnsson (105), *Department of Computer Science, Yale University, New Haven, Connecticut 06520*

Thomas L. Jordan (127), *Computing Division, Los Alamos National Laboratory, Los Alamos, New Mexico 87545*

David R. Kincaid (53, 323), *Center for Numerical Analysis, The University of Texas at Austin, Austin, Texas 78712*

H. T. Kung (141), *Department of Computer Science, Carnegie–Mellon University, Pittsburgh, Pennsylvania 15213*

W. T. Lin (141), *Department of Electrical Engineering, Carnegie–Mellon University, Pittsburgh, Pennsylvania 15213*

R. E. Lynch (427), *Computer Science and Mathematics Departments, Purdue University, West Lafayette, Indiana 47906*

Lois Mansfield (481), *Department of Applied Mathematics and Computer Science, University of Virginia, Charlottesville, Virginia 22904*

Thomas A. Manteuffel (301), *Computing Division, Los Alamos National Laboratory, Los Alamos, New Mexico 87545*

[1]Present address: Princeton Plasma Physics Laboratory, Princeton University, Princeton, New Jersey 08544.

[2]Present address: Department of Mathematics, Fu Jen University, Hsinchuang, 242 Taipei, Taiwan.

Reginald E. Mitchell (557), *Combustion Research Division, Sandia National Laboratories, Livermore, California 94550*

J. Périaux (395), *AMD/BA, 92214 St. Cloud, France*

Wlodzimierz Proskurowski (65), *Department of Mathematics, University of Southern California, Los Angeles, California 90089-1113*

Edna Reiter (161), *California State University, Hayward, Hayward, California 94542*

Werner C. Rheinboldt (345), *Department of Mathematics and Statistics, University of Pittsburgh, Pittsburgh, Pennsylvania 15260*

John R. Rice (3, 467), *Division of Mathematical Sciences, Department of Computer Science, Purdue University, West Lafayette, Indiana 47907*

Garry Rodrigue (161), *Computing Research Group, Lawrence Livermore National Laboratory, University of California, Livermore, California 94550*

Donald J. Rose (493, 531), *AT&T Bell Laboratories, Murray Hill, New Jersey 07974*

A. H. Sameh (175), *Department of Computer Science, University of Illinois at Urbana–Champaign, Urbana, Illinois 61801*

Robert Schreiber (187), *Department of Computer Science, Stanford University, Stanford, California 94305*

N. L. Schryer (531), *AT&T Bell Laboratories, Murray Hill, New Jersey 07974*

Martin H. Schultz (45, 77), *Research Center for Scientific Computation, Department of Computer Science, Yale University, New Haven, Connecticut 06520*

Andrew H. Sherman (45), *Exxon Production Research Company, Houston, Texas 77002*

Mitchell D. Smooke (557), *Combustion Chemistry Division, Sandia National Laboratories, Livermore, California 94550*

Eli Turkel (233), *Tel Aviv University, Tel Aviv, Israel*

John Van Rosendale (93), *ICASE, NASA Langley Research Center, Hampton, Virginia 23665*

Richard S. Varga (197), *Institute for Computational Mathematics, Department of Mathematical Sciences, Kent State University, Kent, Ohio 44242*

M. Vidrascu (23), *INRIA, 78153 Rocquencourt, France*

Eugene L. Wachspress[3] (315), *Department of Mathematics, University of Tennessee, Knoxville, Tennessee 37916*

Olof B. Widlund (245), *Courant Institute of Mathematical Sciences, New York University, New York, New York 10012*

Charles L. Wilson (521), *National Bureau of Standards, Washington, D.C. 20234*

David M. Young (53, 323), *Center for Numerical Analysis, The University of Texas at Austin, Austin, Texas 78712*

[3]Present address: Knolls Atomic Power Laboratory, Schenectady, New York 12301.

Preface

This volume contains the proceedings of the Elliptic Problem Solvers Conference held at the Naval Postgraduate School in Monterey, California, from January 10 to 12, 1983. The conference was sponsored by the Office of Naval Research, through its Mathematics and Mechanics Branch, and was held in conjunction with Garrett Birkhoff's occupancy of the ONR Chair in Applied Mathematics at the Naval Postgraduate School. The program committee consisted of Garrett Birkhoff, John Rice, Donald Rose, Arthur Schoenstadt, Martin Schultz, Richard Varga, and David Young.

These papers, which were presented at the conference, deal with various aspects of the numerical solution of elliptic boundary value problems—a topic central to current large-scale scientific computing. The papers describe continuing advances in the "state of the art" of solving elliptic problems. Especially timely are the discussions of the available software packages, ELLPACK, ITPACK, MODULEF, and the Yale Sparse Matrix Package; the possibilities offered by vector and parallel computers; and the numerical modeling of semiconductors.

This volume is, to some extent, a sequel to a volume "Elliptic Problem Solvers," edited by Martin Schultz and published by Academic Press in 1981, but it differs from that predecessor in that the papers are grouped roughly by topic, as they were at the conference itself, rather than alphabetically. We hope this will structure its contents in a way that is convenient to the reader.

We wish to express our sincere gratitude to the many people whose efforts helped to make the conference a success, especially those who helped arrange the sponsorship of the conference, including Dr. Charles Holland of ONR, Dr. Richard Lau of ONR (Pasadena), and Dean William Tolles, Dean of Research at NPS. We also thank Bob Lande for his assistance in preparing these proceedings. Last, but certainly not least, we thank Pat Paulson and Alesia Mafnas, without whose help the organization of the Elliptic Problem Solvers Conference would have been impossible.

I
Software Packages

BUILDING ELLIPTIC PROBLEM SOLVERS WITH ELLPACK

John R. Rice[1]

I. INTRODUCTION

The origins and general development of the ELLPACK system are given in [Rice, 1981] and a complete documentation is given in the forthcomming book [Rice and Boisvert, 1983]; a concise summary of ELLPACK is given in the next section. The references also contain all the important publications involving the ELLPACK system.

The thesis of this paper is:

Embedding a PDE solving system into a general purpose language allows one easily to build PDE solvers for a much broader range of applications and, often, to achieve high efficiency.

This thesis is supported here using ELLPACK which is embedded in Fortran. ELLPACK is designed to solve linear, second order elliptic problems in two or three dimensions. The geometry in two dimensions can be general, but must be

[1]Mathematical Sciences, Purdue University, West Lafayette, IN 47907

This work supported in part by Department of Energy contract DE-AC02-81ER10997

ISBN 0-12-100560-7

rectangular in three dimensions. In this paper we show how it can be used effectively to solve broad classes of nonlinear elliptic problems, initial value problems and simultaneous nonlinear elliptic problems. Of particular note is the use of Fréchet derivatives and MACSYMA to provide a simple and effective implementation of Newton's method for nonlinear elliptic problems. In [Rice and Boisvert, 1983] it is shown how ELLPACK can be used to solve elliptic problems with interface conditions and in [Dyksen, 1982] it is shown how ELLPACK can be used to solve elliptic problems on three dimensional cylindrical domains.

This study demonstrates the flexibility and programming power of ELLPACK. We believe this power comes from two sources: (1) The use of a natural, very high level language targeted for elliptic problems, (2) The language is built on a foundation of a large number of good elliptic problem solving modules gathered from many sources.

II. THE ELLPACK SYSTEM

The ELLPACK system is a problem statement language designed to handle linear, second order elliptic problems. The principal facilities within ELLPACK are:

2D rectangular domains	-	many problem solving modules
2D general domains	-	several problem solving modules
3D rectangular domains	-	several problem solving modules

These facilities are supported by procedures to:

Initialize solutions and unknowns in several ways
Homogenize the boundary condition in certain cases
Aid in the analysis of methods

All these facilities are embedded in Fortran which provides flexibility in combining modules to solve particular problems. Figure 1 shows a simple ELLPACK program and part of its output; this is an instance where the problem is just stated along with the name of the method (modules) to be used to solve it. No Fortran programming is involved in such a simple application.

Figure 2 presents an "instant ELLPACK guide" which concisely summarizes ELLPACK.

```
*                    EXAMPLE ELLPACK PROGRAM
OPTIONS.             TIME  $  MEMORY
*
EQUATION.            UXX  +  UYY  +  3.0*UX  -  4.0*U  =  EXP(X+Y)*SIN(PI*X)
*
BOUNDARY.            U = 0.0                    ON  X = 0.0
                     U = SIN(PI*X) - X/2.0      ON  Y =-1.0
                     U = Y/2.0                  ON  X = 1.0
                     U = X                      ON  Y = 2.0
*
GRID.                6 X POINTS  $  6 Y POINTS
*
DISCRETIZATION.         5 POINT STAR
SOLUTION.               LINPACK BAND
*
OUTPUT.              TABLE(U)  $  PLOT(U)
*
END.
```

Figure 1. A simple ELLPACK program to solve an elliptic problem using ordinary finite differences and band Gauss elimination.

III. NONLINEAR PROBLEMS AND PICARD'S METHOD

Picard's method is also known as simple iteration, fixed point iteration or successive substitutions and is very simple to implement in ELLPACK. Consider the equation

$$u_{xx} + u_{yy} = u^2(x^2 + y^2)e^{-xy}$$

which has $u(x,y) = e^{xy}$. The ELLPACK program in Figure 3 makes an initial guess of $u = 1.0$ and then iterates six times. The errors in the iteration are as follows:

Iteration	0	1	2	3	4	5	6
Error	1.72	.037	.0025	.00025	.000060	.000073	.000072

The error of .00007 is the discretization error for this problem. If the iteration is initialized by

TRIPLE. SET U BY BLENDING

a much better starting point is obtained and only 3 iterations are required for convergence. Initialization using an Hermite bicubic interpolant requires 5 iterations to converge to the discretization error.

INSTANT ELLPACK GUIDE: BASIC FACILITIES

*** * DRAFT: December 1, 1982 * ***

ELLPACK is an extension of Fortran that allows one to simply state and solve elliptic partial differential equations in 2-dimensions on general domains or in 3-dimensions on rectangular domains. For example, the elliptic problem

$$\begin{aligned} u_{xx} + u_{yy} + 3u_x - 4u &= \exp(x+y)\sin(\pi x) && 0<x<1,\ -1<y<2 \\ u &= 0 && x=0,\ -1<y<2 \\ u &= x && 0<x<1,\ y=2 \\ u &= y/2 && x=1,\ -1<y<2 \\ u &= \sin(\pi x) - x/2 && 0<x<1,\ y=-1 \end{aligned}$$

is solved using ordinary finite differences and Gauss elimination by the ELLPACK program

```
OPTIONS.  TIME $ MEMORY
EQUATION. UXX + UYY + 3.*UX - 4.*U = EXP(X+Y)*SIN(PI*X)
BOUNDARY.
          U = 0.0                ON X =  0.
          U = X                  ON Y =  2.
          U = Y/2.               ON X =  1.
          U = SIN(PI*X) - X/2.   ON Y = -1.
GRID.     6 X POINTS $ 12 Y POINTS
*
DISCRETIZATION.  5 POINT STAR
INDEXING.        AS IS
SOLUTION.        LINPACK BAND
*
OUTPUT.          TABLE(U) $ PLOT(U)
END.
```

GENERAL CONVENTIONS

The independent variables are X, Y (and Z in 3-dimensions), the solution is U with derivatives *UX*, *UY*, *UXX*, etc. If the solution is used as a function (not seen in the above example), then *U(X,Y)*, *UX(X,Y)*, etc. provide the corresponding values at *(X,Y)*. All expressions are in Fortran. The keywords OPTIONS, GRID, etc. start in column 1, otherwise column placement is optional. A long expression may be continued to the next line by ending the line with an ampersand &. The normal structure of an ELLPACK program is

OPTIONS.	Specifies options of the run.
EQUATIONS.	Specifies the elliptic PDE.
BOUNDARY.	Gives domain and boundary conditions.
GRID.	Specifies lines of a rectangular grid.

problem solving, output and Fortran statements

SUBPROGRAMS.	Fortran functions and subroutines used
END.	

OTHER DOMAIN DEFINITION SEGMENTS

HOLE.	Specifies a hole in the domain with boundary conditions. Must follow BOUNDARY segment.
ARC.	Specifies a slit or interface in the domain. Must follow BOUNDARY segment.

The domain and boundary conditions are specified by a list of items like

condition ON *piece*

where *condition* is one of

PERIODIC

*A**UX + *B**UY + *C**U = *D*

with *A*, *B*, *C* and *D* Fortran expressions. Terms with zero coefficients may be omitted. The domain boundary is given in parameterized pieces so that *piece* appears as

X = *Expression(t)*, Y = *Expression(t)* FOR *t* = a TO b

Here, *t* is a parameter with X and Y given as expressions; a and b must be constants. The pieces are given in counter-clockwise order and must join up to form a closed curve. There is a special simple form for lines illustrated by

U = X-Y+2 ON LINE 0.0 TO 1.1 TO 3.6,2.4 TO 1,-1 TO 0,0

The points to be joined by straight lines are listed in order. Finally, there is an even simpler special form for rectangles illustrated by the example above.
The grid is specified by

k variable POINTS *point-list*

where *k* is normally a constant, *variable* is X, Y, or Z and *point-list* is the set of values defining the grid lines. For uniformly spaced grids *point-list* may be *a* TO *b* where *a* and *b* are constants; for uniformly spaced grids on rectangles the *point-list* may be omitted entirely.

BASIC PROBLEM SOLVING MODULES

Most numerical methods consist of the three steps seen in the above example. The basic facilities in ELLPACK are

DISCRETIZATION:

5 POINT STAR	Ordinary finite differences
SPLINE GALERKIN	Galerkin method with piecewise polynomials, rectangles only
HERMITE COLLOCATION	Collocation with Hermite bi-cubic polynomials on rectangles
HODIE ACF	High order finite differences for $a(x,y)u_{xx} + c(x,y)u_{yy} + f(x,y)u$ on rectangles
COLLOCATION	Collocation with Hermite bi-cubics

INDEXING:

AS IS	Ordering not changed
RED-BLACK	Checker board ordering
NESTED DISSECTION	Nested dissection ordering
MINIMUM DEGREE	Minimum degree ordering

SOLUTION:

BAND GE	Band Gauss elimination
LINPACK SPD BAND	Cholesky decomposition for symmetric, positive definite band matrices
JACOBI CG	Jacobi iteration with conjugate gradient acceleration
SOR	SOR iteration
SPARSE	Sparse matrix method

TRIPLE:

FFT-9 POINT	Fast Fourier Transform for rectangles

There are more than 25 other problem solving modules in the complete ELLPACK system.

The ELLPACK system will be available in the spring of 1983 from John R. Rice, Math Science 428, Purdue University.

Figure 2. Instant ELLPACK guide. (A) Basic facilities and modules

OTHER ELLPACK SEGMENTS

OPTIONS: (Fortran option variables given where available)

MAX GRID = NX, NY or NX, NY, NZ to set max grid size for variable grids.
INTERPOLATION = QUADRATICS (default)
= SPLINES
LEVEL = 0 to 5 governs amount of output (I1LEVL).
MEMORY produces memory use estimate.
NO EXECUTION suppresses execution.
PAGE = 0 to 2 sets page advances from none to 1 before each module (I1PAGE).
SELF-ADJOINT = logical switch (L1SELF).
TIME produces time use estimate (L1TIME).
MAX WORKSPACE = Set dimension I1MWRK of R1WORK to indicated value.
MIN WORKSPACE = Dimension I1MWRK of R1WORK is at least indicated value.
CLOCKWISE for boundary pieces given in clockwise order.

OUTPUT:

MAX, RMS, NORM are the same, used as MAX(f) or MAX(f,grid) where f is any Fortran function with 2 (or 3) arguments and grid is like 20, 20.
PLOT gives contour plot, arguments are as in MAX.
PLOT DOMAIN displays domain with grid lines.
TABLE gives table of values, arguments are as in MAX.
SUMMARY = MAX $ TABLE with arguments f or f, grid.

FORTRAN: Fortran code follows.

DECLARATIONS: Fortran declarations follow, must appear at start.

ADVANCED ELLPACK SEGMENTS

GLOBAL: Inserts Fortran declarations in all ELLPACK subprograms Q1PCOE, Q1PRHS, Q1BCOE, Q1BRHS, Q1BCOR.

PROCEDURE: Provides other useful facilities

EIGENVALUES: Computes eigenvalues of discretization matrix.
DISPLAY MATRIX PATTERN: Gives pattern of non-zeros in discretization matrix.
INITIALIZE UNKNOWN FOR FINITE DIFFERENCES (U = *name*)
REMOVE BLENDED BC: Subtract blending function interpolant of boundary conditions from U.
REMOVE BICUBIC BC: Subtract bi-cubic interpolant of boundary conditions from U.
REMOVE (V=*name*) Subtract named function from U.

OPTIONS:

OLDU = k produces functions UOLD(X,Y,J), J=1,2,...,k of previous U(X,Y) functions, also have UXOLD, UYOLD, UXXOLD, etc.
Array dimension control allows one to set Fortran array dimensions independent of what the ELLPACK system thinks they should be.
Problem characteristics may be set by assigning .TRUE. or .FALSE. to the logical variables L1LAPL, L1CSTC, L1POIS, L1HMEQ.
TABLE followed by PROBLEM, EQUATIONS, INDEXES, UNKNOWN, DOMAIN or BOUNDARY provides data about internal ELLPACK variables.

Preprocessor Variables: Certain preprocessor variables can be used in the ELLPACK program, some are $I1KBAN, $I1KWRK, $I1MEND, $I1MIXD, $I1MNCO, $I1MXEQ, $I1MXPT, $I1NBND, $I1MXPT, $I1NGRX, $I1NGRY, $I1NGRZ.

TRIPLE: The following triples directly set U(X,Y) for use in initializing interations or homogenizing boundary conditions. One can use U(X,Y), UX(X,Y), UXX(X,Y), etc. after these triples.

SET U BY BLENDING Use blending functions from interpolating boundary conditions.
SET U BY BICUBICS Use Hermite bicubics from interpolating boundary conditions.
SET (U = *name*) Define U from function named.

REST OF PROBLEM SOLVING MODULES

DISCRETIZATIONS: (for rectangles unless otherwise noted)

HODIE HELMHOLTZ	High order finite differences for $u_{xx} + u_{yy} + f(x,y)u = g(x,y)$
HODIE ACDEF	High order finite differences for variable coefficient PDE $au_{xx} + cu_{yy} + du_x + eu_y + fu = g$
7 POINT 3D	Ordinary finite differences in 3D
INTERIOR COLLOCATION	Variant of HERMITE COLLOCATION which provides better performance for uncoupled boundary conditions

INDEXING:

HERMITE COLLORDER	reorders collocation matrix to have non-zero diagonal
REVERSE CUTHILL MCKEE	reorders equations to possibly help envelope or band solvers

SOLUTION: (GE means Gauss elimination)

BAND GE NO PIVOTING	Band GE without pivoting
LINPACK BAND	LINPACK routine for band GE
JACOBI SI	Jacobi iteration with semi-iteration
SYMMETRIC SOR CG	Symmetric SOR with conjugate gradient acceleration
SYMMETRIC SOR SI	Symmetric SOR with semi-iteration
REDUCED SYSTEM CG	Interation on RED-BLACK with conjugate gradient acceleration
REDUCED SYSTEM SI	Interation on RED-BLACK with semi-iteration
SPARSE GE PIVOTING	GE, sparse matrices with pivoting
SPARSE COMPRESSED	GE, sparse matrices with storage compression
SPARSE FAST	GE, sparse matrices, fast version
SPARSE SYMMETRIC	Symmetric version of SPARSE FAST
ENVELOPE	GE, system in envelope form

TRIPLE: (for rectangles unless otherwise noted)

FISHPAK HELMHOLTZ	Fast solver using ordinary finite differences for $u_{xx} + u_{yy} + \lambda u = f(x,y)$
HODIE 27 POINT 3D	Fast solver for 6th order finite differences for Poisson problem on a cube
MARCHING ALGORITHM	Ordinary finite differences with generalized marching algorithm for separable, self-adjoint problem
DYAKANOV CG	Ordinary finite differences with preconditioned CG iteration for separable, self-adjoint problem
DYAKANOV CG 4	Richardson extrapolation for DYAKANOV CG to obtain fourth order accuracy
P2C0 TRIANGLES	Galerkin method with 6-nodes quadratic triangular elements, **only** for non-rectangular 2D domains.
CAPACITANCE MATRIX	Capacitance matrix method for Poisson problem on non-rectangular 2D domains
MULTIGRID MG00	Multigrid method for self-adjoint problem

Figure 2. Continued. (B) Advanced and other facilities.

```
*                PICARD ITERATION FOR NONLINEAR POISSON PROBLEM
EQUATION.        UXX  +  UYY  =  (X**2 + Y**2) * EXP(-X*Y) * U(X,Y)**2
BOUNDARY.        U = 1.0       ON  X = 0.0
                               ON  Y = 0.0
                 U = EXP(Y)    ON  X = 1.0
                 U = EXP(X)    ON  Y = 1.0
GRID.            7  X POINTS  $  7  Y POINTS
OPTION.          MEMORY    $ TIME
*                    INITIALIZE U
TRIPLE.          SET (U = ONE )
OUTPUT.          MAX(ERROR)
FORTRAN.
                 DO 50 I = 1,8
DISCRETIZATION.            5 POINT STAR
SOLUTION.                  BAND GE
OUTPUT.                    MAX(ERROR)
FORTRAN.
   50            CONTINUE
SUBPROGRAMS.
         FUNCTION TRUE(X,Y)
            TRUE = EXP(X*Y)
         RETURN
         END
         FUNCTION ONE(X,Y)
            ONE = 1.0
         RETURN
         END
END.
```

Figure 3. An ELLPACK program to apply Picard iteration to an equation with $u^2(x,y)$ on the right.

IV. NONLINEAR PROBLEMS AND NEWTON'S METHOD

Picard's method is simple and sometimes very effective, but Newton's method is usually superior. One visualizes the nonlinear elliptic PDE as simply $F(u) = 0$, expands F at the point u_0, drops higher order terms in $u_1 - u_0$ to obtain the following equation for the Newton iterate u_1.

$$F(u_0) + F'(u_0)(u_1 - u_0) = 0$$

For an operator equation, the term $F'(u_0)(u_1 - u_0)$ is the Frechet derivative at u_0 which is a linear operator $L(u_0, u_1)$. Thus the Newton iterate $u_1(x,y)$ satisfies the linear PDE

$$F(u_0) + L(u_0, u_1) = 0$$

This is the linear PDE that one specifies and solves iteratively in ELLPACK. The algorithm is

Compute the Frechet derivative $L(u,v)$
Guess $u^{(0)}$
For $k = 0$ to limit do
    Solve $L(u^{(k)},u^{(k+1)}) = -F(u^{(k)})$
    Exit if convergence test is passed
End-loop
Print results

We consider the simple equation

$$F(u) = u_{xx} + u^2 u_{yy} - e^u - f = 0 \qquad 0 \le x,y \le 1$$

where f is chosen to make the true solution $u(x,y) = sin(x)\,cos(y)$. Set $\delta = u^{(k+1)} - u^{(k)}$ in the Newton expansion to obtain

$$\begin{aligned} &u_{xx}^{(k)} + (u^{(k)})^2 u_{yy}^{(k)} - e^{u^{(k)}} - f + \delta_{xx} + (u^{(k)})^2 \delta_{yy} + (2u_{yy}^{(k)} - e^{u^{(k)}})\delta \\ &= F(u^{(k)}) + L(u^{(k)},u^{(k)} + \delta) = 0 \end{aligned}$$

In the ELLPACK program both $u^{(k)}$ and $u^{(k+1)}$ are denoted by u. The $u^{(k)}$ terms in coefficients or the right side are evaluated before the PDE is solved (and hence use the previous $u(x,y) = u^{(k)}$); then the PDE is solved for $u(x,y)$ which is the next iterate $u^{(k+1)}$. Figure 4 shows a program for this problem using the initial guess $u(x,y) = 0$ and Hermite collocation plus band Gauss elimination to solve the linearized problem. The iteration converges quite rapidly as seen from the following table of the maximum errors

Iteration	Error in u(x,y)
0	.70807
1	.030
2	.00064
3	.0000089
4	.0000035
5	.0000036

If the iteration is initialized by blending it converges to the discretization error in two iterations.

```
                    NEWTONS METHOD TO SOLVE

                         2
                UXX + U  *UYY = EXP(U) + F(X,Y)

*        THE PDE FOR THE NEW U(X,Y) OBTAINED BY LINEARIZATION
EQUATION.
         UXX + (U(X,Y)**2)*UYY + (2.*U(X,Y)*UYY(X,Y)-EXP(U(X,Y)))*U =   &
          2*(U(X,Y)**2)*UYY(X,Y) - EXP(U(X,Y))*(U(X,Y)-1.) + F(X,Y)

BOUNDARY.
                    U = TRUE(X,Y) ON X = 0.
                                  ON X = 1.
                                  ON Y = 0.
                                  ON Y = 1.
GRID.    5 X POINTS $ 5 Y POINTS
*                   INITIALIZE THE NEWTON ITERATION BY GUESSU = 0
TRIPLE.             SET( U = GUESSU )
OUTPUT.             MAX(ERROR)
FORTRAN.
                    NITERS = 5
                    DO 10 NITER = 1, NITERS
DISCRETIZATION.         HERMITE COLLOCATION
SOLUTION.               BAND GE
OUTPUT.                 MAX(ERROR)
FORTRAN.
   10               CONTINUE
SUBPROGRAMS.
      FUNCTION F(X,Y)
C         F IS CHOSEN TO MAKE THE TRUE SOLUTION SIN(X)*COS(Y)
        TRUE   = SIN(X)*COS(Y)
        TRUEXX = -TRUE
        TRUEYY = -TRUE
        F = TRUEXX + TRUE**2*TRUEYY + (2.*TRUE*TRUEYY-EXP(TRUE))*TRUE
     $      - 2.*TRUE**2*TRUEYY + EXP(TRUE)*(TRUE-1.)
        RETURN
      END
      FUNCTION GUESSU(X,Y)
        GUESSU = 0.
      RETURN
      END
      FUNCTION TRUE(X,Y)
        TRUE = SIN(X)*COS(Y)
      RETURN
      END
END.
```

Figure 4. The ELLPACK program to apply Newton's method to a simple problem.

A classical elliptic problem is the Plateau problem

$$(1+u_x^2)u_{xx} - 2\, u_x u_y u_{xy} + (1+u_y^2)u_{yy} = 0$$

Dirichlet boundary conditions.

Newton's method in ELLPACK for this problem is shown in Figure 5, further discussions of the solution of this problem is given in [Birkhoff and Lynch, 1983]. Newton's method has converged rapidly to compute a variety of surfaces for this problem.

```
*                        THE PLATEAU PROBLEM

*          THE PDE FOR THE NEW U(X,Y) OBTAINED BY LINEARIZATION
EQUATION. (1.+UY(X,Y)**2) UXX  +  (1.+UX(X,Y)**2) UYY          &
              - 2.*UX(X,Y)*UY(X,Y)   UXY                        &
          + 2.*(UX(X,Y)*UYY(X,Y) - UY(X,Y)*UXY(X,Y)) UX         &
          + 2.*(UY(X,Y)*UXX(X,Y) - UX(X,Y)*UXY(X,Y)) UY         &
     =      2.*(UX(X,Y)*UYY(X,Y) - UY(X,Y)*UXY(X,Y))*UX(X,Y)    &
          + 2.*(UY(X,Y)*UXX(X,Y) - UX(X,Y)*UXY(X,Y))*UY(X,Y)

BOUNDARY.      U = BOUND(X,Y) ON Y = 0.0
               U = BOUND(X,Y) ON Y = 1.0
               U = BOUND(X,Y) ON X = 1.0
               U = BOUND(X,Y) ON X = 0.0

GRID.          5 X POINTS  $  5 Y POINTS

TRIPLE.        SET ( U = GESSU )
FORTRAN.
               DO 100 IT = 1,5
DISCRETIZATION.         HERMITE COLLOCATION
SOLUTION.               BAND GE
FORTRAN.
  100          CONTINUE
OUTPUT.        PLOT(U)
SUBPROGRAMS.
      FUNCTION GESSU(X,Y)
        GESSU = X*Y
        RETURN
      END
      FUNCTION BOUND(X,Y)
        BOUND = SIN(X+AMAX1(.66,.1+Y**2))*EXP(X-Y)
        RETURN
      END
END.
```

Figure 5. The ELLPACK program to apply Newton's method to the Plateau problem.

As the PDE becomes more complicated, it becomes tedious to carry out the algebra for obtaining the Frechet derivative. For example, we have solved the problem

$$w(u)(u_{xx}+u_{yy})+w_x(u)u_x+w_y(u)u_y=f(x,y)$$

$$u_x=0 \quad \text{on} \quad x=0,1$$

$$u=b(x) \quad \text{on} \quad y=0,1$$

where $w(u)$ varies according to the application. With $a(u)=\sqrt{u_x^2+u_x^2}$ the physically meaningful cases are

$$w(u)=[a(u)]^{\alpha} \qquad w(u)=1/(\alpha+\beta a(u))$$

$$w(u)=e^{[a(u)/(\alpha+\beta a(u))]}/a(u) \qquad w(u)=\alpha\tanh(\beta a(u))/a(u)$$

We have found *MACSYMA* quite useful to carry out the algebra and, instead of showing the ELLPACK problem, we show in Figure 6 the *MACSYMA* program used to compute the Frechet derivative. Note that only the statements (C1) through (C8) depend on the particular PDE being solved; the rest of the *MACSYMA* program can be used for any PDE.

```
       VAXIMA 2.04
(C1)  /*****    LIST THE NONLINEAR PDE COEFFICIENTS    *****/
       A(U)  := W(U)$
(C2)  B(U)  := 0$
(C3)  C(U)  := W(U)$
(C4)  D(U)  := DIFF(W(U),X)$
(C5)  E(U)  := DIFF(W(U),Y)$
(C6)  F(U)  := 0$
(C7)  G(U)  := 0$
(C8)  W(U)  := SQRT ( DIFF(U,X)**2 + DIFF(U,Y)**2 )$

(C9)  /******   DEFINE DERIVATIVES OF U0 AND U1 = U0X, ETC.   *****/
        GRADEF(U0,X,U0X)$
(C10) GRADEF(U0X,X,U0XX)$
(C11) GRADEF(U0X,Y,U0XY)$
(C12) GRADEF(U0Y,X,U0XY)$
(C13) GRADEF(U0,Y,U0Y)$
(C14) GRADEF(U0Y,Y,U0YY)$
(C15) GRADEF(U1,X,U1X)$
(C16) GRADEF(U1X,X,U1XX)$
(C17) GRADEF(U1X,Y,U1XY)$
(C18) GRADEF(U1Y,X,U1XY)$
(C19) GRADEF(U1,Y,U1Y)$
(C20) GRADEF(U1Y,Y,U1YY)$

(C21) /*****    WRITE THE NONLINEAR PDE    *****/
PDE(U)  := A(U)*DIFF(U,X,2) + B(U)*DIFF(DIFF(U,X),Y) + C(U)*DIFF(U,Y,2)
          + D(U)*DIFF(U,X) + E(U)*DIFF(U,Y) + F(U)*U + G(U)$
(C22) /*****       DIFFERENTIATE AND COLLECT TERMS          *****/
        DERIVATIVE:DIFF(PDE(U0+EPS*(U1-U0)),EPS)$
(C23) TSERIES:PDE(U0) + EV(DERIVATIVE,EPS=0)$
(C24) RATSIMP(TSERIES)$
(C25) TSERIES:EXPAND(TSERIES)$
(C26) COEU1XX:COEFF(TSERIES,U1XX)$
(C27) COEU1XY:COEFF(TSERIES,U1XY)$
(C28) COEU1YY:COEFF(TSERIES,U1YY)$
(C29) COEU1X:COEFF(TSERIES,U1X)$
(C30) COEU1Y:COEFF(TSERIES,U1Y)$
(C31) COEU1:COEFF(TSERIES,U1)$
(C32) RS:(COEU1XX*U1XX + COEU1XY*U1XY + COEU1YY*U1YY + COEU1X*U1X +
        COEU1Y*U1Y + COEU1*U1)-TSERIES$
(C33) RATSIMP(RS)$
(C34) FORTRAN(COEU1XX  );
(C35) FORTRAN(COEU1XY  );
(C36) FORTRAN(COEU1YY  );
(C37) FORTRAN(COEU1X   );
(C38) FORTRAN(COEU1Y   );
(D38) ((2*U0Y**3+3*U0X**2*U0Y)*U0YY+U0XX*U0Y**3+2*U0X**3*U0XY)/(U0Y**2+U
     1     0X**2)
(C39) FORTRAN(COEU1    );
(C40) FORTRAN(RS       );
(D40) (2*U0Y**2+U0X**2)*U0YY+U0XX*U0Y**2+2*U0X*U0XY*U0Y+2*U0X**2*U0XX
```

Figure 6. The *MACSYMA* program to compute the Fréchet derivative needed in solving nonlinear PDEs in ELLPACK by Newton's method. For brevity, the output of only statements 38 and 40 is shown.

V. INITIAL VALUE PROBLEMS

Consider the initial value problem (t=time)

$$u_t = Lu + f \qquad u = u0(x,y) \text{ for } t=0$$

$$u = ubound(x,y) \text{ for } (x,y) \text{ on boundary}$$

where L is a linear elliptic operator; for example

$$Lu = 4y^2u_{xx} + u_{yy} + (2 + tan(x+y+t))u_y + u$$

For ELLPACK we use the method of lines in a nonstandard way: time is discretized explicitly and the "lines" are actually planes. Thus one solves boundary value problems at each time step. This approach is not new, it was widely used on analog computers and some felt it was superior to usual method of lines approach where space is discretized explicitly to obtain a system of ordinary differential equations.

The Crank-Nicolson discretization of t is applied to obtain

$$u(t)-u(t-\Delta t) = .5\Delta t\,[Lu(x,y,t) + Lu(x,y,t-\Delta t) + f(x,y,t) + f(x,y,t-\Delta t)]$$

This discretization is always stable, so large Δt values may be used. For each time step one solves the linear elliptic problem

$$\begin{aligned} &Lu(x,y,t) - (2/\Delta t)\,u(x,y,t) \\ &\quad = -(2/\Delta t)u(x,y,t-\Delta t) - Lu(x,y,t-\Delta t) - f(x,y,t) - f(x,y,t-\Delta t) \end{aligned}$$

Figure 7 gives an ELLPACK program to solve this problem for $1 \le t \le 2$, $0 \le x \le 4$, $0 \le y \le 1$ and f chosen so the true solution is $u(x,y,t) = \sin((x+y+t)/4)e^{-y^2-t}$.

A fourth order space discretization (Hermite collocation) is used so that a coarse spatial grid spacing $\Delta x = 4/n$, $\Delta y = 1/n$, can be used, we use $\Delta t = \Delta y^2/2$. The accuracy obtained is summarized as follows:

Maximum Relative Error Times 10,000

time	(x,y)-Grid 2x2	4x4.
1 + 1/8	1.9	.21
1 + 1/4	3.0	.13
1 + 3/8	2.9	.22
1 + 1/2	2.7	.20
1 + 5/8	2.4	.17
1 + 3/4	2.2	.23
1 + 7/8	2.0	.31
2	1.9	.88
2 + 1/4	1.3	6580
2 + 1/2	1.1	$1.7*10^5$

The dominant discretization error is spatial, if one redoes the computation with $\Delta t = \Delta y^2 * 2$, one achieves essentially the same accuracy. The fact that the 4x4 grid computation becomes unstable at t = 2.1 shows that this technique cannot be applied blindly.

The key programming difficulty is how to evaluate the right side function *PDERS*. The function subprogram shown simply writes down the formula using the auxiliary function subprogram *RLUXTT* to evaluate $Lu(x,y,t-\Delta t)$. Both of these subprograms look messier than one expects because special code is used for the first time step to incorporate the initial conditions.

```
*               TIME DEPENDENT PROBLEM USING CRANK-NICOLSON

OPTION. LEVEL = 0
GLOBAL.
         COMMON / GCOMON / T, DELTAT, NSTEP

EQUATION.
     (4.*Y**2)UXX + UYY + (2.+TAN((X+Y+T)/4.))UY + (1.-2./DELTAT)U &
                                    = PDERS(X,Y)
BOUNDARY.            U = UBOUND(X,Y) ON X = 0.
                                    ON X = 4.
                                    ON Y = 0.
                                    ON Y = 1.
GRID.        3 X POINTS $ 3 Y POINTS
FORTRAN.
             TSTART = 1.
             TSTOP  = 2.
C               CHOOSE DELTA T = (DELTA Y)**2 OVER 2
             DELTAT = 1./8.
             DO 10 NSTEP = 1,8
                        T = TSTART + NSTEP*DELTAT
DISCRETIZATION.         HERMITE COLLOCATION
SOLUTION.               BAND GE
OUTPUT.                 MAX(ERROR)

FORTRAN.
   10        CONTINUE
```

Figure 7. The ELLPACK program to solve an initial value problem using the Crank-Nicolson discretization of time. To save space, we omit the body of the initial value functions *U*0, *U*0*Y*, *U*0*XX*, and *U*0*YY*.

```
SUBPROGRAMS.
      FUNCTION PDERS(X,Y)
         COMMON / GCOMON / T, DELTAT, NSTEP
         T = T - DELTAT
         IF (NSTEP .EQ. 1) THEN
               UOFT = U0(X,Y)
            ELSE
               UOFT = U(X,Y)
         ENDIF
         R1 = RLUXYT(X,Y) + F(X,Y,T)
         PDERS = - (2./DELTAT)*UOFT - F(X,Y,T+DELTAT) - R1
         T = T + DELTAT
         RETURN
      END
      FUNCTION RLUXYT(X,Y)
         REAL COEFOF(6)
         COMMON / GCOMON / T, DELTAT, NSTEP
         INTEGER CUXX, CUXY, CUYY, CUX, CUY, CU
         DATA    CUXX, CUXY, CUYY, CUX, CUY, CU
     A        /  1,    2,    3,    4,   5,   6/
C              USE ELLPACK PROGRAM TO EVALUATE PDE COEFFICIENTS
         CALL Q1PCOE(X,Y,COEFOF)
C              APPLY OPERATOR TO U(X,Y,T)
         IF (NSTEP .EQ. 1) THEN
               RLUXYT = COEFOF(CUXX) * U0XX(X,Y)
     A                  + COEFOF(CUYY) * U0YY(X,Y)
     B                  + COEFOF(CUY)  * U0Y(X,Y)
     C                  + (COEFOF(CU) + 2./DELTAT) * U0(X,Y)
           ELSE
               RLUXYT = COEFOF(CUXX) * UXX(X,Y)
     A                  + COEFOF(CUYY) * UYY(X,Y)
     B                  + COEFOF(CUY)  * UY(X,Y)
     C                  + (COEFOF(CU) + 2./DELTAT) * U(X,Y)
         ENDIF
         RETURN
      END

      FUNCTION F(X,Y,T)
         T1 = .25*(X+Y+T)
         T2 = EXP(-Y**2-T)
         F  = - (.25*COS(T1)-2.*Y*SIN(T1))*T2*(TAN(T1)+2.)
     A        + (.0625 - 3.75*Y**2)*T2*SIN(T1)  +  (.25+Y)*T2*COS(T1)
      RETURN
      END
      FUNCTION U0XX(X,Y)
      FUNCTION U0YY(X,Y)
      FUNCTION U0Y(X,Y)
      FUNCTION U0(X,Y)
      FUNCTION UBOUND(X,Y)
         COMMON /GCOMON/ T, DELTAT, NSTEP
         UBOUND = TRUE(X,Y)
         RETURN
      END
      FUNCTION TRUE(X,Y)
         COMMON /GCOMON/ T, DELTAT, NSTEP
         TRUE = SIN((X+Y+T)/4.) * EXP(-Y**2-T)
         RETURN
      END
END.
```

Figure 7. Continued

We also note that multi-step methods can be used with little added difficulty. One needs in this case to obtain values for $Lu(x,y,t-\Delta t)$, $Lu(x,y,t-2\Delta t)$, etc. One can do this by saving values of $RLUXTT(X,Y) + F(X,Y,T)$ in *PDERS* as they are computed. This relies on the fact that the spatial discretization goes through the X,Y values in exactly the same way at each time step; the programmer does not need to know how. Figure 8 shows the function *PDERS* for using a fourth order Adams-Moulton formula (we have suppressed the x,y dependence)

$$Lu(t+\Delta t) - 8/(3\Delta t)\, u(t+\Delta t)$$

$$= -8/(3\Delta t)\, u(t) - f(t)$$

$$- [19(Lu(t)+f(t)) - 5(Lu(t-\Delta t)+f(t-\Delta t)) + Lu(t-2\Delta t)+f(t-2\Delta t)]$$

```
      FUNCTION PDERS(X,Y)
      PARAMETER( NSTORE = 1000 )
      COMMON / GCOMON / T, DELTAT, NSTEP, FIRST
      LOGICAL FIRST
C                STORAGE FOR OLD OPERATOR VALUES, DIMENSION = NSTORE
      REAL R1(NSTORE),R2(NSTORE),R3(NSTORE),X0(NSTORE),Y0(NSTORE)
      DATA NDEX / 0 /
C
C       NEED U, LU AND F AT (X,Y,T-DELTAT).  MOVE TIME T BACK ONE STEP.
      T = T - DELTAT
C
C             FIND U(X,Y,T-DELTAT); IT'S U0(X,Y) AT START, THEN U(X,Y)
      IF (NSTEP .LE. 3) THEN
      UOFT = U0(X,Y)
      ELSE
      UOFT = U(X,Y)
      ENDIF
C
C          SAVE PREVIOUS VALUES OF LU+F IN THE ARRAYS  R1, R2 AND R3
      IF( FIRST ) THEN
            NLOOP = NSTORE
            IF( NSTEP .GT. 1 ) NLOOP = NDEX
            DO 10 I = 1,NLOOP
                  R3(I) = R2(I)
                  R2(I) = R1(I)
   10       CONTINUE
            NDEX = 0
            FIRST = .FALSE.
      ENDIF
      NDEX = NDEX + 1
C      EVALUATE RIGHT SIDE USING RLUXYT FOR LU AT PREVIOUS TIME STEP
      R1(NDEX) = RLUXYT(X,Y) + F(X,Y,T)
      PDERS    = - (8./DELTAT)*UOFT/3. - F(X,Y,T+DELTAT)
     *           - (19.*R1(NDEX) - 5.*R2(NDEX) + R3(NDEX))/9.
      T = T + DELTAT
      RETURN
      END

      FUNCTION RLUXYT(X,Y)
C     THIS FUNCTION EVALUATES LU(X,Y,T).  ITS BODY IS NOT GIVEN HERE.
      END
```

Figure 8. The subprogram *PDERS* needed for using a fourth order Adams-Moulton multi-step formula to solve initial value problems using ELLPACK.

VI. SYSTEMS OF ELLIPTIC PROBLEMS

Iteration can also be used to solve systems of equations in ELLPACK. In order to define a system we write, for example with two equations,

```
EQUATION. A(X,Y)UXX + B(X,Y)UYY = 0
```

along with the Fortran FUNCTION program

```
FUNCTION A(X,Y)
COMMON/EQNUM/KEQN
IF(KEQN .EQ.1) THEN
   A = u_xx COEFFICIENT OF FIRST EQUATION
  ELSE
   A = u_xx COEFFICIENT OF SECOND EQUATION
END IF
RETURN
END
```

Thus changing *KEQN* changes the coefficients of the equation. This technique is used for all coefficients and functions which vary with *KEQN* .

The next task is to create more than one solution. The function $u(x,y)$ always refers to the current PDE, so one must create copies of it. One way is to use the Fortran function

```
FUNCTION U1(X,Y)
COMMON/SAVEU1/UNKNU(1)
COMMON/INITIAL/ISWIT
IF(ISWIT .EQ. 0) THEN
   U1 = FIRST GUESS FOR U(X,Y)
  ELSE
   U1 = R3 COPY(X,Y,UNKNU1)
ENDIF
RETURN
END
```

The function R3COPY is a copy of the ELLPACK function that evaluates the solution at (x,y) from the unknowns *UNKNU*1 saved in the COMMON block

SAVEU 1. The ELLPACK function that is copied depends on the discretization module used; to carry this out the programmer must learn something of how ELLPACK operates internally. The function $U1(x,y)$ is then used in the coefficient functions. Thus if the PDE has $(1+u^2)u_{xx}$, then the $A(X,Y)$ function has

$$A = 1. + U1(X,Y) ** 2$$

Figure 9 shows the ELLPACK program to solve the system

$$\begin{aligned} u_{xx} + (1+v)\, u_{yy} + (1-v)yu &= 0 \\ (1+u)\, u_{xx} + u_{xx} + (1+u)yv &= 0 \end{aligned}$$

The iteration is initialized here by using the ELLPACK initialization facility

TRIPLE. SET U BY BI-CUBICS

which interpolates the boundary conditions by Hermite bicubic polynomials.

The effectiveness of this iteration is problem dependent. We have used it for a number of examples and found that it works more often than not; it, of course, helps to have a good initial guess. Examples of systems where this iteration works well are

$$\begin{aligned} u_{xx} + (1+v)u_{yy} + (v+y)u_x + y(1-v)u &= f_1 \\ (1+u)v_{xx} + v_{yy} + (u+x)v_x + y(1+u)v &= f_2 \end{aligned}$$

$$\begin{aligned} (1+2v)u_{xx} + (4-v)u_{xy} + (v+4)U_x + xvu &= f_1 \\ (1+3u)v_{xx} + (1-u/40)v_{xy} + (4u+x)v_x + x^2yv &= f_2 \end{aligned}$$

$$\begin{aligned} u_{xx} + (2v+u)u_{yy} + (v+sin(u))u_x + vxyu &= f_1 \\ v_{xx} + (1+u)v_{yy} + (u+sin(v))v_x + x^2uv &= f_2 \end{aligned}$$

$$\begin{aligned} (1+v/4)u_{xx} + u_{yy} + (u+v/4)u_x + (v+(1+x)^2)u_y + 6xsin(v/5)u &= f_1 \\ (1+vu^2)v_{xx} + v_{yy} + (4u-v)v_x + (2u+v)v_y + 8\cos^2(x+u)v &= f_2 \end{aligned}$$

The second system has also been solved with the coupled, nonlinear boundary condition on one side of

$$(1+v)u + u_x = g_1 \qquad (1+u)v + uv_x = g_2$$

Several similar systems of three equations have also been solved by this technique.

The use of this iteration raises an interesting question: what are the properties of the system that make this Gauss-Siedel type iteration converge? Are there properties (e.g. diagonal dominance) similar to those for algebraic systems which are relevant?

```
*            TWO SIMULTANEOUS, SIMPLY COUPLED EQUATIONS
*        L1(U,V)  :     UXX + (1+V)UYY + (1-V)*Y*U = F1(X,Y)
*        L2(U,V)  :     (1+U)VXX + VYY + (1+U)*Y*V = F2(X,Y)
*                          U = (X+Y)**4           V = EXP(X-Y)
DECLARE.
                   COMMON / EQNUM / KEQN
                   COMMON / SAVEU1   / UNKNU1($I1MNEQ)
                   COMMON / SAVEV1   / UNKNV1($I1MNEQ)

EQUATION.      A(X,Y)UXX + B(X,Y)UYY + E(X,Y)U = G(X,Y)

BOUNDARY.    UX = TRUEX(X,Y)   ON X=0.0  $  UX = TRUEX(X,Y)   ON X=1.0
             U  = TRUE (X,Y)   ON Y=0.0  $  U  = TRUE (X,Y)   ON Y=1.0

GRID.              4 X POINTS $ 4 Y POINTS
FORTRAN.
C                        INITIALIZE V(X,Y)
                   KEQN = 2
TRIPLE.            SET U BY BICUBICS
OUTPUT.            MAX(ERROR)
FORTRAN.
C                        SAVE THE COEFFICIENTS OF INITIAL V(X,Y)
                   CALL SAVE(UNKNV1)
                   DO 100 NITER = 1, 6
                         KEQN = 1
DISCRETIZATION.          HERMITE COLLOCATION
SOLUTION.                BAND GE
FORTRAN.
                         CALL SAVE(UNKNU1)
OUTPUT.                  MAX(ERROR)
FORTRAN.
                         KEQN = 2
DISCRET.                 HERMITE COLLOCATION
SOLUTION.                BAND GE
FORTRAN.
                         CALL SAVE(UNKNV1)
OUTPUT.                  MAX(ERROR)
FORTRAN.
  100              CONTINUE
```

Figure 9. The ELLPACK program to solve a pair of nonlinear, simultaneous elliptic problems.

```
SUBPROGRAM.
      FUNCTION A(X,Y)
         COMMON / EQNUM / KEQN
         IF( KEQN .EQ. 1 ) THEN
                 A = 1.
            ELSE
                 A = 1. + U1(X,Y)
         ENDIF
      RETURN
      END
      FUNCTION B(X,Y)
      FUNCTION E(X,Y)
      FUNCTION TRUE(X,Y)
         COMMON / EQNUM / KEQN
         IF( KEQN .EQ. 1 ) THEN
                 TRUE = (X+Y)**4
            ELSE
                 TRUE = EXP(X-Y)
         ENDIF
         RETURN
      END
      FUNCTION TRUEXX(X,Y)
      FUNCTION TRUEYY(X,Y)
      FUNCTION AT(X,Y)
C                AT(X,Y) GIVES TRUE COEFFICIENTS OF UXX
         COMMON / EQNUM / KEQN
      FUNCTION BT(X,Y)
      FUNCTION ET(X,Y)
      FUNCTION G(X,Y)
         G = AT(X,Y)*TRUEXX(X,Y) + BT(X,Y)*TRUEYY(X,Y) + ET(X,Y)*TRUE(X,Y)
         RETURN
      END
      SUBROUTINE SAVE(UNKN)
         REAL UNKN(1)
         COMMON / C1UNKN / R1UNKN(1)
         COMMON / C1IVDI / I1NEQN
         DO 10 I = 1,I1NEQN
   10          UNKN(I) = R1UNKN(I)
         RETURN
      END
      FUNCTION  U1(X,Y)
         COMMON / SAVEU1   / UNKNU1(1)
               U1 = R3COPY(X, Y, UNKNU1)
         RETURN
      END
      FUNCTION  V1(X,Y)
         COMMON / SAVEV1   / UNKNV1(1)
               V1 = R3COPY(X, Y, UNKNV1)
         RETURN
      END
      FUNCTION R3COPY (X,Y,R1UNKN)
C  R3COPY IS A COPY OF THE ELLPACK FUNCTION TO RETURN THE VALUE
C  OF THE COLLOCATION APPROXIMATION EVALUATED AT THE POINT (X,Y).
END.
```

Figure 9. Continued.

REFERENCES

Birkhoff, G. and Lynch, R.E. (1983). *Numerical Solution of Elliptic Problems,* SIAM Publications.

Boisvert, R.F., Houstis, E.N. and Rice, J.R. (1978). A system for performance evaluation of partial differential equations software, IEEE Trans. Software Engineering 5, pp. 418-425.

Coriell, S.R., Boisvert, R.F., Rehm, R., and Sekerka, R. (1981). Solute segregations during unidirectional solidification of a binary alloy with curved solid-liquid interface II — large departure from planarity, J. Crystal Growth 5, pp 167-175.

Gherson, P., Lykoudis, P.S. and Lynch, R.E. (1980). Analytic study of end effects in liquid metal MHD generators, Seventh Inter. Conf. MHD Electrical Power Generation, II, M.I.T., pp. 590-594.

Houstis, E.N. and Rice, J.R. (1980). An experimental design for the computational evaluation of elliptic partial differential equation solvers, in *Production and Assessment of Numerical Software* (M. Hennell and L. Delves, eds) Academic Press, pp. 55-65.

Houstis, E.N. and Rice, J.R. (1982). High order methods for elliptic partial differential equations with singularities, Inter. J. Numer. Meth. Engin. 18, pp. 737-754.

Lynch, R.E. and Rice, J.R. (1978). The Hodie method and its performance for solving elliptic partial differential equations, in *Recent Advances in Numerical Analysis* (C. deBoor, ed.) Academic Press, pp. 143-176.

Rice, J.R. (1977). ELLPACK: A research tool for elliptic partial differential equations software, in *Mathematical Software III* (J. Rice, ed). Academic Press, pp. 319-341.

Rice, J.R. (1979). Programming effort analysis of the ELLPACK Language, SIGNUM Newsletter 14, pp. 109-111.

Rice, J.R., (1981). ELLPACK: Progress and plans, in *Elliptic Problems Solvers* (M. Schultz, ed.) Academic Press, pp. 135-162.

Rice, J.R. (1982). Machine and compiler effects on the performance of elliptic PDE software, in Proc. 10th World Congress, IMACS 1, pp. 446-448.

Rice, J.R. (1983). Performance analysis of 13 methods to solve the Galerkin method equations. Linear Alg. Appl., to appear.

Rice, J.R. and Boisvert, R.F. (1984). *Solving Elliptic Problems with ELLPACK,* Springer-Verlag.

Rice, J.R., Houstis, E.N. and Dyksen, W.R. (1980). A population of linear, second order, elliptic partial differential equations on rectangular domains, Math. Comp. 36, pp. 475-484.

PRESENTATION AND EVOLUTION OF THE CLUB MODULEF: A LIBRARY OF COMPUTER PROCEDURES FOR FINITE ELEMENT ANALYSIS

D. Bégis[1], F. Hecht[1] and M. Vidrascu[1]

I. THE CLUB MODULEF

In this paper, we present the goals, structure, working methods and reasons for the existence of the Modulef Club and the achievements of the Modulef library. Some Modulef productions are described in the second part of this paper, and in the last part, we indicate the orientations of the Club Modulef.

A. Goals

The Modulef Club, created in 1974, brings together French or foreign universities and private or public industrial companies in order to design and realize a library of "finite elements" modules. The library is formed of contributions from individual members which then are combined and redistributed among the members.

B. Objectives

The objectives of the Modulef Club are:

- To expand a finite element library using the latest results of theoretical and numerical research in applied mathematics, and industrial experience in solving large and practical problems.
- To bring together laboratories interested in the same topic, in order to determine the optimal program and methods.
- To promote the exchange and sharing of each members' knowledge.
- To validate new methods and to compare them with those already implemented.

[1]INRIA - B.P. 105, 78153 Rocquencourt, France.

ISBN 0-12-100560-7

The realization of these objectives requires:

- standardization of programming
- portability of the programs
- development of computer tools in order to realize a coherent multi-computer set.
- full documentation and maintenance of the Modulef library.

C. Structure of the Club

The rules of the Modulef Club dictate that all exchanges between the members are only scientific and not financial. At this moment, there are 87 members in the Club: 27 industrial companies and 60 university laboratories.

D. Achievements

The Modulef library currently makes possible the numerical solution of:

- steady state or time-dependent heat conduction problems, whether linear or nonlinear, and two- or three-dimensional,
- static or dynamic elasticity problems with the hypotheses of small elastic displacements or large deformations,
- fluid mechanics problems (two-dimensional flow of fluids, ...).

The different steps of computation are separated; each step corresponds to a module. A module is a set of procedures that transforms input data structures into output data structures. The different modules allow:

- the generation and display of meshes,
- the specification of characteristics of materials or external forces by sub-domains or pieces of boundary,
- the choice of type of the finite element method (e.g., conforming or not conforming, hybrid, mixed, etc.). For non-conforming methods, the flow or the stresses can be computed together with the temperature or the displacement, or separately,
- the solution of linear systems using a choice of several techniques to manage storage, depending upon the size of the problem and that of main memory, i.e.,
    - Direct methods (Gauss, Crout, Cholesky) using skyline storage in main memory or secondary memory, or generalized array storage,
    - Iterative methods (relaxation, conjugate gradient algorithms with or without preconditioning (SSOR, ICCG,

etc.) or multigrid method),

- the computation of eigenvalues of two or three dimensional operators using a choice of subspace iteration, Lanczos or QR methods,
- the integration of time-dependent heat equations using multistep, Runge Kutta or Gear methods. These last two select automatically the step size and the order of the scheme (for GEAR) with respect to the stiffness of the data,
- the solution of dynamic problems by means of free vibration modes or multistep methods,
- the determination of plastic or elastic zones of a structure,
- the solution of thermo-elastic problems,
- the computation of velocity and pressure of a plane incompressible fluid flow (Navier-Stokes equations),
- the computation of magnetic fields,
- the computation of the effective elastic moduli of composite materials,
- the display of isovalues, main stresses, or deformed configuration.

E. Present Programming and Use

The methodology used to build the library of procedures is a combination of the abstract approach with modular programming. Each algorithm is a succession of mathematical operators, which becomes a computer module after choosing the data structure [1].

New problems can be easily solved. The chosen algorithm is simply split into a succession of mathematical operators.

Generally, most of these operators are already programmed in the Modulef library. The others are being developed consistent with the existing norms and data structures.

If a problem must be solved very often, a "black box" can easily be made so that only a few data must be specified.

The Modulef norms have been defined in order to make the software as portable as possible [2]. The programming language used is Fortran. The Modulef library includes about 2200 procedures making 250,000 cards. This set can be implemented on BURROUGHS, CDC, CII-HB, IBM, NORSK-DATA, UNIVAC, VAX and CRAY computers.

More details about the Modulef Club are given in [3].

II. SOME MODULEF PRODUCTIONS

Here we present some sample problems solvable with the Modulef library.

A. *Navier-Stokes Equations for Incompressible Viscous Fluids*

Several methods are implemented in the Modulef library:

i) a non conforming P1 method using a zero divergence basis [4],

ii) a method which combines finite elements for the space discretization and alternating directions for the time discretization. The splitting effectively decouples the two main difficulties, the nonlinearity and the incompressibility. The nonlinearity is solved by a least squares and conjugate gradient method [5],

iii) a Q^2 conforming method for the flow velocity, where the pressure is a discontinuous function linear by quadrilateral, and we build a zero-divergence local basis [6].

Two numerical methods have been implemented: the least squares method for the steady equations and the method of characteristics for the time dependent equations. The main value of this last method is that it makes possible the solution of the Navier Stokes equations with large Reynolds number in the least CPU run time. We used these methods to compute the flow and the pressure in the following problem of a channel with a step.

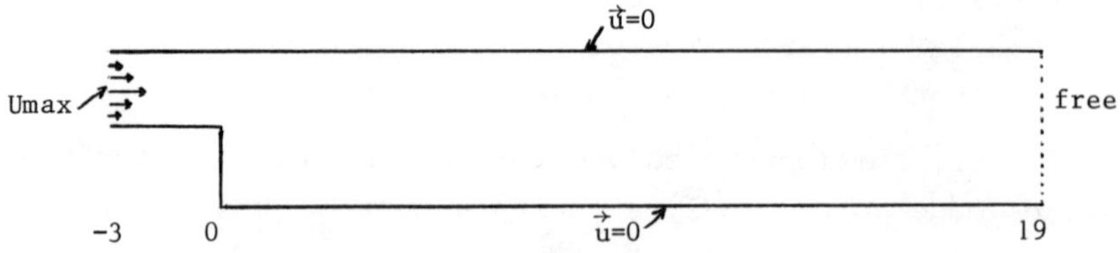

At the intake, we have a fully developed laminar flow profile with Umax = 1.

The two methods yield the same good results for a Reynold's number up to 200, but only the method of characteristics converges at Re = 500. Figure 1 shows the line at Re = 500 computed by the method of characteristics.

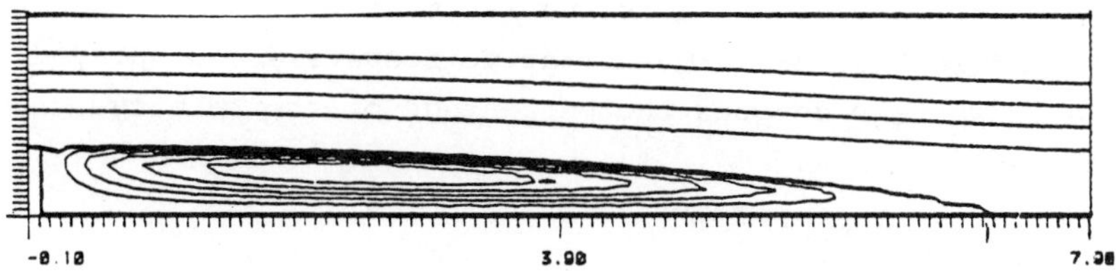

Characteristics method
REYNOLDS = 500.0

Figure 1.

B. Crack Problems

We next consider two different crack problems, which have been studied with Babuska [7]:

i) A plane stress elasticity problem, with the unit square as domain and loading as shown in figure 2.

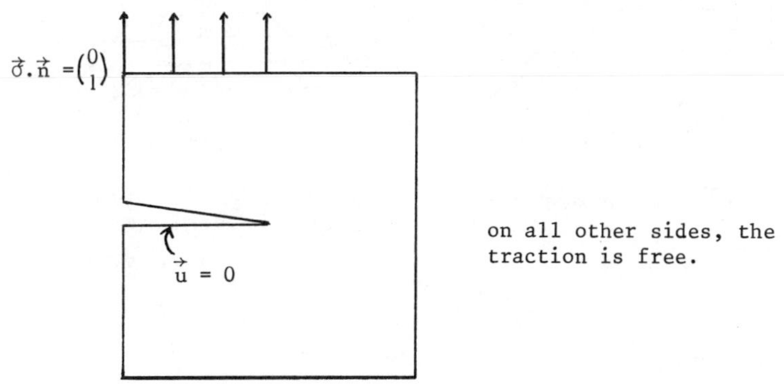

Figure 2.

For Young's modulus E = 1 and Poisson ratios of ν = .3 or .49, we know the potential energy. Thus we can compare the accuracy of each finite element solution. Several classical approximations are used, the linear Lagrangian elements P1 (triangular) and Q1 (quadrilateral), and the isoparametric quadratic Lagrangian elements P2 and Q2.

In the Modulef library, the mesh is generated by a tree structured up down technique. We generate only an elementary subdomain and then build the total mesh by linear mappings. For each elementary domain, we give some characteristic points. From these points, characteristic edges are generated, using points generated according to a geometrical ratio. From these edges we automatically generate the mesh of the subdomain.

A quadrilateral and a triangular mesh thus generated are displayed in Fig. 3/4. Quantitative results are displayed in Table I, where P denotes the percentage of the relative error of the energy, i.e.,

$$P = [(E_c - E_e)/E_c]^{1/2} ,$$

where E_c is the computed energy and E_e is the exact energy.

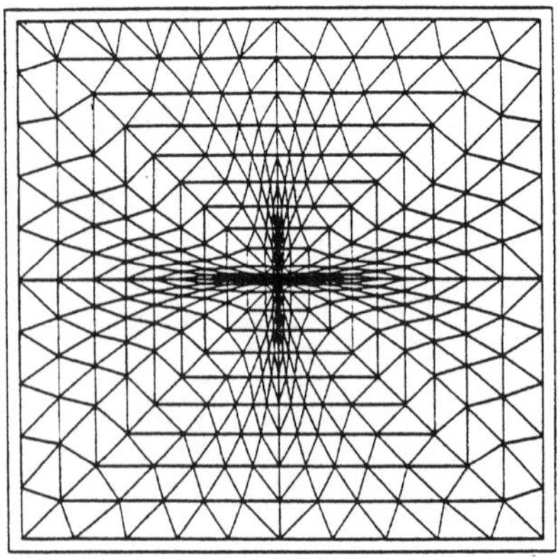

Triangular mesh

Figure 3.

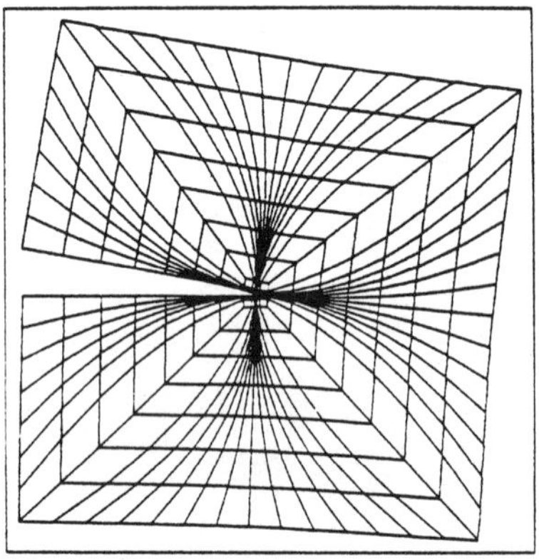

A deformed configuration quadrilateral mesh.

Figure 4.

TABLE I.

f.e.m.	ν = .3 Energy	ν = .3 P	ν = .49 Energy	ν = .49 P
Q1	1.54	15.6	1.224	27
P1	1.56	11.0	1.264	20
P2	1.575	5.5	1.305	10
Q2	1.575	5.6	1.305	10
Exact	1.58	-	1.320	-

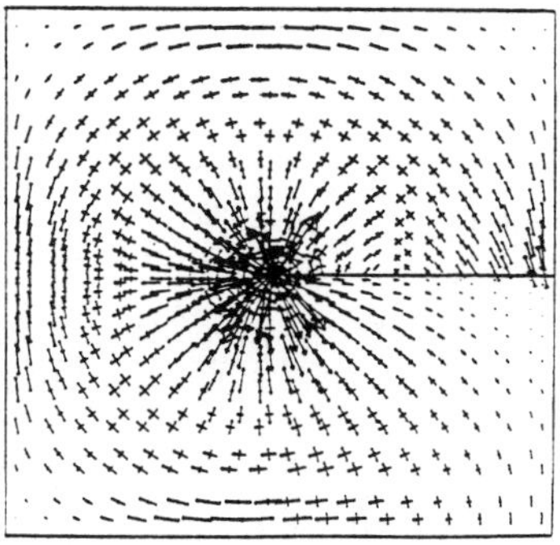

An elarging of the mesh around the tip of the crack

Figure 5.

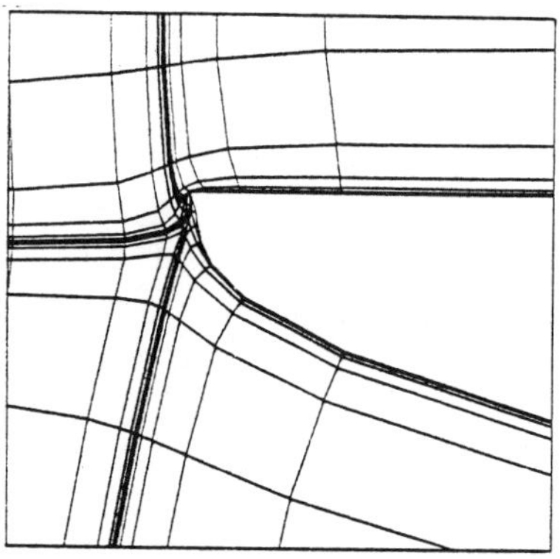

A stress distribution

Figure 6.

ii) The plane stress elasticity problem with $E = 3 \times 10^7$ and $\nu = .3$, with domain and loading as shown:

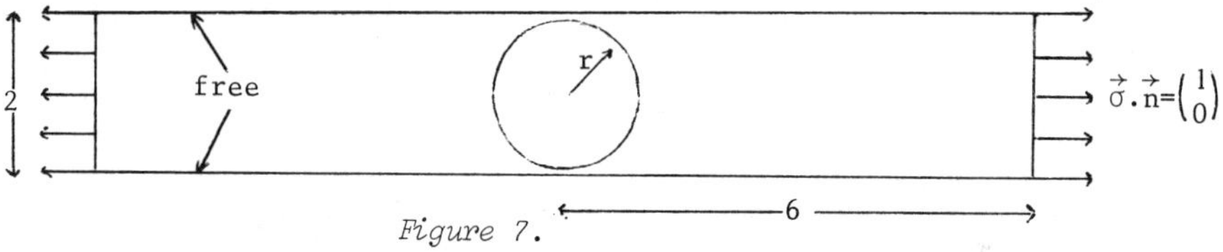

Figure 7.

Note the bar contains a hole whose diameter is almost equal to the width of the bar. Five different radii were considered:

$$r = .85, .95, .98, .99, .999 .$$

Note also that because of symmetry, we can need study a quarter part of the bar.

As with the previous problem, the solution using Q1 elements proves to be the worst and that with Q2 the best. For each radius, Table II gives the computed energy and the percentage of the relative error. Even with a radius equal to .999, the percentage of the relative error is less than 5% for about 2,000 degrees of freedom. This shows the accuracy of the energy stresses.

TABLE II.

r	D.O.F.	E_e	Q1 E_c	Q1 P	Q2 E_e	Q2 P
.85	1234	.189410^{-6}			.189410^{-6}	0.87
	1562	"	.18710^{-6}	10.8		
.98	1234	.521310^{-6}			.519610^{-6}	5.78
	1562	"	.47510^{-6}	29.7		
	5278	"			.521310^{-6}	0.67
	8974	"	.51910^{-6}	12.2		
.99	1234	.7387510^{-6}			.728310^{-6}	11.92
	1900	"	.60510^{-6}	42.5		
	5258	"			.738610^{-6}	1.62
	6900	"	.66210^{-6}	32.1		
.999	1078	2.3409610^{-6}			2.320410^{-6}	9.37
	4982	"			2.340810^{-6}	0.85

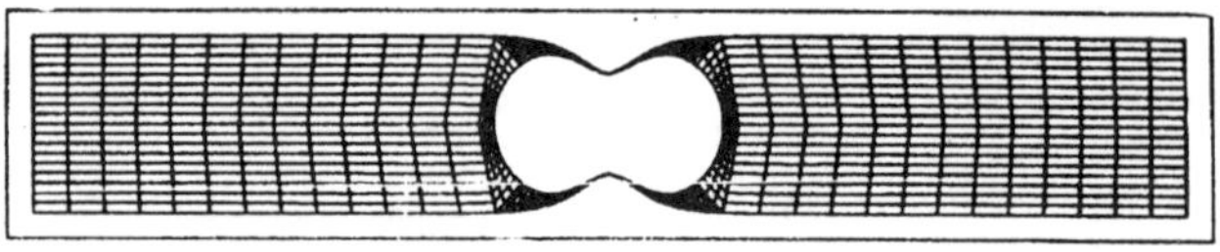

A deformed configuration.
Figure 8.

C. *Problems in Incompressible and Compressible Finite Elasticity*

Problems of this type have been considered by P. Letallec and M. Vidrascu, [8]-[10]. The problem involves computation of the deformation of a hyperelastic body. The solution is found using an augmented Lagrangian technique which combines a penalty function and Lagrange multipliers. We can reduce the solution of such complicated nonlinear, nonconvex problems to a sequence of linear unconstrained elasticity problems and to a sequence of elementary nonlinear problems of small dimension.

For the compressible case [10], the numerical method was first tested on Rivlin's cube, a classical problem whose analytical solution is known. This is the problem of computing the deformation of a cube uniformly loaded on all its faces. The computed results agree closely with the analytical solution. We next considered a three dimensional problem with bifurcation. This problem illustrates the ability of the method to compute the stable post buckling solutions of non-linear elasticity problems. We computed the equilibrium positions of a thick elastic beam submitted to a very small horizontal surface load and a compression ratio of 10%. Two solutions are obtained:

1) a symmetric solution with small displacements which is unstable (Figure 9)
2) a stable bifurcated solution with large horizontal displacements (Figure 10)

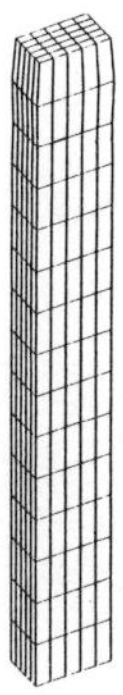

Symmetric solution
Figure 9.

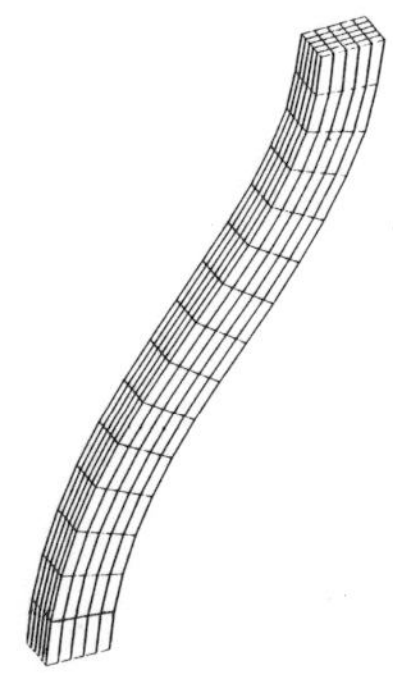

Post buckling solution
Figure 10.

Starting from the reference configuration, the algorithm approached first the symmetric solution (17 iterations, energy = 1.695), then diverged and finally converged towards the stable post buckling solution (510 iterations, energy = 0.425).

III. MODULEF: ADVANCES AND DEVELOPMENT

Currently, problems in various engineering fields can be solved using finite element methods available in the MODULEF library.

Modular programming is the key concept of the library. Results of a significant computation step form a structured, standardized data set called a "Data Structure", to be used in subsequent steps. Data structure handling is transparent to the user, but must fit into some computer file organization. Hence some data is duplicated and the user must specify the file environment. A relational data base, fitted to numerical software, would advantageously replace this file organization, but available systems seem too limited or unsuitable. Future programming standards will take into account latest developments in computer technology (vector programming) and languages (Fortran 77, Ada, etc.), [11].

Another area of current concern is to simplify user interface requirements for new problems. The user must be assisted in expressing his needs, perhaps by using predefined patterns (menus); and when interpreting and evaluating results by the use of 3D graphic representations on color display screens, or animation techniques. (Graphic representation is already provided with Modulef [12], and Figure 11.)

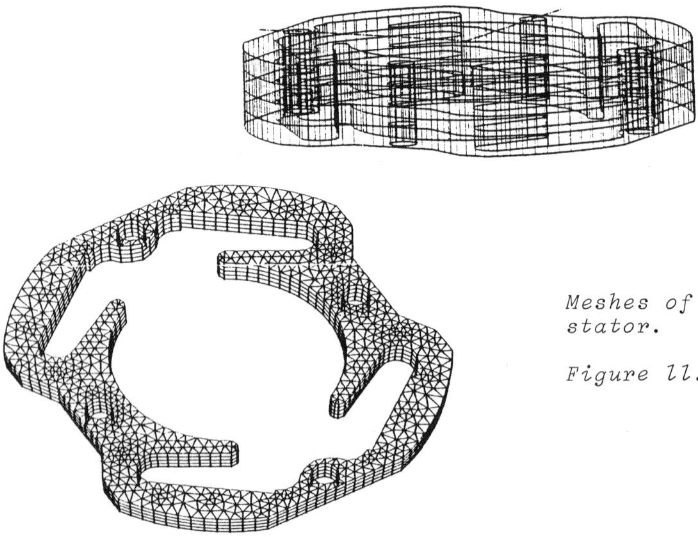

Meshes of the stator.

Figure 11.

The user or programming language is indeed important, but we believe a good choice of the programming methodology is the key ingredient in the success of numerical software. This software must be portable and adapted to the distributed computing environment like the CAD/CAM systems, since the programming environment is assuming a larger role in the efficacious and safe development and maintenance of large scientific software.

The design and realization of a CAE (Computer Aided Engineering) system is a key future goal of the Modulef library.

REFERENCES

1. Perronnet, A., "Description des structures de données du Club Modulef," Février 1979.
2. Latil, J., Perronnet, A., "Normes d'appel, Normes de programmation, Normes d'insertion des modules du Club Modulef," Octobre 1977.
3. "Présentation du Club Modulef," Version 3.3, Mai 1982.
4. Thomasset, F., "Equations de Navier Stokes bidimensionnelles: modules NSNCEV, NSNCST, TRSD," Version 3.1, Novembre 1979.
5. Glowinski, R., "Numerical methods for nonlinear variational problems," (Second Edition), Springer, New York, 1983.
6. Hecht, F., "Construction d'une base de fonctions P_1 non conforme à divergence nulle dans R^3," RAIRO, Analyse numerique/Numérical Analysis, vol. 15, No. 2, 1981, p. 119-150.
7. Babuska, I., Bégis, D., Szabo, B., to appear.
8. Glowinski, R., Letallec, P., "Elasticité non linéaire, formulation mixte et méthode numérique associée," 5ème colloque international sur les méthodes de calcul scientifique et technique, INRIA Dec. 1981.
9. Letallec, P., Vidrascu, M., "Elasticité non linéaire: quelques modules de calcul en grandes déformations, Modules: COBDNO, COTAE, GDEFIN, GRAFEL, Super-module: PRELAS," Version 3.3, Rapport Modulef Mars 1982.
10. Letallec, P., Vidrascu, M., "Une méthode numérique pour les problèmes d'équilibre des corps hyperélastiques en grandes déformations (to appear).
11. Laug, P., Perronnet, A., "Modulef: une expérience de programmation en éléments finis," Conf. Internationale Outils, Méthodes et Langages Adaptés au Calcul Scientifique, Mai 1983.
12. Bégis, D., George, P.L., Pistre, F., "Présentation générale des mailleurs de la bibliothèque Modulef," Journées sur la Génération Automatique et la Visualisation des Maillages, Décembre 1982.

A FOURTH ORDER ACCURATE FAST DIRECT METHOD FOR THE HELMHOLTZ EQUATION

Ronald F. Boisvert[1]

I. INTRODUCTION

Fast direct methods have proven to be very effective techniques for the solution of separable elliptic problems. One reason for their widespread use is the availablity of high quality mathematical software such as the FISHPAK collection [12]. These successes have generated much interest in exploring the range of discrete elliptic problems amenable to such treatment [8], [6], [7].

Here I describe software which uses a fourth order accurate finite difference scheme to discretize the Helmholtz equation on rectangular domains in two and three dimensions with Dirichlet, Neumann, or periodic boundary conditions. The discrete problem is solved using Fourier transforms. This software differs from [8] and [6] in the boundary conditions handled as well as in the details of the direct solution. It is roughly twice as fast as the fourth order solver in FISHPAK (version 3) which is based on deferred corrections.

---

[1]Center for Applied Mathematics
National Bureau of Standards
Washington, DC

 ISBN 0-12-100560-7

We first consider the solution to the problem

$$Lu = u_{xx} + u_{yy} + fu = g(x,y) \tag{1}$$

(f constant $\leqslant 0$) on the open rectangle R. On the boundary $\partial R = \partial R_1 \cup \partial R_2 \cup \partial R_3$ we have

(a) u prescribed (Dirichlet condition) on ∂R_1

(b) u_n prescribed (Neumann condition) on ∂R_2

(c) the solution is periodic on ∂R_3.

(∂R_1, ∂R_2, and ∂R_3 are each unions of edges of R.) We place a uniform grid with mesh width h over the domain, and first consider the most common difference stencil,

$$\begin{array}{|c|c|c|} \hline c & b & c \\ \hline b & a & b \\ \hline c & b & c \\ \hline \end{array}\; u = dh^2 g$$

where either $a=-4+h^2f$, $b=d=1$, $c=0$, or $a=-20+6h^2f$, $b=4$, $c=1$, $d=6$. Both are second order accurate. After incorporating the boundary conditions one obtains the linear system $Mu=g$, which we write in block form as (the natural ordering is used)

$$\begin{bmatrix} A/\mu & B & & & & & \Theta B \\ B & A & B & & & & \\ & B & A & B & & & \\ & & \cdot & \cdot & \cdot & & \\ & & & \cdot & \cdot & \cdot & \\ & & & & \cdot & \cdot & \cdot \\ & & & & B & A & B \\ \Theta B & & & & & B & A/\nu \end{bmatrix} \begin{bmatrix} u_1 \\ u_2 \\ u_3 \\ \cdot \\ \cdot \\ \cdot \\ u_{m-1} \\ u_m \end{bmatrix} = \begin{bmatrix} g_1/\mu \\ g_2 \\ g_3 \\ \cdot \\ \cdot \\ \cdot \\ g_{m-1} \\ g_m/\nu \end{bmatrix} \tag{2}$$

$$A = \begin{bmatrix} a/\eta & b & & & & & \tau b \\ b & a & b & & & & \\ & b & a & b & & & \\ & & \cdot & \cdot & \cdot & & \\ & & & \cdot & \cdot & \cdot & \\ & & & & b & a & b \\ \tau b & & & & & b & a/\sigma \end{bmatrix} \qquad B = \begin{bmatrix} b/\eta & c & & & & & \tau c \\ c & b & c & & & & \\ & c & b & c & & & \\ & & \cdot & \cdot & \cdot & & \\ & & & \cdot & \cdot & \cdot & \\ & & & & c & b & c \\ \tau c & & & & & c & b/\sigma \end{bmatrix}$$

A and B are both n by n. The scalars μ, ν, η, σ depend on the boundary conditions on the top, bottom, left, and right sides of the domain respectively. They are 2 for Neumann conditions and 1 otherwise. The scalars Θ and τ are 1 when u is periodic in y and x respectively and 0 otherwise.

A number of very efficient direct methods for solving Mu=g have been developed [3], [11]. Our software is based on the fact that high order accurate discretizations may be obtained by a proper choice of a, b, and c, and a perturbation of the right hand side. The resulting matrix problem is then exactly the same as (2). We use the Fourier method since software for the discrete Fourier transform is readily available, and the algorithm is easily adapted for use on parallel or vector processors [13].

II. THE FOURIER ALGORITHM

This technique relies on the fact that A and B have a common set of linearly independent eigenvectors (see Table 1). Let Q be the matrix of these eigenvectors. If we premultiply each block equation in (2) by Q^{-1} and scale each block of unknowns by Q^{-1}, the system decouples to the solution of n tridiagonal systems of size m by m. The algorithm may be summarized as follows (see [3] for details).

Step 1. Compute $g_j \leftarrow Q^{-1}g_j$, $j=1,..,m$

Step 2. Solve $T_i u_i \leftarrow g_i$, $i=1,..,n$

Step 3. Compute $u_j \leftarrow Qu_j$, $j=1,..,m$

The vectors u_i and g_i are composed of the ith components of each of the u_j and g_j subvectors respectively. T_i has the same form as M, except that A and B are replaced by their ith eigenvalues. The vectors u and g may coincide in memory.

Multiplication by Q^{-1} is equivalent to performing a discrete Fourier transform, and hence may be computed using FFT techniques. Hence the algorithm has $O(mn\log_2 n)$ complexity on a scalar computer. We use the FFT software of [13] which contains all the special transforms required and places no restriction on vector lengths (although the transform will be fast only when NX-1 is highly composite).

III. HIGH ORDER COMPACT DIFFERENCES

We obtain high order accurate discretizations using a compact 9-point stencil by augmenting the right side of the difference equations. The precise form of these equations is

TABLE 1. Eigenvalues/eigenvectors of A and B*

	Case : Nonperiodic in x ($\tau=0$)	
	$\sigma=1$	$\sigma=2$
$\eta=1$	$\rho_j=\cos[j\pi/(n+1)]$ $Q_{ij}=\sin[ij\pi/(n+1)]$ $n=NX-2$	$\rho_j=\cos[(2j-1)\pi/(2n)]$ $Q_{ij}=\sin[i(2j-1)\pi/(2n)]$ $n=NX-1$
$\eta=2$	$\rho_j=\cos[(2j-1)\pi/(2n)]$ $Q_{ij}=\cos[(i-1)(2j-1)\pi/(2n)]$ $n=NX-1$	$\rho_j=\cos[(j-1)\pi/(n-1)]$ $Q_{ij}=\cos[(i-1)(j-1)\pi/(n-1)]$ $n=NX$
	Case : Periodic in x ($\eta=\sigma=\tau=1$)	
	$\rho_j = \cos(2k\pi/n)$ where $k=\text{floor}(j/2)$ $Q_{ij} = \cos[j(i-1)\pi/n]$, j even; $-\sin[(j-1)(i-1)\pi/n]$, j odd $n = NX-1$	

* Eigenvalues of A and B are $a+2b\rho_j$ and $b+2c\rho_j$, $j=1,..,n$ respectively. NX is the number of grid lines in x.

$$L_h u = h^2 I_h g + h J_h u_n \tag{3}$$

$$L_h u = \sum_{i=0}^{8} \alpha_i u(p_i) \qquad I_h g = \sum_j \beta_j g(q_j) \qquad J_h u_n = \sum_k \gamma_k u_n(r_k)$$

Here p_0 is the central stencil point. For $p_0 \varepsilon R$ the discretization points $p_1,..,p_8$ are the nearest neighbors of p_0 and J_h is dropped. The auxiliary evaluation points q_j are taken near p_0, but not necessarily at grid points. For $p_0 \varepsilon \partial R_2$ we drop the p_i and q_j outside $R \cup \partial R$ and the points r_k are taken on ∂R_2 near p_0. The number and location of the points q_j and r_k may be chosen to increase the order of accuracy of the formula. To obtain $O(h^N)$ accuracy one must select the coefficients α_i, β_j, and γ_k such that the truncation error $L_h v - h^2 I_h L v - h J_h v_n = 0$ for all $v \varepsilon P^{N+1}$ (or $v \varepsilon P^N$ when p_0 is a boundary point where u_n is given). Such discretizations are studied in [9] and [2] where they are called HODIE methods. They are related to both Mehrstellen [5] and OCI [4] methods.

Our discretization of (1) results in the matrix equation (2) with $a = -(480-118F+5F^2)/24$, $b = (192+8F+F^2)/48$, $c = (48+5F)/48$, $F = h^2 f$, and right hand side given in Table 2. $O(h^4)$ convergence is shown in [1]. Note that some q_j are taken off the grid. A matrix of the form (2) results provided L_h at $p_0 \varepsilon \partial R_2$ is obtained from L_h at $p_0 \varepsilon R$ by reflecting u through ∂R_2. This requirement restricts our choice of $\{q_j\}$ as the following theorem shows.

THEOREM. No difference equation of the form (3) for $\nabla^2 u$ at a point $p_0 \varepsilon \partial R_2$ can simultaneously (a) be exact on P^4, and (b) have the reflection property, when the evaluation points q_j are the same as the discretization points p_i.

TABLE 2. Right Side of 2D Finite Difference Formulae*

Type	$I_h g + J_h u_n$
Interior points	$\beta_0 g(0,0) + \beta_1[g(h,0)+g(0,h)+g(-h,0)+g(0,-h)]$ $+ \beta_2[g(k,k)+g(-k,k)+g(-k,-k)+g(k,-k)]$
Neumann sides (right)	$0.5\{\beta_0 g(0,0) + \beta_1[g(0,h)+g(0,-h)+2g(-h,0)]$ $+ 2\beta_2[g(-k,k)+g(-k,-k)]$ $+ \gamma_0 u_x(0,0) + \gamma_1[u_x(0,h)+u_x(0,-h)]\}$
Neumann corners (top r.)	$0.25\{\beta_0 g(0,0) + 2\beta_1[g(-h,0)+g(0,-h)] + 4\beta_2 g(-k,-k)$ $+ \delta_0[u_x(0,0)+u_y(0,0)] + \delta_1[u_x(0,-h)+u_y(-h,0)]$ $+\delta_2[u_x(0,-2h)+u_y(-2h,0)] +\delta_3[u_x(0,-3h)+u_y(-3h,0)]\}$

*$F=h^2 f$, $k=h/2$, $\beta_0=(48-5F)/24$, $\beta_1=F/48$, $\beta_2=1$, $\gamma_0=-12+13F/12$, $\gamma_1=F/12$, $\delta_2=2-F/48$, $\delta_3=((6-F)/3-4\delta_2)/18$, $\delta_1=8\delta_2+33\delta_3-(10-F)\beta_2$, $\delta_0=-2(\beta_0+2\beta_1)-(\delta_2+4\delta_3)-(4-F/2)\beta_2$

IV. THE THREE-DIMENSIONAL CASE

The extension of the Fourier method to three-dimensional problems is straightforward. Here the matrix has the same form as M, but A and B are themselves block tridiagonal. If there are ℓ grid lines of unknowns in z, then taking nm discrete Fourier transforms decouples the problem to a set of ℓ two-dimensional problems to which the algorithm of Section II may be applied. Our discretization is based on the following fourth order HODIE formula for $p_0 \varepsilon R$.

$$L_h u = h^2 I_h g$$

$$\begin{aligned} L_h u = {} & au(0,0,0) + \\ & b[u(h,0,0)+u(0,h,0)+u(0,0,h)+ \\ & \quad u(-h,0,0)+u(0,-h,0)+u(0,0,-h)] + \\ & c[u(h,h,0)+u(-h,h,0)+u(-h,-h,0)+u(h,-h,0)+ \\ & \quad u(h,0,h)+u(-h,0,h)+u(-h,0,-h)+u(h,0,-h)+ \\ & \quad u(0,h,h)+u(0,-h,h)+u(0,-h,-h)+u(0,h,-h)] \end{aligned}$$

$$\begin{aligned} I_h g = {} & \beta_0 g(0,0,0) + \\ & \beta_1[g(k,k,k)+g(-k,k,k)+g(k,-k,k)+g(k,k,-k)+ \\ & \quad g(k,k,-k)+g(-k,k,-k)+g(k,-k,-k)+g(-k,-k,-k)] \end{aligned}$$

TABLE 3. Right Side of 3D Finite Difference Formulae*

Type	$I_h g + J_h u_n$
Interior	given in text
Neumann sides	$0.5\{\beta_0 g(0,0,0) + 2\beta_1[g(-k,k,k)+g(-k,-k,k)+g(-k,k,-k)+g(-k,-k,-k)] - 72u_x(0,0,0)\}$
Neumann edges	$0.25\{\beta_0 g(0,0,0) + 4\beta_1[g(-k,-k,k)+g(-k,-k,-k)] +\gamma_0[u_x(0,0,0)+u_y(0,0,0)]+\gamma_1[u_x(0,-h,0)+u_y(-h,0,0)] +\gamma_2[u_x(0,-2h,0)+u_y(-2h,0,0)]+\gamma_3[u_x(-3h,0,0)+u_y(0,-3h,0)]+\gamma_4[u_x(0,0,h)+u_x(0,0,-h)+u_y(0,0,h)+u_y(0,0,-h)]+\gamma_5[u_x(0,-h,h)+u_x(0,-h,-h)+u_y(-h,0,h)+u_y(-h,0,-h)]\}$
Neumann corners	$0.125\{\beta_0 g(0,0,0) + 8\beta_1 g(-k,-k,-k) +\delta_0[u_x(0,0,0)+u_y(0,0,0)+u_z(0,0,0)] +\delta_1[u_x(0,-h,0)+u_x(0,0,-h)+u_y(-h,0,0)+u_y(0,0,-h)+u_z(-h,0,0)+u_z(0,-h,0)]+\delta_2[u_x(0,-2h,0)+u_x(0,0,-2h)+u_y(-2h,0,0)+u_y(0,0,-2h)+u_z(-2h,0,0)+u_z(0,-2h,0)]\}$

*These formula apply only in the case of f=0 or $u_n=0$. $F=h^2f$, $k=h/2$, $\beta_0=2-3F/8$, $\beta_1=1/2$, $\gamma_0=23/3$, $\gamma_1=6$, $\gamma_2=-2$, $\gamma_3=1/3$, $\gamma_4=1/2$, $\gamma_5=-1/2$, $\delta_0=-6$, $\delta_1=-4$, $\delta_2=1$

where $k = h/2$, $F = h^2f$, $a = -24+6F-3F^2/8$, $b = 2-F/8$, $c = 1+F/16$, $\beta_0 = 2-3F/8$, and $\beta_1 = 1/2$. For $p_0 \varepsilon \partial R_2$ the right hand side of the discrete system is augmented as shown in Table 3.

V. SOFTWARE AND COMPUTATIONAL EXAMPLES

Two Fortran subprograms, HFFT2D and HFFT3D, implement the algorithms described above for two and three-dimensional problems respectively. We compare the performance of HFFT2D with one widely distributed high order fast solver, the subprogram SEPX4 from FISHPAK (version 3). SEPX4 uses standard second order differences, solving the discrete problem using recursive cyclic reduction [11], which is of the same complexity as the Fourier method. Fourth order accuracy is obtained using deferred corrections [10], which requires a second fast solve using a modified right hand side.

All our tests are run in single precision on a Univac 1100/82 computer using the FTN 10R1 compiler (options OZ). We collect data for SEPX4 both with and without the deferred corrections step (denoted by SEPX4(2) and SEPX4(4) respectively). Timing data includes time to evaluate the forcing function g and the boundary conditions. Since HFFT2D requires only one fast solve we expect its basic execution time to be roughly half that of SEPX4(4). This is verified in Table 4 where we present timings for the homogeneous Dirichlet problem for Laplace's equation on a sequence of square grids. Figure 1 displays the results obtained on four additional test problems.

PROBLEM A. $Lu = \nabla^2 u-100u$, u prescribed on x=1, y=1, u_n prescribed (=0) on x=0, y=0, $u = [\cosh(10x)/\cosh(10) + \cosh(20y)/\cosh(20)]/2$, solution has mild boundary layer along x=1 and y=1.

PROBLEM B. $Lu = \nabla^2 u-20u$, u prescribed on y=0, u_n prescribed on y=1, u periodic in x, $u = \cos[4\pi y] + \sin[4\pi(x-y)]$, solution is smooth and slowly varying.

PROBLEM C. $Lu = \nabla^2 u$, u prescribed on y=0,1, u_n prescribed on x=0,1, $u = (xy)^{5/2}$, solution has singular third derivatives along x=0 and y=0.

PROBLEM D. $Lu = \nabla^2 u$, u prescribed on x=0,1, y=0,1, $u = (x^{3/4}-x)(y^{3/4}-y)$, solution has singular first derivatives along x=0 and y=0.

Since the solutions to problems A and B are entire the convergence rates of each method is as expected (four for SEPX4(4) and HFFT2D, two for SEPX4(2)). For a given grid both the running time and the resulting error are much smaller for HFFT2D, however. The observed convergence rates for problems C and D were 1.5 and 0.7 for all methods. In contrast to SPEX4(4) which shows no advantage over SEPX4(2) for these problems, HFFT2D again produces a smaller error.

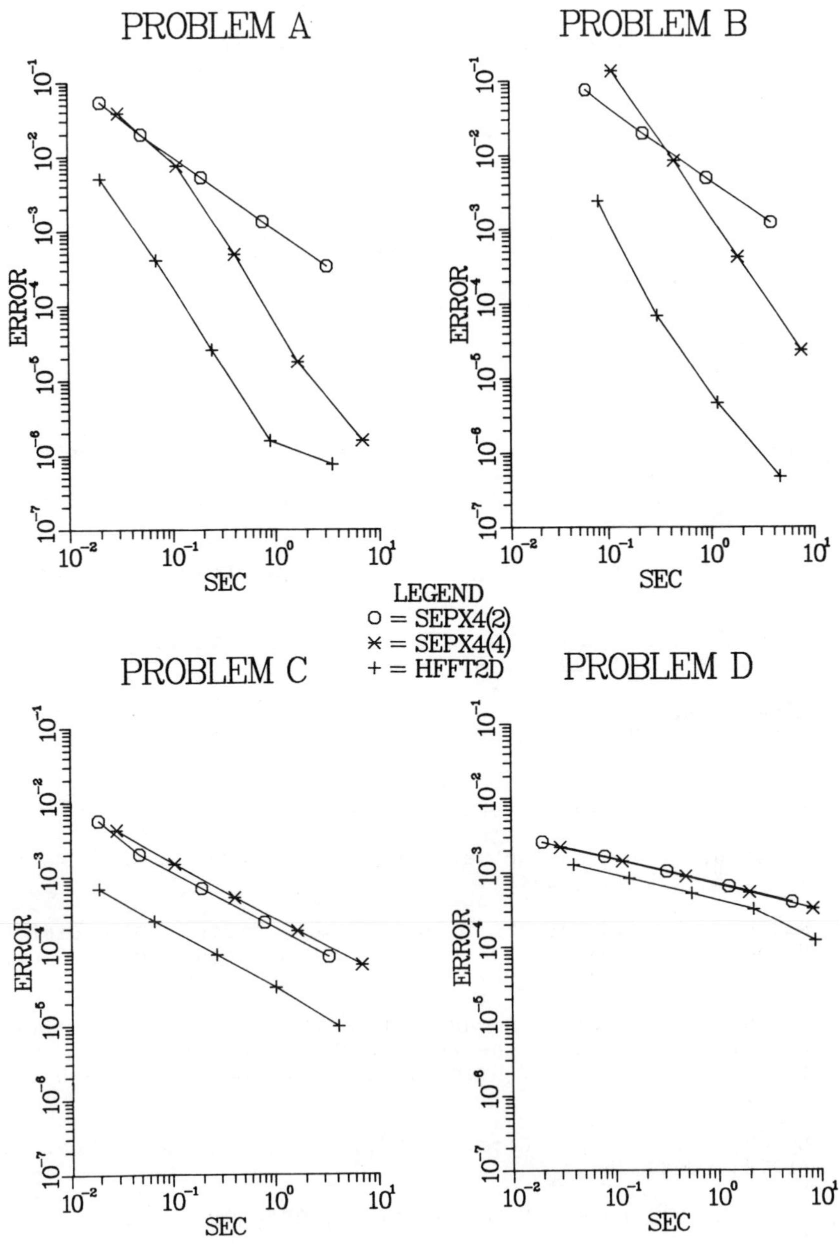

FIGURE 1. Performance of HFFT2D, SEPX4(2), and SEPX4(4) on four sample problems. Graphs show maximum error on grid vs. computing time on a log-log scale. Marker symbols (+,*,o) show results on grids with h=1/8, 1/16, 1/32, 1/64, and 1/128 (except Problem B where h=1/8 is dropped.)

TABLE 4. Raw Timing Data For HFFT2D and SEPX4*

n	SEPX4(2)	SEPX4(4)	HFFT2D
8	0.01	0.02	0.01
16	0.02	0.06	0.04
32	0.10	0.27	0.14
64	0.43	1.14	0.52
128	1.91	4.94	2.10

*in seconds on Univac 1100/82, n=NX-1.

REFERENCES

1. Boisvert, R. F., in "Advances in Computer Methods for Partial Differential Equations," Vol. IV (R. Vichnevetsky and R. S. Stepleman, eds.), IMACS, Rutgers Univ. (1981), 193-199.

2. Boisvert, R. F., SIAM J. Sci. Stat. Comp. 2 (1981), 268-284.

3. Buzbee, B. L., Golub, G. H., and Nielson, C. W., SIAM J. Numer. Anal. 7 (1970), 627-655.

4. Ciment, M., Leventhal, S. H. and Weinberg, B. C., J. Comp. Phys. 28 (1978), 135-166.

5. Collatz, L., "The Numerical Treatment of Differential Equations," Springer-Verlag, Berlin (1960).

6. Houstis, E. N. and Paptheodorou, T. S., ACM Trans. Math. Soft. 5 (1979), 431-441.

7. Kaufman, L. and Warner, D. D., "High Order, Fast-Direct Methods for Separable Elliptic Equations," Tech. Rept. #402, Dept. of Math. Sci., Clemson Univ. (1982).

8. Lynch, R. E., "$O(h^6)$ Accurate Finite Difference Approximation to Solutions of the Poisson Equation in Three Variables," CSD-TR 221, Dept. of Comp. Sci., Purdue Univ. (1977).

9. Lynch, R. E. and Rice, J. R., Proc. Nat. Acad. Sci. 75 (1978), 2541-2544.

10. Pereyra, V., Numer. Math. 8 (1966), 376-391.

11. Swarztrauber, P. N., SIAM Rev. 19 (1977), 490-501.

12. Swarztrauber, P. N. and Sweet, R. A., ACM Trans. Math. Soft. 5 (1979), 352-364.

13. Swarztrauber, P. N., in "Parallel Computation" (G. Rodrique, ed.), Academic Press, New York (1982).

The (New) Yale Sparse Matrix Package[1]

Stanley C. Eisenstat[2]
Howard C. Elman[2]
Martin H. Schultz[2]
Andrew H. Sherman[3]

1. Introduction

The *Yale Sparse Matrix Package* is a collection of routines for solving the $n \times n$ system of linear equations

$$Mx = b \tag{1}$$

when the coefficient matrix M is large and sparse. The current release features direct methods based on Gaussian elimination without pivoting, and is described in Section 2. The next release will also include a number of preconditioned conjugate-gradient-like methods as described in Section 3.

2. The (Old) Yale Sparse Matrix Package

The routines included in the current release of the *Package* [3, 4] are all implementations of Gaussian elimination without pivoting. If M is symmetric, they form the $U^t DU$ decomposition of M, where U is unit upper triangular and D is diagonal, and then successively solve the triangular systems

---

[1] The work presented in this paper was supported in part by the Office of Naval Research under contract N00014-82-K-0184 and by the National Science Foundation under grant MCS-81-04874.

[2] Department of Computer Science and Research Center for Scientific Computation, Yale University, New Haven, Connecticut.

[3] Exxon Production Research Company, Houston, Texas.

ISBN 0-12-100560-7

$$U^t y = b, \qquad Dz = y, \qquad Ux = z \ . \tag{2}$$

If M is nonsymmetric, they form the LU decomposition of M, where L is lower triangular and U is unit upper triangular, and then successively solve the triangular systems

$$Lz = b, \qquad Ux = z \ . \tag{3}$$

Since M is sparse, most entries of M and the triangular factors are zero, and the routines take advantage of this fact by forming the decomposition and solving the triangular systems without storing or operating on zero entries.

Storage of Sparse Matrices

The coefficient matrix M is stored using a row-by-row storage scheme which requires three one-dimensional arrays IA, JA, and A:

- The nonzero entries of M are stored row-by-row in the array A.
- To identify the individual nonzero entries in each row, the column in which each entry lies is stored in the array JA; i.e., if $m_{i,j}$ is stored in A(k), then JA(k) $= j$.
- To identify the individual rows, pointers to the start of each row are stored in the array IA; i.e., if $m_{i,j}$ is stored in A(k) and is the first nonzero entry from the ith row stored in A, then IA(i) $= k$.
- IA(n+1) contains a pointer to the first location following the last entry (stored) in the last row.

Thus, the number of entries in the ith row is IA(i+1) − IA(i), the nonzero entries in the ith row are stored consecutively in

A(IA(i)), A(IA(i)+1), ..., A(IA(i+1)−1),

and the corresponding column indices are stored consecutively in

JA(IA(i)), JA(IA(i)+1), ..., JA(IA(i+1)−1).

Since the coefficient matrix is symmetric, only the nonzero entries in the upper triangle need be stored. For example, the matrix

$$M = \begin{bmatrix} 1 & 0 & 2 & 3 & 0 \\ 0 & 4 & 0 & 0 & 0 \\ 2 & 0 & 5 & 6 & 0 \\ 3 & 0 & 6 & 7 & 8 \\ 0 & 0 & 0 & 8 & 9 \end{bmatrix}$$

could be stored as

	1	2	3	4	5	6	7	8	9	10	11	12	13
IA	1	4	5	8	12	14							
JA	1	3	4	2	1	3	4	1	3	4	5	4	5
A	1	2	3	4	2	5	6	3	6	7	8	8	9

or (symmetrically) as

	1	2	3	4	5	6	7	8	9
IA	1	4	5	7	9	10			
JA	1	3	4	2	3	4	4	5	5
A	1	2	3	4	5	6	7	8	9

Reordering

The work and storage required to solve (1) clearly depend on the zero-nonzero structure of M. Thus it may be advantageous to solve the symmetrically permuted system

$$PMP^t y = Pb, \qquad Px = y$$

for some permutation matrix P (this corresponds to symmetrically reordering the variables and equations).

The ordering driver ODRV selects P by applying the *minimum degree algorithm* to the matrix $V+V^t$, where V denotes the upper triangular part of M. This heuristic algorithm effectively performs a symbolic elimination, at each stage choosing a pivot element from among those uneliminated diagonal entries which require the fewest arithmetic operations to eliminate. Thus it locally optimizes the elimination process with respect to the number of arithmetic operations performed. If the zero-nonzero structure of M is not symmetric, this can yield a bad ordering.

The Solution Process

The solution process is broken into three stages:

1. *Symbolic Factorization*: The coefficient matrix is factored symbolically to determine the zero-nonzero structure of the triangular factors.
2. *Numerical Factorization*: The coefficient matrix is factored numerically taking advantage of the structure information generated by the symbolic factorization to avoid storing or operating on zero entries.
3. *Numerical Solution*: The system (1) is solved by solving a sequence of triangular systems as in (2) or (3).

The advantage of this approach is two-fold:

- *Flexibility*: To solve several systems with the same coefficient matrix, the symbolic and numerical factorization need be done only once (for the first system), and the numerical solution is done once for each right-hand side. To solve several systems whose coefficient matrices have the same zero-nonzero structure, the symbolic factorization need be done only once (for the first system), and the numerical factorization and solution are done once for each system.

- *Efficiency*: Handling fill-in within the innermost loop of the factorization procedure is extremely inefficient (requiring the use of dynamic data structures), unless the elements which fill in are known in advance (allowing the use of a static data structure).

There are four drivers for the solution phase, depending on whether the matrix is symmetric (SDRV) or nonsymmetric (NDRV, CDRV, TDRV), and, in the latter case, whether run-time or storage is more critical.

SDRV computes the U^tDU factorization of a symmetric positive (or negative) definite M, using a compressed storage scheme for U. If M is indefinite (*there is no check for definiteness*), SDRV may break down or produce incorrect results.

NDRV, CDRV, and TDRV all compute the LU factorization of a nonsymmetric M, assuming that such a factorization exists (*there is no pivoting for stability*). NDRV uses the same (uncompressed) storage scheme for L and U as for M. CDRV uses a compressed storage scheme for L and U which can take advantage of structural symmetry; it requires somewhat less storage (but somewhat more run-time) than NDRV. TDRV, which interleaves the symbolic factorization with a combined numerical factorization and solution, uses a compressed storage scheme for U and does not store L; it requires the least storage (but the most run-time).

Availability

Listings and machine-readable versions of the *Package* may be obtained by sending a written request to YSMP LIBRARIAN, BOX 2158 YALE STATION, NEW HAVEN, CT 06520 .

3. The (New) Yale Sparse Matrix Package

The (preconditioned) conjugate gradient and conjugate residual methods have proven to be extremely effective iterative methods for solving large sparse symmetric positive definite systems. However, in order to extend these methods to apply to general nonsymmetric systems, one must give up either the finite termination property (convergence to the exact solution in a finite number of steps) or the fixed term recursion (e.g., the two-term and three-term formulations) for either the solution or the direction vectors [6]. In this section we will describe the planned expansion of the *Package* to include a wide class of such methods.

The Basic Methods

For symmetric positive definite systems, the only method implemented is the conjugate gradient method (the more efficient two-term formulation [9]). Neither the conjugate residual method nor any of the extensions of the conjugate gradient method to symmetric indefinite problems is implemented.

For nonsymmetric systems where the coefficient matrix is positive real (i.e., the symmetric part of M is positive definite), the following conjugate-gradient-like methods are implemented:

- ORTHOMIN [15, 5, 2]
- GCR (Generalized Conjugate Residual) [5, 2]
- IOM/ORTHORES (Incomplete Orthogonalization Method) [13, 17]
- HYBRID (a combination of the Chebyshev semi-iterative method and GMRES [14])

For more general nonsymmetric systems, the following methods are implemented:

- CGNE/CGNR (the Conjugate Gradient method applied to the Normal equations for minimizing the Error/Residual) [5]
- GMRES [14]
- BICG (the BiConjugate Gradient method) [7].

Preconditionings

Instead of solving $Mx = b$, it is often advantageous to solve one of the preconditioned systems

$$[Q^{-1}M]\ [x] = [Q^{-1}b] \qquad \textit{(left preconditioning)}$$

$$[MQ^{-1}]\ [Qx] = [b] \qquad \textit{(right preconditioning)}$$

$$[Q_L^{-1}MQ_R^{-1}]\ [Q_Rx] = [Q_L^{-1}b] \qquad \textit{(split preconditioning)}$$

instead, where $Q\ (= Q_LQ_R)$ is the preconditioning matrix and systems of the form

$$Qz = d, \qquad Q_Lz = d, \qquad Q_Rz = d$$

are easy to solve [5]. Note that if M is symmetric and positive definite, then one must take $Q_L = Q_R^t$ and use the split preconditioning to maintain symmetry.

The following preconditionings are implemented:

- DIAGONAL (the diagonal of M)
- SSOR (Symmetric Successive Over-Relaxation) [1]
- IC/ILU (Incomplete Cholesky / LU) [12]
- MIC/MILU (Modified Incomplete Cholesky / LU) [8]
- SIC/SILU (Shifted Incomplete Cholesky / LU) [11]

In the last three ***approximate factorization*** preconditionings, one specifies the zero-nonzero structure of the lower triangular matrix Q_L and the upper triangular matrix Q_R and forces $Q_L Q_R$ to be close to M. The *Package* will allow the zero-nonzero structure to be specified either explicitly or implicitly (as the maximum order of fill-in) [16].

In order to avoid having a separate routine for each combination of method, preconditioning, and means of applying the preconditioning as in ITPACK [10], the *Package* is organized in an ***orthogonal*** manner. Each preconditioning is implemented by a single driver which is then passed as a parameter to another driver which implements the basic method on the preconditioned system.

Preprocessors

The preconditionings listed in the preceding section can all be implemented implicitly; i.e., the basic method applied to the preconditioned system can be reformulated in terms of the original solution and residual. Another class of preconditionings, which we prefer to call ***preprocessors***, are more easily implemented by forming the new system explicitly and then applying a (possibly preconditioned) basic method to that system. The two preprocessors which are implemented are

- REORDER (symmetric reordering of the variables and equations)
- REDUCE (cyclic reduction of a two-cyclic matrix).

Stopping Criteria

The stopping criterion for the iteration can be based on

- the true residual;
- the preconditioned residual;
- an estimate of the error in the approximate solution.

Availability

The new release of the *Package* exists but will not be available for general distribution before the end of 1983.

References

[1] O. Axelsson. A generalized SSOR method. *BIT* 13:443-467 (1972).

[2] Stanley C. Eisenstat, Howard C. Elman, and Martin H. Schultz. Variational Iterative Methods for Nonsymmetric Systems of Linear Equations. *SIAM Journal on Numerical Analysis* 20:345-357 (1983).

[3] S. C. Eisenstat, M. C. Gursky, M. H. Schultz, and A. H. Sherman. Yale Sparse Matrix Package I: The Symmetric Codes. *International Journal for Numerical Methods in Engineering* 18:1145-1151 (1982).

[4] S. C. Eisenstat, M. C. Gursky, M. H. Schultz, and A. H. Sherman. *Yale Sparse Matrix Package II: The Nonsymmetric Codes.* Research Report #114, Department of Computer Science, Yale University, July 1977.

[5] Howard C. Elman. *Iterative Methods for Large, Sparse, Nonsymmetric Systems of Linear Equations.* Ph. D. Thesis, Department of Computer Science, Yale University, 1982. Also available as Research Report #229.

[6] Vance Faber and Thomas Manteuffel. *Conjugate Gradient Methods.* Technical Report LA-UR-82-1967, Los Alamos National Laboratory, 1982. Submitted to *SIAM Journal on Numerical Analysis.*

[7] R. Fletcher. Conjugate gradient methods for indefinite systems. In G. A. Watson, Editor, *Numerical Analysis Dundee 1975*, Springer, 1976, pp. 73-89.

[8] Ivar Gustafsson. A class of first order factorization methods. *BIT* 18:142-156 (1978).

[9] Magnus R. Hestenes and Eduard Stiefel. Methods of conjugate gradients for solving linear systems. *Journal of Research of the National Bureau of* Standards 49:409-436 (1952).

[10] David R. Kincaid, John R. Respess, David M. Young, and Roger G. Grimes. ITPACK 2C: A FORTRAN package for solving large sparse systems by accelerated iterative methods. *ACM Transactions on Mathematical Software* 8:302-322 (1982).

[11] T. A. Manteuffel. An incomplete factorization technique for positive definite linear systems. *Mathematics of Computation* 34:473-497 (1980).

[12] J. A. Meijerink and H. A. van der Vorst. An iterative solution method for linear systems of which the coefficient matrix is a symmetric M-matrix. *Mathematics of Computation* 31:148-162 (1977).

[13] Y. Saad. Krylov subspace methods for solving large unsymmetric linear systems. *Mathematics of Computation* 37:105-126 (1981).

[14] Youcef Saad and Martin H. Schultz. *GMRES: A Generalized Minimal Residual Algorithm for Solving Nonsymmetric Linear Systems.* Research Report 254, Department of Computer Science, Yale University, 1983.

[15] P. K. W. Vinsome. Orthomin, an iterative method for solving sparse sets of simultaneous linear equations. In *Proceedings of the Fourth Symposium on Reservoir Simulation*, Society of Petroleum Engineers of AIME, February 1976, pp. 150-159.

[16] James W. Watts III. A conjugate gradient-truncated direct method for the iterative solution of the reservoir simulation pressure equation. *Society of Petroleum Engineers Journal* 21:345-353 (1981).

[17] David M. Young and Kang C. Jea. Generalized conjugate-gradient acceleration of nonsymmetrizable iterative methods. *Linear Algebra and Its* Applications 34:159-194 (1980).

THE ITPACK PROJECT: PAST, PRESENT, AND FUTURE

David R. Kincaid & David M. Young*

I. INTRODUCTION

The ITPACK project involves the development of research-oriented mathematical software, based on iterative algorithms, for solving large systems of linear algebraic equations with sparse coefficient matrices. The emphasis is on linear systems arising in the solution of partial differential equations by discretizations such as finite difference or finite element methods. Among other things this project has resulted in the development of a package of subroutines known as ITPACK which is available to the scientific community [14]. Important features of the ITPACK routines include the adaptive determination of the iteration parameters and realistic procedures for terminating the iterative processes. An important application of ITPACK is for use as solution modules in the ELLPACK package of computer routines for solving a class of partial differential equations [18].

We review the objectives of the ITPACK project, summarize the progress made to date and outline plans for future work. A summary is given of work on both software and research that has contributed to the present mathematical software package. The 2C version of ITPACK for scalar computers and a version for use on high performance vector computers are discussed. An overview is given of the developmental work on other software involving more general preconditioners and additional nonsymmetric procedures.

* University of Texas at Austin, Austin, Texas

Work on this project was supported in part by National Science Foundation grant MCS 79-19829, by Department of Energy grant DE-AS05-81ER10954, and by Control Data Corporation grant 81T01 to The University of Texas at Austin.

ISBN 0-12-100560-7

Among the objectives of the ITPACK project are: to write quality software as a research and teaching tool, to provide linear equation solution modules for ELLPACK, to contribute to research on mathematical software methodology, to conduct comparisons between various iterative algorithms, and to facilitate comparisons between iterative methods and direct methods [17,20]. Since flexibility rather than maximum efficiency is stressed in ITPACK, it should successfully handle only moderate-size industrial problems that fit into high-speed memory [1,9]. Nevertheless, its usage might well provide a guide for the construction of large tailor-made production software.

II. PAST

The current version of ITPACK is the result of several years of research and development. The ITPACK project at The University of Texas at Austin dates from the early 1970's when Professor Garrett Birkhoff of Harvard University suggested that a general-purpose software package be written for solving linear systems using iterative methods. While many computer programs had been written in order to conduct basic research on iterative methods, little if any general-purpose software was available. Initially, prototype programs were written based on preliminary iterative algorithms. In 1975, David Kincaid and David Young wrote a paper [6] on the development of a computer package for solving a class of partial differential equations using iterative methods. The algorithms used in the current ITPACK were based in part on technical reports, lecture notes, and books written at the Center for Numerical Analysis. In particular, we used the preliminary draft of a monograph by Louis Hageman and David Young on the effective use of iterative methods for solving large linear systems which was revised and published in 1981 under the title Applied Iterative Methods [5]. A survey of the iterative methods in ITPACK is presented in [10]. The first ITPACK report [7] was written by David Kincaid and Roger Grimes on the numerical studies of several adaptive iterative algorithms. Programs were designed for solving self-adjoint elliptic partial differential equations with Dirichlet boundary conditions on a region with a uniform grid and with horizontal, vertical, or forty-five degree sides. In ITPACK 1.0 [2,8], a major change of direction was introduced by starting from the linear system rather than the partial differential equation. Seven adaptive iterative algorithms for solving large sparse linear systems were included. The matrix manipulation in ITPACK 1.0 was handled by a sub-package written by Bill MacGregor which allowed three interchangeable symmetric sparse storage schemes to be used. While this created a great deal of flexibility and made it possible to study different storage modes, the resulting execution times were rather slow. The next version, ITPACK 2.0 [3] , was significantly faster since it was restricted to only one sparse symmetric storage format. Most of the iterative algorithms utilized in the 2.0 version of this package assume that the coefficient matrix of the linear system is symmetric positive definite. As with many packages, the need to handle a slightly larger class of problems, such as nearly symmetric systems, soon became evident. This required modifying the software to allow either symmetric or nonsymmetric sparse storage in ITPACK 2A [4]. In addition, a modification of the conjugate gradient algorithms was developed for nearly symmetric systems [11-12]. More efficient versions of several key subroutines were incorporated into ITPACK 2B [13]. This package was finalized for publication [14] as ITPACK 2C in 1982 with authors David Kincaid, John Respess, David Young, and Roger Grimes.

III. PRESENT

The ITPACK package includes routines for solving, by several alternative iterative algorithms, large sparse linear systems. In all but one case, each algorithm of ITPACK consists of four components: (1) a basic iterative method such as the Jacobi method, the Reduced System method, or the Symmetric Successive Overrelaxation method, (2) an acceleration method such as Chebyshev acceleration or conjugate gradient acceleration, (3) an automatic procedure for computing acceleration parameters where needed, and (4) a stopping procedure to determine whether the approximate solution is sufficiently close to the exact solution.

Each of the seven main subroutines in ITPACK corresponds to one of the following iterative methods:

NAME	ITERATIVE METHOD
JCG()	Jacobi Conjugate Gradient
JSI()	Jacobi Semi-iteration
SOR()	Successive Overrelaxation
SSORCG()	Symmetric SOR Conjugate Gradient
SSORSI()	Symmetric SOR Semi-iteration
RSCG()	Reduced System Conjugate Gradient
RSSI()	Reduced System Semi-iteration

The calling sequence is of the form

```
CALL <name>(N,IA,JA,A,RHS,U,IWKSP,NW,WKSP,IPARM,RPARM,IER)
```

where the parameters in the calling sequence correspond to the order of the system (N), the coefficient matrix in sparse-storage format (IA, JA, A), the right-hand side

(RHS), the initial and final solution vector (U), integer and real workspace arrays of a given length (IWKSP, WKSP, NW), integer and real parameter arrays (IPARM, RPARM), and an error flag (IER). The adaptive iterative procedures can be user-guided to some degree by specifying values in each of the 12-component parameter arrays IPARM and RPARM after assigning the default values with a call to subroutine DFAULT (IPARM, RPARM) and before calling one of the ITPACK iterative methods. The initial approximation to the solution of the linear system can be set to a constant vector with a call to subroutine VFILL (N, U, VAL). Hence, a standard example of the usage of ITPACK might be:

```
...
ITMAX = 125
ZETA = 5.E-6
...
CALL DFAULT (IPARM, RPARM)
IPARM(1) = ITMAX
RPARM(1) = ZETA
CALL VFILL (N, U, 0.E0)
CALL JCG (N,IA,JA,A,RHS,U,IWKSP,NW,WKSP,IPARM,RPARM,IER)
...
```

Here the default values are set in the parameter arrays IPARM and RPARM. Then the maximum number of iterations allowed is reset to ITMAX and the stopping criterion is changed to ZETA. The zero vector is used as the starting solution in the call to subroutine JCG(). (See [14] for additional details.)

IV. ITPACK IN ELLPACK

ITPACK can be used as a stand-alone package, as described above, or as part of the ELLPACK system [18] for solving elliptic partial differential equations. Parts of ITPACK 2C were submitted as solution modules for this system after changing them to use the ELLPACK data structure. In the most recent version of ELLPACK, linear systems are generated by discretization modules and the ITPACK solution modules solve them without translation to a different storage format. Thus, a considerable savings in the memory allocation is achieved.

The use of ITPACK within ELLPACK can best be illustrated by an example. Here a partial differential equations on the unit square is solved using central differences (DIS. 5 POINT STAR) with mesh spacing 1/40. The resulting symmetric positive definite linear system is of order 1521 and is solved by the Jacobi Conjugate Gradient method with the natural ordering (INDEX. AS IS) of the grid points and by the Reduced System Conjugate Gradient method with the red/black ordering

(INDEX. RED-BLACK). The algorithmic procedures in ITPACK can be controlled to some degree within ELLPACK by the use of private parameters which are enclosed in parentheses following the solution module name. The initial guess to the solution vector is the zero vector if not user specified and an ITPACK module iterates producing a sequence of approximate solution vectors until the convergence criterion is satisfied. The sample ELLPACK program is:

```
*
*       ELLPACK EXAMPLE PROGRAM
*
OPT.    TIME     $        LEVEL = 1
EQ.     (X*UX)X + (Y*UY)Y = X + Y
BOUND.  U = TRUE(X,Y)     ON       X = 0.0
                          ON       X = 1.0
                          ON       Y = 0.0
                          ON       Y = 1.0
GRID.   41 X-POINTS
        41 Y-POINTS
DIS.    5 POINT STAR
INDEX.  AS IS
SOL.    JACOBI CG ( ITMAX = 125, ZETA = 5.E-6 )
INDEX.  RED-BLACK
SOL.    REDUCED SYSTEM CG ( ITMAX = 125, ZETA = 5.E-6 )
SUB.    REAL FUNCTION TRUE (X,Y)
        TRUE = 1.E0 + X*Y
        RETURN
        END
END.
```

Part of the ELLPACK output is:

```
                ----------------
                 ELLPACK OUTPUT
                ----------------

                        +++++++++++++++++++++++++
                        +    EXECUTION  TIMES   +
                        +++++++++++++++++++++++++

                     MODULES NAME                     SECONDS
                     -----------------------------------

                     5-POINT STAR                        .53
                     AS IS                               .01
                     JACOBI CG SETUP                    0.00
                     JACOBI CG                          2.86
                     RED-BLACK                           .05
                     REDUCED SYSTEM CG SETUP            0.02
                     REDUCED SYSTEM CG                  1.22
                     TOTAL TIME                         4.70
        . . .
        JACOBI CG  HAS CONVERGED IN     107   ITERATIONS
        . . .
        REDUCED SYSTEM CG  HAS CONVERGED IN     55   ITERATIONS
        . . .
```

V. FUTURE

We plan to carry out a major expansion of the ITPACK package based on the results of research on iterative algorithms. The expanded ITPACK will be written in a modular form so that any one of several basic iterative methods (preconditioners) can be used in combination with any one of several acceleration procedures. This will provide greater flexibility than is possible with the current software in ITPACK where there is a separate subprogram for each basic method/acceleration procedure combination. Moreover, the expanded package will include iterative procedures specifically designed for nonsymmetric linear systems whereas the algorithms in the present package were designed primarily for symmetric positive definite systems. It should be noted that the modular approach is somewhat complicated by the fact that in some cases several iteration parameters are involved. Thus, for example, the SSOR method with Chebyshev acceleration needs both an estimate for the largest eigenvalue of the iteration matrix and for the relaxation factor.

For the expanded package, modules will be prepared for several basic iterative preconditioners including methods based on approximate matrix factorization. Modules will also be written for various acceleration procedures which can be applied to nonsymmetrizable iterative methods. Such procedures include the generalized conjugate gradient acceleration procedures ORTHODIR, ORTHOMIN, & ORTHORES, as well as the Lanczos method. In the symmetrizable case, the conjugate residual acceleration and the two- and three-term conjugate gradient acceleration will be available. The routines of the expanded ITPACK will be tested over a wide variety of problems including many generated by ELLPACK. We plan to prepare complete documentation by writing a guide which will describe the usage of this package as well as the algorithms used.

We plan to continue to study the properties of the generalized conjugate gradient acceleration procedures for solving nonsymmetrizable iterative methods. In particular, there is a need for a better theoretical understanding of the properties of truncated generalized conjugate gradient methods. We plan to continue our studies of breakdown, convergence, and rate of convergence under various assumptions on the properties of the coefficient matrix and the preconditioning matrix. Another primary focus of our future work is to conduct a comparative study of a number of preconditioned matrices and their effectiveness when used with various acceleration procedures, especially for problems arising from the solution of partial differential equations. Studies will be made for the cases where the coefficient matrix is symmetric positive definite, nearly symmetric positive definite, symmetric indefinite, positive real, and for the general case. (For additional details, see the companion article [21] in this volume on accelerating nonsymmetrizable iterative methods.)

Another area of research involves the question of when to stop an iterative process. Realistic stopping tests have been developed for the symmetrizable case and are incorporated into the present ITPACK. However, the procedures which we have developed for the nonsymmetrizable case involve the determination of eigenvalues of certain nonsymmetric matrices which can be somewhat costly. We plan to establish more efficient procedures.

VI. VECTOR ITPACK

An area for additional studies in both research and software is the development of vectorizable algorithms so that the procedures can be effectively used on high performance computers.

The ITPACK 2C software has been adapted for efficient use on supercomputers such as the CDC CYBER 205. Since this is a rather large package of code, we wanted to determine what obvious changes were needed in the software to produce vectorizable code without completely rewriting the entire package. (See reports by Kincaid, Oppe, and/or Young [15-16].) In this effort, we decided to use a minimum amount of the vector syntax available in the CYBER 200 Fortran since we wanted the resulting vector version of ITPACK 2C to run on other supercomputers such as the CRI CRAY 1. We shall call the version of ITPACK 2C published [14] "scalar ITPACK 2C" and the version for supercomputers "vector ITPACK 2C" in order to clarify the discussion.

In adapting scalar ITPACK 2C for high performance computers, two major changes in the software were required. First, DO loops which had been unrolled in the original package for scalar optimization were not recognized by optimizing vector compilers as vectorizable. Hence, these loops were rewritten as simple tight DO loops so that they would be executed in vector mode. Second, the data structure used to store the iteration matrix was found to be of critical importance in vectorization. The data structure used in scalar ITPACK 2C is row-oriented which inhibits vectorization of many matrix operations. For increased vectorization, vector

ITPACK 2C uses a column-oriented data structure similar to the one in ELLPACK.

As an example of these sparse storage schemes, consider the matrix-vector product

$$\begin{vmatrix} a_{11} & a_{12} & 0 & a_{14} \\ a_{21} & a_{22} & a_{23} & 0 \\ a_{31} & 0 & a_{33} & 0 \\ 0 & a_{42} & 0 & a_{44} \end{vmatrix} \begin{vmatrix} x_1 \\ x_2 \\ x_3 \\ x_4 \end{vmatrix}$$

In scalar ITPACK, the coefficient matrix might be stored as follows:

Scalar ITPACK - Nonsymmetric Sparse Storage

$$\begin{aligned} A &= (a_{11}, a_{12}, a_{14}, a_{21}, a_{22}, a_{23}, a_{31}, a_{33}, a_{42}, a_{44}) \\ JA &= (1, 2, 4, 1, 2, 3, 1, 3, 2, 4) \\ IA &= (1, 4, 7, 9, 11) \end{aligned}$$

The coefficient matrix is stored in a 2 dimensional array in vector ITPACK as follows:

Vector ITPACK - Sparse Storage

$$\text{COEF} = \begin{vmatrix} a_{11} & a_{12} & a_{14} \\ a_{22} & a_{21} & a_{23} \\ a_{33} & a_{31} & 0. \\ a_{44} & a_{42} & 0. \end{vmatrix}$$

$$\text{JCOEF} = \begin{vmatrix} 1 & 2 & 4 \\ 2 & 1 & 3 \\ 3 & 1 & 1 \\ 4 & 2 & 1 \end{vmatrix}$$

As an illustration of the type of recoding that was done for vector ITPACK 2C to accommodate the COEF/JCOEF data structure, notice that the matrix-vector product given previously can be computed in a vector sense as follows:

$$\begin{vmatrix} a_{11} \\ a_{22} \\ a_{33} \\ a_{44} \end{vmatrix} \begin{vmatrix} x_1 \\ x_2 \\ x_3 \\ x_4 \end{vmatrix} + \begin{vmatrix} a_{12} \\ a_{21} \\ a_{31} \\ a_{42} \end{vmatrix} \begin{vmatrix} x_2 \\ x_1 \\ x_1 \\ x_2 \end{vmatrix} + \begin{vmatrix} a_{14} \\ a_{23} \\ 0. \\ 0. \end{vmatrix} \begin{vmatrix} x_4 \\ x_3 \\ x_1 \\ x_1 \end{vmatrix}$$

The matrix associated with the Jacobi method is of the form $B = L + U$ where L and U are strictly lower and upper triangular matrices, respectively. When the red/black ordering is used it assumes the block form

$$B = L + U = \begin{vmatrix} 0 & H \\ K & 0 \end{vmatrix}$$

Depending on the iterative method and the ordering to be used, the elements of COEF/JCOEF are automatically moved around, when possible, into one or the other of the following forms to increase vectorization of the basic step.

Storage for Natural Ordering

```
       | d11 |       .         |       .         |
       |  .  |       .         |       .         |
       |  .  |       .         |       .         |
COEF = |  .  | nonzeros of U   | nonzeros of L   |
       |  .  |       .         |       .         |
       |  .  |       .         |       .         |
       | dnn |       .         |       .         |
```

Storage for Red/Black Ordering

```
       | d11 |       .         |
       |  .  | nonzeros of H   |
       |  .  |       .         |
COEF = |  .  | -------------   |
       |  .  |       .         |
       |  .  | nonzeros of K   |
       | dnn |       .         |
```

where the d_{ii} are diagonal elements. (See [15-16] for additional details.)

VII. ACKNOWLEDGMENT

We wish to thank Thomas C. Oppe and John R. Respess for reading a draft of this paper and offering helpful suggestions.

VIII. REFERENCES

1. Eisenstat, S., George, A., Grimes, R., Kincaid, D., and Sherman, A., "Some Comparisons of Software Packages for Large Sparse Linear Systems," in Advances in Computer Methods for Partial Differential Equations III, (R. Vichnevetsky and R. Stepleman, eds.), publ. IMACS, Dept. Computer Science, Rutgers Univ., New Brunswick, New Jersey, 08903, 1979, pp. 98-106.

2. Grimes, R., Kincaid, D., MacGregor, W., and Young, D., "ITPACK Report: Adaptive Iterative Algorithms Using Symmetric Sparse Storage," CNA-139, Center for Numerical Analysis, Univ. of Texas, Austin, Texas, 78712, August 1978.

3. Grimes, R., Kincaid, D., and Young, D., "ITPACK 2.0 User's Guide," CNA-150, Center for Numerical Analysis, Univ. of Texas, Austin, Texas, 78712, August 1979.

4. Grimes, R., Kincaid, D., and Young, D., "ITPACK 2A: A Fortran Implementation of Adaptive Accelerated Iterative Methods for Solving Large Sparse Linear Systems," CNA-164, Center for Numerical Analysis, Univ. of Texas, Austin, Texas, 78712, October 1980.

5. Hageman, L. , and Young, D., Applied Iterative Methods, Academic Press, New York, 1981.

6. Kincaid, D., and Young, D., "The Development of a Computer Package for Solving a Class of Partial Differential Equations by Iterative Methods," CNA-102, Center for Numerical Analysis, Univ. of Texas, Austin, Texas, 78712, July 1975.

7. Kincaid, D. , and Grimes, R., "Numerical Studies of Several Adaptive Iterative Algorithms," CNA-126, Center for Numerical Analysis, Univ. of Texas, Austin, Texas, 78712, August 1977.

8. Kincaid, D., Grimes, R., MacGregor, W., and Young, D., "ITPACK - Adaptive Iterative Algorithms Using Symmetric Sparse Storage," in Symposium on Reservoir Simulation, SPE-7687, Soc. Pet. Engr. of AIME, 6200 North Central Expressway, Dallas, Texas, 75206, February 1979, pp. 151-160.

9. Kincaid, D., Grimes, R., and Young, D., "The Use of Iterative Methods for Solving Large Sparse PDE-Related Linear Systems," Mathematics and Computers in Simulation XXI, North-Holland Publ. Co., New York, 1979, pp. 368-375.

10. Kincaid, D., and Young, D., "Survey of Iterative Methods," in Encyclopedia of Computer Sciences and Technology, Vol. 13 (J. Belzer, A. Holzman, and A. Kent, eds.), Marcel Dekker, Inc., New York, 1979, pp. 354-391.

11. Kincaid, D., and Young, D., "Adapting Iterative Algorithms Developed for Symmetric Systems to Nonsymmetric Systems," in Elliptic Problem Solvers, (M. Schultz, ed.), Academic Press, New York, 1981, pp. 353-359.

12. Kincaid, D., "Acceleration Parameters for a Symmetric Successive Overrelaxation Conjugate Gradient Method for Nonsymmetric Systems," in Advances in Computer Methods for Partial Differential Equations IV, (R. Vichnevetsky and R. Stepleman, eds.), publ. IMACS, Dept. Computer Science, Rutgers Univ., New Brunswick, New Jersey, 08903, 1981, pp. 294-299.

13. Kincaid, D., Grimes, R., Respess, J., and Young, D., "ITPACK 2B: A Fortran Implementation of Adaptive Accelerated Iterative Methods for Solving Large Sparse Linear Systems," CNA-173, Center for Numerical Analysis, Univ. of Texas, Austin, Texas, 78712, September 1981.

14. Kincaid, D., Respess, J., Young, D., and Grimes, R., "Algorithm 586 ITPACK 2C: A FORTRAN Package for Solving Large Sparse Linear Systems by Adaptive Accelerated Iterative Methods," ACM Transactions on Mathematical Software, Vol. 8, No. 3, September 1982, pp. 1302-322.

15. Kincaid, D., Oppe, T., Young, D., "Adapting ITPACK Routines for Use on a Vector Computer," CNA-177, Center for Numerical Analysis, Univ. of Texas, Austin, Texas, August, 78712, August 1982.

16. Kincaid, D., and Oppe, T., "ITPACK on Supercomputers," CNA-178, Center for Numerical Analysis, Univ. of Texas, Austin, Texas, 78712, September 1982.

17. Lawson, C., Hanson, R., Kincaid, D., and Krogh, F., "Basic Linear Algebra Subprograms for Fortran Usage," ACM Transactions on Mathematical Software, Vol. 5., No. 3, September 1979, pp. 308-323.

18. Rice, J., "ELLPACK User's Guide," CSD-TR 372, Computer Science Department, Purdue Univ., West Lafayette, Indiana, 47907, July 1981.

19. Young, D., Iterative Solution of Large Linear Systems, Academic Press, New York, 1971.

20. Young, D., and Kincaid, D., "The ITPACK Package for Large Sparse Linear Systems," in Elliptic Problem Solvers ,(M. Schultz, ed.), Academic Press, New York, 1981, pp. 163-185.

21. Young, D., Jea, K., and Kincaid, D., "Accelerating Nonsymmetrizable Iterative Methods," this volume.

CMMPAK - THE CAPACITANCE MATRIX SOFTWARE PACKAGE

Wlodzimierz Proskurowski[1]

In this paper we are looking at fast and efficient numerical solvers. In order to quantify the efficiency let us define Fast Elliptic Solvers (FES), in the manner of Detyna (1981), as solvers that produce the solution in s operations per unknown, where s is a constant or a very slowly (logarithmically) growing function of the number of unknowns. Let N be the number of mesh points inside a rectangular region (in some coordinate system). The Helmholtz equation: $-\Delta u + cu = f$, $c = \text{const}$, in such regions can be solved efficiently at a cost proportional to NlogN arithmetic operations (and essentially using N storage locations), or logN operations per unknown. See Swarztrauber-Sweet (1979) as an example of such Helmholtz solvers. These solvers satisfy our definition of FES since logN is a slowly growing function of N.

The capacitance matrix method (CMM) extends the usefulness of such solvers to non-rectangular regions. An iterative variant of CMM developed in O'Leary-Widlund (1979) and Proskurowski (1979) makes use of the fact that only a product of the capacitance matrix C and a given vector is required

---

[1]Department of Mathematics, University of Southern California, Los Angeles, California, 90089-1113.

Supported in part by a National Science Foundation Grant MCS-8ØØ3383 at the University of Southern California

ISBN 0-12-100560-7

(without any explicit knowledge of the matrix C) and this product can be obtained essentially at the cost of a Fast Solver. In addition, the discrete potential theory developed in Proskurowski-Widlund (1976) guarantees that the conjugate gradient iteration with a properly constructed matrix C will converge almost independently of the size of the mesh, see also Shieh (1978, 1979). All in all, in such implicit CMM a solution can be obtained at a cost proportional to several calls of a FES (of the order of 10). Therefore the CMM can be regarded as satisfying the definition of FES, with the reservation that here the constant s is large: of one order higher than for the solvers on rectangular regions. For details on the efficiency of the implicit CMM see Proskurowski (1981). In the explicit CMM, in addition to calling a FES once or twice, one needs to generate and solve a large and dense system of equations, called the capcitance matrix C, representing the modifications imposed by the irregular boundary. The operations connected with this matrix are very costly and require additional storage of kN locations, where k is of the order of 10, see Buzbee-Dorr-George-Golub (1971). In the explicit CMM the capacitance matrix C can be generated and decomposed into triangular factors only once (in the preprocessing stage). Therefore for problems with multiple right hand sides, only the backsolving of the system with matrix C is needed, which substantially reduces the cost per single solution. On the other hand, if p, the number of the mesh points adjacent to the boundary is large then the size of the matrix C increases rapidly and the fast memory is easily saturated.

Here we present the outline of the existing CMM package in 2D. For details see Proskurowski (1983). The package consists of three solvers: CMMIMP, CMMEXP and CMMSIX. Each of these solvers computes the finite difference approximation to the solution of the Helmholtz equation in cartesian coordinates $-\Delta u(x,y) + cu(x,y) = f(x,y)$ on a non-rectangular 2D region. Here Δ is the Laplacian and c is a real constant. On the boundary of this region either the solution (Dirichlet condition) or the normal derivative of the solution is specified (Neumann condition).

CMMIMP - is an implicit (iterative) solver well suited for very large meshes; up to approximately N=5ØØ*5ØØ and p=15ØØ (assuming 65K core storage), and for problems with only one or very few right hand sides that are solved in the same irregular region.

CMMEXP - is an explicit (direct) solver that should be used whenever the capacitance matrix C can be stored in the fast memory, i.e., roughly for meshes up to N=64*64 and p=15Ø assuming 65K core storage. This solver can be used efficiently for problems with multiple right hand sides.

Both CMMIMP and CMMEXP produce a solution with second order accuracy for the Dirichlet problem and with first order accuracy for the Neumann problem.

For the Neumann problem, a solution may not exist if the consistency condition is not satisfied. In such cases CMMIMP and CMMEXP compute a least squares solution to the original approximation. This solution is not unique and is unnormalized.

CMMSIX - is a variant of CMMEXP producing the solution to the Dirichlet problem with higher order accuracy. Fourth or sixth order accuracy is obtained by using one or two deferred corrections in an inexpensive way; see Pereyra-Proskurowski-Widlund (1977). Note that CMMSIX is limited to constants c=Ø and the Dirichlet boundary conditions.

All these programs use a special fast solver, STRIP, with periodic boundary conditions on an infinite strip. This choice of solver has its advantages and its limitations.

STRIP exploits the translation invariancy of the solution with these boundary conditions, which significantly decreases the cost for the generation of the matrix C. This is efficiently exploited in CMMEXP and CMMSIX, see Proskurowski-Widlund (1976).

Moreover, STRIP is designed so that either sparse or fast Fourier transform is performed for the analysis and for the synthesis of the data. In this way it can operate in four different modes: sparse-sparse, sparse-fast, fast-sparse or fast-fast transforms. This design makes STRIP very flexible, which is advantageously exploited in the package. The increased efficiency whenever transformation is performed from or to a sparse set is documented in Proskurowski (1979).

On the other hand, if the irregular region has one or more sides parallel to the coordinate axes, the use of STRIP limits the flexibility of the solvers insomuch as it cannot match the prescribed boundary conditions on such sides. In such cases it then leads to the unnecessary increase of the size of the capacitance matrix.

The COMMENT cards at the beginning of each subroutine give detailed description of the calling sequence, parameters on input and output, and some additional useful information, e.g. memory space requirement.

Input data for the package (variables and vectors specified in the drivers) are defined through subroutines DOMAIN, BNDRY and CHARGE. These subroutines are supplied in the package (for the test example described below). The user can easily test the package with different parameters, such as DIR,LAPL,CON,N,M or H, and for regions with different radii and centra.

Assume that we wish to solve the Poisson equation in a circular region with center at (Ø.5,Ø.5) and radius Ø.375. The solution specified at the boundary of the region (Dirichlet conditions) is chosen so that the exact solution to this problem is

$$u(x,y) = x^5 + y^5.$$

The circular region is imbedded in a unit square and we choose a square grid of length h=1/16 for our approximation.

On the next two pages, we list the driver program DRIMP that defines the necessary input parameters for the subroutine CMMIMP.

```
C
C     DRIVER PROGRAM TO ILLUSTRATE THE USE OF CMMIMP
C
      REAL F(2Ø,2Ø),W(3ØØ),HX(4Ø),HY(4Ø),
     *           TX(4Ø),TY(4Ø),BX(4Ø),BY(4Ø)
      INTEGER IW(8ØØ),IC(4Ø),JC(4Ø)
      LOGICAL DIR,LAPL,TEST
C
C     NDIM IS THE FIRST DIMENSION OF THE ARRAY F.
C     THIS PARAMETER IS USED TO SPECIFY THE VARIABLE
C     DIMENSION OF F.  NDIM MUST NOT BE LESS THAN N.
C
C     LEW IS THE TOTAL DIMENSION OF THE WORKSPACE W.
C     FOR THE DIRICHLET PROBLEM
C             LEW MUST NOT BE LESS THAN 6*IP;
C     FOR THE NEUMANN PROBLEM
C             LEW MUST NOT BE LESS THAN 7*IP.
C
C     LEIW IS THE TOTAL DIMENSION OF THE WORKSPACE IW.
C     FOR THE DIRICHLET PROBLEM
C             LEIW MUST NOT BE LESS THAN 19*IP;
C     FOR THE NEUMAN PROBLEM
C             LEIW MUST NOT BE LESS THAN 13*IP.
C
C     IMPAX IS THE DIMENSION OF VECTORS IC,JC,HX,HY,
C     TX,TY,BX,BY AND IS AN UPPER ESTIMATE OF IP,
C     THE NUMBER OF IRREGULAR MESH POINTS.
C
C                         ********************
C                         *    INITIALIZE    *
C                         ********************
C
      NDIM=2Ø
      LEW=3ØØ
      LEIW=8ØØ
      IMPAX=4Ø
      CON=Ø.Ø
      DIR=.TRUE.
      LAPL=.FALSE.
      TEST=.TRUE.
      ACCY=1E-3
      ITMAX=2Ø
      N=16
      M=N
      IOUT=6
C
C     THE TEST IS PERFORMED IN THE CIRCULAR REGION
C
C     ( X - Ø.5 ) ** 2 + ( Y - Ø.5 ) ** 2 = R ** 2,
C
C     WHERE R=Ø.375,
C
C     WITH THE EXACT SOLUTION BEING U(X,Y) = X**5 + Y**5.
```

```
C
C                         ********************
C                         *    INPUT DATA    *
C                         ********************
C
C       PROCESS THE INFORMATION
C       ABOUT THE REGION AND ITS BOUNDARY.
C
C       INPUT DATA ABOUT THE REGION:  DEFINE IP, IPP, H,
        IC,JC,HX,HY, AND TX,TY (FOR THE NEUMANN PROBLEM).
C
        CALL DOMAIN(N,M,IC,JC,HX,HY,TX,TY,IPMAX,IP,IPP,
     *  H,DIR)
C
C       INPUT BOUNDARY DATA:  DEFINE BX AND BY.
C
        CALL BNDRY(IC,JC,HX,HY,BX,BY,IP,IPP,H,DIR)
C
C       INPUT RIGHT SIDE:  DEFINE F
C
        CALL CHARGE (F,NDIM,N,M,H,CON)
C
C                         ********************
C                         *   CALL CMMIMP    *
C                         ********************
C
        CALL CMMIMP(IC,JC,HX,HY,TX,TY,BX,BY,F,NDIM,N,M
     *  H,CON,DIR,LAPL,TEST,ACCY,ITMAX,IPMAX,IP,IPP,
     *  W,LEW,IW,LEIW,IOUT,IERROR)
C
        IF(IERROR.GT.Ø) WRITE(IOUT,9999) IERROR
C
C                         ********************
C                         *     PRINTOUT     *
C                         ********************
C       FOR THE TEST EXAMPLE
C       PRINT OUT THE NUMBER OF CORRECT DIGITS
C       FOR ALL POINTS INSIDE THE IRREGULAR REGION.
C
        IF(TEST.AND.IERROR.EQ.Ø)
     *  CALL OUTPUT(F,NDIM,N,M,H,DIR,IOUT)
C
C       NOTE:  SUBROUTINES DOMAIN, BNDRY, CHARGE AND OUTPUT
C       ARE USED ONLY IN THIS DRIVER
C       AND ARE NOT REQUIRED IN CMMIMP.
C
9999    FORMAT(//1X,19H DETECTED ERROR NO.,I3//)
        STOP
        END
```

An acknowledged deficiency of CMMPAK is that the user is required to input the information about the region (and the rest of the data) in a rather sophisticated manner.

The use of the 2D domain processor designed by Rice (1982) can remedy this problem in a satisfactory way. All the user must do in order to describe the irregular region is define the number of boundary pieces, the range of the parameter along each piece, and a routine that returns (x,y) for each value of the parameter (the boundary can also be described discretely by the set of boundary-grid intersections). The domain processor then generates automatically all the necessary information, such as the number and coordinates of boundary points, the data structures connected with them, etc.

In recent months, I have been working on linking this domain processor with CMMPAK. The task consisted in transforming the information generated by the processor into the form that is required by CMMPAK. As a result codes for subroutines DOMAIN, BNDRY and CHARGE have been developed and incorporated into the body of CMMPAK, and the user no longer needs to supply these routines. The final product will be included in the ELLPAK system.

Here follows a simulated data input for the same example as above, i.e. a circular region with the center at (ø.5,ø,5) and radius ø.375, imbedded in a unit square with a square grid of 1/16.

Box: ∅.∅,1.∅, ∅.∅,1.∅

Number of grids: 16,16

Ncase: 1 (circle)

Center and Radius: ∅.5,∅.375

Number of boundary pieces 1

Range of the parameter: ∅.∅,2.∅

The user-supplied routine to define the boundary looks now as follows:

```
      SUBROUTINE BCOORD(P,X,Y,IPIECE)
      DATA PI/3.14159265359/
      COMMON/CIRCLE/CENTER,RAD
      GOTO(1∅∅∅),IPIECE
1∅∅∅  CONTINUE
      X=CENTR+COS(PI*P)*RAD
      Y=CENTR+SIN(PI*P)*RAD
      RETURN
      END
```

The convenience of the user is at the expense of some additional computational time and storage required for the domain processor. The detailed results will be presented elsewhere.

REFERENCES

1. Buzbee, B.L., Dorr, F.W., George, J.A. and Golub, G.H. SIAM J.Numer.Anal, 8 (1971), 722-736.

2. Detyna, E. pp. 245-265 in I.S.Duff (ed.), Sparse matrices and their uses, Academic Press, 1981.

3. O'Leary, D.P. and Widlund O. Math.Comp., 33 (1979), 849-879.

4. O'Leary, D.P. and Widlund, O. Algorithm 572, ACM Trans.Math.Software, 7 (1981), 239-246.

5. Pereyra, V., Proskurowski, W. and Widlund, O. Math.Comp., 31 (1977), 1-16.

6. Proskurowski, W. Report LBL-7516, Lawrence Berkeley Laboratory, 1978.

7. Proskurowski, W. ACM Trans.Math.Software, 5 (1979), 36-49.

8. Proskurowski, W. pp. 274-280 in Vichnevetsky R. and Stepleman R.S. (eds.), Advances in computer methods for PDEs - IV, IMACS Publ., 1981.

9. Proskurowski, W. pp. ACM Trans.Math.Software, 9 (1983), 117-124.

10. Proskurowski, W. and Widlund, O. Math.Comp., 30 (1976), 433-468.

11. Rice, J.R. Report CSD-TR 417, Purdue Univ. (1982).

12. Shieh, A.S. Numer.Math., (1978) 29, 307-327.

13. Shieh, A.S. Numer.Math., 31(1979), 405-429.

14. Swarztrauber, P. and Sweet, R. Algorithm 541, ACM Trans.Math.Software, 5 (1979), 352-364.

II

Vector and Parallel Processing

SOLVING ELLIPTIC PROBLEMS ON AN ARRAY PROCESSOR SYSTEM

Martin H. Schultz[1]

I. INTRODUCTION

Elliptic problems not only are very important in their own right, but often occur as the compute-intensive innermost loop of large-scale scientific computations. It is thus incumbent upon us to fully understand the algorithm-software-architecture issues for such problems. In the Proceedings of the 1980 meeting on Elliptic Problem Solvers [4], we gave a hypothetical discussion of the application of array processor systems to solving elliptic problems. While such systems have been successfully applied to cost-effective signal processing, it is by no means obvious that they would be effective in this application. In this paper we continue our discussion (on a less hypothetical plane), summarizing our progress to date and describing potential future developments.

In Section II, we discuss array processor systems and our experience in using one, the Floating Point Systems AP-164, to solve the Poisson equation. We discuss three standard algorithms based on FFTs, their implementation and algorithmic complexity on the AP-164, and numerical experiments. We show that for these algorithms the AP-164 is essentially as fast as published benchmarks for the CDC-7600.

---

[1]Research Center for Scientific Computation, Department of Computer Science, Yale University, New Haven, Connecticut 06520. This work was supported in part by ONR Grant #N00014-82-K-0184, NSF Grant #MCS-8106181 and by External Research Grants from Digital Equipment Corp. and Floating Point Systems.

ISBN 0-12-100560-7

In Section III, we consider future developments for very large three-dimensional problems. In particular, we consider the effect of using secondary storage. We compare three typical systems: (1) the AP-164 with a disk; (2) the Cray-1 with a disk; and (3) the AP-164 with a bulk memory system under development. We show that the AP-164 with bulk memory system is the fastest of the three systems for three dimensional problems with less than one billion unknowns. Furthermore, we show how sufficiently fast FFT devices could be integrated into the system to produce an order of magnitude speedup. Finally we will briefly discuss the ELI-512, a new array processor being designed at Yale; cf. [5] for more details.

II. ELLIPTIC PROBLEM SOLVERS

It has been observed for several years that Very Long Instruction Word (VLIW) computers and specialized processors are especially cost effective for signal processing. We have been seeking to determine their effectiveness for solving elliptic problems. It has also been observed that the programming effort for these machines is significantly greater in general than for serial computers and we will address that issue.

What are VLIW computers? They are computers with three basic properties:

(1) there is one central control unit issuing a single wide instruction per cycle;

(2) each wide instruction consists of many coupled independent operations; and

(3) each operation requires a small, statically predictable number of cycles to execute (operations may be pipelined). The Floating Point Systems AP-164 is the example of such a machine which we will consider in detail. The ELI-512 is another example.

If we are going the route of specialized processors, one may ask why we bother with VLIW computers at all. Our argument is very simple. Suppose on the one hand that we could build a specialized processor that would do 80% of a large scientific computation with unlimited parallelism, i.e., it would do 80% of the computation in zero time. Then the overall computing time would be reduced by a factor of 5. On the other hand, suppose we could find an average 10-fold parallelism over the whole computation for a VLIW computer. Then we could get a speed-up of a factor of 10 on a VLIW computer, which is clearly better. However, combining the best of both approaches would be even better. In fact, if we used a specialized processor on 80% of the computation and a VLIW machine on the remaining 20%, we could get a reduction of a factor of 50. In Section III, we explore the use of specialized FFT processors in conjunction with the AP-164 for solving the Poisson equation in three dimensions.

As a test of feasibility, we consider solving the Poisson equation on the AP-164. We made this choice for a number of reasons:

(1) the AP-164 is a commercially available, "off-the-shelf" VLIW computer;

(2) the Poisson equation is the simplest nontrivial elliptic equation and if we can't solve it fast then we can't solve anything fast; and

(3) the Poisson equation occurs in many important applications such as image enhancement, geoacoustics, semiconductor device simulation, fluid mechanics, and mechanical CAD/CAM.

As a specific example, we mention work we are doing with Dr. Ding Lee of NUSC on underwater acoustics. We are investigating the use of a fast Poisson solver as a preconditioner to be used in conjunction with the conjugate gradient method (applied to the preconditioned normal equations), for solving discrete approximations to nonseparable Helmholtz equations. We have been able to prove that this algorithm requires at most order log n iterations to reduce the error in a difference approximation with n grid points in each coordinate by a factor of n^{-2}.

This brings us to the question of how well a VLIW computer like the AP-164 will solve the Poisson equation. In order to put our discussion into context we note the following information about the AP-164:

(1) the word length of the AP-164 is 64 bits;

(2) the cycle time is 167 nanoseconds;

(3) 10 instructions can be executed per cycle, including a floating point multiplication and addition;

(4) floating point multiplications and additions are pipelined (requiring 3 and 2 cycles respectively);

(5) a division requires 22 cycles, but vector division requires 6 cycles per component;

(6) the machine can be either microprogrammed in APAL or programmed in FORTRAN; and

(7) there exists a good scientific APAL subroutine library.

The model discrete problem we wish to solve is the standard five point finite difference approximation to the Poisson equation, $-\Delta u = f$, on the unit square, i.e.,

$$-\Delta_h u_{ik} = f_{ik} \equiv f(x_i, y_k) \ , \quad 1 \leq i,k \leq n \ , \tag{1}$$

subject to Dirichlet boundary conditions

$$u_{0k} = u_{n+1,k} = u_{i0} = u_{i,n+1} = 0 \ , \ 1 \leq i,k \leq n \ , \tag{2}$$

where

$$n = 2^{t-1}, \text{ and}$$

$$\Delta_h u_{ik} = h^{-2}[-4u_{ik} + u_{i-1,k} + u_{i+1,k} + u_{i,k-1} + u_{i,k+1}] .$$

Problem (1)-(2) can be written in matrix form as

$$M_h \underset{\sim}{u}_h = \underset{\sim}{f}_n \tag{3}$$

where

$$M_h \equiv h^{-2} \begin{bmatrix} 4 & -1 & & -1 & 0 \\ -1 & \ddots & \ddots & 0 & -1 \\ & \ddots & \ddots & \ddots & \\ -1 & & 0 & \ddots & -1 \\ 0 & -1 & & -1 & 4 \end{bmatrix} \tag{4}$$

Since array processors are excellent at doing FFTs, we consider three standard FFT based algorithms for solving the discrete Poisson equation (1)-(2), cf. [1], [3], [6], [7], [9] and [10]. The basic idea of all these methods is to extend $\underset{\sim}{u} \equiv \{u_{ik}\}$ and $\underset{\sim}{f} \equiv \{f_{ij}\}$ to real, odd, periodic sequences of period 2(n+1) in both dimensions.

If $\underset{\sim}{d} \equiv \{d_{ik}\}$ is the real, doubly periodic sequence of period 2(n+1) with

$$d_{0,0} = 4, \quad d_{-1,0} = d_{1,0} = d_{0,-1} = d_{0,1} = -1, \text{ and}$$

$$d_{ik} = 0 \quad \text{otherwise,}$$

then the discrete Poisson equation (1)-(2) can be represented as

$$\underset{\sim}{d} * \underset{\sim}{u} = h^2 \underset{\sim}{f} \quad \text{or} \quad \hat{\underset{\sim}{\alpha}} \cdot \hat{\underset{\sim}{u}} = h^2 \hat{\underset{\sim}{f}}$$

or

$$\hat{u}_{ij} = h^2 \, \hat{d}_{ik}^{\;-1} \, \hat{f}_{ik} \, , \quad 1 \leq i,k \leq n \, , \tag{5}$$

where $\hat{z}$ denotes the discrete, double, complex Fourier transform of z. We note that in this example the real Fourier transform or sine transform would suffice, cf. [2]. We will return to the impact of this observation later. The solution to (1)-(2) is obtained by taking the inverse transform of the normalized vector $\hat{u}$ of (5). We refer to this method as the (double) FOURIER method.

It is also possible to view the discrete Poisson equations as being block tridiagonal, i.e., if we partition the solution vector $\underset{\sim}{u}$ as

$$\underset{\sim}{u} \equiv \begin{bmatrix} \underset{\sim}{u}_1 \\ \underset{\sim}{u}_2 \\ \vdots \\ \underset{\sim}{u}_n \end{bmatrix}, \quad \text{where} \quad \underset{\sim}{u}_i \equiv \begin{bmatrix} u_{i1} \\ u_{i2} \\ \vdots \\ u_{in} \end{bmatrix}, \quad \text{then}$$

$$M_n = \begin{bmatrix} A & I & & 0 \\ I & A & I & \\ & \ddots & \ddots & \ddots \\ & & \ddots & \ddots & I \\ 0 & & I & A \end{bmatrix}, \quad \text{where} \quad A = \begin{bmatrix} -4 & 1 & & 0 \\ 1 & \ddots & \ddots & \\ & \ddots & \ddots & 1 \\ 0 & & 1 & -4 \end{bmatrix}$$

has the eigenvalues $\lambda_k = -4 + 2\cos\left(\frac{k\pi}{n+1}\right)$ and

eigenvectors $\underset{\sim}{V}_k \equiv \left[V_{kj}\right] \equiv \left[\frac{1}{\sqrt{n+1}} \sin\left(\frac{jk\pi}{n+1}\right)\right]$.

If $\underset{\sim}{u}_i = \Sigma\, \hat{u}_{ij}\underset{\sim}{v}_j$, then

$$A\underset{\sim}{u}_1 + \underset{\sim}{u}_2 = \underset{\sim}{y}_1,$$

$$\underset{\sim}{u}_{i-1} + A\underset{\sim}{u}_i + \underset{\sim}{u}_{i+1} = \underset{\sim}{y}_i, \quad 2 \le i \le n-1,$$

$$\underset{\sim}{u}_{n-1} + A\underset{\sim}{u}_n = \underset{\sim}{y}_n$$

or for each component $1 \le k \le n$

$$\left.\begin{aligned} &\lambda_k\, \hat{u}_{1,k} + \hat{u}_{2,k} = \hat{y}_{1,k} \\ &\hat{u}_{i-1,k} + \lambda_k\, \hat{u}_{i,k} + \hat{u}_{i+1,k} = \hat{y}_{i,k}, \quad 2 \le i \le n-1, \\ &\hat{u}_{n-1,k} + \lambda_k\, \hat{u}_{n,k} = \hat{y}_{n,k}\, , \end{aligned}\right\} \tag{6}$$

which is an easily solved tridiagonal system. Once we have solved for all the transformed vectors $\{\hat{u}_i\}$, we obtain the solution by doing n one-dimensional inverse transforms. We call this the FOURIER/TRID method.

The third method which we call the ODD/EVEN method consists of first doing an odd/even reduction, cf. [3] and [9], to eliminate all the unknowns along vertical lines with even index and then applying the FOURIER/TRID method to the remaining unknowns. Table I summarizes the asymptotic complexity of these three methods.

TABLE I

	FOURIER	FOURIER/TRID	ODD/EVEN
# transforms of length 2n+1*	4n	2n	n
# tridiagonal systems of order n**	0	n	n/2
# tridiagonal systems of order n/2**	0	0	n

*requires 0 (n log n) work

**requires 0(n) work

It is clear that on a serial computer that the ODD/EVEN method is asymptotically fastest and the FOURIER/TRID method is asymptotically second fastest. On a vector machine, the hierarchy is not so clear. It would seem that if we could do FFTs sufficiently fast it wouldn't pay to trade them off for tridiagonal solves which are basically serial. In addition, there are a variety of transforms which may be employed. As shown in [2], the ratio of work for the transforms is given in Table II.

TABLE II

complex FFT	1
real FFT*	1/2
sine transform*	1/4

*requires low order pre- and post-processing

On a serial machine, it is clear that we would choose the sine transform. On a VLIW machine, the choice may not be so obvious because an optimal microcoded implementation can be so much faster than a compiled one. The AP-164 has very efficient complex and real FFTs in the APAL subroutine library while it does not have a comparable sine transform at present. A sine transform on the AP-164 must be done by doing the pre- and post-processing [2] in FORTRAN and calling the APAL real transform for the inner-loop.

Vectorization is not the only important factor in algorithm selection for a machine like the AP-164 which has a very slow division (especially scalar) operation. In the FOURIER/TRID and ODD/EVEN methods we must solve linear systems with tridiagonal matrices of the form

$$B \equiv \begin{bmatrix} \lambda & 1 & & 0 \\ 1 & \ddots & \ddots & \\ & \ddots & \ddots & 1 \\ 0 & & 1 & \lambda \end{bmatrix} \qquad \text{where } |\lambda| > 2.$$

It would seem that this would involve quite a few divisions.

Following [6] and [8] we can factor

$$B = \begin{bmatrix} 1 & & & 0 \\ \ell_1 & \ddots & & \\ & \ddots & 1 & \\ 0 & & \ell_{n1} & \end{bmatrix} \begin{bmatrix} u_1 & 1 & & 0 \\ & \ddots & \ddots & \\ & & \ddots & 1 \\ 0 & & & u_n \end{bmatrix} \equiv LU$$

where $u_i^{-1} = \ell_i$, $1 \leq i \leq n-1$. Moreover, $\ell_i \to \ell$ as $i \to \infty$, which means we can avoid most of the divisions and storage in calculating and using the LU-factorization. If we are in the regime where we are solving several Poisson equations, then we can view the factorization as preprocessing and we need do only the forward and backward substitution. Moreover, since $u_i^{-1} = \ell_i$, we need store only L and forward and backward substitution can be carried out without any divisions.

There are a number of important vectorization "tricks" or techniques which can be employed on these problems, cf. [1], [6], and [10]:

(1) The normalization in the FOURIER method can be vectorized. In fact, if we are solving at least two Poisson equations, it pays to form the reciprocals of the normalization constants by means of a vector divide and then to do a normalization by means of a vector multiplication.

(2) In the FOURIER/TRID and ODD/EVEN methods, the collection of tridiagonal systems can be solved in parallel. Thus on the AP-164, the time required to solve the tridiagonal systems is about twice the time required to do the normalization assuming preprocessing of the normalization constants and it probably requires less time assuming no preprocessing.

(3) The reduction step in the ODD/EVEN method can be done in parallel.

(4) The pre- and post-processing necessary to compute all the sine transforms, given the existence of a real FFT, can be vectorized. Thus on the AP-164, we should not have to pay too large a penalty for not having a hand-optimized APAL sine transform code.

These points have some implications for our algorithm selection. Points (2) and (3) imply that the FOURIER method is very unlikely to be the most efficient on the AP-164 no matter how fast we can do an FFT. Point (4) implies that we should use the vectorized sine transform whenever possible even if it isn't implemented entirely in APAL.

We have run a large number of experiments for these methods on the AP-164. The results are too voluminous to reproduce here and we refer to [6]. We do reproduce one interesting graph (Figure I) showing the predicted running times of these algorithms based on a sine transform hand-coded in APAL. We approximated the running time of such a transform by taking the product of the running time of the APAL real FFT and the quotient of the running time of the FORTRAN sine transform divided by the running time of the FORTRAN real FFT. The curve labeled D is the FOURIER method without preprocessing or vectorized normalization, the curve labeled T is the FOURIER/TRID method using the fact that the entries in the factors converge but without preprocessing or vectorization, and the curve labeled R is the ODD/EVEN method without preprocessing or vectorization. Curves 1 and 2 from [10] are for algorithms similar to those in T and R except implemented on a CDC-7600.

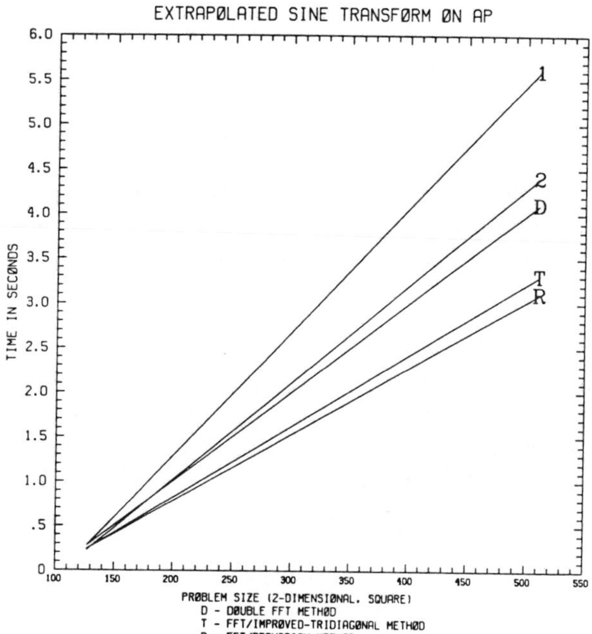

FIGURE I

Since we didn't exhaust our possible improvements in the benchmarks of Figure I, it seems quite clear that even if we were slightly optimistic on the running time of our sine transform, Poisson solvers on a AP-164 are competitive with comparable solvers on a CDC-7600. This proves our contention that we can effectively solve nontrivial elliptic problems on a VLIW computer.

III. VERY LARGE PROBLEMS

We now switch our attention to solving very large three dimensional Poisson problems. To get a handle on things we consider a simple out-of-core method based on the FOURIER method in three dimensions. Our goal is to analyze the impact of different architectures on the running time of a fixed algorithm (as distinct from Section II in which we fixed the architecture and varied the algorithm).

In this section, we assume that the source term, i.e., the right-hand side of the equation, is initially stored as a 3-dimensional block array in secondary storage (so far unspecified)

<u>OUT-OF-CORE ALGORITHM</u>:

```
FOR EACH DIMENSION:
    BRING EVERY BLOCK ROW INTO MAIN MEMORY
    DO 1-DIMENSIONAL FFTs
    STORE TRANSFORMS BACK IN SECONDARY STORAGE

FOR EACH DIMENSION:
    BRING EVERY TRANSFORMED BLOCK ROW INTO MAIN MEMORY
    IF DIMENSION = 1:
        BRING NORMALIZATION CONSTANTS INTO MAIN
        MEMORY and NORMALIZE
    DO 1-D INVERSE FFTs
    STORE ANSWER BACK IN SECONDARY STORAGE
```

The required I/O for the source, solution, and normalization constants is $13n^3$ words. We consider the following three systems:

(1) a AP-164 with a disk having a transfer rate of 1.2 megabytes/sec;

(2) a CRAY-1 with a disk having a transfer rate of 5 megabytes/sec; and

(3) a AP-164 with a bulk memory having a transfer rate of 48 megabytes/sec.

To simplify the analysis and favor the disks as much as possible we ignore latency. As an aside we mention that disk latency is on the order of milliseconds, while bulk memory latency is on the order of microseconds.

The I/O time in microseconds required for this algorithm on systems (1)-(3) is given by the following estimates: (1) $87n^3$; (2) $21n^3$; and (3) $2.2n^3$. The compute time on the AP-164 required for this algorithm is asymptotically equal to the time to do n^3 multiplications and $6n^2$ one-dimensional FFTs which is approximately $2n^3 \log n$ microseconds.

Comparing these numbers, we can reach several conclusions:

(1) the relative times needed for I/O on systems (1)-(3) are approximately 40:10:1;

(2) if $n \leq 2^{11}$, disk I/O on the Cray-1 takes more time than I/O <u>and</u> computation on the AP-164 with bulk memory; and

(3) a AP-164 with a disk takes about $1+(44/\log n)$ times longer than a AP-164 with a bulk memory to solve this problem. If $n = 2^7$, this works out to about 7 times longer.

It is interesting to note that this algorithm is one of the most favorable for systems for slow I/O. Most iterative methods for solving elliptic problems involve work proportional to the amount of data (and not a higher power of the amount of data) for every sweep through the data. For these algorithms, data movement is as important asymptotically as floating point computation. In this sense, solving elliptic problems is quite different than combinatorial computing for which the number of operations often grows superlinearly (or even exponentially) with the amount of data.

The time and memory requirements for solving the 3-dimensional Poisson equation for $n = 2^7$, 2^8, and 2^{10} by the FOURIER OUT-OF-CORE method on the AP-164 with bulk memory are given in Table III.

TABLE III

n	time (secs)	memory (megabytes)
2^7	28	16
2^8	256	128
2^{10}	20,000	8,000

Clearly $n = 2^7$ is viable on this system, $n = 2^8$ is marginal because of the large amount of memory required, and $n = 2^{10}$ is too large.

What can be done to reduce this time? Even though the AP-164 does FFTs very rapidly, it can be significantly improved by the addition of special hardware. Moreover, if that hardware is interfaced to the bulk memory, we may be able to get data to and from it sufficiently fast.

Consider the situation shown in Figure II.

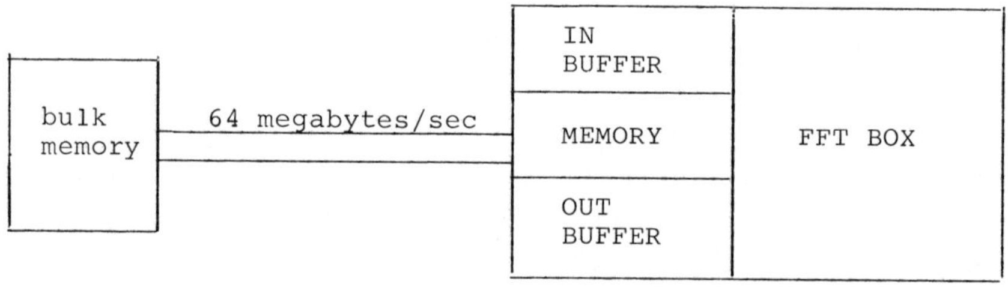

FIGURE II

To simplify our analysis, we make the assumption that all the 1-dimensional FFT's on the data in the memory of the FFT-box can be completed in the time, T, necessary to fill the in-buffer and empty the out-buffer. Thus the computation in the FFT-box is totally overlapped with I/O. If we have an $n \times n \times n$ grid and a $k \times k \times k$ block size and a block row fits into each buffer, then

$$T = k^2 n/(4 \cdot 2^{20}) \text{ secs},$$

which implies that the flops rate of the FFT box must be at least

$$(k^2 \cdot 2 \cdot 2n \log n)/(k^2 n/(4 \cdot 2^{20})) \approx 2^{24} \log n$$

or 16 log n megaflops. Thus for $n = 2^7$, the FFT box must achieve a computational rate of 112 megaflops, which is clearly achievable with today's technology.

Since we need to do $6 \cdot (\frac{n}{k})^2$ block transforms, each requiring time T, the total time for the transforms is $6 \cdot (\frac{n}{k})^2 \cdot \frac{k^2 n}{4 \cdot 2^{20}} \approx .75 n^3 \cdot 10^{-6}$ seconds. For the normalization, we need:

(1) to transfer both the normalization constants and the data to the AP-164, which takes $2n^3$ cycles;

(2) to carry out the normalization, which requires $3n^3$ cycles; and

(3) to output the normalized data back to the bulk memory, which takes n^3 cycles.

This analysis implies that the normalization requires $6n^3$ cycles or about n^3 microseconds. Thus the total computation would require about 1.75 n^3 microseconds. We note that while an FFT box saves a log n factor in the run time of the FFT portion of the algorithm, a normalization box (attached to the bulk memory) would save at most a factor of two in the run time of the normalization portion of the algorithm.

With multiple bulk memory boxes, it is possible to interleave I/O. For example, with 4 memory boxes we could get a 4-fold interleave which would speed I/O by a factor of 4. If we increased the computing speed of the FFT box accordingly, the transforms could be done 4 times faster. However, if we are using the AP-164 for the normalization, it probably isn't worth the effort.

Of course, we could build a normalization box with a 4-fold interface. In fact, at 256 megabytes/second, we could input two words and output one word approximately every 80 nanoseconds. In order to be I/O bound such a box would have to be able to do a 64 bit floating point multiply every 80 nanoseconds. These new timings are summarized in Table IV.

TABLE IV

n	AP-164 and bulk memory	AP-164, bulk memory, and FFT box
2^7	28	2.8
2^8	256	22
2^{10}	20,000	1,333

So far, we have talked only about the Poisson solver. Clearly we want something faster for the rest of the code. We are developing a replacement for the AP-164 which pushes the VLIW architecture to its limit, cf. [5]. We call it the ELI (Enormously Long Instruction)-512. The 512 comes from the instruction word length. The goals of the ELI project are:

(1) to run compiled general purpose scientific code written in FORTRAN;

(2) yield high speed at low cost without microcoding; and

(3) to first build an "optimizing" compiler called BULLDOG which will help determine how much parallelism to build into ELI.

Very roughly speaking, the machine could have a cycle time 4 times faster (if built out of ECL) than the AP-164 and 8 times as much parallelism. Thus it is potentially 32 times as powerful.

Of course, ELI has its limitations. We want to make sure that everyone understands that the ELI computer:

(1) is an attached processor;

(2) is only optimized for running compute bound scientific code; and

(3) will perform poorly for "dynamic" code.

IV. SUMMARY

In this paper, we have shown the following main points about VLIW computers, like the FPS-164:

(1) They can be used very effectively to solve nontrivial elliptic partial differential equations such as the Poisson equation in a square domain;

(2) When coupled with very fast, large bulk memories, they can be used very effectively to solve Poisson equations in three dimensions with arbitrarily fine grids. The resulting algorithms are CPU, rather than I/O, bound. Such a bulk memory system for the FPS-164 has been designed, built, and tested at Yale; and

(3) For very large problems, an order of magnitude speedup can be achieved by the use of specialized FFT hardware in conjunction with a VLIW computer with high speed bulk storage.

ACKNOWLEDGMENTS

We acknowledge the assistance of our colleagues at Yale, especially Peter Geiger and Susan O'Donnell who aided with the research described in Section II.

REFERENCES

[1] Buzbee, B.L. A Fast Poisson solver amenable to parallel computation. IEEE Transactions on Computers C-22:793-796 (1973).

[2] Cooley, J.W., Lewis, P.A.W. and Welch, P.D. The Fast Fourier Transform Algorithm: Programming considerations in the calculation of Sine, Cosine and Laplace transforms. J. Sound Vib 12:315-337 (1970).

[3] Dorr, F.W. The direct solution of the discrete Poisson equation on a rectangle. Siam Rev 12:248-263 (1970).

[4] Eisenstat, S.C. and Schultz, M.H. On some trends in elliptic problem solvers. In "Elliptic Problem Solvers" (Martin H. Schultz, ed.), pp. 99-114. Academic Press, New York (1981).

[5] Fisher, J.A. Very long instruction word architectures and the ELI-512. Technical Report 253, Department of Computer Science, Yale University, 1982.

[6] Geiger, P., O'Donnell, S., Schultz, M.H. Some fast Poisson solvers on an array processor. In preparation.

[7] Hockney, R.W. A fast direct solution of Poisson's equation using Fourier analysis. Journal of the ACM 12:95-113 (1965).

[8] Malcolm, M.A. and Palmer, J. A fast method for solving a class of tridiagonal linear systems. Communications of the ACM 17:14-17 (1974).

[9] Swartztrauber, P.N. Themethods of cyclic reduction, Fourier analysis and the FACR algorithm for the discrete solution of Poisson's equation on a rectangle. SIAM Rev 19:490-501 (1977).

[10] Swartztrauber, P.N. Vectorizing the FFTs. In "Parallel Computations" (Garry Rodrigue, ed.), pp. 51-83. Academic Press, New York (1982).

Parallel Architectures for Iterative Methods on Adaptive, Block Structured Grids

Dennis Gannon

John Van Rosendale

I. INTRODUCTION

The development of larger and faster computers is being driven by a number of application areas where todays computers are not adequate. One of the most important of these areas is the numerical solution of partial differential equations. No computer in existence could tackle the problem of accurately modeling air flow over realistic aircraft configurations. This is so even if one neglects all time dependent effects. Similar remarks could be made about partial differential equations arising in a number of applications areas.

Faced with these needs, one could hope for computers similar to present designs, but with memory size and computation speeds thousands of times greater than current models. While no major technological barriers impede the development of large memories, the speed of the switching elements used for sequential computation seems to have reached a plateau.

The solution must lie in the use of parallelism. With device densities rising rapidly, and prices dropping correspondingly, computers having hundreds or thousands of processors will be practical within a decade. Such computers will present a challenge to both the computer architect and numerical analyst. From the architects point of view, current problems like processor communication and synchronization will become more troublesome. But the changes facing the numerical analyst may be much more sweeping.

Authors' address: Dennis Gannon, Department of Computer Sciences, Purdue University, West Lafayette, IN. 47907. John Van Rosendale, ICASE, NASA Langley Research Center, Hampton, Va. 23665.

This research was supported by the National Aeronautics and Space Administration under NASA contracts no. NAS1-15810 and no. NAS1-16394 while the authors were in residence at ICASE, NASA Langley Research Center, Hampton Va. 23665. Addional support for the first author was provided by NSF Grant MCS-8108512.

ISBN 0-12-100560-7

In the past, the programmer or numerical analyst was able to design algorithms taking relatively little cognizance of the architectures on which they would run. With the advent of highly parallel computation, this situation may be drastically altered. It may be that the available parallelism can be exploited only by special purpose architectures tuned to a given problem or problem class. As a result, the algorithms employed will probably be highly architecture dependent, chosen as often for architectural reasons as for traditional numerical analysis reasons. More precisely, the impact of error analysis on algorithm design should change little, but complexity questions will tend to revolve around architectural considerations.

With the exception of the FEARS project (Zave, 1979) much of the recent work in computer architectures has been stimulated by a small set of simple numerical problems. Great attention has been paid to such problems as finding efficient architectures for Fourier transforms or band solution algorithms. Less tractable, and equally important problems, such as Lagrangian hydrodynamics, have received scant attention.

One class of numerical algorithms which is well understood, yet offers a great deal of the complexity frequently present in real world applications, is adaptive iterative solution of elliptic partial differential equations. This paper proposes an algorithm-architecture pair for the adaptive solution of elliptic problems in complicated domains. An attempt is made to point out and extract as much of the parallelism present as possible. We also attempt to retain as much flexibility in both the algorithm and architecture as possible.

The plan of the paper is as follows. Section II and III present the data structure and numerical algorithm being considered. Section IV describes our proposed architecture. The final section draws some tentative conclusions and suggests directions for further research.

II. DATA STRUCTURE

In this section we present our data structure for the finite element resolution of complex three dimensional domains. The data structure chosen permits the use of adaptive refinement and multigrid solution techniques. Our focus is on simple scalar elliptic partial differential equations, such as the Poisson or Helmholtz equations, and finite element grids consisting of trilinear isoparametric elements.

The proposed data structure is based on body fitted finite element grids made of macro-elements or cells. Each cell is an 8 x 8 x 8 grid of trilinear isoparametric elements. The use of such cells has advantages from both the architectural and data structure point of view. From the data structure side, it creates an attractive homogeneity, and requires less storage than completely general grids. From the architectural side, it enables the use of fast special purpose hardware designed to handle the fixed size blocks of array data.

To create such cellular grids, we begin by assuming the domain is covered by curved *rectangular* or hexahedral regions having disjoint interiors. Associated with each curved rectangular region, we have a mapping from the canonical rectangle [0,1] x [0,1] x [0,1] onto the given curved rectangular region, as in Figure 1.

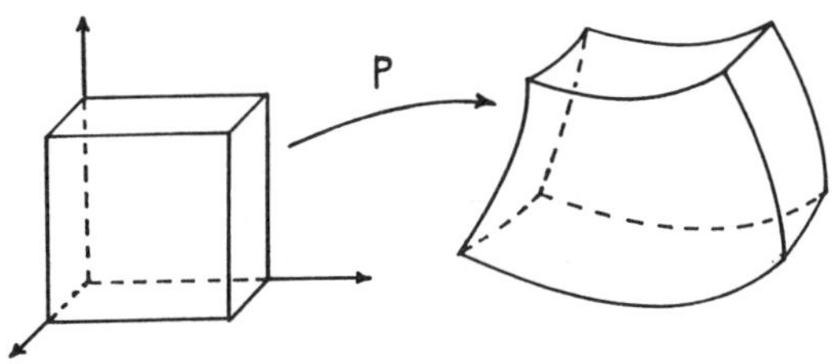

Figure 1. Mapping from Reference Cube to Physical Domain

The *triangulation* of a given domain into curved rectangular regions, defined by mappings like that in Figure 1, is a complex computer aided design problem. The necessity of having well defined and invertible Jacobians, usual in finite element theory, applies here as well, greatly complicating this problem. However, since a tetrahedron can be decomposed into curved rectangular regions, there is, in principle, no loss of generality here.

Associated with each of these curved rectangular regions we have an *ancestral cell*. The collection of all ancestral cells form the initial grid used by the multigrid solution algorithm. This initial grid will be refined adaptively during the course of the computation. Figure 2 shows the refinement of an ancestral cell into eight children.

Associated with each cell, which is really an 8 x 8 x 8 macro-element composed of 512 trilinear elements, we need three kinds of information:

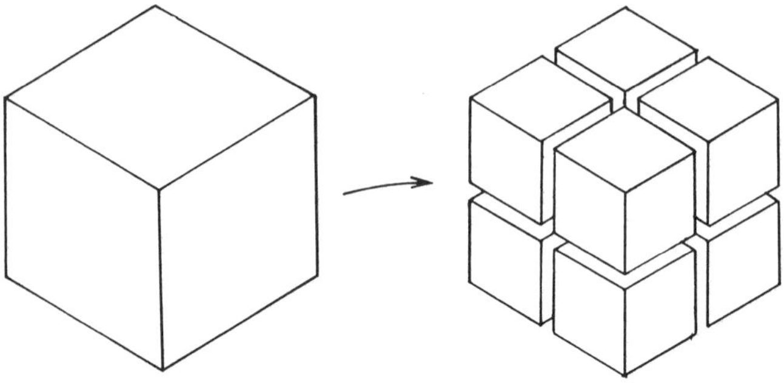

Figure 2. Refinement of a Cell into Eight Children

1. geometry information describing the region covered by the cell and the grid inside it.
2. numerical information, such as the local stiffness matrix.
3. data structure information to keep track of neighborhood relations between cells.

One of the advantages of the use of cells of this type is that items 1 and 3 here need be stored only once per cell, rather than separately for each of the trilinear finite elements.

We turn now to a closer examination of these three types of information associated with each cell.

A. Geometric Information

The geometry information associated with each cell is the region in the physical domain it occupies, and the location of all mesh points within it. Each cell must be either an ancestral cell or the child of an ancestral cell created by adaptive refinement. If it is ancestral, it covers the curved rectangular region p([0,1] x [0,1] x [0,1]), where p is its associated mapping, as in Figure 1. If this cell is a child it covers the region p($[r_0, r_1]$ x $[s_0, s_1]$ x $[t_0, t_1]$), for parameters r_0, r_1, s_0, s_1, t_0, t_1, and where p is the coordinate transformation mapping associated with its ancestor.

B. Numerical Information

Associated with each cell we need solution, residual and data vectors, and also the set of matrix coefficients. The three vectors here can be stored as 9 x 9 x 9 arrays of reals. This implies redundant storage along the boundaries, since boundary mesh points will be shared with other cells. The matrix coefficients are best stored as 27 point difference formulas associated with each of the 729 mesh points in the 9 x 9 x 9 grid. The alternative, storing the element stiffness matrices for each element is wasteful of storage.

C. Data Structure Information

The data structure must permit dynamic allocation of cells, in order to allow adaptive refinement and must also keep track of a number of kinds of relationships between cells. We must be able to extract neighborhood relationships, including the relative size and orientation of neighbors, in order to perform relaxation iterations. We must also maintain parent-child relationships, as these are required in multigrid iteration. Finally, we must keep track of storage allocation in the computer memory, since this is not done automatically at the moment.

These data structure issues for locally refined grids are inherently complex, and made more so here by the need to rapidly access data structure information in the course of the computation. The data structure chosen is based on a global numbering of nodes, and requires that mesh size vary by at most a factor of two from a cell to its neighbor. Space does not permit detailed description of this data structure here, but it will be described in detail when we present the results of the completed study.

III. ADAPTIVE SOLUTION ALGORITHM

An adaptive multigrid solution algorithm can be based on the data structure described in the last section. Such a solution algorithm requires six basic numerical operations.

1. Form the stiffness matrix on a cell.

2. Form properly scaled data vectors on all cells.

3. Perform a Jacobi smoothing iteration on a multigrid level.

4. Perform a multigrid projection from one multigird level to the next coarser level.

5. Perform a multigrid injection from a multigird level to the next finer level.

6. Form an estimate of the local truncation error on a cell.

Designing multigrid algorithms built from these operations is reasonably straight forward and is discussed in detail in Gannon and Van Rosendale (1983). Mathematically, operation 6, estimating the truncation error, is the most subtle. But from a data structure or architectural point of view, operations 2, 3 and 4 are the most difficult, because of the complex data movements involved. The basic structure of operations 3, 4 and 5 are similar. Operation 3 is described in more detail in section IV.C.

IV. ARCHITECTURE

The computations described in the previous section contain two distinct levels of parallelism which may be exploited. These computations are, at their lowest level, composed of array operations on the cells of the global mesh. The matrix product Au and the multigrid injection and projection operators are highly parallel array computations on each cell. These matrix computations form a core of operators that dominate the serial complexity of our adaptive multigrid algorithm. Since the cells here are relatively small and since these computationally intensive operations are quite simple, a large systolic array (Kung and Leiserson, 1980) or an SIMD processor like the Burroughs BSP would be well suited to these computation.

Parellelism is also available at a higher level, since operations on different cells in the global mesh can be done concurrently. A virtual machine that would be ideal for the algorithms here consists of a network of processors organized as a complete graph with one cell per processor. More realistically, a system of processors communicating through a high bandwidth switching network, such as in the TRAC system (Kapur, Lipovski, Premkumar 1980) or PASM (Siegel, 1979) could be used, with grid cells mapped onto these processors in a many to one manner. Then each processor would be responsible for computation on each of its assigned cells.

The primary difficulty with this scheme is that an operation, such as a relaxation iteration, introduces data dependencies in the form of partially computed residuals along shared faces or edges of adjacent cells. If these adjacent cells are assigned to different processors, then the partial result data must be passed from one processor to the other over the switching network. The problem of optimally scheduling the computations on a general grid can be extremely difficult if network delays are significant. If on the other hand, computation, task scheduling and communication can be sufficiently overlapped, then it would be possible to use arrival of partial result data and the availability of hardware (such as a free floating point unit) as a mechanism for dynamically scheduling local cell computations. This is know as the "data driven" approach to task scheduling.

Both levels of parallelism can be simultaneously exploited if the algorithms and architecture are carefully matched. As one approach to achieving this, we propose a system architecture based on low level SIMD or systolic array computations, governed by a data flow synchronized, distributed operating system. The proposed hardware is organized as 16 "local clusters" communicating with each other through a message switching network. A global control processor also communicates with all local clusters via a shared bus as illustrated in Figure 3. The overall system organization is similar to that of the Cedar project (Gajski, Kuck, Lawrie, Sameh 1983).

Each local cluster consists of four units.

1. A local control processor (LCP) which communicates with the global controller and coordinates the activities of the other devices in the cluster.
2. A configurable mesh of floating point processing elements (PEs) known as the array unit (AU).
3. A network input processor called the receive unit (RU).
4. A network output processor referred to as the send unit (SU).

The global controller is viewed as a conventional, but reasonably powerful, front end processor which communicates with the local clusters by a system of message passing. The global controller originates and distributes the data structure information to the rest of the system. Upon command from the global controller the local controllers initiate tasks such as Jacobi smoothing or truncation error estimation. Decisions about local refinement are made by the global controller based on information returned by the local controllers.

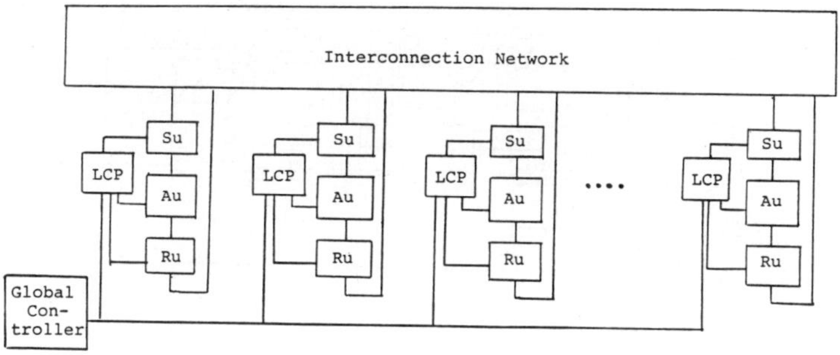

Figure 3. System Organization.

A. Local Cluster Hardware

The local controller and processing element array unit are modeled after one configuration of the CHiP system (Snyder, 1982) which is being designed for implementation on a single silicon wafer. The processors in the array are organized as a "nearest neighbor", toroidal lattice - each processor P_{ij} is connected to the 8 processors defined by $P_{(i \pm 1 \, mod \, 9)(j \pm 1 \, mod \, 9)}$. As shown in Figure 4 (for a 4 by 4 sublattice), the PE array has two other networks interwoven with it. These networks are used for "parallel by row" loading and unloading of the local memories of the processing elements. A moderately intelligent memory interface to the I/O networks permits the system to overlap the computation on one cell with data input on another cell and data output on a third.

Each processing element is assumed to have access to approximately 8K words of local storage which is sufficient to store the matrix components, solution vector and several temporary vectors for up to 20 cells per AU. This is also sufficient space to house the "microprograms" for each of the distinct computational tasks required in the multigrid algorithm. The array unit can be preprogrammed to perform any of the matrix operations required for computations on one cell. The 2 dimensional processor array can process a

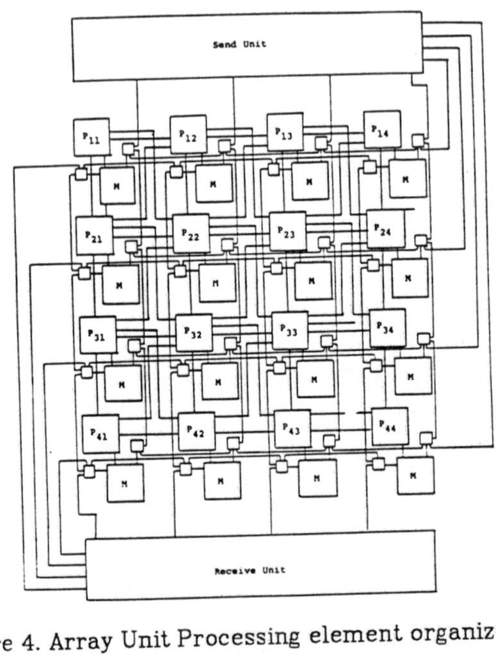

Figure 4. Array Unit Processing element organization

3 dimensional "cube" of data as a sequence of 2 dimensional slices. Assuming an instruction cycle of 1 microsecond, a well designed PE can be shown to compute the matrix product Au on one cell in about 500 microseconds. These timing issues will be covered in more detail in a report (in preparation) describing our architecture simulation results in detail.

If a grid structure is composed of cells C_i, $i = 1..m$, then any linear operator A defined as a sum of cell integrals can be expressed as a nondisjoint sum of matrix products of the form

$$Au = \sum_{i=1}^{m} A^i u^i$$

where u^i is the vector of coefficients of u on C_i and A^i is the corresponding matrix defined by the integrations of A restricted to C_i. The summation takes place over the vector components corresponding to vertices of the grid that are shared by two or more cells. To extend the cell product computed as systolic array operations for individual cells to a "grid wide" product requires that the above summation be completed.

Each local cluster contains, for each of its cells, a list of adjoining cells and a description of the corresponding intersections (faces, edges and vertices). The send unit, upon command from the local controller, accesses the partial result data from the array unit memories one row at a time until a complete face or edge has been buffered. It then forms a network message using the adjoining cell record as a network address tag.

The receive unit at the destination local cluster stores the data in the array unit memories and signals the local controller that a message has arrived for the targeted cell.

B. The Switching Network

The processor clusters communicate asynchronously through a crossbar message switching network. In spite of the large number of processor elements, the demands on the switching network turn out to be quite modest: one can easily show that over 95 percent of the communication takes place within processor clusters, rather than between them. Even so, transmitting the remaining 5 percent is a non-trivial task if the machine is expected to perform in the gigaflop range.

C. The Local Cluster System Software.

The local controller has the responsibility for managing the resources and scheduling the computations on each of the cells assigned to it. Each cell is viewed as having a virtual processor which may execute any of several "cell processes".

A cell process consists of a sequence of tasks. For example, the cell process for a relaxation iteration that would be executed on each cell of a specified grid level goes roughly as follows:

Relaxation_process{ x: cell# }

1. request the A.U. to do a matrix multiply on cell x.
2. request the send unit to transmit the partial result face data to all adjoining cells.
3. wait for the arrival of partial result face data from adjoining cells.
4. request the A.U. to add the partial results and to compute $u := u + \lambda(f - Au)$.
5. signal the global controller that an iteration is complete on cell x.

In a similar manner cell processes are constructed for the other basic solution operations, such as the matrix assembly and interpolations between grid levels.

The operating system for each local control processor is interrupt driven and is organized as a set of resource monitors (Hoare, 74) for the A.U., the send unit, the receive unit and for messages from the global controller. Each cell has a cell descriptor record which, in addition to its geometric attributes, lists its current process and task state. Each resource monitor maintains a queue of waiting cells (a list of descriptor records) and a pointer to a descriptor for the cell currently in control of the device. When a cell process of a cell requests service from a device, it invokes the appropriate monitor, which places that cell's descriptor on the queue of waiting cells.

The global control message monitor maintains a queue of "jobs" for each cell. Messages that arrive from the global controller cause an interrupt and the LCP puts the message in the appropriate queue. When a cell completes a process it enters an idle state and the system initiates a new process by fetching the next command from the queue.

V. EXPERIMENTAL PERFORMANCE AND CURRENT PROJECT STATUS.

Using simulation techniques, we have investigated several questions about system performance issues. The first of these is the problem of estimating the system overhead due to local controller task scheduling. Preliminary simulation results indicate that a system with a 1 microsecond cycle time for all local cluster operations would have software overhead that would consume 10-25% of total execution time, depending on the degree of multiprogramming. At this clock rate, a system with 16 array units, each consisting of a 9 by 9 array of processor elements, has a possible throughput of 1.3 gigaflops. Accounting for software overhead and a 100 microsecond message delay, the achieved performance is in excess of 800 megaflops. This result was obtained using a multigrid program on a model problem having 93 cells. The switching network needed to communicate 12 words per clock cycle on average. As the complexity of the problem increased the performance increased to over 900 megaflops with an increase of message traffic to only 14 words per clock cycle. The amount of data transmitted by the switching network turns out to be only about 2 percent as large as the amount of data transmitted between the processor elements and their attached memory. This low message traffic justifies our claim that a reasonably simple network will provide all the global communication required in this class of algorithms. A more complete analysis will be published when the study is complete.

Among the questions under current study are:

1. The effect of different degrees of multiprogramming on system performance. As the number of cells per local cluster increases, the utilization of the array units rises until the system is saturated. The exact form of this performance curve is being studied.

2. The design of efficient I/O algorithms compatible with parallel architectures. As computers become faster, extracting meaningful information from the reams of output produced becomes an increasingly difficult problem. Distributed graphics algorithms are being considered.

VI. REFERENCES.

Gajski, D, Kuck, D., Lawrie, D., Sameh, A., (1983) "CEDAR, A Large Scale Multiprocessor," Technical Report, Cedar project, Dept. of Comp. Science University of Illinois at Urbana-Champaign.

Gannon, D., Van Rosendale, J. (1983), "Highly Parallel Multigrid Solvers for Elliptic PDEs: An Experimental Analysis," ICASE Report. ICASE, NASA Langley Research Center, Hampton Va., 1983.

Hoare, C. A. R. (1974), "Monitors: An Operating System Structuring Concept,", Comm. ACM 17, 10, pp.549-57, Oct.

Kapur, R., Premkumar, U., Lipovski, J.(1980), "Organization of the TRAC Processor-Memory Subsystem," AFIPS Conf. Proc. pp.632-629.

Kung, H. T., Leiserson, C. E. (1980), "Algorithms for VLSI Processor Arrays," in C. Mead and L. Conway, Introduction to VLSI Systems, Addison-Wesley, Reading, Ma., pp.271-292.

Siegel, H. J., et. al. (1979), "An SIMD/MIMD Multimicroprocessor System for Image Processing and Pattern Recognition," IEEE Conf. on Pattern. Recog. Image Proc. pp. 214-224.

Snyder, L. (1982), "CHiP: The Configurable Highly Parallel Computer," Computer, Jan. 1982.

Zave, P., Rheinboldt, W. (1979), "Design of an Adaptive, Parallel Finite-Element System," ACM Trans. on Math. Software. Vol. 5(1), 1979, pp.1-17.

HIGHLY CONCURRENT ALGORITHMS FOR SOLVING LINEAR SYSTEMS OF EQUATIONS

Lennart Johnsson [1]

I. INTRODUCTION

Concurrency in computation has the potential of offering vastly increased computational power. With the emerging VLSI technology multiprocessor systems with thousands of processors are realistic. The smaller the processor the larger is the number of processors that economically can be built into a system. The term ensemble architecture, [Seitz 82], is often used. Such architectures typically have a significantly improved price/performance ratio compared to high performance vector processors.

The ensemble architectures for which algorithms are discussed consist of a thousand to tens of thousands of sparsely interconnected processors with local storage. There is no shared storage. A node consists of a processor with its local storage, and necessary communication circuitry. Two construction projects based on this concept are currently under way at Caltech: the Tree Machine project, [Browning 80], [Browning, Seitz 81], [Li 81], and the Homogeneous Machine project, [DeBenedictis, Seitz 82], [Fox, Seitz 83]. These two projects are architectural experiments with a significant difference in processor architecture, size, machine topology, and programming model. Both machines use message passing to synchronize computations in neighboring processors.

The Tree Machine processors are intentionally kept very simple, the idea being to achieve high performance by numerous processors and efficient communication. The processor is a 16-bit processor with floating point in software and 2-4k words of local storage. Even though such a processor has very

[1] Computer Science, Yale University, New Haven, CT 06520

ISBN 0-12-100560-7

limited capabilities, a 1000 chip machine could in 2 micron technology contain 8000 processors and 16-32M 16-bit words of storage. Such a machine would have an estimated capacity of about 0.5 Gflops.

The Homogeneous Machine node is currently based on standard components, an Intel 8086 processor with an 8087 floating point processor, with additional chips for handling the communication. The node that today requires a small PC board is expected to fit on a single chip at 1/2 micron technology. At 5 MHz clock rate each node has been benchmarked as having 1/6 of the capacity of a VAX 11/780 for certain PDE calculations.

Processors of a significantly smaller size than those used in the Tree Machine are being studied at Caltech for ensembles aimed for special purposes, such as computations frequently encountered in numerical linear algebra. Programming those special ensembles consists of changing the data flow within the network of processing elements. Instructions as well as data are pipelined through the network.

To be able to use the processing power of ensemble architectures such as the Tree Machine, the Homogeneous Machine, or any other machine with numerous processors, e.g., the CHiP machine [Snyder 82], or the Ultra Computer, [Schwartz 80], [Gottlieb, Schwartz 82], a good match is necessary between the communication requirements of the algorithms and the topology and bandwidth of communication paths of the machine. The quality of an algorithm can no longer be judged by the number of arithmetic operations.

Much of the analysis of concurrent algorithms has been based on the assumption of unrestricted communication. Excellent surveys are given by Sameh, [Sameh 77], and Heller, [Heller 78]. Results derived under such assumptions can serve as lower bounds against which the merits of various interconnection networks in multiprocessor systems can be judged.

Recent work has focused on devising algorithms with a high degree of concurrency under the restriction of limited communication, [Kung, Leiserson 80], [Kung 80], [Johnsson 81], [Heller, Ipsen 82], [Schreiber 82], etc. Indeed the topology has typically been restricted to regular graphs with interior nodes of degree 4 or 6 (hexagonal arrays). Algorithms for binary tree topologies have been devised by, e.g., Bentley and Kung, [Bentley, Kung 79], Browning, [Browning 80], and Presnell and Pargas, [Presnell, Pargas 81].

Systolic algorithms, like many tree machine algorithms, are very communication intensive. The goal is to maximize concurrency. One communication action per channel per multiply/add operation is often the case. A few registers are typically sufficient for data storage in each node. A few examples are given below. Such algorithms are efficient whenever the time for a communication action is comparable to a local storage reference.

For large problems the assumption of unbounded concurrency, for example a processor per matrix element, is often unrealistic. Computations must be sequenced through a smaller number of processors. Storage can either be external to the network of processors, each with very limited capacity for data storage, or distributed among the processing nodes. The allocation of storage and its organization affect what performance is possible to achieve. The order of algorithm complexity is dependent on the machine organization, [Johnsson 81], [Johnsson 82a], [Heller 82], [Schreiber, Keukes 82].

In the following we will focus on some concurrent algorithms for the conjugate gradient method. The effect of preconditioning on concurrency is discussed and a few algorithms presented. First, a brief, by no means complete, survey of some recent work on concurrent algorithms for the solution of linear systems of equations through factorization of the system matrix will be given. We write a linear system of equations as Ax = y where A is a N by N nonsingular matrix and x and y vectors of size N. Algorithms for LU- as well as QR-decomposition are discussed. Solution methods based on fast Fourier Transforms or cyclic reduction are not included. In the following logx denotes $\log_2 x$.

Detailed analysis of direct hardware (VLSI) implementations of FFT algorithms have been carried out in [Despain 79] and [Johnsson, Cohen 82]. The latter reference contains an algebraic treatment of FFT algorithms partially implemented in space, partially in time. Stone, [Stone 71], has described an O(logN) time FFT algorithm for shuffle networks, as well as O(logN) time algorithms for tridiagonal equations, [Stone 73], [Stone 75]. An $O(\log^2 N)$ cyclic reduction algorithm on a tree is given in [Presnell, Pargas 81].

II. LU-DECOMPOSITION

Concurrent algorithms for Gaussian elimination on unbounded hexagonally interconnected arrays have been presented by Kung, [Kung, Leiserson 80] and others, e.g., [Kung 80]. These systolic algorithms for hexagonal arrays use diagonals as a data abstraction for sets of elements. Diagonals as a data abstraction arise naturally when the stucture of band matrices is exploited and has earlier efficiently been used for vector machines, [Madsen, Rodrigue, Karushi 76]. The systolic algorithms only demand that the processing nodes have a few data registers and an arithmetic and logic capability. All interior nodes of the array compute the same function. The boundary nodes compute different functions, but sets of them compute the same function. For banded matrices all computations except those in a window of size uv+w are trivial, [Johnsson 81]. With u nonzero subdiagonals, v nonzero superdiagonals and bandwidth w=u+v+1, uv+w processors are needed to compute L and U in 3N+min(u,v) time.

For a full matrix N^2 processors are required. In that case computations can be organized so that a network with interior processors of degree 4 offers sufficient communication capabilities. Several such algorithms can be devised. An algorithm with A, L, and U resident in a mesh of interconnected processors is implicit in the inversion algorithm described in [Flanders et. al. 77]. An algorithm computing the factorization while the matrix enters a mesh is easily devised for a data flow similar to that employed by Gentleman and Kung for Givens rotations, [Gentleman, Kung 81]. Algorithms that match a mesh of degree 4 are easily mapped to execute efficiently on Boolean n-cubes. It is also possible to devise band matrix algorithms for a mesh with interior nodes of degree 4, but explicit control is typically required. Partial products are accumulated in place.

The methods of Cholesky, Crout, and Doolittle are closely related to Gaussian elimination. Concurrent algorithms have been devised that for these methods have the same communication needs as concurrent algorithms for Gaussian elimination. An algorithm for Cholesky's method based on matrix splitting, called hyperbolic Cholesky, is presented in [Delosme, Morf 81], [Ahmed, Delosme, Morf 82]. A direct, concurrent, factorization is described in [Johnsson 82b], that also contains algorithms for the methods of Crout and Doolittle. The node complexities are comparable. A graphical illustration of the function of an interior node in an array for Gaussian elimination is given in Figure 1, [Johnsson 81]. The node contains features for partial pivoting and matrix-matrix multiplication. Figure 2, [Johnsson 82b], shows an interior node for Cholesky's method.

All the above algorithms can be considered as vectorized versions of common sequential algorithms. The speed-up is

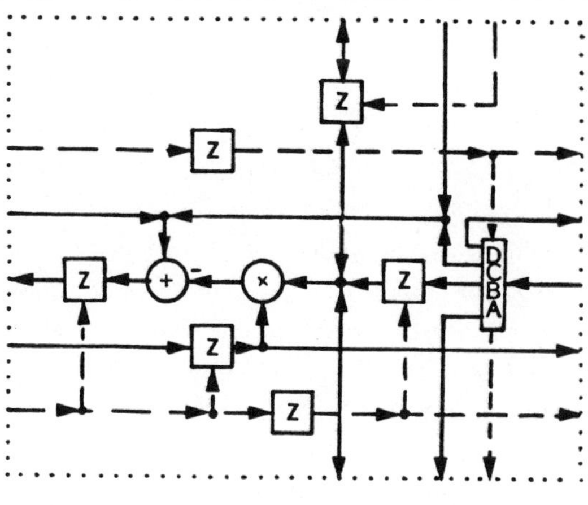

Figure 1: An interior node in an array for Gaussian elimination

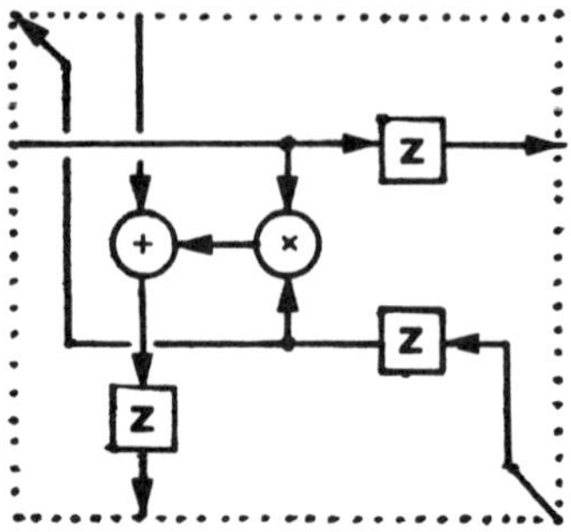

Figure 2: An interior node in an array for Cholesky's method

proportional to the number of processors, but the concurrent algorithms do not necessarily compute a solution in the minimum time possible. For instance, for banded matrices the above algorithms require O(N) time irrespective of the bandwidth. A uniprocessor can also compute the solution in O(N) time, see, e.g., [Stone 75]. However, algorithms for solving tridiagonal equations in O(logN) time on N processors interconnected in a shuffle network are known, [Stone 73], [Stone 75], as well as $O(\log^2 N)$ algorithms for 2N binary tree interconnected processors, [Presnell, Pargas 81].

Pipelining is a term used to denote the subdivision of an operation into suboperations, each of which is executed independently in a separate stage. The data moves from a stage performing one suboperation to another until all suboperations have been performed. The data flows through a "pipe". Traditionally pipelining is applied to operations such as addition, multiplication, and division. However, the technique can also be applied at higher levels, in particular in regular ensembles of processors. For instance, the computation of a succession of inner products can also be pipelined by providing as many multiply/add elements as there are terms in the inner product. Each element computes a product and adds it to a partial product sum. Partial product sums are passed between elements in such a way that the inner product is completed by the last element in the direction of the data flow, [Huffman 56], [Cohen 78], [Kung, Leiserson 80]. The processing elements can be organized to form many different topologies, such as paths, trees, and two dimensional meshes. A fully pipelined ensemble architecture is defined as one in which only communication to nearest neighbors is possible within one computational step. Global communication is implemented as a sequence of local communications.

Concurrent algorithms are often described in their fully pipelined versions. In factoring a matrix, computations on row elements are pipelined, as are operations on rows for the elimination of successive column entries below the diagonal in one column. In addition, the elimination operations for successive columns are pipelined. Pivoting on the diagonal results in the elements of columns and rows moving in arrowhead like formation through a rectangularly organized array of processing elements, Figure 3, [Kung 80], [Johnsson 81]. Without pipelining in the row and column directions the number of time steps for factoring a matrix is reduced to N+min(u,v) instead of 3N+min(u,v). However, the duration of a step in a partially pipelined array will be significantly longer, except for very small arrays, [Johnsson 81].

A fully pipelined mesh does not perform well when partial pivoting is needed. The elements in a column are computed at successive instances of time. The time to find the next pivot element and restart the computations goes from O(N) for the first column to O(1) for the last. For a band matrix the time for pivoting is O(u). Hence, a linear array will perform as

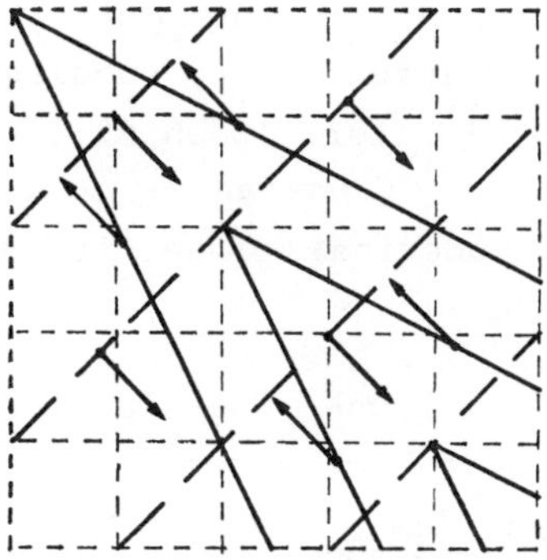

Figure 3: Wave fronts in a fully pipelined array performing Gaussian elimination

well as a two-dimensional mesh connected array, if partial pivoting is used. Interprocessor communication limits the performance. Adding a binary tree interconnection within columns of the array would allow Gaussian elimination with partial pivoting to be performed in O(NlogN) time on $O(N^2)$ processors. The performance of vector processors is less sensitive to partial pivoting.

If the problem to be solved is so large that it is impossible to associate one processor with each element within the active window, different algorithms are required. The total amount of storage is determined by the problem size. The storage can either be distributed among the network of processing elements or be external to it to keep the processing nodes small. Block algorithms using external storage are described in [Johnsson 81]. These block algorithms perform a factorization of a diagonal block followed by the proper matrix-matrix multiplications on the block row in which the diagonal element was just factorized. The blocks below the diagonal block are then eliminated and the remaining submatrix finally updated. The computations for different steps of the block algorithm can be pipelined, rendering a third level of pipelining.

The report, [Johnsson 81], also describes an algorithm in which several matrix elements are stored in each node. This algorithm performs sweeps over successive columns and/or rows. Both the block and sweeping algorithms have the same numerical properties as Gaussian elimination with pivoting on the diagonal elements. Elimination by subdiagonals as proposed for Givens rotations by [Heller, Ipsen 82], instead of by columns, can also be made for Gaussian elimination, but will have entirely different numerical properties irrespective of which row is used for pivoting.

An algorithm for block Gaussian elimination, with and without partial pivoting, on a torus is described by Fox, [Fox 82]. This algorithm communicates a vector in each communication action. The algorithm is intended for ensembles with an interprocessor communication bandwidth that is significantly less than the bandwidth to local storage. The time complexity given by Fox for Gaussian elimination with partial pivoting on a torus can be improved by a factor of $O(2^n/n)$ using the richer interconnect of a Boolean n-cube.

III. QR-DECOMPOSITION

If the numerical properties of the problem are such that partial pivoting is required in Gaussian elimination, stable methods such as Givens rotations and Householder reflections are often preferred. Systolic algorithms for Givens method have been proposed by [Gentleman, Kung 81], [Delosme, Morf 81], [Heller, Ipsen 82], [Johnsson 82c], and [Ahmed, Delosme, Morf 82]. In the algorithm described by Gentleman and Kung the upper triangular matrix R resides in the array after the computations. All rotations required to eliminate the first column are handled by the first row of the array, i.e., the row of the array which accepts input data. Similarly, the rotations needed for the second column is performed by the second row, etc. The array size is N(N+1)/2. A processor on the input side of the array receives one column of the matrix.

In the algorithm of Delosme and Morf a processor on the input side of the array instead receives one row of the matrix. Heller and Ipsen supplies an input processor with a diagonal. For the elimination of each subdiagonal w processors are needed, and hence uw processors for the complete factorization. Johnsson uses the column elimination strategy frequently used in Gaussian elimination. The number of processors used is uv+w. The latter two approaches both have an input processor for each diagonal. The computation time is 3N for a single matrix factorization according to [Gentleman, Kung 81] and [Delosme, Morf 81], 3N+min(u,v) for [Johnsson 82c], and 2(N+u+v) for [Heller, Ipsen 82]. By pipelining several factorizations the factor 3 in the time complexity can be reduced for the algorithms in [Gentleman, Kung 81] and [Delosme, Morf 81].

Givens method is used by Schreiber, [Schreiber 82],

[Schreiber, Keukes 82], and [Kung, Gal-Ezer 82], in the QR algorithm for eigenvalue computation. Methods of partitioning large problems to fit on a small array are discussed in [Heller 82], and [Schreiber, Keukes 82].

Givens rotations are based on local operations. The amount of interconnection present in a mesh (nodes of degree 4 or 6) suffice to reduce the computation time with a factor proportional to the number of processors. Householder's method, that has numerical properties similar to Givens, and is often preferred over Givens on sequential machines, requires that a vector norm be computed in every step of the algorithm. A vector norm is a global quantity. A time complexity of O(NlogN) is the best that can be expected. An $O(N^2)$ algorithm is presented in [Johnsson 82d] which also contains a discussion of an O(NlogN) algorithm.

IV. CONJUGATE GRADIENT METHODS

The conjugate gradient method, [Hestenes, Stiefel 52], combined with preconditioning is an interesting alternative to the above methods in solving linear systems of equations, in particular sparse systems [Meijerink, van der Vorst 77], [Meijerink, van der Vorst 78], [Kershaw 78], [Munksgaard 79], [Jennings, Malik 78], [Manteuffel 80], [Glowinski et. al. 80], [van der Vorst 82].

The conjugate gradient (CG) method, [Hestenes, Stiefel 52], is defined as follows:

$$a_i = (r_i, r_i)/(p_i, Ap_i)$$

$$x_{i+1} = x_i + a_i p_i$$

$$r_{i+1} = r_i - a_i A p_i$$

$$b_i = (r_{i+1}, r_{i+1})/(r_i, r_i)$$

$$p_{i+1} = r_{i+1} + b_i p_i$$

with initial condition $r_0 = p_0 = y - Ax_0$

and where A is an N by N positive definite matrix, Ax = y, r residuals and p a direction vector. The vectors x, y, p and r have dimension N, and a and b are scalars. The subscript is the iteration index, and $x_N = x$, where $Ax = y$.

In the following we will only concern ourselves with the iteration phase of the computations.

The scalar a_i has to be computed before any of the components of x_{i+1} and r_{i+1} can be computed. Similarly b_i has to be known before any of the components of p_{i+1} can be computed. The coefficient a_i can be computed in O(logN) time steps using a tree structure, or in O(N) time using a linear array. The variables a_i and b_i are global. They require information from the entire N-space for their computation. Every step in every direction is dependent on information from every dimension of the space.

For a matrix of bandwidth w, N binary trees of 2w-1 processors each can be used to compute the product Ap_i in logw steps. Each tree contains a row of A. The quadratic form (p_i,Ap_i) can be computed in another logN steps by making the N trees into subtrees of a tree of 2Nw-1 processors. The leaf nodes each contain one element of A. The products of Ap_i are formed in the leafs. Hence, each of the N subtrees of 2w-1 processors contains w components of p_i, and each component of p_i is needed in w subtrees. Also, one component of p_i is needed in the root of each subtree. The products of (r_i,r_i) can be computed at distance logN from the root one time step prior to the computation of the products of (p_i,Ap_i) and the inner product obtained by performing the summations in passing partial sums towards the root. The value of a_i can be sent down the tree logN levels and x_{i+1} and r_{i+1} computed at that level. For the next iteration it is also necessary to update the direction vector p_i, which requires the computation of b_i. If b_i were calculated directly from the equation above (r_{i+1}, r_{i+1}) would be needed, i.e., another logN steps. These steps can be avoided by rewriting the equation for b_i.

Substituting the equation for r_{i+1} into the expression for b_i and observing the property that $(r_i,r_j) = 0$, $i \neq j$, and

$(r_i, Ap_i) = (p_i, Ap_i)$ since $(p_i, Ap_j) = 0$, $i \neq j$, the following equation can be derived

$$b_i = -1 + a_i(Ap_i, Ap_i)/(p_i, Ap_i)$$

The values of a_i and (p_i, Ap_i) are already available in the root of the tree. The value of (Ap_i, Ap_i) can be available in the root one time step after (p_i, Ap_i) by computing the products of this additional inner product after those of (p_i, Ap_i) and passing the terms towards the root. Hence, b_i can be available in the root two steps after a_i, sent down to level logN and p_{i+1} computed at that level immediately after x_{i+1} and r_{i+1}. Finally, the values of p_{i+1} in the leaves need to be updated.

Each subtree contains w consecutive components of p_i. Each component of p_{i+1} is computed from corresponding components of r_{i+1} and p_i. The proper components of p_i are aleady available in the subtree, but those of r_{i+1} are not. They are available at the roots of adjacent subtrees. Computing w components of r_{i+1} in each subtree from r_i offers no easy escape from the communication problem, since w components of Ap_i are needed. Those components are only available in the same nodes that hold r_{i+1}. A left shift of information by one node at level logN can be accomplished in time 2logN, in a way similar to the rotate operator of Presnell and Pargas, [Presnell, Pargas 81]. Hence, for w left shifts a total of 2wlogN time steps are required. After an additional logw steps the leaves have received all needed values of r_{i+1}, and p_{i+1} can be computed. The tree is ready for the next iteration. The total time for an iteration of the CG method is 2wlogN + 2log(Nw) + 2.

The dominating term in this estimate can be reduced to w if a path at level logN from the root is available. The left shifts could then be accomplished in w time steps. With a path through the leaves the time would be reduced to 2wlogw. In both these cases the time complexity is reduced by providing increased communication capability, but without any additional processors. Providing a path through the leaves do not require any processor modification, since spare ports can be used. A path at any other level would require additional processor ports.

If A is the result of a finite difference approximation of differential operators, its nonzero elements will with typical node orderings fall on diagonals. In such a case it is possible to replace w in the previous complexity estimates with the number of nonzero diagonals of A. The trees can be reduced correspondingly.

The processor utilization of this algorithm is fairly low. Increased processor utilization at the expense of a slightly increased time complexity is obtained by performing the computations for a row of A in one processor. The N subtrees of 2w-1 processors are reduced to a single node. The resulting algorithm will require 2(w+1)logN + w + 4 time on a tree of 2N-1 processors. The time complexity can be reduced to 2logN + w + 5 by providing a path through the leaves, [Johnsson 82a]. Notice, the algorithm using a path through the leafs has a lower time complexity than all of the 2Nw-1 processor algorithms.

If A is of size 1000x1000 with a bandwidth of 30, then one iteration of the CG method would require about 6-7 ms on the Tree Machine, assuming it was fabricated in 2 micron technology. A path through the leaves would reduce the time needed by approximately one order of magnitude. The tree would consist of 256 chips, which indeed is a physically small machine. In 1 micron technology only 64 chips would be needed for the same machine, that would solve the problem in approximately half the time. The Tree Machine would easily fit on a small PC board.

The tree algorithms described above can be mapped onto Boolean n-cubes with a number of processors equal to the number of leaf processors in the tree. It should be noticed that the tree nodes participating in the summation required in computing inner products, only perform additions during three cycles of every CG iteration. Three inner products are computed. Three wave fronts are going up the tree. Two values are broadcast from the root towards the leaves. These operations are easily mapped onto a Boolean n-cube without loss of performance. A Boolean n-cube is a Hamiltonian graph. Hence, a cyclic shift of length w at any level of the binary tree described above is easily accomplished in time w in an n-cube. An n-cube with N processors can perform the computations of one iteration of the CG method in time 2logN+w+const, i.e., the same time complexity as the algorithm for 2N-1 tree interconnected processors with a path through the leafs.

If there are fewer than N processors available a different form of mapping of the problem onto the network of processors is required. If the matrix A is of band type a reduction in the number of required processors is easily obtained by exploiting the structure. A processor is associated with each non-zero diagonal, (instead of rows or columns), [Madsen, Rodrigue, Karushi 76], [Kung, Leiserson 80], [Kung 80], [Johnsson 82c], [Johnsson 82a]. An algorithm for the computation of matrix-vector products in 2N+w steps using w processors is given in [Kung, Leiserson 80]. Another algorithm requiring N+u steps using w processors is described in [Johnsson 82a]. With a few additional processors for the computation of inner products and the variables a_i and b_i it is possible to complete the computations for a step of the CG method in N+u+4 steps, using w+6 processors, [Johnsson 82a]. Of the w+6 processors, the w that are linearly interconnected corresponds to the non-zero diagonals of A, while the remaining 6 processors are sparsly interconnected and perform the other computations of the CG method. Storage of O(N) is associated with w+3 processors.

Algorithms for different partitionings of the computations needed when the number of processing elements does not exactly match any of the characteristic dimensions of the problem can be found in [Johnsson 82a].

V. PRECONDITIONED CONJUGATE GRADIENT METHODS

The preconditioned conjugate gradient method can be defined as:

$$x_{i+1} = x_i + a_i p_i$$

$$r_{i+1} = r_i - a_i A p_i$$

$$a_i = (r_i, z_i)/(p_i, Ap_i)$$

$$Mz_{i+1} = r_{i+1}$$

$$b_i = (r_{i+1}, z_{i+1})/(r_i, z_i)$$

$$p_{i+1} = z_{i+1} + b_i p_i$$

with initial conditions $r_0 = p_0 = y-Ax_0$

Preconditioning implies that a linear system of equations, $Mz_i = r_i$, must be solved in each step of the conjugate gradient method. With proper preconditioning the number of steps to reach an approximate solution is reduced to the extent that the total time to compute the approximate solution decreases. Most recent work on preconditioning has been concerned with the effect various choices of M have on convergence rate, storage requirement and solution time on seqential machines. Obtaining M through incomplete Cholesky factorization of A has been studied by, [Meijerink, van der Vorst 77], [Manteuffel 80], [Kershaw 78], and [van der Vorst 82], incomplete Crout factorization by [Meijerink, van der Vorst 77], SSOR preconditioning by [Jennings, Malik 78] and [Axelsson 76], and partial elimination depending on size of the elements of A by [Jennings, Malik 78] or the factor L, ($A = LL^T$), by [Glowinski et. al. 80]. Recently, the feasibility of different preconditioners from a computational point of view has been studied also for vector and multiprocessor machines, [van der Vorst 82], [Jordan 83], and [Rodrigue 83], [Sameh 83].

In the context of the machine topologies presented here, a diagonal preconditioning matrix obviously only demands one extra time step per iteration in the tree algorithms and one extra processor in the solution based on O(w) processors. In the following it is assumed that $M=LL^T$. The best known algorithm for solving dense triangular linear systems requires $3+(1/2)(3+\log N)\log N$ steps and at most $(15/1024)N^3+O(N^2)$ processors, [Sameh, Brent 77]. For the banded case the results are: $3+(2+\log u)\log N-1/2(\log u+1)\log u$ steps using $(1/2)u^2N+O(uN)$ processors. These results are based on a model where any processor can access any information at any time without delay. The tree, the n-cube and the O(w) processor solution do neither provide this number of processors nor unrestricted communication. Hence, preconditioning on a tree must be expected to have a time complexity that at best is comparable to that of one iteration of the CG method.

The system $Mz=r$ is for $M=LL^T$ solved as

$$Lq = r$$

$$L^Tz = q$$

A column sweep algorithm for the forward and backward substitution requires that $N^2/4$ data items pass through the root of a tree in each substitution phase, with the data structures used above, and if L were dense. For the banded case the lower bound is easily seen to be $u(u+1)/2+\log N$, and if only nonzero elements in L are considered is $c(N)+\log N$, where $c(N)$ is the number of nonzero elements in the lower lefthand N/2 by N/2 submatrix of L. The logN term is due to the allocation of the vector r at that distance from the root.

An O(N) tree algorithm is as follows. Compute q_i in the processor that stores r_i. Column i of L is assumed to be stored in the same processor. To compute q_i the sum of products of q_j and L_{ij}, $j = 1,\ldots,i-1$ is needed. A leaf node receives this sum from its parent. Having computed its own component of q, q_i, the node changes mode and computes the products of column i of L and q. Product sums of the product of L and q are computed at successive levels of the tree, i.e., columns properly multiplied by q are added pairwise in processors at distance one from the leafs, such sums added again pairwise at distance two, etc. The computations start from the left leaf cell and terminates at the right leaf cell. Nodes are activated in inorder tree traversal. An analysis of the time complexity of this algorithm yields $2(2N-1-\log N)$ steps. This time is independent of the bandwidth of L. The time equals the path length of inorder tree traversal. This algorithm offers a speed up of O(N) using O(N) processors.

If there is a path of length N through the nodes at distance logN from the root, then the algorithms used for forward and backward substitution on linear arrays, [Kung, Leiserson 80], [Kung 80], [Johnsson 81], can be employed and the number of time steps equals 2N, independent of the bandwidth.

The time complexity of the above two preconditioning algorithms is the same even if L is sparse. Exploiting sparsity leads to reduced storage needs. Reduced time complexity can be obtained if different data structures are employed. For instance, if L is of bordered block diagonal form with block size m, a tree algorithm can be devised that concurrently computes all but the last m components of q in 2(2m-1-logm) time steps. To compute the last m components of q an additional 2logN+2(2m-1-logm) steps are needed. These complexity estimates illustrate the importance of the data structures used. This point is further underscored by the $O(\log^2 N)$ cyclic reduction tree algorithm of Presnell and Pargas, [Presnell, Pargas 81].

VI. CONCLUSIONS

Several concurrent algorithms for the solution of linear systems of equations were surveyed. They are all feasible for machines with a large number of sparsely interconnected processors with local storage only. The communication topology of the algorithms have been restricted to meshes with internal nodes of degree 4 or 6 (hexagonal arrays), trees and Boolean n-cubes.

For ensembles with fewer processors than what the fully concurrent algorithms demand, requires additional sequencing of the computations, and additional storage management, compared to algorithms fully instantiated in space. Block algorithms and sweeping algorithms are two possibilities that were briefly discussed.

New algorithms for the conjugate gradient method on binary tree interconnected processors as well as Boolean n-cubes have been presented. The algorithms range in time complexity from O(logN+w) to O(N) with the corresponding number of processors ranging from 2Nw-1 to O(w). It was shown that a path through the leafs reduces the time complexity significantly. Indeed, a tree of 2N-1 processors with the leafs interconnected to a ring performs the computations of one iteration of the CG method in less time than a pure binary tree with 2Nw-1 processors. The time complexity of concurrent CG algorithms on a Boolean n-cube of N processors is the same as the time complexity of algorithms on a tree of 2N-1 processors with a path through the leafs.

Preconditioned conjugate gradient methods include the solution of a linear system of equations in each step of the conjugate gradient algorithm. An O(N) algorithm for the forward and backward substitution was presented for a binary tree. This algorithm makes no particular assumption of the structure of the factors of the preconditioning matrix, but assumes the components of the given vector and the solution vector to be located in successive cells at distance logN from the root. An O(logN+m) algorithm was also presented for the case where the factors of the preconditioning matrix are of bordered block diagonal form with block size m.

ACKNOWLEDGMENT

The author wish to thank Carver A. Mead and Pey-yun Peggy Li for their careful reading of the manuscript, and many valuable suggestions.

This work was in part supported by the Defense Advanced Project Agency (DARPA) under Contract N00014-79-C-0597 with the California Institute of Technology. Views and conclusions contained in this paper are the author's and should not be interpreted as representing the official opinion or policy of DARPA, the U.S. Government, or any person or agency connected with them.

Partial support was also provided by the System Development Foundation.

REFERENCES

Ahmed, H.M., Delosme, J.-M., Morf, M. Highly Concurrent Computing Structures for Matrix Arithmetic and Signal Processing. IEEE Computer (1): 65-82, January, 1982.

Axelsson, O. A Class of Iterative Methods for Finite Element Equations. Computer Methods in Applied Mechanics and Engineering 9: 123-137, 1976.

Bentley, J.L., Kung, H.T. A Tree Machine for Sorting Problems. In International Conference on Parallel Processing, pp. 257-266. IEEE Computer Society, 1979.

Browning, S.A., Seitz, C.L. Communication in a Tree Machine. In Proceedings of the Second Caltech Conference on VLSI. Computer Science, Caltech, 1981.

Browning, S.A. The Tree Machine: A Highly Concurrent Computing Environment. Technical Report 3760, Computer Science, California Institute of Technology, January, 1980.

Cohen, D. Mathematical Approach to Iterative Computational Networks. In Proceedings of the Fourth Symposium on Computer Arithmetic, pp. 226-238. IEEE Computer Society, October 1978.

DeBenedictis, E., Seitz, C.L. Homogeneous Machine--Technical Plan. Technical Report 4705:DF:82, California Institute of Technology, 1982.

Delosme, J.-M., Morf, M. Scattering Arrays for Matrix Computations. In Real-Time Signal Processing IV, pp. 74-83. SPIE--The International Society for Optical Engineering, August 1981.

Despain, A.M. Very Fast Fourier Transform Algorithms Hardware for Implementation. IEEE Transaction on Computers C-28 (5): 333-341, May 1979.

Flanders, P.M., Hunt, D.J., Reddaway, S.F., Parkinson, D. Efficient High Speed Computation with the Distributed Array Processor. In High Speed Computer and Algorithm Organization, pp. 113-128. Academic Press, 1977.

Fox, G.C., Seitz, C.L. Concurrent Processing and the Decomposition of Problems. Technical Report, California Institute of Technology, 1983.

Fox, G.C. Matrix Operations on the Homogeneous Machine. Technical Report CALT-68-939, California Institute of Technology, August 1982.

Gentleman, S. Morven, Kung, H.T. Matrix Triangulation by Systolic Arrays. In Real-Time Signal Processing, pp. 19-26. SPIE, August 1981.

Glowinski, R., Lions, Mantel, B., Periaux, J., Pironneau, O., Poirier, G. An Efficient Preconditioned Conjugate Gradient Method Applied to Nonlinear Problems in Fluid Dynamics via Least Square Formulations. In Computing Methods in Applied Science and Engineering, pp. 445-487. North-Holland, 1980. Eds. R. Glowinski, J.L. Lions.

Gottlieb, A., Schwartz, J.T. Networks and Algorithms for Very-Large-Scale Parallel Computation. IEEE Computer 15(1): 27-36, January 1982.

Heller, Don E., Ipsen, I.I.P. Systolic Networks for Orthogonal Equivalence Transformations and Their Applications. In P. Penfield Jr., editor, Proc., Conf. on Advanced Research in VLSI, pp. 113-122. Artech House, 1982.

Heller, D.E. A Survey of Parallel Algorithms in Numerical Linear Algebra. SIAM Review 20: 740-777, 1978.

Heller, D.E. Partitioning Big Matrices for Small Systolic Arrays. In VLSI & Modern Signal Processing, p. 109. Computer Science, University of Southern California, November 1982. Also, Report CS-83-02, Computer Science Dept., Pennsylvania State Univ., University Park, PA.

Hestenes, M.R., Stiefel E. Methods of Conjugate Gradient for Solution of Linear Systems. J. Res. Nat. Bur. Standards 49: 409-436, 1952. 21

Huffman, D.A. The Synthesis of Linear Sequential Coding Networks. In Information Theory, London Symposium on Information Theory. Butterworth and Company, Ltd., 1956.

Jennings, A., Malik, G.M. The Solution of Sparse Linear Equations by the Conjugate Gradient Method, Int. J. Numer. Methods 12: 141-158, 1978.

Johnsson, S.L., Cohen, D. A Formal Description of Array Implementations of FFT Algorithms. In VLSI & Modern Signal Processing. Computer Science, University of Southern California, 1982.

Johnsson, S.L. Computational Arrays for Band Matrix Equations. Technical Report 4287, Computer Science, California Institute of Technology, May 1981.

Johnsson, S.L. Concurrent Algorithms for the Conjugate Gradient Method. Technical Report 5040:TR:82, Computer Science, California Institute of Technology, September 1982.

Johnsson, S.L. VLSI Algorithms for Doolittle's, Crout's and Cholesky's Methods. In International Conference on Circuits and Computers 1982, ICCC82, pp. 372-377. IEEE, Computer Society, September 1982.

Johnsson, S.L. Pipelined Linear Equation Solvers and VLSI. In Microelectronics '82, pp. 42-46. Institution of Electrical Engineers, Australia, May 1982.

Johnsson, S.L. A Computational Array for the QR-method. In P. Penfield, Jr., editor, _Proc., Conf. on Advanced Research in VLSI_, pp. 123-129, Artech House, January 1982.

Jordan, T.L. Conjugate Gradient Preconditioners for Vector and Parallel Processors. In _Elliptic Problem Solvers II_. Academic Press, 1983.

Kershaw, D.S. The Incomplete Cholesky Conjugate Gradient Method for the Iterative Solution of Systems of Linear Equations, _J. Comput. Phys._ 26: 43-65, 1978.

Kung, S.Y., Gal-Ezer, R.J. Linear or Square Array for Eigenvalue and Singular Value Decomposition? In _VLSI & Modern Signal Processing_, pp. 89-98. Computer Science, University of Southern California, November 1982.

Kung, H.T. and Leiserson, C.E. Algorithms for VLSI Processor Arrays. In _Introduction to VLSI Systems_, pp. 271-294. Addison-Wesley, 1980. C.A. Mead, and L.A. Conway.

Kung, S.Y. VLSI Matrix Computation Array Processor. In _MIT Conference on Advanced Research in Integrated Circuits_, MIT, February 1980.

Li, P. _The Tree Machine Operating System_. Technical Report 4618, Computer Science, California Institute of Technology, July 1981.

Madsen, K., Rodrigue, G.H., Karushi, J.I. Matrix Multiplication by Diagonals on a Vector/Parallel Processor. _Information Processing Letters_ 5(2): 41-45, June 1976.

Manteuffel, T.A. Solving Structures Problems Iteratively with a Shifted Incomplete Cholesky Preconditioning. In R. Glowinski, J.L. Lions, editor, _Computing Methods in Applied Science and Engineering_, pp. 427-444. North-Holland, 1980.

Meijerink, J.A., van der Vorst. An Iterative Solution Method for Linear Systems of which the Coefficient Matrix is a Symmetric M-Matrix. _Math. Comp._ 31: 148-162, 1977.

Meijerink, J.A., van der Vorst. _Guidelines for the usage of of incomplete decompositions in solving sets of linear equations that occur in practical problems_. Technical Report TR-9, ACCU, Utrecht, 1978.

Munksgaard, N. _Solving Sparse Symmetric Sets of Linear Equations by Preconditioned Conjugate Gradients_. Technical Report CSS 67, Harwell, 1979.

Presnell, H.A., Pargas, R.P. Communication Along Shortest Paths in a Tree Machine. In _Functional Programming Languages and Computer Architecture_, pp. 107-114, ACM, 1981.

Rodrigue, G.H. Incomplete Factorization and Parallel Computations. In _Elliptic Problem Solvers II_. Academic Press, 1983.

Sameh, A.H., Brent, R.P. Solving Triangular Systems. SIAM J. Numerical Analysis 14(6): 1101-1113, December 1977.

Sameh, A.H. Numerical Parallel Algorithms--A Survey. In High Speed Computer and Algorithm Organization, pp. 207-228. Academic Press, 1977.

Sameh, A.H. Parallel Algorithms for Elliptic Problems. In Elliptic Problem Solvers II. Academic Press, 1983.

Schreiber, R., Keukes, P. Systolic Linear Algebra Machines in Digital Signal Processing. In VLSI & Modern Signal Processing, p. 201. Computer Science, University of Southern California, November 1982.

Schreiber, R. Systolic Arrays for Eigenvalue Computation. In Real-Time Signal Processing, Vol. 341, pp. 27-34. SPIE, May 1982.

Schwartz, J.C. Ultracomputers. ACM Transactions on Programming Languages and Systems 2(4): 484-521, October 1980.

Seitz, C.L. Ensemble Architectures for VLSI--A Survey and Taxonomy. In P. Penfield Jr., editor, Proc., Conf. on Advanced Research in VLSI, pp. 130-135. Artech House, 1982.

Snyder, L. Introduction to the Configurable, Highly Parallel Computer. IEEE Computer 15(1): 47-56, January 1982.

Stone, H.S. Parallel Processing with the Perfect Shuffle. IEEE Trans. Computers C-20: 153-161, 1971.

Stone, H.S. An Efficient Parallel Algorithm for the Solution of a Tridiagonal Linear System of Equations. JACM 20 (1): 27-38, January 1973.

Stone, H.S. Parallel Tridiagonal Equation Solvers. ACM Transactions on Mathematical Software 1(4): 289-307, December 1975.

van der Vorst, H.A. A Vectorizable Variant of Some ICCG Methods. SIAM, J. Scientific and Statistical Computing 3(3): 350-357, 1982.

CONJUGATE GRADIENT PRECONDITIONERS FOR VECTOR AND PARALLEL PROCESSORS

Thomas L. Jordan [1/]

I. INTRODUCTION

Vector and parallel processors have caused program designers to select or develop algorithms that allow efficient use of these computers. The relatively simple tasks of solving linear systems of equations (Fong and Jordan, 1977) or performing matrix multiplication (Madsen, et al., 1976) require, as a minimum, a reformulation of existing methods. At another extreme, radically new approaches, such as chaotic relaxation (Chazan and Miranker, 1969; Miellou, 1974; and Baudet, 1978;), are invented to accommodate large numbers of asynchronous processors, commonly referred to as MIMD computers.

We shall examine the conjugate gradient (CG) algorithm and some preconditioners for suitability on both vector and parallel computers. Preconditioners considered include Jacobi, incomplete Cholesky, polynomial approximation, block Jacobi, and block cyclic reduction. Performance data on Cray-1 is provided for finite 5- and 9-point difference

[1/] Computing Division, Los Alamos National Laboratory
Los Alamos, New Mexico

ISBN 0-12-100560-7

approximations to elliptic PDEs. Problems solved include the model problem (LaPlace's equation with Dirichlet boundary conditions) and a variable coefficient matrix. The variable elements are generated randomly such that the diagonal element of a row is the negative sum of the off-diagonal elements except for the boundary equations. In all cases, we start with a 0 guess, a solution of 1, and a convergence criterion of 10^{-6} error in the residual.

The problem is to solve Ax = y, where A is symmetric, positive definite, $N = n^2$ is the order, and n is the number of nodes on one side of a square grid. Let

$$A = [-B_j \ A_j \ -B_{j+1}^T] \tag{1}$$

represent the blocks of the jth row of A. Then A_j is also symmetric, positive definite, and tridiagonal. The B_j are either diagonal (5-point) or tridiagonal (9-point). Designate M as approximations to A such that $M^{-1}A \simeq I$, where M^{-1} is sparse and easy to compute.

II. CONJUGATE GRADIENT ALGORITHM

The iterative portion of the CG algorithm is described by

$$x_{i+1} = x_i + \alpha_i p_i \qquad \alpha_i = (r_i M^{-1} r_i)/(p_i^T A p_i)$$

$$r_{i+1} = r_i - \alpha_i p_i$$

$$p_{i+1} = M^{-1} r_{i+1} + \beta_i p_i \qquad \beta_i = (r_{i+1} M^{-1} r_{i+1})/(r_i M^{-1} r_i) \ ,$$

where p is a descent direction (initially arbitrary), r is a residual vector, and x is an approximation to the solution vector (initially a guess supplied by the user).

The sequential and parallel complexity of CG is tabulated in Table I. We conjecture that the three reduction operators of CG will play an increasingly important role in determining the time of the CG algorithm.

TABLE I. Complexity of CG With $M^{-1} = I$

	Sequential	Parallel
3 triads	3N	0(1)
3 reductions[a]	3N	$0(\log_2 N)$
1 matrix multiply	$n_d N$	$0(\log_2 n_d)$[b]

[a] *Two dot products and a convergence norm.*

[b] n_d *is 5 or 9 for our problem.*

III. PRECONDITIONERS

A. *Incomplete Cholesky (ICCG)*

Meijerink and Van Der Vorst (1977) developed a popular family of incomplete Cholesky factorizations for solving elliptic problems with the CG algorithm where $M = \tilde{L}\tilde{L}^T$ contains only a limited number of off-diagonals of the complete Cholesky factorization of $A = LL^T$. For example, ICCG(0) contains elements in L only where elements in the lower half of A are present.

1. *Mostly Scalar.* We include an implementation of ICCG(0) in which all of the Fortran code was compiled in a scalar mode. However, two triads and the three reduction operations used vectorized library subroutines. Thus the times for this implementation are considerably more favorable than a completely scalar implementation.

2. L^{-1} Scalar. Next we allowed vectorization whenever possible but implemented the forward and back substitution in the classical manner. Hence the application of $\tilde{L}^{-1}$ and $\tilde{L}^{-T}$ were the only scalar components.

3. Linear Recurrences Scalar. Greenbaum and Rodrigue (1977) observed that most of the work in solving $\tilde{L}^{-1}w = y$ could be vectorized except for a first-order linear recurrence. Let $M = \tilde{L}D\tilde{L}^T$, where $L = [C_j \; B_j \; 0]$ is a block bidiagonal matrix and $B_j = [b_{ij} \; 1 \; 0]$ is a bidiagonal matrix with unit diagonal. Then the forward substitution,

$$w = \tilde{L}^{-1}y$$

can be described by

$$w_1 = B_1^{-1}y \quad ,$$

$$w_j = B_j^{-1}(y_j - C_j y_{j-1}), \; j = 2,n \quad .$$

Hence, the only scalar code is in applying B_j^{-1}. This linear recurrence is, of course, vectorizable using cyclic reduction. However, it is not very profitable on the Cray-1 (Jordan, 1981).

4. All Vector. Van Der Vorst (1982) modified the preceding algorithm to achieve a fully vectorizeable procedure for ICCG.

Since B_j^{-1} can be approximated by

$$B_j^{-1} \simeq I + N_j + N_j^2 + N_j^3 = (I + N_j^2)(I + N_j) \quad ,$$

where $N_j = I - B_j$ is a diagonal matrix, then N^2 does not fill in and may either be precomputed and stored or another multiply may be added to each iteration. Our times reflect the latter choice. This amounts to two steps of cyclic reduction with an approximation error generally within that due to the incompleteness of the $\tilde{L}\tilde{L}^T$ factorization.

Finally, we have arrived at an ICCG algorithm that is fully vectorizable with vectors of length n. Although this is marginally satisfactory for vector computers, note that the parallel complexity, as described, is O(n) and requires n steps of vector recursion. While it is possible to apply block cyclic reduction to this block linear recursion and thereby obtain $O(\log_2 n)$ complexity, the algorithm remains unattractive to us for parallel computation.

5. *ICCG Results.* Table II gives the Cray-1 times for the four implementations just described. We attribute the extra iterations required to convergence in the "all vector" algorithm to a relatively less accurate solution for two steps of cyclic reduction, because $\log_2 n$ is 1 greater in those cases.

TABLE II. ICCG Times (s) and Iteration Count - Model Problem

n	Mostly Scalar	L^{-1} Scalar	B^{-1} Scaalar	All Vector	Iteration Count
30	0.11	0.057	0.034	0.010	20(20)[a]
50	0.40	0.21	0.12	0.080	27(27)
70	1.03	0.53	0.31	0.21	36(38)
90	2.10	1.09	0.63	0.48	45(57)

[a]*Iteration count for the all-vector code.*

B. Block Methods

Block methods are more attractive for both vector and parallel processors. Kowalik and Kumar (1981) studied the efficiency of a form of block Jacobi as a conjugate gradient preconditioner, labeled BCG by Malik and Jenning (1976), on the DENELCOR HEP MIMD computer. They determined that they could achieve close to the maximum possible speed-up with BCG. We programmed the algorithm for the Cray-1 and show the iteration count in Sec. IV.

1. BCG. Scale A, as defined in (1), such that

$$A = [-\tilde{B}_j \ \ I \ \ -\tilde{B}_{j+1}^T],$$

where

$$\tilde{B}_j = L_j^{-1} B_j L_{j-1}^{-T}$$

and

$$A_j = L_j L_j^T.$$

Since $\tilde{B}_j$ is full if formed explicitly, sparsity may be preserved by computing $\tilde{B}_j$ implicitly. Also, observe that even though L^{-1} is a first-order linear recurrence, n of these may be computed simultaneously. Hence vector lengths are n for this phase of the computation. The remainder of the calculation can be done with vector lengths N. The convergence rate is not attractive, but the scaling does provide an improvement over Jacobi as shown in Sec. IV.

2. Incomplete Block Cyclic Reduction. Block cyclic reduction can be described in terms of $\log_2 n$ pairs of steps such that scaling as described in the BCG algorithm is one step, and matrix multiplication is a second. Let

$$A = [-B_j \ 1 \ -B_{j+1}^T].$$

Step 1. Perform the matrix multiplication

$$[B_j \ 1 \ B_{j+1}^T]A \quad \text{to obtain}$$

$$[-B_jB_{j-1} \ 0 \ 1-B_jB_j^T-B_{j+1}^TB_{j+1} \ 0 \ -B_{j+1}^TB_{j+2}^T] \quad .$$

Step 2. Scale symmetrically by L_j, where

$$\tilde{L}_j\tilde{L}_j^T = 1 - B_jB_j^T - B_{j+1}^TB_{j+1} \ , \quad \text{to obtain}$$

$$[-\tilde{B}_j \ 0 \ 1 \ 0 \ -\tilde{B}_{j+2}] \quad ,$$

and

$$\tilde{B}_j = \tilde{L}_j^{-1}B_j \ B_{j-1} \ \tilde{L}_{j-2}^{-T} \quad .$$

The matrices $\tilde{L}_j^{-1}$ and $\tilde{B}_j$ have become full, and some form of incompleteness or approximation must be introduced to prevent fill-in. We carried out a scaling step followed by a multiplication step to produce the results labeled BCG 1 in Sec. IV. The convergence rates have halved compared to BCG; however, the times were about the same. We attribute this to an inconsistent implementation in the sense of Lambiotte and Voigt (1974). We eliminated variables from every equation rather than just the even ones.

The reader is referred to the works of Rodrigue and Wollitzer (1981) and Lichnewsky (1982) that describe their approaches to, and results from, incomplete block cyclic reduction. The times given by Lichnewsky are impressive.

C. Polynomial Approximations.

Polynomial approximations provide the highest level of parallelism, because the vector lengths are all of order N. Also, vector hardware supports the evaluation of them very well.

1. Neuman Series. To exploit the above features for the STAR-100, Dubois, et al. (1979) explored the feasibility of using s terms of the Neuman series

$$M^{-1} = I + N + N^2 + \cdots + N^{s-1} \quad ,$$

where N = I - A. They found that for s > 2 there was no improvement in running time. Our results are given in Table III for the Cray-1 and agree qualitatively with theirs.

TABLE III. Neuman Times (s) and Iteration Count - Model Problem

n/s	1	2	4	6	8
30	0.020(46)	0.017(26)	0.019(19)	0.022(16)	0.023(13)
50	0.082(74)	0.062(38)	0.073(27)	0.081(22)	0.091(19)
70	0.22(103)	0.16(52)	0.19(37)	0.21(30)	0.24(26)
90	0.46(130)	0.34(66)	0.39(46)	0.45(38)	0.49(33)

Johnson and Paul (1981) considered a third-order polynomial

$$M^{-1} = uI + vN + wN^2 \quad ,$$

where u,v and w were based upon optimal relaxation factors of the matrix A. They claim convergence rates equivalent to ICCG(3).

2. Chebychev Polynomials. One can evaluate polynomials of matrices on the Cray-1 at a rate of two to three times faster than the work done in the conjugate gradient portion of this code when implemented in assembly language. The megaflop rates for matrix multiplication and polynomial evaluation over matrices for 3, 9, and 27 diagonal matrices are 80, 95, and 100, respectively. We conjectured, wrongly, that these rates would permit us to economically take more terms of the Neuman series. The results in Table III are just those. However, T. Manteuffel made the constructive suggestion that we try Chebychev coefficients shifted to the range of the eigenspectrum. The results shown in Table IV demonstrate that if the polynomials can be evaluated at a somewhat higher rate, then polynomials of considerably higher order are economical, provided "good" coefficients are available.

TABLE IV. Chebychev Times (s) and Iteration Count - Model Problem

n/s	1	2	4	6	8
30	0.020(46)	0.017(26)	0.015(14)	0.015(10)	0.015(8)
50	0.083(74)	0.065(38)	0.058(21)	0.056(15)	0.058(12)
70	0.22(103)	0.17(52)	0.15(28)	0.14(19)	0.14(15)
90	0.46(130)	0.35(66)	0.30(34)	0.28(24)	0.29(19)

A drawback to the shifted Chebychev method is the determination of the shift that generates the "best" coefficients. To better understand the sensitivity of the shift on a harder problem, we took as our matrix A a variable coefficient problem in which the off-diagonal elements are generated randomly and the diagonal elements are the negative sum of the off-diagonal elements. The matrix is then rescaled symmetrically to unit diagonal (as in the model problem), and the coefficients generated for the model problem are used. The results given in Table V show "good" convergence rates with coefficients that are not "best" and exhibit the same economy in time as the model problem.

TABLE V. Chebychev Times (s) and Iteration Count Random

n/s	1	2	4	6	8
30	0.034(78)	0.026(40)	0.023(22)	0.023(16)	0.022(12)
50	0.14(126)	0.11(66)	0.093(35)	0.090(24)	0.091(19)
70	0.36(162)	0.30(90)	0.25(47)	0.24(33)	0.24(26)
90	0.71(200)	0.55(107)	0.49(56)	0.46(39)	0.45(30)

IV. OTHER COMPARISONS

Times given in Sections II.A. and II.B. use the fast routines for the 9-point operators, even though the problems solved are 5-point problems. An optimal 5-point code would not improve times appreciably because of limitations on memory referencing. However, using a 9-point operator for the model problem reduces the iteration count and the total solution time (compare Tables IV and VI). Hence, a "broader" operator appears to be worthwhile.

A comparison of the iteration counts for the various methods solving the model problem (5-point operator) is made in Table VII.

TABLE VI. Chebychev Times (s) and Iteration Count Model Problem (9-point Operator)

n/s	1	2	4	6	8
30	0.015(33)	0.012(19)	0.013(12)	0.012(8)	0.014(7)
50	0.061(54)	0.053(31)	0.050(18)	0.049(13)	0.050(10)
70	0.16(74)	0.14(43)	0.12(24)	0.12(17)	0.12(13)
90	0.33(93)	0.29(54)	0.25(29)	0.24(20)	0.24(16)

TABLE VII. Iteration Count - Model Problem

N	J1[a] S1	BCG	J2 S2	BCG1	All Vector	ICCG	S4	S6	S8
30	46	41	26	25	20	20	14	10	8
50	74	65	38	39	28	27	21	15	12
70	103	85	52	50	38	36	28	19	15
90	130	105	66	63	57	45	34	24	19

[a]*Ji and Si denote i Neuman and Chebychev terms, respectively. Also, J1 is equivalent to Jacobi.*

V. CONCLUSIONS

We see that ICCG, block, and polynomial methods are desirable, from the standpoint of parallelism, in that order. All appear satisfactory for vector computers. ICCG is even less attractive for MIMD computers. Also we may have a glimpse of the conjugate gradient portion of the codes becoming the slow component of the calculation as we move toward truly parallel computation. Because polynomial preconditioners have 0(1) parallel complexity, more work is needed to simplify the derivation of "good" coefficients.

REFERENCES

Baudet, G. M. (1978). "The Design and Analysis of Algorithms for Asynchronous Computers," Rep. CMU-CS-78-116, Carnegie-Mellon University, Pittsburg, Pennsylvania.

Chazan, D., and Miranker, W. (1969). "Chaotic Relaxation," Linear Algebra and Its Applications, Vol. 2, pp. 199-222.

Dubois, P. F., Greenbaum, A., and Rodrigue, G. H. (1979). "Approximating the Inverse of a Matrix for Use in Iterative Algorithms on Vector Processors," Computing, 22, pp. 257-268.

Fong, K. W., and Jordan, T. L. (1977). "Some Linear Algebraic Algorithms and Their Performance on Cray-1," Rep. LA-6744, Los Alamos National Laboratory, Los Alamos, New Mexico.

Greenbaum, A., and Rodrigue, G. H. (1977). "The Incomplete Cholesky Conjugate Gradient for the STAR (5-point) Operator," Rep. UCID 17574, Lawrence Livermore National Laboratory, Livermore California.

Jennings, A., and Malik, G. M. (1978). "The Solution of Sparse Linear Equations by the Conjugate Gradient Method," Intern. J. for Numerical Methods in Engineering, 12, pp. 141-158.

Johnson, O. G., and Paul, G. (1981). "Optimal Parameterized Incomplete Inverse Preconditionings for Conjugate Gradient Calculations," Res. Rep. RC 8644, IBM-Research Center, Yorktown Heights, New York.

Jordan, T. L. (1982). " A Guide to Parallel Computation and Some Cray-1 Experiences," Parallel Computations, Academic Press, pp. 1-50.

Kowalik, J. S., and Kumar, S. P. (1982). "An Efficient Parallel Block Conjugate Method for Linear Equations," Department of Computer Science, Washington State University, Pullman, Washington.

Lambiotte, J. J., and Voigt, R. G. (1974). "The Solution of Tridiagonal Linear Systems on the CDC-STAR-100 Computer," ICASE Rep., Institute for Computer Applications in Systems Engineering, NASA, Langley Research Center, Hampton, Virginia.

Lichnewsky, A. (1982). "Solving Some Linear Systems Arising in Finite Element Methods on Parallel Processors," Universite de Paris - Sud and Inria, Paris, France.

Madsen, N., Rodrigue, G., and Karush, J. (1976). "Matrix Multiplication by Diagonals on a Vector/Parallel Processor," J.of Info. Proc. Letters, Vol. 5, No. 2.

Meijerink, J. A., and Van Der Vorst, H. A. (1977). "An Iterative Solution Method for Linear Systems of Which the Coefficient Matrix Is a Symmetric M-Matrix," Math. Comp., Vol. 31.

Miellou, J. C. (1974). "Iterations Chaotiques a Retards," Comptes Rendus de L'Academie des Sciences de Paris, Series A., Vol. 278, pp. 957-960.

Rodrigue, G., and Wollitzer, D. (1981). "Incomplete Block Cyclic Reduction," Rep. UCRL-85837, Lawrence Livermore National Laboratory, Livermore, California.

Van DerVorst, H. A. (1982). "A Vectorizable Variant of Some ICCG Methods," SIAM J. Sci. Stat. Comput., Vol. 3, No. 3, pp. 350-356.

AN ALGEBRA FOR SYSTOLIC COMPUTATION

H.T. Kung[1] and W.T. Lin[2]

1. INTRODUCTION

Over the past several years, many systolic algorithms have been proposed as solutions to computation-bound problems (see, e.g., Fisher and Kung (1982), Kung (1979), Kung (1982), Kung and Leiserson (1979)). By exploiting the regularity and parallelism inherent to given problems and by employing high degrees of parallelism and pipelining, systolic algorithms implemented in VLSI achieve high performance with regular communication structures and low I/O requirements (see Kung (1982) for detailed discussions of advantages of systolic structures). A number of prototype machines for implementing systolic algorithms, ranging from single-purpose chips (see Corry and Patel (1983), Foster and Kung (1980), and Kung and Song (1982)), through application-oriented yet programmable systems (Blackmer, Kuekes and Frank (1981), Yen and Kulkarni (1982)), to very general systems with reconfigurable interconnections (see Bromley (1981), Symanski (1981), and Symanski (1982)), have been designed and built. Somewhat more recently, building-block chips for systolic architectures have also been proposed or designed (see Fisher, et al. (1983), Kuekes and Shen (1983), Sorasen and Solberg (1983), and Weste, et al. (1982)), including the CMU programmable systolic chip (PSC) (Fisher, et al. (1983), Fisher, et al. (1983)). The general question of automatically deriving systolic arrays and verifying their correctness, however, remains open, although several significant attempts have been made in this direction (see, e.g., Chen and Mead (1983), Lam and Mostow (1983), Leiserson and Saxe (1981), Weiser (1981)). Instead of suggesting methods for deriving or verifying systolic designs, we provide in this paper an algebra for manipulating algorithm designs for VLSI implementation in general. With this algebra a designer is able to manipulate designs by "pushing symbols," in order to conveniently meet desirable design criteria such as locality and regularity of data communication.

---

[1]Department of Computer Science, Carnegie-Mellon University, Pittsburgh, Pennsylvania 15213

[2]Department of Electrical Engineering, Carnegie-Mellon University, Pittsburgh, Pennsylvania 15213

ISBN 0-12-100560-7

Section 2 illustrates the notation and basic principles by considering the hardware implementation of a finite impulse response (FIR) filter. Two representations are proposed to specify a design with the property that from either representation we can derive the other. The *z-graph representation* is close to a hardware or VLSI specification of a design, and the *algebraic representation* is convenient for performing algebraic transformations on a design. Starting with a design that corresponds directly to the mathematical definition of the filtering problem (and thus its correctness is obvious), we perform a set of algebraic transformations on its algebraic representation and obtain the algebraic representation of a systolic design, from which a systolic filtering array can be derived automatically. Section 3, the heart of this paper, provides a mathematical foundation for the algebraic transformations used in Section 2. These transformations are formally justified with respect to a proposed semantics for design. Once justified, they become "legal" transformations that can be applied freely to any design without impairing correctness. Section 4 presents another application of the algebra, namely, the derivation of a systolic infinite impulse response (IIR) filtering array. The last section contains some concluding remarks.

2. BASIC PRINCIPLES AND NOTATION--ILLUSTRATED BY A FIR FILTERING EXAMPLE

To illustrate the basic idea and notation of this paper, this section considers a concrete example--the FIR filtering problem. We will use many diagrams to make the presentation as clear as possible, although algebraic transformations of this paper rely only on the algebraic representation. We will perform algebraic transformations formally here and postpone their justification to Section 3.

2.1. FIR Filtering and z-Notation

Consider the following FIR filter with weights w_i:

$$y_i = w_1 x_i + w_2 x_{i+1} + w_3 x_{i+2} + w_4 x_{i+3}. \tag{1}$$

Figure 1(a) depicts a straightforward design, called design S, for the hardware implementation of the filter. In the diagram, each ⊗ and ⊕ represent a multiplier and adder, respectively and each ▭ or ▯ represents a register capable of latching incoming data for one cycle time. Note that the cycle time must be long enough to allow data flow from register to register, possibly performing some computations in between. One of the objectives of systolic designs is to minimize the cycle time by avoiding long communications and large numbers of computations done inside each cycle, and thus maximize the throughput of the resulting system.

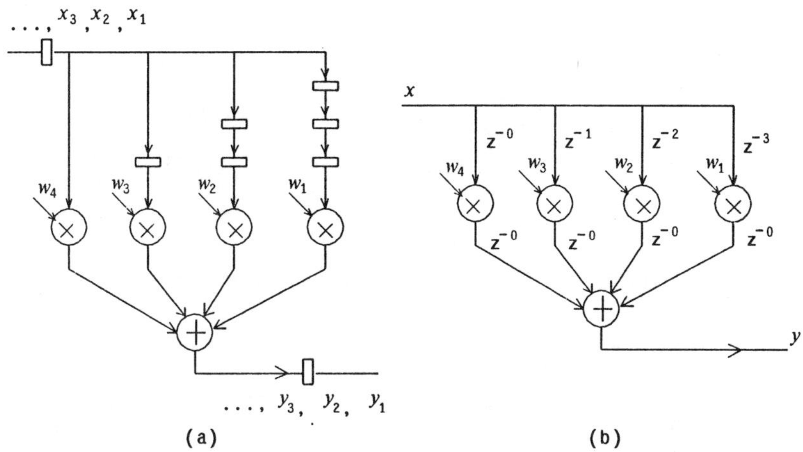

Figure 1. Design S: straightforward design (a) and in the z-notation (b).

Figure 1(b) describes design S (ignoring the input and output registers □) with the usual z-notation, where a delay of k cycles is indicated by z^{-k}. We see that in the z-notation the minimum cycle time is the time to perform all the operations connected by edges with label z^{-0}. Thus for design S the cycle time is at least the time to perform one multiplication (assuming that four hardware multipliers are available) and one 4-input addition. In the next section we show a systolic design for which only one multiplication and one 2-input addition will have to be done in each cycle.

2.2. Systolic FIR Filtering and z-Graph Representation

Figure 2 depicts a typical systolic design for FIR filtering, called design W2 in Kung (1982). In this design the w_i stay and x_i and y_i both move systolically from left to right, but the x_i move twice as slowly as the y_i. Note that each x value passes from cell to cell without changing. Figure 3(a) depicts the systolic array in the z-notation.

By grouping every pair of multiplication and addition as one node to be executed by a separate processor, we derive the *z-graph representation* of the design (Figure 3(b)). The z-graph representation of a systolic design has the ''systolic property'' that the input (the x in Figure 3(b)) is distributed to all the nodes (v_1, v_2, v_3, v_4) at dif-

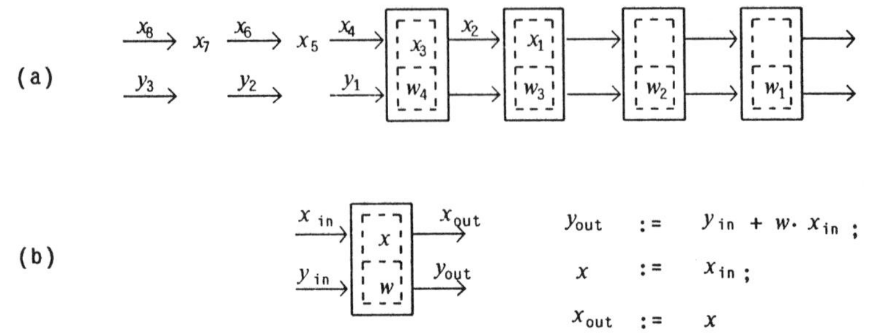

Figure 2. Design W2: systolic FIR filtering array (a) and cell (b).

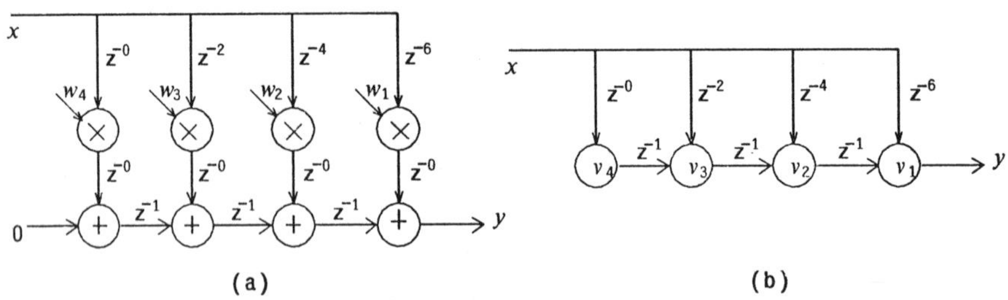

Figure 3. Design W2: in the z-notation (a) and in the z-graph representation(b).

ferent time instants and edges between nodes have labels z^{-k} with $k \geq 1$. One of objectives of this paper is to introduce an algebra for deriving designs whose z-graph representations will have the systolic property (see Section 4 below for precise conditions for a systolic design). Given a design like Figure 3(b), whose z-graph representation enjoys the systolic property, a corresponding systolic array design is readily obtained by simply passing the input x through the nodes with appropriate delays as depicted in Figure 4. It is instructive to examine the correspondence between Figures 2(a) and 4.

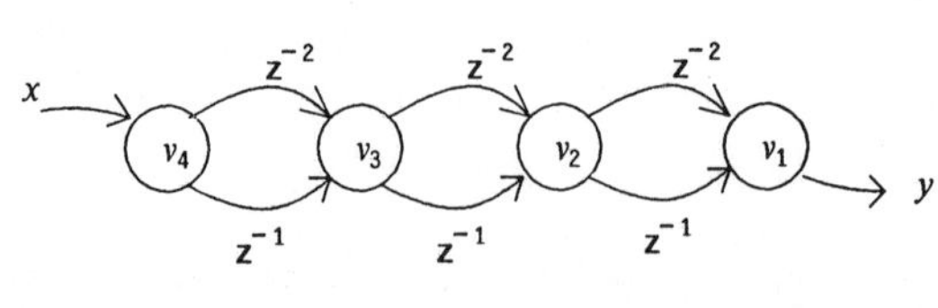

Figure 4. Design W2 in the graph representation.

2.3. Algebraic Representation of Design

In this and next sections we show that the systolic design W2 of the preceding section can be derived systematically by algebraic transformation analogous to some typically used in linear algebra. Our starting design is design C of Figure 5(a), which is a variant of the straightforward design, design S, of Figure 1(b). In design C the summation is distributed over a cascade of four 2-input adders as shown in Figure 5(a). Figure 5(b) describes design C in the z-graph representation.

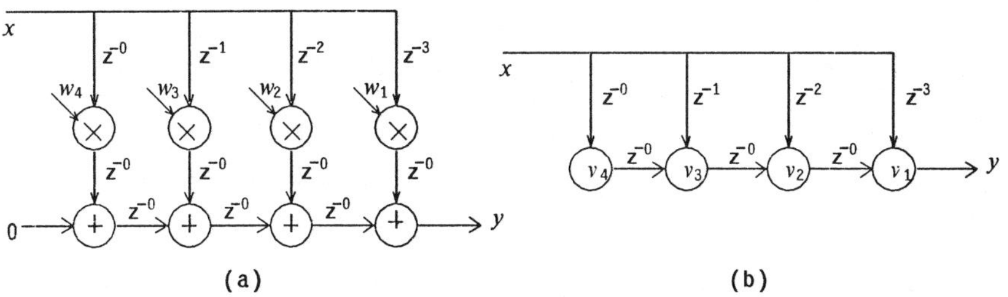

Figure 5. Design C: (a) a variant of design S of Figure 1(b) and (b) in the z-graph notation.

Design C relies on the fact that in the filter computation (1) there are as many multiplications as additions. Similar designs apply to many other inner-product-like computations of this kind. Note that in design C of Figure 5(b) the edges linking nodes v_1, v_2, v_3 and v_4 all have labels z^{-0} and therefore the cycle time must be long enough to perform computations associated with all the nodes in sequence. Thus design C is not systolic. Assuming that design C in the z-graph representation (Figure 5(b)) is given, our task is to transform it to the systolic design, design W2, of Figure 3(b) by linear algebra techniques. To this end, we formally associate the z-graph representation of design C of Figure 5(b) with an algebraic representation. To see the correspondence between the two representations, consider for example that

$$(a) \quad v_2 \leftarrow z^{-0}v_3 + z^{-2}x \quad \text{and} \quad (b) \quad y = z^{-0}v_1. \tag{2}$$

The complete algebraic representation is

$$\begin{bmatrix} v_1 \\ v_2 \\ v_3 \\ v_4 \end{bmatrix} \longleftarrow \begin{bmatrix} 0 & z^{-0} & 0 & 0 \\ 0 & 0 & z^{-0} & 0 \\ 0 & 0 & 0 & z^{-0} \\ 0 & 0 & 0 & 0 \end{bmatrix} \begin{bmatrix} v_1 \\ v_2 \\ v_3 \\ v_4 \end{bmatrix} + \begin{bmatrix} z^{-3} \\ z^{-2} \\ z^{-1} \\ z^{-0} \end{bmatrix} x \tag{3}$$

$$y = \begin{bmatrix} z^{-0} & 0 & 0 & 0 \end{bmatrix} \begin{bmatrix} v_1 & v_2 & v_3 & v_4 \end{bmatrix}^{T}$$

Consistently with Figure 5(b), (2.a) states that at any time t, the value of node v_2, $v_2(t)$, depends on the values of node v_3 at time t, $v_3(t)$, and the value of input x at time $t-2$, $x(t-2)$, and (2.b) states that the value of output y is the same as the value of node v_1 at any time. More precisely,

$$v_2(t)=f_2[v_3(t), x(t-2)], \tag{4}$$

where f_2 is a 2-variable function associated with v_2 such that

$$f_i[a,b]=a+w_i b.$$

This defines a one-to-one correspondence between the z-graph representation of a design and its algebraic representation, in the sense that from either representation one can derive the other. Note that the plus sign in (2.a) represents some combination of information by (4) rather than the usual arithmetic addition. In Section 3.2 below semantics for algebraic expressions involving the ''←'' symbol such as (2.a) will be given.

It is readily seen from Figure 3(b) that the algebraic representation of design W2 is that shown below:

$$\begin{bmatrix} v_1 \\ v_2 \\ v_3 \\ v_4 \end{bmatrix} \longleftarrow \begin{bmatrix} 0 & z^{-1} & 0 & 0 \\ 0 & 0 & z^{-1} & 0 \\ 0 & 0 & 0 & z^{-1} \\ 0 & 0 & 0 & 0 \end{bmatrix} \begin{bmatrix} v_1 \\ v_2 \\ v_3 \\ v_4 \end{bmatrix} + \begin{bmatrix} z^{-6} \\ z^{-4} \\ z^{-2} \\ z^{-0} \end{bmatrix} x \tag{5}$$

$$y = \begin{bmatrix} z^{-0} & 0 & 0 & 0 \end{bmatrix} \begin{bmatrix} v_1 & v_2 & v_3 & v_4 \end{bmatrix}^{T}$$

2.4. Deriving Systolic Designs by Algebraic Transformations

In this section we demonstrate that the algebraic representation of the systolic design, design W2, can be obtained from that of design C through formal algebraic transformations; in the next section we will provide a mathematical foundation for these transformations. To simplify notation, we denote the algebraic representation of design C by

$$(a)\quad v \leftarrow Av + bx \quad \text{and} \quad (b)\quad y = c^T v. \tag{6}$$

where matrix A and vectors b, c are defined according to (3). Consider the diagonal matrix

$$D = \begin{bmatrix} z^{-3} & 0 & 0 & 0 \\ 0 & z^{-2} & 0 & 0 \\ 0 & 0 & z^{-1} & 0 \\ 0 & 0 & 0 & z^{-0} \end{bmatrix} \quad \text{and its inverse} \quad D^{-1} = \begin{bmatrix} z^{3} & 0 & 0 & 0 \\ 0 & z^{2} & 0 & 0 \\ 0 & 0 & z^{1} & 0 \\ 0 & 0 & 0 & z^{0} \end{bmatrix}$$

Let $u = Dv$. Then $v = D^{-1}u$. Multiplying the equation for v by D, we have $Dv \leftarrow DAv + Dbx$. Then by the relation between v and u, we can reduce (6) to:

$$(a)\quad u \leftarrow (DAD^{-1})u + (Db)x, \quad \text{and} \quad (b)\quad y = (c^T D^{-1})u. \tag{7}$$

Through formal calculation, one can check that

$$DAD^{-1} = \begin{bmatrix} 0 & z^{-1} & 0 & 0 \\ 0 & 0 & z^{-1} & 0 \\ 0 & 0 & 0 & z^{-1} \\ 0 & 0 & 0 & 0 \end{bmatrix}, \quad Db = \begin{bmatrix} z^{-6} \\ z^{-4} \\ z^{-2} \\ z^{-0} \end{bmatrix}, \quad \text{and} \quad c^T D^{-1} = \begin{bmatrix} z^{3} \\ 0 \\ 0 \\ 0 \end{bmatrix}^T.$$

Thus (7.a) and (7.b) are the algebraic representation of the design whose z-graph representation is shown in Figure 6.

We have transformed design C of Figure 5(b) to the design of Figure 6. After renaming the value of output y at time t to be that of output y at time $t+3$, the design becomes exactly the systolic design W2 of Figure 3(b). In conclusion, we have derived a systolic design by applying a transformation D to the algebraic representation of a non-

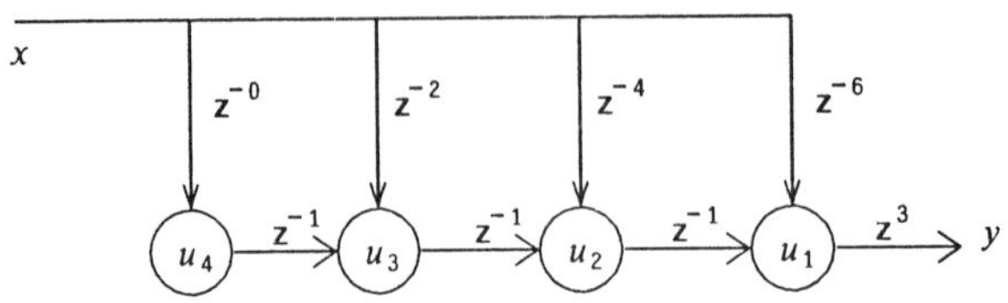

Figure 6. Design corresponding to (7.a) and (7.b) in the z-graph representation.

systolic design.

3. FOUNDATION FOR ALGEBRAIC TRANSFORMATIONS

In Section 2.4 we illustrated that a systolic design could be derived by formal algebraic manipulations similar to those used in linear algebra. This section provides a mathematical foundation for these formal manipulations. To do so, we first need to give a semantics for VLSI algorithm design.

3.1. Semantics of Design

We define the semantics of a design to be a function of time that the design implements. More precisely, the semantics of some basic design constructs given in either the z-graph representation or the algebraic representation are summarized in the table of Figure 7 with the following comments:

1. Each node v_i in the z-graph representation or each variable v_i in the algebraic representation is a function of time defined in terms of some *implicit function* f_i associated with v_i.

2. The value of node or variable v_1 at time t, $v_1(t)$, is $f_1[v_2(t-i), x(t-j), v_3(t-k)]$, where $v_2(t-i)$ is the value of v_2 at time $t-i$, $x(t-j)$ is the value of input x at time $t-j$, and $v_3(t-k)$ is the value of v_3 at time $t-k$.

3. The value of output y at time t is the same as the value of v_i at time $t-k$. (If $k=0$, symbol z^{-k} can be omitted from the z-graph representation as in Figures 3(b) and 5(b).)

Note that for designs of Figures 5(b) and 6, implicit function f_i, $i=1,2,3$, associated with node v_i or u_i with weight w_i, is defined by $f_i[a,b]=a+w_ib$, and implicit function f_4 associated with node v_4 or u_4 is defined by $f_4[b]=w_4b$, where a and b are the left and top inputs to the node, respectively. Note that implicit functions f_i are functions independent of time. As far as the algebraic transformations of this paper are concerned, the semantics of implicit functions need not be specified, as they are invariant under these transformations. This is the reason why we call them implicit functions.

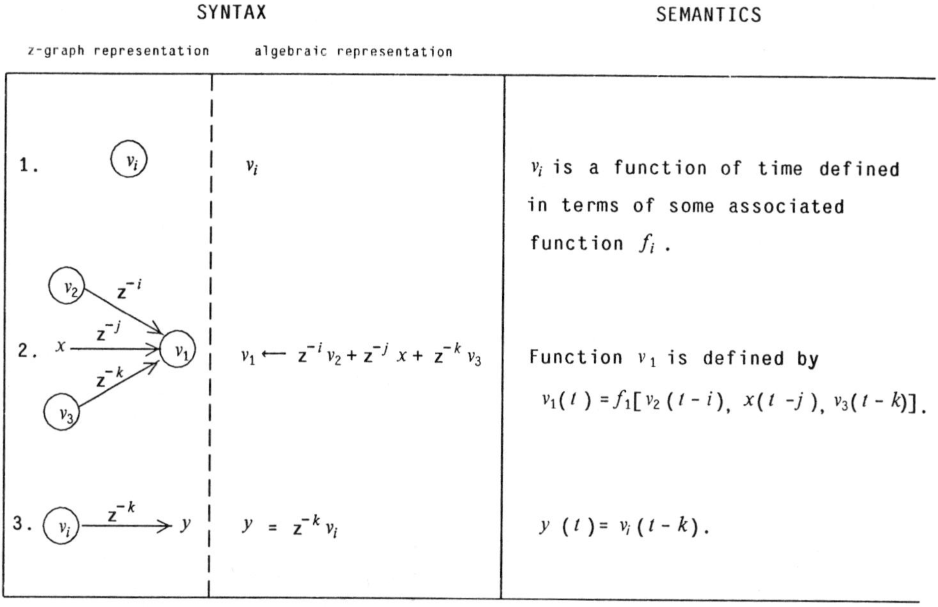

Figure 7. Semantics of basic design constructs.

3.2. Canonical Algebraic Representation

As shown in Figures 3(b) and 5(b), a general design in the z-graph representation has input x, output y and nodes $v_1, \cdots, v_n$. By grouping multiple expressions for defining individual functions $v_1, \cdots, v_n$ into a single matrix expression, the algebraic representation of a general VLSI algorithm design often has the form:

$$(a) \quad v \leftarrow Av + bx, \quad (b) \quad y = c^T v, \tag{8}$$

where $A = (z^{-a_{ij}})$ is an $n \times n$ matrix, $b = (z^{-b_1}, \cdots, z^{-b_n})^T$, $v = (v_1, \cdots, v_n)^T$, and $c^T = (z^{-c_1}, \cdots, z^{-c_n})$ with only one nonzero entry. This *canonical form* of algebraic representation has been illustrated by (3) and (5), and will be assumed in the rest of the paper except the concluding remarks section.

3.3. Well-Defined Design and Equivalent Designs

For $i = 1, \cdots, n$, the i-th component of (8.a) is

$$v_i \leftarrow z^{-a_{i1}} v_1 + z^{-a_{i2}} v_2 + \cdots + z^{-a_{in}} v_n + z^{-b_n} x. \tag{9}$$

That is, (8.a) is a collection of expressions (9) for $i=1,\cdots,n$. For defining the semantics of design (8), (9) means that function v_i satisfies

$$v_i(t)=f_i[v_1(t-a_{i1}), v_2(t-a_{i2}), \cdots, v_n(t-a_{in}), x(t-b_i)] \tag{10}$$

for some implicit function f_i associated with node v_i, and (8.b) means that

$$y(t)=v_j(t-c_j),$$

where $-c_j$ is the exponent of the only nonzero entry in vector c^T. (Mechanically, we can think that in the transformation from (9) to (10) "$\leftarrow$" is replaced with "$=f_i$.") Here we use the convention that a zero entry of A, b or c^T is $z^{-\infty}$ and it is omitted from expressions (9) and (10).

We say that a design is *well-defined* starting from t_0, for some t_0, if for $i=1,\cdots,n$ and $t\geq t_0$, $v_i(t)$ is completely determined by values in the sets $\{x(t'): t'\leq t\}$ and $\{v_i(t'): t'<t\}$, $i=1,\cdots,n$, and this property holds for *any* implicit functions. In view of (10) a sufficient condition for design (8) to be well-defined is that a_{ij}'s are all positive. This is, however, not a necessary condition. It is instructive to see that design C of Figure 5(b) is well-defined in spite of the fact that for this design $a_{12}=a_{23}=a_{34}=0$. From (3), we have

$$v_1(t)=f_1[v_2(t), x(t-3)],\quad v_2(t)=f_2[v_3(t), x(t-2)],$$

$$v_3(t)=f_3[v_4(t), x(t-1)],\quad v_4(t)=f_4[x(t)].$$

Therefore

$$v_1(t)=f_1[f_2[f_3[f_4[x(t)], x(t-1)], x(t-2)], x(t-3)],\quad v_2(t)=f_2[f_3[f_4[x(t)], x(t-1)], x(t-2)],$$

$$v_3(t)=f_3[f_4[x(t)], x(t-1)],\quad v_4(t)=f_4[x(t)].$$

We see that for $i=1,\cdots,4$, $v_i(t)$ is completely determined by values in the set $\{x(t'): t'\leq t\}$ for any implicit functions f_i, and thus design C is well-defined. It is easy to prove that a sufficient and necessary condition for a design to be well-defined is that in its z-graph representation there does not exist any cycle whose edges all have label z^{-0}. Verifying this condition for a design can be done in linear time. Hereafter we only be interested in designs that are well-defined.

Consider a well-defined design (8) with some implicit function associated with each node. Given an input function (of time) x and *initial values* $v_i(t)$ for $t<t_0$, by (10) design (8.a) defines a unique vector function (of time) $v=(v_1,\cdots,v_n)^T$,[3] and together with (8.b), defines a unique output function (of time) y. We say two output functions A and B are essentially the same if $A(t)=B(t+\alpha)$, where α is some constant, for all t greater than certain time intant.

> Definition 3.1: Two given designs are *equivalent*, if for any initial values given for one design, there exist initial values for the other design such that with the same input function the two designs produce essentially the same output function.

In the following section we will show that design defined by (6) and (7) are equivalent.

3.4. Fundamental Results

To express our results on algebraic transformations, we need the following definitions. Let $D=(z^{-d_i})$ be an $n\times n$ diagonal matrix.

1. For $v=(v_1,\cdots,v_n)^T$, define Dv to be $u=(u_1,\cdots,u_n)^T$ such that for $i=1,\cdots,n$, $u_i(t)=v_i(t-d_i)$ for all t for which $v_i(t-d_i)$ is defined. Thus, D can be viewed as an operator that maps a vector function v to another vector function Dv.

2. For $b=(z^{-b_1},\cdots,z^{-b_n})^T$, define Db to be $e=(z^{-e_1},\cdots,z^{-e_n})^T$ where $e_i=d_i+b_i$, for $i=1,\cdots,n$.

3. Let $A=(z^{-a_{ij}})$ be an $n\times n$ matrix. Define DA to be an $n\times n$ matrix $B=(z^{-b_{ij}})$ where $b_{ij}=d_i+a_{ij}$ for $i,j=1,\cdots,n$. Product AD is defined similarly. We can easily check that $(DA)D^{-1}=D(AD^{-1})$ and thus we can simply denote them by DAD^{-1}.

Here we use the convention that $\infty=d_i+\infty$ for any d_i. Thus zero entries of b or A remain to be zero entries in Db or DA, respectively.

> Lemma 3.1: Suppose that v and u are defined by well-defined designs

$$v\leftarrow Av+bx \tag{11}$$

and

$$u\leftarrow(DAD^{-1})u+(Db)x, \tag{12}$$

respectively, with their initial values satisfying

$$u_i(t+d_i)=v_i(t) \quad \text{for} \quad t<t_0. \tag{13}$$

---

[3] In the semantics literature, function v such defined is called the "fixpoint solution" of "fixpoint equation" (8.a).

Then $u=Dv$.

Proof: Let v_i and u_i be the i-th components of v and u, respectively. Note that $DAD^{-1}=(z^{-d_i+d_j-a_{ij}})$ and $Db=(z^{-d_1-b_1}, z^{-d_2-b_2}, \cdots, z^{-d_n-b_n})^T$. Thus, u_i defined by (12) satisfies

$$u_i(t)=f_i[u_1(t-d_i+d_1-a_{i1}), u_2(t-d_i+d_2-a_{i2}), \cdots, u_n(t-d_i+d_n-a_{in}), x(t-b_i-d_i)],$$

Replacing t with $t+d_i$ in the above equation, we have, successively

$$u_i(t+d_i)=f_i[u_1(t+d_1-a_{i1}), u_2(t+d_2-a_{i2}), \cdots, u_n(t+d_n-a_{in}), x(t-b_i)]. \tag{14}$$

By (10),

$$v_i(t)=f_i[v_1(t-a_{i1}), v_2(t-a_{i2}), \cdots, v_n(t-a_{in}), x(t-b_i)]. \tag{15}$$

We prove by induction on t that for $i=1, \cdots, n$,

$$u_i(t+d_i)=v_i(t) \tag{16}$$

for $t=t_0, t_0+1, t_0+2, \cdots$. By (13), (16) holds for $t<t_0$. Thus,

$$u_j(t_0+d_j-a_{ij})=v_j(t_0-a_{ij})$$

for any j for which $a_{ij}>0$. Since designs (11) and (12) are well-defined, (14) and (15) imply that $u_i(t_0+d_i)=v_i(t_0)$, that is, (16) holds for $t=t_0$. By induction (16) holds for $t=t_0+1, t_0+2, \cdots$, and so on. □

The following lemma can be proven by a similar method:

Lemma 3.2: If

$$y=c^Tv \quad \text{and} \quad u=Dv,$$

then

$$y=(c^TD^{-1})v.$$

Immediately following from Lemmas 3.1 and 3.2, we have the following result:

Theorem 3.1: Design (8) is equivalent to design (7), assuming that both designs are well-defined.

The above theorem is essentially the "retiming lemma" by Leiserson and Saxe (1981). Not using the algebraic notation and approach taken here, they had to rely on a very long (4 pages) and rather unclean proof.

In the following, we introduce another transformation whose function is to scale down the throughput of an existing design. Consider a well-defined design M with input function x and output function y, and another design M′ with input function x' and output function y'. We say that design M′ is a *k-slowed* design of M for some positive integer k, if the following holds for some integer p: for any initial values for M, there exist initial values for M′ such that if $x'(kt+p)=x(t)$ for all t, then $y'(kt+p)=y(t)$ for all t where $y(t)$ is defined. Therefore as far as the outside world is concerned, the function of a k-slowed design is the same as that of the original design, except that input and output are taken in and out, respectively, once every k time units. The usefulness of k-slowed designs in the derivation of systolic designs was first pointed out by Leiserson and Saxe (1981), and it will become clear in the next two sections. The following lemma shows a simple way to implement a well-defined, k-slowed design.

Lemma 3.3: If design (8) is well-defined then the design

$$(a) \quad v' \leftarrow A'v' + b'x', \tag{17}$$

$$(b) \quad y' = c'^T v',$$

with $A'=(z^{-ka_{ij}})$, $b'=(z^{-kb_1}, \cdots z^{-kb_n})^T$, and $c'^T=(z^{-kc_1}, \cdots z^{-kc_n})$ is a well-defined, k-slowed design.

Proof: Since in their z-graph representations the two designs have the same set of edges with label z^{-0}, well-defineness of one design implies that of the other. Let v_i and v_i' be the i-th components of v and v', respectively. Without loss of generality, assume that the output functions y and y' of the two designs satisfy

$$y(t)=v_i(t-c_i), \text{ and } y'(t')=v_i'(t'-kc_i),$$

respectively. Suppose that the original design is well-defined starting from t_0. It suffices to prove that if $x'(kt)=x(t)$ for all t, and $v_i'(kt)=v_i(t)$ for $i=1,\cdots,n$ and $t<t_0$, then $y'(kt)=y(t)$ for $t \geq t_0$, for all t for which $y(t)$ is defined. The proof is similar to that of Lemma 3.1 and is omitted.

4. DETERMINING ALGEBRAIC TRANSFORMATIONS

Given a well-defined design, we want to determine a k-slowed design and $D=(z^{-d_i})$ such that design

$$(a) \quad u \leftarrow (DA'D^{-1})u + (Db')x', \tag{18}$$

$$(b) \quad y' = (c'^T D^{-1})u$$

will be well-defined and systolic. This imposes the following conditions on the entries of $DA'D^{-1}$ and Db':

C1. For $i=1,\cdots,n$, $d_i + ka_{ij} - d_j \geq 1$.
{This assures not only that the design is well-defined, but also that the cycle time only has to be long enough to perform the computation of at most one node.}

C2. All nonzero entries of any column of $DA'D^{-1}$ and Db' must be distinct.
{This assures that the value of a node never has to be sent to more than one node simultaneously, and thus no broadcasting or fanout of data is needed.}

It is an easy exercise to show that if the original design is well-defined, that is, in its z-graph representation there does not exist cycles whose edges all have label z^{-0}, then there exist k and D for which conditions C1 and C2 are satisfied. To maximize throughput we are interested in a solution which has the smallest-possible k. It turns out that for some designs to satisfy C1 and C2, k must be greater than one, as to be illustrated by IIR filtering example in the next section. This is the reason why we perform transformations on a k-slowed design, with $k \geq 1$, rather than the original design.

5. IIR FILTERING--A FURTHER EXAMPLE

Consider the implementation of the following infinite impulse response (IIR) filter with weights w_i:

$$y_i = w_1 y_{i-1} + w_2 y_{i-2} + w_3 x_i + w_4 x_{i-1}. \tag{19}$$

The above equation states that at any given time t, the value of output y depends on the values of y at times $t-1$ and $t-2$, and input x at times t and $t-1$. Figure 8(a) depicts a straightforward design for the IIR filter in the z-notation. Similar to the FIR design of Figure 5(a), the 4-input adder of Figure 8(a) can be distributed over a cascade of four 2-input adders. This forms a design with four identical nodes, whose z-graph representation is described in Figure 8(b). The algebraic representation of the design is

$$\begin{bmatrix} v_1 \\ v_2 \\ v_3 \\ v_4 \end{bmatrix} \leftarrow \begin{bmatrix} z^{-1} & z^{-0} & 0 & 0 \\ z^{-2} & 0 & z^{-0} & 0 \\ 0 & 0 & 0 & z^{-0} \\ 0 & 0 & 0 & 0 \end{bmatrix} \begin{bmatrix} v_1 \\ v_2 \\ v_3 \\ v_4 \end{bmatrix} + \begin{bmatrix} 0 \\ 0 \\ z^{-0} \\ z^{-1} \end{bmatrix} x \tag{20}$$

$$y = \begin{bmatrix} z^{-0} & 0 & 0 & 0 \end{bmatrix} \begin{bmatrix} v_1 & v_2 & v_3 & v_4 \end{bmatrix}^T$$

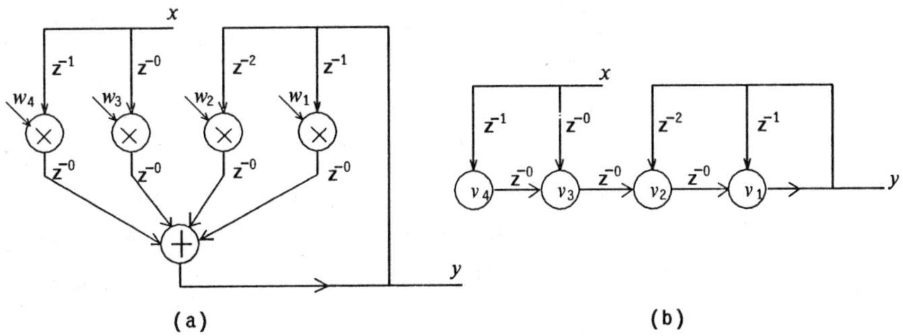

Figure 8. Straightforward design for the IIR filter: (a) in the z-notation and (b) in the z-graph representation.

According to Lemma 3.3, a k-slowed design can be obtained by changing labels z^{-h} to z^{-kh} for any h. The algebraic representation of the k-slowed IIR filter is

$$\begin{bmatrix} v_1' \\ v_2' \\ v_3' \\ v_4' \end{bmatrix} \leftarrow \begin{bmatrix} z^{-k} & z^{-0} & 0 & 0 \\ z^{-2k} & 0 & z^{-0} & 0 \\ 0 & 0 & 0 & z^{-0} \\ 0 & 0 & 0 & 0 \end{bmatrix} \begin{bmatrix} v_1' \\ v_2' \\ v_3' \\ v_4' \end{bmatrix} + \begin{bmatrix} 0 \\ 0 \\ z^{-0} \\ z^{-k} \end{bmatrix} x \tag{21}$$

$$y = \begin{bmatrix} z^{-0} & 0 & 0 & 0 \end{bmatrix} \begin{bmatrix} v_1' & v_2' & v_3' & v_4' \end{bmatrix}^T$$

or,

$$(a) \quad v \leftarrow A'v + b'x, \tag{22}$$

$$(b) \quad y = c'^T v.$$

We seek a diagonal matrix D such that the design described by (18) will be well-defined and systolic. By condition C1 of Section 4, $k \geq 1$; $2k+d_2-d_1 \geq 1$; $d_2-d_3 \geq 1$; $d_3-d_4 \geq 1$; and by condition C2, $k \neq 2k+d_2-d_1$; $d_3 \neq k+d_4$. One can check that a solution with the minimum-possible value for k is that $k=2$ and

$$D = \begin{bmatrix} z^{-2} & 0 & 0 & 0 \\ 0 & z^{-1} & 0 & 0 \\ 0 & 0 & z^{-0} & 0 \\ 0 & 0 & 0 & z^{1} \end{bmatrix}, \quad DA'D^{-1} = \begin{bmatrix} z^{-2} & z^{-1} & 0 & 0 \\ z^{-3} & 0 & z^{-1} & 0 \\ 0 & 0 & 0 & z^{-1} \\ 0 & 0 & 0 & 0 \end{bmatrix},$$

$$Db' = \begin{bmatrix} 0 & 0 & z^{-0} & z^{-1} \end{bmatrix}^T, \quad \text{and} \quad c'^T D^{-1} = \begin{bmatrix} z^2 & 0 & 0 & 0 \end{bmatrix}.$$

Thus the resulting systolic IIR filtering array in the z-graph representation is shown in Figure 9. This systolic array was previously described in Kung (1980).

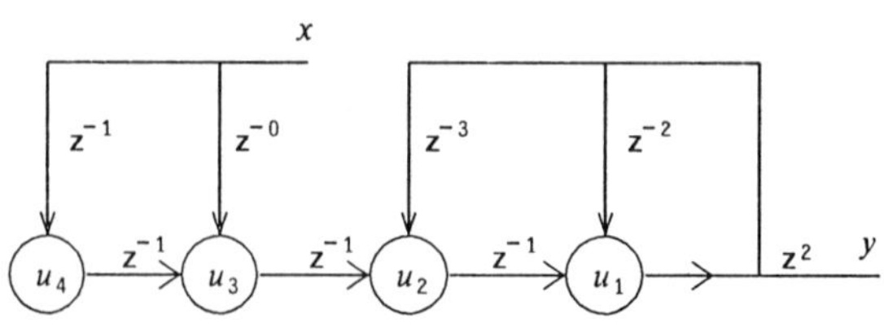

Figure 9. Systolic IIR filter in the z-graph representation.

6. CONCLUDING REMARKS

We proposed two representations for specifying a design--the z-graph representation and the algebraic representation. From either representation we can derive the other. The z-graph representation is readily mappable to a hardware or VLSI implementation, whereas the algebraic representation is suitable for algebraic transformations. For algebraic transformations, only algebraic representations of designs are needed. By working within an algebraic framework, rather than a network or graph-theoretic framework, one can use powerful algebraic operators to manipulate designs and can deal with abstractions conveniently. For example, using matrix notation, a simple algebraic expression such as (8) can represent design of arbitrary size.

A more general algebraic representation than the one described in (8) is:

$$(a) \quad v \leftarrow Av + Bx, \quad (b) \quad y = C^T v, \tag{23}$$

where input x and output y are vectors rather than scalars, and B and C are matrices rather than vectors b and c. This general form of representation seems to cover all the interesting VLSI algorithm designs that we know of and can anticipate. For example, for the design of Figure 10(a) for multiplying a bidiagonal upper triangular matrix with a bidiagonal lower triangular matrix, we have

$$A = \begin{bmatrix} 0 & 0 & 0 & 0 \\ 0 & 0 & 0 & 0 \\ 0 & 0 & 0 & 0 \\ z^{-0} & 0 & 0 & 0 \end{bmatrix}, \quad B = \begin{bmatrix} 0 & z^{-0} & z^{-0} & 0 \\ z^{-0} & 0 & z^{-1} & 0 \\ 0 & z^{-0} & 0 & z^{-0} \\ z^{-0} & 0 & 0 & z^{-1} \end{bmatrix}, \quad C^T = \begin{bmatrix} 0 & z^{-0} & 0 & 0 \\ 0 & 0 & 0 & z^{-0} \\ 0 & 0 & z^{-0} & 0 \end{bmatrix}.$$

Without loss of generality we can always assume that there is only one nonzero entry in each row of C^T, that is, at any time the value of each output y_i is equal to that of some node at that time or earlier. Results and definitions of this paper can all be extended in a straightforward way to this general form of the algebraic representation (23). For example we can show that starting with the non-systolic design of Figure 10(a), a systolic solution with the minimum-possible value for k is that $k=1$ and

$$D = \begin{bmatrix} z^{-0} & 0 & 0 & 0 \\ 0 & z^{-0} & 0 & 0 \\ 0 & 0 & z^{-1} & 0 \\ 0 & 0 & 0 & z^{-1} \end{bmatrix}.$$

The resulting systolic array is illustrated in Figure 10(b), which is precisely the systolic design for band matrix multiplication proposed in Weiser (1981). Detailed discussions of this and other results including the use of the proposed algebra in the derivation of two-level pipelined systolic arrays Kung (1983) and systolic arrays for LU-decomposition of matrices will appear in forthcoming papers.

We view that major contributions of this paper are at the proposed

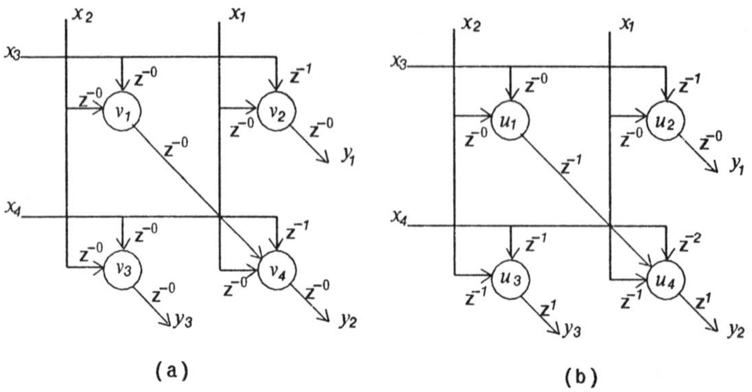

Figure 10. Designs for band matrix multiplication in the z-graph representation: (a) a non-systolic design, and (b) a systolic design.

semantics for VLSI algorithm design, algebraic representation and transformations, and the mathematical foundation for these transformations. With these algebraic tools, we are able to manipulate designs by ''pushing symbols'' as we do in algebra, and to prove theorems about design transformations (e.g., Theorem 3.1) without relying on any drawings. Deriving systolic design is just of one of many potential applications of the proposed algebra.

ACKNOWLEDGEMENT

This research was supported in part by the Office of Naval Research under Contracts N00014-76-C-0370, NR 044-422 and N00014-80-C-0236, NR 048-659. We want to thank Allan Fisher and Monica Lam for reading early drafts of the paper.

REFERENCES

Blackmer, J., P. Kuekes and Frank, G. (1981). "A 200 MOPS Systolic Processor." In "Proceedings of SPIE Symposium, Vol. 298, Real-Time Signal Processing IV," Society of Photo-Optical Instrumentation Engineers, August.

Bromley, K., Symanski, J.J., Speiser, J.M., and Whitehouse, H.J. (1981). "Systolic Array Processor Developments." In Kung, H.T., Sproull, R.F., and Steele, G.L., Jr. (editors), "VLSI Systems and Computations," pages 273-284, Computer Science Department, Carnegie-Mellon University, Computer Science Press, Inc., October.

Chen, M.C. and Mead, C.A. (1983). "A Hierarchical Simulator Based on Formal Semantics." In Bryant, R. (editor), "Proceedings of the Third Caltech Conference on Very Large Scale Integration," pages 207-223, California Institute of Technology, Computer Science Press, Inc., March.

Corry, A. and Patel, K. (1983). "A CMOS/SOS VLSI Correlator." In "Proceedings of 1983 International Symposium on VLSI Technology, Systems and Applications," pages 134-137.

Fisher, A.L. and Kung, H.T. (1982). "Special-Purpose VLSI Architectures: General Discussions and a Case Study." In "Proceedings of the USC Workshop on VLSI and Modern Signal Processing," November.

Fisher, A.L., Kung, H.T., Monier, L.M. and Dohi, Y. (1983). "Architecture of the PSC: A Programmable Systolic Chip." In "Proceedings of the Tenth International Symposium on Computer Architecture," June.

Fisher, A.L., Kung, H.T., Monier, L.M., Walker, H. and Dohi, Y. (1983). "Design of the PSC: A Programmable Systolic Chip." In Bryant, R. (editor), "Proceedings of the Third Caltech Conference on Very Large Scale Integration," pages 287-302, California Institute of Technology, Computer Science Press, Inc., March.

Foster, M.J. and Kung, H.T. (1980). "The Design of Special-Purpose VLSI Chips" Computer 13(1):26-40, January.
Reprint of the paper appears in *Digital MOS Integrated Circuits*, edited by Elmasry, M.I., IEEE Press Selected Reprint Series, 1981, pp. 204-217. A preliminary version of the paper, entitled "Design of Special-Purpose VLSI Chips: Example and Opinions," also appears in *Proceedings of the 7th International Symposium on Computer Architecture*, pp. 300-307, La Baule, France, May 1980.

Kuekes, P. and Shen, J. (1983). "One-Gigaflop VLSI Systolic Processor." In "Proceedings of SPIE Symposium, Vol. 431, Real-Time Signal Processing VI," Society of Photo-Optical Instrumentation Engineers, August.

Kung, H.T. (1979). "Let's Design Algorithms for VLSI Systems." In "Proceedings of Conference on Very Large Scale Integration: Architecture, Design, Fabrication," pages 65-90, California Institute of Technology, January.
Also available as a CMU Computer Science Department technical report, September 1979.

Kung, H.T. (1980). "Special-Purpose Devices for Signal and Image Processing: An Opportunity in VLSI." In "Proceedings of the SPIE, Vol. 241, Real-Time Signal Processing III," pages 76-84, Society of Photo-Optical Instrumentation Engineers, July.

Kung, H.T. (1982). "Why Systolic Architectures?" Computer Magazine 15(1):37-46, January.

Kung, H.T., Ruane, L.M., and Yen, D.W.L. (1983). "Two-Level Pipelined Systolic Array for Multidimensional Convolution" Image and Vision Computing 1(1):30-36, February.
A preliminary version appears in *VLSI Systems and Computations*, edited by H. T. Kung, G. L. Steele, Jr. and R. F. Sproull, Computer Science Press, Maryland, 1981, pp. 255-264.

Kung, H.T. and Leiserson, C.E. (1979). "Systolic Arrays (for VLSI)." In Duff, I. S. and Stewart, G. W. (editors), "Sparse Matrix Proceedings 1978," pages 256-282, Society for Industrial and Applied Mathematics.
A slightly different version appears in *Introduction to VLSI Systems* by C. A. Mead and L. A. Conway, Addison-Wesley, 1980, Section 8.3.

Kung, H.T. and Song, S.W. (1982). "A Systolic 2-D Convolution Chip." In Preston, K., Jr. and Uhr, L. (editor), "Multicomputers and Image Processing: Algorithms and Programs," pages 373-384, Academic Press.
An extended abstract appears in *Proceedings of 1981 IEEE Computer Society Workshop on Computer Architecture for Pattern Analysis and Image Database Management*, November 11-13, 1981, pp. 159-160.

Lam, M. and Mostow, J. (1983). "A Transformational Model of VLSI Systolic Design." In Uehara, T. and Barbacci, M. (editors), "Proceedings of the 6th International Symposium on Computer Hardware Description Languages and their Applications," pages 65-77, IFIP, May.

Leiserson, C.E. and Saxe, J.B. (1981). "Optimizing Synchronous Systems." In "Proceedings of the 22nd Annual Symposium on Foundations of Computer Science," pages 23-36, IEEE Computer Society, October.
The final version of the paper is to appear in *Journal of VLSI and Computer Systems*.

Sorasen, O. and Solberg, B. (1983). "VLSI Implemented Systolic Array Processor." In "Proceedings of VLSI 83 International Conference," Trondheim, Norway, August.

Symanski, J.J. (1981). "A Systolic Array Processor Implementation." In "Proceedings of SPIE Symposium, Vol. 298, Real-Time Signal Processing IV," Society of Photo-Optical Instrumentation, August.

Symanski, J.J. (1982). "Progress on a Systolic Processor Implementation." In "Proceedings of SPIE Symposium, Vol. 341, Real-Time Signal Processing V," pages 2-7, Society of Photo-Optical Instrumentation, May.

Weiser, U. and Davis, A. (1981). "A Wavefront Notation Tool for VLSI Array Design." In Kung, H.T., Sproull, R.F., and Steele, G.L., Jr. (editors), "VLSI Systems and Computations," pages 226-234, Computer Science Department, Carnegie-Mellon University, Computer Science Press, Inc., October.

Weste, N.H.E., Burr, D.J. and Ackland, B.D. (1982). "A Systolic Processing Element for Speech Recognition." In "Proceedings of 1982 IEEE International Solid-State Circuits Conference," pages 274-275, February.

Yen, D.W.L. and Kulkarni, A.V. (1982). "Systolic Processing and an Implementation for Signal and Image Processing" IEEE Transactions on Computers C-31(10):1000-1009, October.

AN INCOMPLETE-CHOLESKY FACTORIZATION BY A MATRIX PARTITION ALGORITHM

Edna Reiter* and Garry Rodrigue**

I. INTRODUCTION

Matrix-partition algorithms for solving tridiagonal systems of equations are highly adaptable to parallel processors. In this paper, a matrix-partition algorithm for generating a block-Cholesky factorization of a permuted form of a block tridiagonal system is presented. A preconditioning system based on an incomplete application of this algorithm is then described. Under certain dominance conditions, it is shown that the computations within a partition can be performed independently thus yielding a highly parallel incomplete-Cholesky-factorization particularly suitable for multiprocessing architectures.

A fundamental process in the numerical solution of elliptic partial differential equations is the solution of

*California State University, Hayward

**Lawrence Livermore National Laboratory, University of California

ISBN 0-12-100560-7

the block-tridiagonal system of symmetric positive-definite equations:

$$(1.1)\quad Ax = \begin{bmatrix} A_1 & B_1 & & & \\ B_1^t & A_2 & B_2 & & \\ & \ddots & \ddots & \ddots & \\ & & \ddots & \ddots & B_{L-1} \\ & & & B_{L-1}^t & A_L \end{bmatrix} \begin{bmatrix} x_1 \\ x_2 \\ \vdots \\ \vdots \\ x_L \end{bmatrix} = \begin{bmatrix} b_1 \\ b_2 \\ \vdots \\ \vdots \\ b_L \end{bmatrix} = b$$

where the A_i, B_i are K x K tridiagonal matrices, i = 1, 2, .., L, and x_i, b_i are K-dimensional vectors, i = 1, ..., L, cf. Varga [1], Rodrigue, Hendrickson, Pratt [2].

Direct algorithms for solving (1.1) involve three processes:

1) The Decomposition Process: Construct a finite sequence of matrices $\{Q_i\}_{i=1}^{m}$ and form the matrix product

$$(1.2)\quad \prod_{i=1}^{m} [I - Q_i]A = T;$$

2) The Forward - Substitution Process: Form

$$(1.3)\quad \prod_{i=1}^{m} [I - Q_i]\, b = b';$$

3) The Backward-Substitution Process: Solve

$$(1.4)\quad Tx = b'$$

When a Gaussian Elimination process is used, the Q_i matrices in the Decomposition Process are constructed from sums of elementary block multiplier matrices $M_iE_j^t$ where E_j is a column block L-vector having zero block entries except for the j entry which is the K x K identity matrix, and M_i is a column block L-vector such that

$$(1.5)\quad E_k^tM_i = 0,\ k \neq i,$$

cf. Stewart [3]. It follows that the only non-zero block of M_i is the k entry.

For the familiar block - LU - decomposition algorithm, $Q_i = M_i E^t_{i+1}$ and the non-zero (i+1) - st block of M_i is a matrix product $B A^{-1}$ where B is a tridiagonal matrix. In this case, the matrix T is a block-upper lower triangular matrix. Further, $\prod_{i=1}^{n-1} (I - Q_i) = L^{-1}$ is a unit triangular matrix. By the symmetry of A, $T = DL^t$ where D is a block diagonal matrix so that the block-Cholesky-decomposition, $A = LDL^t$, is formed.

Symmetry and positive-definiteness are preserved under permutations of the matrix A. Consequently, a decomposition process (i.e. the Cholesky decomposition) (1.2) can be constructed so that

$$\prod_{i=1}^{n-1} (I-Q_i)PAP^t = U \qquad (1.6)$$

where P is a permutation matrix and U is a block upper triangular matrix. In this case, $[\prod (I - Q_i)] = L^{-1}$ is a unit lower triangular, $U = DL^t$, and D is block-diagonal, so that LDL^t is a block-Cholesky factorization of PAP^t. The choice of the permutation matrices P in (1.6) depends on the computational hardware and the available parallelism, (see Ehrel, Lichnewsky, and Thomasset, [4], and Rodrigue and Wolitzer, [5], for examples).

II. A MATRIX-PARTITION ALGORITHM

Matrix-partition algorithms are a class particularly suitable for solving (1.1) on multi-processors (see Kuck and Sameh, [6], Wang [7], and Jordan [9] for examples). In this section, we describe a matrix-partition algorithm for generating the block-Cholesky factorization of a permuted form of (1.1). To describe the algorithm, we assume $K = 9$ for simplicity. The number of partitions will be 3.

The Decomposition Process:

Let $Q^t = M_1E_2^t + E_4E_4^t + M_7E_7^t$ where the non-zero blocks of M_i are $B_i^tA_i^{-1}$. Then, $(I - Q_1)A$ takes the form

$$\left[\begin{array}{ccc|ccc|ccc}
A_1 & B_1 & & & & & & & \\
0 & A_2^{(1)} & B_2 & & & & & & \\
 & B_2^t & A_3 & B_3 & & & & & \\
\hline
 & & B_3^t & A_4 & B_4 & & & & \\
 & & R_4 & 0 & A_5^{(1)} & B_5 & & & \\
 & & & & B_5^t & A_6 & B_6 & & \\
\hline
 & & & & & B_6^t & A_7 & B_7 & \\
 & & & & & R_7 & 0 & A_8^{(1)} & B_8 \\
 & & & & & & & B_8^t & A_9
\end{array}\right]$$

Similarly, $Q_2, Q_3, \ldots$ are constructed so that $\prod_{i=1}^{j} (I - Q_i)A$

$$\left[\begin{array}{ccc|ccc|ccc}
x & x & & & & & & & \\
 & x & x & & & & & & \\
 & 0 & A_3^{(1)} & x & & & & & \\
\hline
 & & x & x & x & & & & \\
 & & x & & x & x & & & \\
 & & R_5 & & 0 & A_6^{(1)} & & & \\
\hline
 & & & & & x & x & x & \\
 & & & & & x & & x & x \\
 & & & & & R_8 & & 0 & A_9^{(1)}
\end{array}\right]
\qquad
\left[\begin{array}{ccc|ccc|ccc}
x & x & & & & & & & \\
 & x & x & & & & & & \\
 & & A_3^{(2)} & 0 & S_4 & & & & \\
\hline
 & & x & x & x & & & & \\
 & & x & & x & x & & & \\
 & & x & & & A_6^{(2)} & 0 & S_7 & \\
\hline
 & & & & & x & x & x & \\
 & & & & & x & & x & x \\
 & & & & & x & & & x
\end{array}\right]$$

(j = 2) (j = 3)

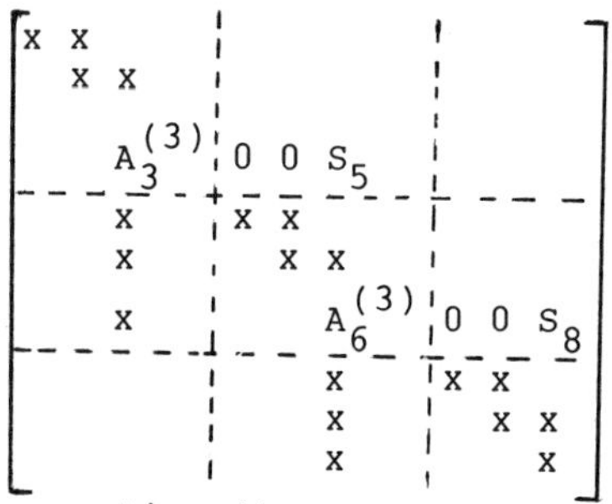

(j = 4)

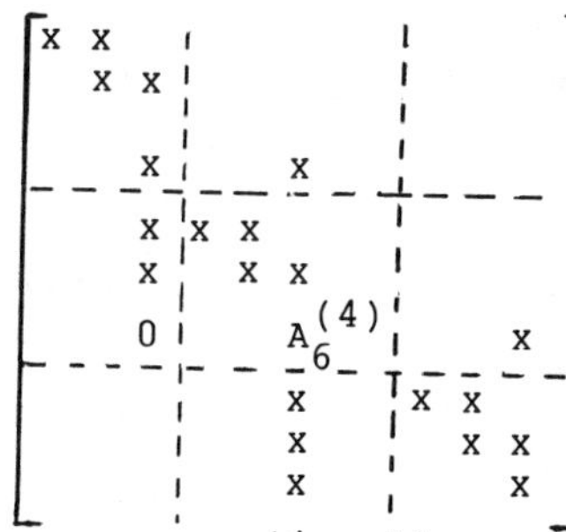

(j = 5)

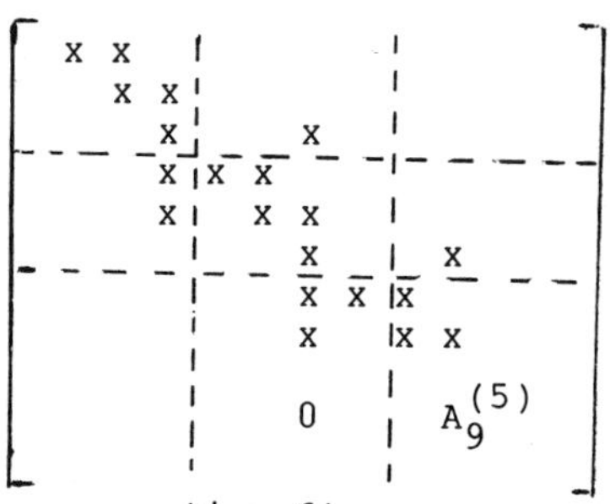

(j = 6)

The above figures illustrate the independence of the computations within a partition. Likewise, the Forward Substitution process is also partition independent. Unfortunately, independence is lost in the Backward Substitution process because of the coupling between partitions from the S_5 and S_8 matrices. However, as the following theorem asserts, if the matrix A has certain diagonal dominance properties, this coupling can be very weak. (We will assume a consistent matrix norm is given, i.e., $\| PQ \| < \| P \| \| Q \|$.)

<u>Theorem 2.1</u> Let A be an L x L block tridiagonal matrix as in (2.1). Suppose

$$\| A_1^{-1} B_1 \| < 1,$$

$$(\| A_i^{-1} B_{i-1}^t \| + \| A_i^{-1} B_i \|) < 1, \qquad 2 < i < L - 1,$$

$$\| A_L^{-1} B_{L-1}^t \| < 1.$$

Let

$$\alpha = [L/p] = \min \{n : n > L/p \text{ and } n \text{ an integer}\}.$$

Then for m = 1, .., p-1,

$$\| R_{i+m\alpha} \| < \| R_{i-1+m\alpha} \|, \ i = 1, .., \alpha,$$

$$\| S_{i+m\alpha} \| < \| S_{i-1+m\alpha} \|, \ i = 1, .., \alpha,$$

where $R_{m\alpha} = B^t_{m\alpha}$ and $S_{m\alpha} = B_{M\alpha}$,

Proof: Fix m throughout the proof. We first show by induction that for i = 1,.., α,

$$\| [A^{(1)}_{i + m\alpha}]^{-1} B_{i+m\alpha} \| < 1.$$

It is true for i = 1 since $A^{(1)}_{1+m\alpha} = A_{1+m\alpha}$.

Suppose it is true for i - 1. Then,

$$A^{(1)}_{i+m\alpha} = A_{i+m\alpha} - B^t_{i+m\alpha-1}[A^{(1)}_{i+m\alpha-1}]^{-1} B_{i+m\alpha-1}$$

$$= A_{i+m\alpha} [I - R]$$

where

$$R = [A_{i+m\alpha}]^{-1} B^t_{i+m\alpha-1}[A^{(1)}_{i+m\alpha-1}]^{-1} B_{i+m\alpha-1}$$

By both the theorem and induction hypothesis,

$$\| R \| < 1 - \| [A_{i+m\alpha}]^{-1} B_{i+m\alpha}\|.$$

Hence,

$$[A^{(1)}_{i+m\alpha}]^{-1} B_{i+m\alpha} = [I - R]^{-1}[A_{i+m\alpha}]^{-1} B_{i+m\alpha}$$

so that

$$\| [A^{(1)}_{i+m\alpha}]^{-1} B_{i+m\alpha} \| < [1 - \| R \|]^{-1} \| [A_{i+m\alpha}]^{-1} B_{i+m\alpha} \| < 1$$

We now prove the theorem;

$$\| S_{i+m\alpha} \| = \| [A^{(1)}_{i+m\alpha-1}]^{-1} B_{i+m\alpha} S_{i+m\alpha-1} \| < \| S_{i+m\alpha-1} \|$$

and

$$\| R_{i+m\alpha} \| = \| B^t_{i+m\alpha} A^{-1}_{i+m\alpha} R_{i+m\alpha-1} \| < \| R_{i+m\alpha-1} \|.$$

Note that the matrix-partition algorithm described above is actually a block-Cholesky decomposition on the permuted form of A:

$$(2.1)\quad PAP^t = \begin{bmatrix} x & & & x & & & & & \\ & x & & & x & & x & & \\ & & x & & & x & & x & \\ x & & & x & & & x & & \\ & x & & & x & & & x & \\ & & x & & & x & & & x \\ & x & & x & & & x & & \\ & & x & & x & & & x & \\ & & & & & x & & & x \end{bmatrix}$$

Hence, the matrix-partition algorithm as described here is well defined.

III. INCOMPLETE-CHOLESKY

Because of numerical fill-in and slowness of a direct algorithm, iterative methods are commonly used to solve (1.1). Of late, second order schemes such as the Chebychev and conjugate gradient methods have become popular because of their rapid rates of convergence. The diminutiveness of the condition number of the matrix A contributes considerably to this rapid convergence rate. An effective means of reducing the condition number of A is to precondition the matrix. This amounts to the construction of a matrix $M = LL^t$ so that

$$\text{cond } (L^{-1}AL^{-t}) < \text{cond } (A) .$$

When used within an iterative method, the product $L^{-1}AL^{-t}$ is not explicitly computed but rather is brought in through a residual equation

$$(3.1)\quad LL^t\, z^{(i)} = b - Ax^{(i)}$$

where $x^{(i)}$ is the current iterative approximation to the solution of (1.1). The system (3.1) must be solved at each iteration.

A preconditioning system can be defined from an incomplete application of the decomposition processes used in the block-Cholesky factorization of the matrix A of (1.1) or in the block-Cholesky factorization of a permuted form of A. Specifically, once a sequence of matrices $\{Q_i\}_{i=1}^{m}$ have been defined in (1.2), a similar sequence of matrices $\{\tilde{Q}_i\}$ are constructed so that

i) the zero blocks of $\tilde{Q}_i$ are the same as those of Q_i, $1 < i < m.$;

ii) the non-zero blocks of $\tilde{Q}_i$ are formed from the tridiagonal portion of a matrix product BA^{-1} where B and A are K x K tridiagonal matrices;

iii) the pre-multiplication by $(I - \tilde{Q}_i)$ is done so that the resultant blocks of the product are constrained to be tridiagonal.Thus, preconditioning systems for specific computational hardware can be defined, (see [4], [5] for examples).

An incomplete application of the matrix-partition algorithm can be defined under the conditions stated above. This follows the same pattern as that given in section 2.

The final decomposed matrix is

$$\left[\begin{array}{ccc|ccc|ccc}
A_1 & B_1 & & & & & & & \\
 & \tilde{A}_2^{(1)} & B_2 & & & & & & \\
 & & \tilde{A}_3^{(2)} & & & \tilde{S}_5 & & & \\
\hline
 & & B_3^t & A_4 & & & & & \\
 & & \tilde{R}_4 & & \tilde{A}_5^{(1)} & & & & \\
 & & & & & \tilde{A}_6^{(4)} & & & \tilde{S}_8 \\
\hline
 & & & & & B_6^t & A_7 & & \\
 & & & & & \tilde{R}_7 & & \tilde{A}_8^{(1)} & \\
 & & & & & & & & \tilde{A}_9^{(5)}
\end{array}\right]$$

where the resultant blocks are now all tridiagonal. This incomplete algorithm yields an incomplete-Cholesky factorization of the matrix PAP^t of (2.1).

The matrices $\tilde{Q}_i$ in the incomplete-matrix-partition algorithm are constructed from a tridiagonal approximation of the K x K block matrices BA^{-1} where B and A are K x K tridiagonal matrices and A is symmetric. This is done by letting $A = LDL^t$ where L is a unit lower bidiagonal matrix and D is a diagonal matrix. A^{-1} is approximated by $(2I - L^t)D^{-1}(2I - L)$. The approximation to BA^{-1} is then the tridiagonal part of $B(2I - L^t)D^{-1}(2I - L)$. If the original system (1.1) is an M-Matrix, no zero-division will occur. We can now prove a theorem similar to Theorem 2.1.

Theorem 3.1 With the same hypothesis as Theorem 2.1, then for $m = 1, .., p - 1$,

$$\| \tilde{R}_{i+m\alpha} \| < \| \tilde{R}_{i-1+m\alpha} \|, \quad i = 1, \ldots, \alpha,$$

$$\| \tilde{S}_{i+m\alpha} \| < \| \tilde{S}_{i-1+m\alpha} \|, \quad i = 1, .., \alpha,$$

where $\tilde{R}_{m\alpha} = B^t_{m\alpha}$ and $\tilde{S}_{m\alpha} = B_{m\alpha}$.

Theorem 3.1 gives the possibility of defining many preconditioners from the matrix partition algorithm when dominance is present in the matrix A of (1.1). Since the $\tilde{R}_{i+m\alpha}$, $\tilde{S}_{i+m\alpha}$ are decaying to zero as i increases, it is possible to discard them prematurely. In particular, the $\tilde{S}_5$ and $\tilde{S}_8$ can be discarded so that no coupling exists between the partitions and the Backward substitution process can proceed in a partition independent fashion. Thus, unlike the direct matrix partition algorithm described in section 2, the decomposition, forward, and backward processes of the incomplete algorithm are completely partition independent, making it a highly desirable preconditioning method for use on a multi-processor. Further, many computations within the partitions can be avoided if many of the $\tilde{R}_{i+m\alpha}$ and $\tilde{S}_{i+m\alpha}$ matrices can be discarded without the causing an increase in the number of iterations within an iterative algorithm. Also, variations of the incomplete decomposition process such as those given in Gustafsson, [8], and Evans [10] can also be used.

IV. APPLICATIONS

The incomplete-matrix-partition algorithm was used to generate preconditioning systems for use within the generalized conjugate-gradient method, (cf. Golub, Concus, and O'Leary [11]). Setting $\tilde{R}_{i+m\alpha} = \tilde{S}_{i+m\alpha} = 0$ for each $i = 0, 1, \ldots, \alpha - 1$ gives different preconditioning systems. Their performances are compared in the first problems against a block-Jocobi preconditioning system where $M = LL^t$ = block-diagonal (A).

For all runs, the components of the exact solution x were randomly generated and then $Ax = b$ is formed. The initial iterate $x^{(0)}$ is zero and the conjugate gradient algorithm is terminated when

$$\frac{\| x^{(k)} - x \|_2}{\| x \|_2} < 10^{-7}$$

In all cases K = L. For problems 1 and 2, the number of partitions was taken to be 5.

Problem 1 We solved the Dirichlet problem $\Delta U = f$ with zero boundary conditions. The 5-point approximation to the Laplacian was used to generate the system (1.1) with $row_i(A) = [0...0, -1, 0..0, -1, 4, -1, 0..0, -1, 0..0]$.

Problem 2 The off-diagonal elements of A in (1.1) were negative random numbers. The diagonal of A was then taken to be the negative of the sum along the rows. In this case, A is a singular M-matrix. To solve a non-singular problem, we solved the system $(A + \sigma I)\, x = b$ where $\sigma > 0$. In Table 2, $\sigma = .0001$.

To see the effects of the number of partitions on the iteration count, the Dirichlet problem was tested with K = L fixed at 100. Here α is number of partitions. The results are recorded in Table 3.

K = L	i = 0	i = 1	i = 2	i = 3	B - Jac
10	19	13	8		25
30	37	25	23	23	72
50	54	40	34	34	107
75	70	55	50	50	157
100	92	72	67	67	208

Table 1: Dirichlet Matrix

K = L	i = 0	i = 1	i = 2	i = 3
10	35	27	25	
30	73	63	62	61
50	116	105	102	101
75	181	169	168	167

Table 2: Random Matrix

α	i = 0	i = 1	i = 2	i = 3
5	92	72	67	67
10	98	73	67	66
20	104	73	64	64
50	152	104	65	

Table 3: Dirichlet Matrix with α partitions

ACKNOWLEDGEMENTS

The authors would like to thank the U.S. Department of Energy, Office of Basic Energy Sciences, Applied Mathematical Sciences Division, for support.

This work was performed under the auspices of the U.S. Department of Energy by Lawrence Livermore Laboratory under contract No. W-7405-Eng-48.

REFERENCES

1. Varga, R.S., Matrix Iterative Analysis, Prentice Hall, Inc., New Jersey, 1962.

2. Rodrigue, G., Hendrickson, C., Pratt, M., "An Implicit Solution of the Dimensional Diffusion Equation and Vectorization Experiments", in Parallel Computations, G. Rodrigue, ed., Academic Press, 1982.

3. Stewart, G.W., Introduction to Matrix Computations, Academic Press, 1973.

4. Ehrel, J., Lichnewsky, A., Thomasset, F., "Parallelism in Finite Element Computation", Proceedings of I.B.M. Symposium on Vector Computers and Scientific Computation, Rome, 1981.

5. Rodrigue, G. and Wolitzer, D., "Preconditioning by Incomplete Block Cyclic Reduction", to appear in Mathematics of Computation.

6. Sameh, A. and Kuck, D., "On Stable Parallel Linear System Solvers", JACM, 25, 1978, pp. 81-91.

7. Wang, H.H., "A Parallel Method for Tridiagonal Systems", ACM-TOMS, Vol. 7, No. 2, June 1981, pp. 170-183.

8. Gustafsson, I., "A Class of First-Order Factorization Methods", BIT 18, (1978), pp. 142-156.

9. Jordan, T., "A Guide to Parallel Computation and Some CRAY-1 Experiences", Parallel Computations, G. Rodrigue (ed.), Academic Press.

10. Evans, D. and Lipitakis, E.A., "A Normalized Implicit Conjugate Gradient Method For the Solution of Large Sparse Systems of Linear Equations", Computer Methods in Applied Mechanics and Engineering 23 (1980), pp. 1-19.

11. Concus, P., Golub, G., O'Leary, O., "A Generalized Conjugate Gradient Method for the Numerical Solution of Elliptic Partial Differential Equations", in Sparse Matrix Computation, J.R. Bunch and D.J. Rose (ed.), Academic Press, New York, 1976.

A FAST POISSON SOLVER FOR MULTIPROCESSORS[1]

A. H. Sameh[2]

I. INTRODUCTION

Efficient direct methods for solving the finite-difference approximation of the Poisson equation on the unit square have been developed by Buneman [2], Hockney [5, 6], and Golub [1]. The most effective sequential algorithm combines the block cyclic reduction and Fourier analysis schemes. This is Hockney's $FACR(l)$ algorithm, [6]. Excellent reviews of these methods on sequential machines have been given by Swarztrauber [14] and Temperton [15, 16]. In [14] it is shown that the asymptotic operation count for $FACR(l)$ on an $n \times n$ grid is $O(n^2 \log_2 \log_2 n)$, and is achieved when the number l of the block cyclic reduction steps preceding Fourier analysis is taken approximately as $(\log_2 \log_2 n)$. Using only cyclic reduction, or Fourier analysis, to solve the problem on a sequential machine would require $O(n^2 \log_2 n)$ arithmetic operations.

Buzbee [3] observed that Fourier analysis, or the matrix decomposition Poisson solver (MD-Poisson solver), is ideally suited for parallel computation. It consists of performing a set of independent sine transforms, and solving a set of independent tridiagonal systems. On a parallel computer consisting of n^2 processors, with an arbitrarily powerful interconnection network, the MD-Poisson solver for the two-dimensional case requires $O(\log_2 n)$ parallel arithmetic steps [10]. It can be shown,

[1]Research supported in part by NSF grant MCS 81-17010.

[2]Department of Computer Science, University of Illinois at Urbana-Champaign, Urbana, Illinois 61801.

ISBN 0-12-100560-7

[9] and [13], that a perfect shuffle interconnection network is sufficient to keep the communication cost to a minimum. Ericksen [4] considered the implementation of $FACR(l)$, [6], and $CORF$, [1], on the ILLIAC IV; and Hockney [7] compared the performance of $FACR(l)$ on the Cray-1, Cyber-205, and the ICL-DAP.

In this paper we consider the MD-algorithm for solving:

(i) the 5-point finite difference approximation of the two-dimensional Poisson equation on a uniform $n \times n$ grid using a ring of $p < n$ processors, and

(ii) the 7-point finite difference approximation of the three-dimensional Poisson equation on a uniform $n \times n \times n$ grid using a mesh of n^2 processors (n rings each of n processors).

Given such limited processor interconnection, we show that problems (i) and (ii) are solved with speedups of $O(p)$ and $O(n^2)$, respectively, provided that the following assumption holds.

> *Each processor is capable of simultaneously performing an arithmetic operation, receiving a floating-point number from an immediate neighbor, and transmitting a previously computed floating-point number to another immediate neighbor in one time step.*

Such an assumption is realistic and the multiprocessors considered in this paper can be realized using present-day technology.

For the sake of illustration, we assume Dirichlet boundary conditions for both the 2-D and 3-D cases. The multiprocessor algorithms presented below can be readily modified to accommodate Neumann and periodic boundary conditions.

II. THE TWO-DIMENSIONAL MD-POISSON SOLVER ON A LINEAR ARRAY OF PROCESSORS

Consider the Poisson problem on the unit square with Dirichlet boundary conditions,

$$u_{xx} + u_{yy} = \phi(x,y), \quad x,y \in R ,$$
$$u(x,y) = \psi(x,y), \ x,y \in \partial R .$$

Using the 5-point finite difference approximation on a uniform mesh of width $h = 1/(n+1)$ with natural ordering of the grid points, we obtain the well-known linear system of order n^2

$$\begin{vmatrix} T & -I & & & & \\ -I & T & -I & & & \\ & \cdot & \cdot & \cdot & & \\ & & \cdot & \cdot & \cdot & \\ & & & -I & T & -I \\ & & & & -I & T \end{vmatrix} \begin{vmatrix} u_1 \\ u_2 \\ \cdot \\ \cdot \\ u_{n-1} \\ u_n \end{vmatrix} = \begin{vmatrix} f_1 \\ f_2 \\ \cdot \\ \cdot \\ f_{n-1} \\ f_n \end{vmatrix}, \tag{2.1}$$

where $T = [-1, 4, -1]$ is a tridiagonal matrix of order n. We wish to solve (2.1) on the ring of $p < n$ processors shown in Fig. 1 below.

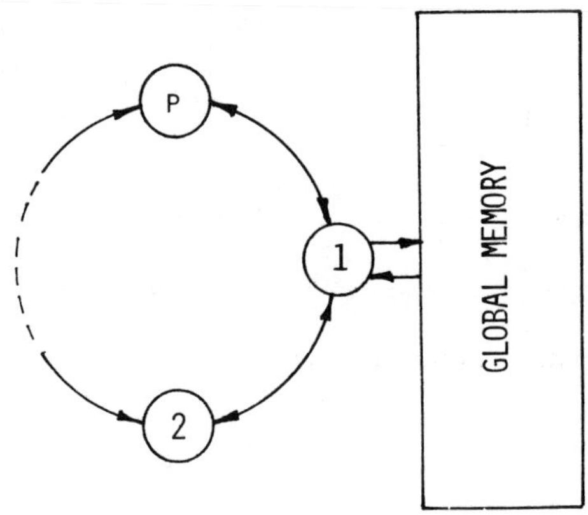

Fig. 1

We assume that n and p are powers of 2 with $q = n/p \geq 4$, and that each processor has a local memory of $O(qn)$ words. Supplying the functions $\phi(x,y)$ and $\psi(x,y)$ from the global memory to processor 1 which in turn transmits them to processor 2, and so on, until each of the p processors contains both functions, the MD-algorithm may be organized as follows:

Stage 1: (Parallel)

1. Each processor j, $1 \leq j \leq p$, forms the subvectors $f_{(j-1)q+1}$, $f_{(j-1)q+2}, \ldots, f_{jq}$ of the right-hand side of (2.1) using $\phi(x,y)$ and $\psi(x,y)$.

2. Next, each processor j obtains $\hat{g}_j^T = (g_{(j-1)q+1}^T, \ldots, g_{jq}^T)$, where $g_k = Qf_k$, in which $Q = [(2/[n+1])^{1/2} \, sin(lm\pi/[n+1])]$, $l,m = 1, 2,\ldots, n$, is the eigenvector matrix of T. This amounts to performing, in each processor, q sine transforms each of length n at a cost of $O(qn\log_2 n)$ arithmetic operations.

3. Now we have the system

$$\begin{vmatrix} M & E & & & \\ E^T & M & E & & \\ & \cdot & \cdot & \cdot & \\ & & \cdot & \cdot & \cdot \\ & & E^T & M & E \\ & & & E^T & M \end{vmatrix} \begin{vmatrix} \hat{v}_1 \\ \hat{v}_2 \\ \cdot \\ \cdot \\ \hat{v}_{p-1} \\ \hat{v}_p \end{vmatrix} = \begin{vmatrix} \hat{g}_1 \\ \hat{g}_2 \\ \cdot \\ \cdot \\ \hat{g}_{p-1} \\ \hat{g}_p \end{vmatrix}, \tag{2.2}$$

where each processor contains one block row. Here, $\hat{v}_j^T = (v_{(j-1)q+1}^T, \ldots, v_{jq}^T)$ with $v_k = Qu_k$, $M = [-I, \Lambda, -I]$ is a block tridiagonal matrix of order qn, and

$$E = \begin{vmatrix} & O \\ -I_n & \end{vmatrix}.$$

If both sides of (2.2) are premultiplied by $diag(M^{-1}, \ldots, M^{-1})$, system (2.2) is reduced to

$$\begin{bmatrix} I & F & & & \\ G & I & F & & \\ & \cdot & \cdot & \cdot & \\ & & \cdot & \cdot & \cdot \\ & & G & I & F \\ & & & G & I \end{bmatrix} \begin{bmatrix} \hat{v}_1 \\ \hat{v}_2 \\ \cdot \\ \cdot \\ \hat{v}_{p-1} \\ \hat{v}_p \end{bmatrix} = \begin{bmatrix} \hat{h}_1 \\ \hat{h}_2 \\ \cdot \\ \cdot \\ \hat{h}_{p-1} \\ \hat{h}_p \end{bmatrix}, \tag{2.3}$$

where $I \equiv I_{qn}$, $\hat{h}_j^T = (h_{(j-1)q+1}^T, \ldots, h_{jq}^T)$, F and G are given by: $M\hat{h}_j = \hat{g}_j$, $1 \leq j \leq p$, $MF = E$, and $MG = E^T$. Observing that M consists of n independent tridiagonal matrices $T_k = [-1, \lambda_k, -1]$ each of order q, where $\lambda_k = 4\text{-}2cos(k\pi/[n+1])$, $k = 1, 2, \ldots, n$, the right-hand side of (2.3) is obtained by solving in each processor j the n independent systems

$$T_k r_k = s_k, \quad k = 1, 2, \ldots, n, \tag{2.4}$$

where $\hat{e}_i^T s_k = e_k^T g_{(j-1)q+i}$, and $\hat{e}_i^T r_k = e_k^T h_{(j-1)q+i}$, $i = 1, 2, \ldots, q$, and $j = 1, 2, \ldots, p$. Here, $\hat{e}_i$ and e_i are the i-th columns of I_q and I_n, respectively.

The matrices F and G can be similarly obtained by solving, in each processor j, the independent systems $T_k c_k = \hat{e}_1$, and $T_k d_k = \hat{e}_q$, for $k = 1, 2, \ldots, n$. Since T_k is a Toeplitz matrix, however, we have $c_k = Jd_k$, where $J = [\hat{e}_q, \ldots, \hat{e}_2, \hat{e}_1]$, see [8] for example. As a result, in order to obtain F and G we need only solve in each processor the n systems

$$T_k d_k = \hat{e}_q, \quad k = 1, 2, \ldots, n. \tag{2.5}$$

Hence, F and G are of the form,

$$F = \begin{vmatrix} \Gamma_q & \\ \vdots & O \\ \Gamma_1 & \end{vmatrix}, \quad G = \begin{vmatrix} & \Gamma_1 \\ O & \vdots \\ & \Gamma_q \end{vmatrix},$$

where $\Gamma_i = -diag(\gamma_i^{(1)}, \gamma_i^{(2)}, \ldots, \gamma_i^{(n)})$, in which $\gamma_i^{(k)} = \hat{e}_i^T c_k$, $i = 1, 2, \ldots, q$, $k = 1, 2, \ldots, n$.

The 2n positive definite systems, (2.4) and (2.5), may be solved at a cost of $O(qn)$ arithmetic operations.

Stage 2: (Pipeline)

From the structure of (2.3) we can see that the (p-1) pairs of n equations above and below each partition are completely decoupled from the rest of the n^2 equations [11]. This reduced system, of order $2(p-1)n$, consists of interlocking blocks of the form:

$$\begin{array}{cccccccc} I & \Gamma_1 & 0 & 0 & & & & \\ \Gamma_1 & I & 0 & \Gamma_q & & & & \\ \Gamma_q & 0 & I & \Gamma_1 & 0 & 0 & & \\ & & \Gamma_1 & I & 0 & \Gamma_q & & \\ & & \Gamma_q & O & I & \Gamma_1 & & \\ & & & & & & \ddots & \end{array}$$

The first n rows of this block-pentadiagonal system are contained in processor 1, the next 2n rows in processor 2, and so on with the last n rows residing in processor p. This system, in turn, constitutes n independent pentadiagonal systems each of order $2(p-1)$. Further, it can be verified that these systems may be solved via Gaussian elimination without pivoting. The forward and backward sweeps for these n systems may be pipelined across the p processors at a cost of $O(n)$ time steps. This cost includes both the arithmetic and the largely masked interprocessor communication, see for example [12]. This pipelining feature is made possible by our main assumption in section 1.

Stage 3: (Parallel)

1. Now, that the subvectors v_{kq}, v_{kq+1}, $k = 1, 2,..., p-1$, are available, each processor $j = 1, 2,..., p$ obtains

$$v_{(j-1)q+i} = h_{(j-1)q+i} - (\Gamma_i v_{(j-1)q} + \Gamma_{q-i+1} v_{jq+1})$$

for $i = 2, 3,..., q-1$, where $v_0 = v_{pq+1} = 0$, at a cost of $O(qn)$ arithmetic operations.

2. Finally, each processor j retrieves the q subvectors $u_{(j-1)q+i} = Qv_{(j-1)q+i}$, $i = 1, 2,..., q$, of the solution of (2.1) via q sine transforms, each of length n, at a cost of $O(qn\log_2 n)$.

Clearly, the algorithm consumes $O((n^2/p)\log_2 n)$ time steps on our ring of p processors. If $p = n$, however, step 3 of stage 1, and step 1 of stage 3 are not needed and the cost of the fast solver is reduced to $O(n\log_2 n)$.

III. THE THREE-DIMENSIONAL MD-POISSON SOLVER ON A MESH OF n^2 PROCESSORS

Here, we consider the Poisson problem on the unit cube with Dirichlet boundary conditions. Using the usual 7-point finite difference approximation on a uniform $n \times n \times n$ grid, with natural ordering of the grid points, we obtain the linear system of order n^3,

$$\begin{vmatrix} A & -I_\nu & & & \\ -I_\nu & A & -I_\nu & & \\ & \cdot & \cdot & \cdot & \\ & & \cdot & \cdot & \cdot \\ & & -I_\nu & A & -I_\nu \\ & & & -I_\nu & A \end{vmatrix} \begin{vmatrix} \hat{w}_1 \\ \hat{w}_2 \\ \cdot \\ \cdot \\ \hat{w}_{n-1} \\ \hat{w}_n \end{vmatrix} = \begin{vmatrix} \hat{a}_1 \\ \hat{a}_2 \\ \cdot \\ \cdot \\ \hat{a}_{n-1} \\ \hat{a}_n \end{vmatrix} \tag{3.1}$$

where $A = [-I_n, \hat{T}, -I_n]$ is a block tridiagonal matrix of order n^2 in which $\hat{T} = [-1, 6, -1]$ is a tridiagonal matrix of order n, and $\nu \equiv n^2$.

We wish to solve (3.1) on the mesh of n^2 processors shown in Fig. 2. Let each of the n rings contain one plane with each of its n processors containing one line of n mesh points. In other words each ring i generates $\hat{a}_i^T = (a_{(i-1)n+1}^T, \ldots, a_{in-1}^T, a_{in}^T)$ and at completion of the algorithm contains $\hat{w}_i^T = (w_{(i-1)n+1}^T, \ldots, w_{in-1}^T, w_{in}^T)$. As before, our main assumption regarding the capability of each processor holds, and each ring is connected to the global memory through one processor only.

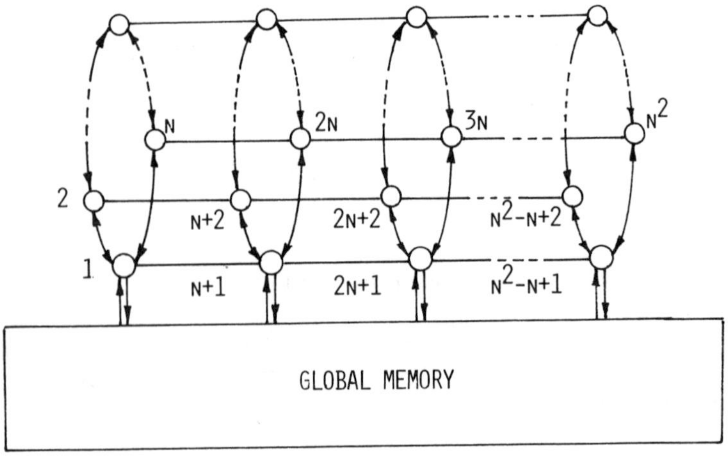

Fig. 2

Assuming that each processor has a local memory of $O(n)$ words, the three-dimensional MD-Poisson solver may be organized on our mesh as follows.

Stage 1: (Parallel among all processors)

After each processor $1 \leq k \leq n^2$ has generated a_k, its corresponding part of the right-hand side of (3.1), we simultaneously perform n^2 sine transforms, $b_k = Qa_k$, each of length n at the cost of $O(n\log_2 n)$, where Q is the eigenvector matrix of T and $\hat{T}$. Now, we have the system

$$\begin{vmatrix} B & -I_\nu & & & & \\ -I_\nu & B & -I_\nu & & & \\ & \cdot & \cdot & \cdot & & \\ & & \cdot & \cdot & \cdot & \\ & & & -I_\nu & B & -I_\nu \\ & & & & -I_\nu & B \end{vmatrix} \begin{vmatrix} \hat{x}_1 \\ \hat{x}_2 \\ \cdot \\ \cdot \\ \hat{x}_{n-1} \\ \hat{x}_n \end{vmatrix} = \begin{vmatrix} \hat{b}_1 \\ \hat{b}_2 \\ \cdot \\ \cdot \\ \hat{b}_{n-1} \\ \hat{b}_n \end{vmatrix}, \tag{3.2}$$

where $B = [-I_n, \tilde{\Lambda}, -I_n]$ is a block tridiagonal matrix of order n^2 in which $\tilde{\Lambda} = diag(\tilde{\lambda}_1, \ldots, \tilde{\lambda}_n)$ with $\tilde{\lambda}_k = 6 - 2\cos(k\pi/[n+1])$, and $\hat{x}_i = \hat{Q}\hat{w}_i$. Here, $\hat{Q} = diag(Q, \ldots, Q)$ is of order n^2, and each ring i contains B and $\hat{b}_i$.

Stage 2: (Parallel among rings--pipelined within each ring)

First, we observe that the matrix B consists of n independent tridiagonal matrices $C_k = [-1, \tilde{\lambda}_k, -1]$, $1 \le k \le n$, i.e., there exists a permutation matrix P such that $PBP = diag(C_1, \ldots, C_n)$. In ring i, each processor $(i-1)n+j$, $1 \le j \le n$, contains the j-th rows of $C_1, \ldots, C_n$. Observing that $QC_kQ = \Sigma_k = diag(\sigma_1^{(k)}, \ldots, \sigma_n^{(k)})$, where $\sigma_l^{(k)} = \tilde{\lambda}_k - 2\cos(l\pi/[n+1])$, we can obtain within each ring i the vector $\hat{c}_i = P\hat{Q}P\hat{b}_i$ via n independent sine transforms, each of length n, pipelined across the n processors of the ring. This consumes $O(n\log_2 n)$ time steps for both the arithmetic and interprocessor communication.

The system (3.2) is now reduced to

$$\begin{vmatrix} \Sigma & -I_\nu & & & & \\ -I_\nu & \Sigma & -I_\nu & & & \\ & \cdot & \cdot & \cdot & & \\ & & \cdot & \cdot & \cdot & \\ & & & -I_\nu & \Sigma & -I_\nu \\ & & & & -I_\nu & \Sigma \end{vmatrix} \begin{vmatrix} \hat{y}_1 \\ \hat{y}_2 \\ \cdot \\ \cdot \\ \hat{y}_{n-1} \\ \hat{y}_n \end{vmatrix} = \begin{vmatrix} \hat{c}_1 \\ \hat{c}_2 \\ \cdot \\ \cdot \\ \hat{c}_{n-1} \\ \hat{c}_n \end{vmatrix}, \tag{3.3}$$

where $\Sigma = diag(\sigma_1^{(1)}, \ldots, \sigma_1^{(n)};\ldots;\sigma_n^{(1)}, \ldots, \sigma_n^{(n)})$, $\hat{c}_i^T = (c_{(i-1)n+1}^T, \ldots, c_{in}^T)$, and $\hat{y}_i = P\hat{Q}P\hat{x}_i = (y_{(i-1)n+1}^T, \ldots, y_{in}^T)^T$ are contained in ring i, $i = 1, 2,\ldots, n$.

Stage 3: (Parallel within each ring--pipelined across rings)

The system (3.3) consists of n^2 independent tridiagonal systems each of order n, with ring i containing the i-th row of each system. Specifically, processors $j, n+j,\ldots, n(n-1)+j$, for $j = 1, 2,\ldots, n$, contain the n tridiagonal systems,

$$\begin{vmatrix} \sigma_j^{(k)} & -1 & & & \\ -1 & \sigma_j^{(k)} & -1 & & \\ & \cdot & \cdot & \cdot & \\ & & \cdot & \cdot & \cdot \\ & & -1 & \sigma_j^{(k)} & -1 \\ & & & -1 & \sigma_j^{(k)} \end{vmatrix} \begin{vmatrix} \eta_1 \\ \eta_2 \\ \cdot \\ \cdot \\ \eta_{n-1} \\ \eta_n \end{vmatrix} = \begin{vmatrix} \phi_1 \\ \phi_2 \\ \cdot \\ \cdot \\ \phi_{n-1} \\ \phi_n \end{vmatrix}, \quad k = 1, 2,\ldots, n \ , \tag{3.4}$$

where $\phi_i = e_k^T c_{(i-1)n+j}$, and $\eta_i = e_k^T y_{(i-1)n+j}$. The n systems (3.4) may be solved via Gaussian elimination without pivoting, pipelined across the processors $j, n+j,\ldots, n(n-1)+j$. This stage consumes $O(n)$ time steps, including the cost of interprocessor communication (communication across the rings).

Finally, stages 4 and 5 are analogous to stages 2 and 1, respectively. In stage 4, we retrieve $\hat{x}_i = P\hat{Q}P\hat{y}_i$ in each ring i via pipelining n sine transforms, each of length n, across the processors $(i-1)n+1,\ldots, in$. In stage 5, each processor k retrieves $w_k = Qx_k$, $k = 1, 2,\ldots, n^2$, via n^2 independent sine transforms of length n. Hence, the total cost of this three-dimensional MD-Poisson solver is $O(n\log_2 n)$.

If the multiprocessor consists of $r < n$ rings with each ring consisting of $p < n$ processors, then an obvious modification of the algorithm organization of the two-dimensional case described in section 2 will lead to a speedup of $O(pr)$.

References

[1] B. Buzbee, G. Golub, and C. Nielson, On direct methods for solving Poisson's equation, *SIAM J. Numer. Anal. 7,* 627-656, 1970.

[2] O. Buneman, A compact non-iterative Poisson solver, Report 294, Stanford University Institute for Plasma Research, Stanford, California, 1969.

[3] B. Buzbee, A fast Poisson solver amenable to parallel computation, *IEEE Trans. Comput. C-22,* 793-796, 1973.

[4] J. Ericksen, Iterative and direct methods for solving the Poisson's equation and their adaptability to Illiac IV, CAC Document No. 60, University of Illinois, Urbana, Illinois, December 1972.

[5] R. Hockney, A fast direct solution of Poisson's equation using Fourier analysis, *JACM 12,* 95-113, 1965.

[6] R. Hockney, The potential calculation and some applications, in *Methods of Computational Physics,* (Alder, Fernback and Rotenberg, eds.), Vol. 9, 135-211, Academic Press, 1970.

[7] R. Hockney, Optimizing the *FACR(l)* Poisson solver on parallel computers, *Proc. 1982 Conf. Parallel Processing,* 62-71, IEEE Computer Society, 1982.

[8] T. Kailath, A. Vieira, and M. Morf, Inverses of Toeplitz operators, innovations, and orthogonal polynomials, *SIAM Rev. 20,* 106-119, 1978.

[9] M. Pease, An adaptation of the fast Fourier transform for parallel processing, *JACM 15,* 252-264, 1968.

[10] A. Sameh, S. Chen, and D. Kuck, Parallel Poisson and biharmonic solvers, *Computing 17,* 219-230, 1976.

[11] A. Sameh and D. Kuck, On stable parallel linear system solvers, *JACM 25,* 81-91, 1978.

[12] A. Sameh, Solving the linear least squares problem on a linear array of processors, Purdue Workshop on *Algorithmically-Specialized Computer Organizations,* September 1982 (proceedings to be published by Academic Press).

[13] H. Stone, Parallel processing with the perfect shuffle, *IEEE Trans. Comput. C-20,* 153-161, 1971.

[14] P. Swarztrauber, The methods of cyclic reduction, Fourier analysis, and the FACR algorithm for the discrete solution of Poisson's equation on a rectangle, *SIAM Rev.* *19,* 490-501, 1977.

[15] C. Temperton, Direct methods for the solution of the discrete Poisson equation: some comparisons, *J. Comput. Phys.* *31,* 1-20, 1979.

[16] C. Temperton, On the *FACR(l)* algorithm for the discrete Poisson equation, *J. Comput. Phys.* *34,* 314-329, 1980.

SYSTOLIC ARRAYS: HIGH PERFORMANCE PARALLEL MACHINES FOR MATRIX COMPUTATION

Robert Schreiber[1]

I. INTRODUCTION

In this paper we shall summarize the recent development of systolic array methods for some of the important standard problems in numerical linear algebra. We shall discuss LU and QR factorizations, eigenvalue problems, and the singular value decomposition. All the work we shall describe has been done since 1981. Our aim is to introduce the reader to this rapidly developing branch of numerical computation.

Systolic arrays were introduced and named by Kung and Leiserson [12]. Although their ideas had antecedents (see [8] for example), they first fully realized and proclaimed that these designs--highly parallel computing networks with regular data flows, two-dimensional lattice form, simple identical cells, regular input and output patterns, and heavy reuse of data--were an ideal way to use the emerging VLSI technology to obtain very high performance for suitable computations.

After these first promising designs, which implemented a few relatively simple matrix computations, two important questions were open. Could systolic arrays be found for a broader class of matrix problems, including the more important and difficult matrix decompositions? And could they be integrated into real computing systems without introducing incapacitating losses of efficiency?

---

[1]Department of Computer Science, Stanford University, Stanford, California 94305

ISBN 0-12-100560-7

The first of these questions has now been answered and the answer is a definite "yes." This research has done more than answer the question of the applicability of the systolic array idea. It is, rather, the beginning of an intriguing new approach to numerical computation. In this new approach the flow of data in both time and space is important, whereas traditionally only the sequence of computations in time was considered. An interesting facet is that the standard algorithms have not always proven suitable for systolic array realization. For instance, for symmetric eigenvalue problems, Brent and Luk [5] have found that Jacobi's method is better than QR. Their method uses a permutation of the off-diagonal elements that seems to be better than the usual cyclic-by-rows permutation [5].

The second question is being answered now. Several projects for building VLSI based systolic hardware are currently in progress or complete ([1], [19]). An effort at ESL, Inc., should produce a systolic machine with performance in the 100-1000 megaflop range in the next few years.

It appears at the moment that digital signal and image processing [15], rather than standard scientific computing and elliptic equation solving in particular, offers the sort of problem best suited for systolic array solution. Nevertheless these ideas can be usefully applied to elliptic equations. In Section VI we shall discuss an implementation of multigrid methods by highly parallel computing networks.

II. LU AND QR FACTORIZATIONS

The first such array was an $m \times m$ hexagonally connected array for LU factorization of $n \times n$ banded matrices with bandwidth m in time $O(3n)$ [12]. Unfortunately, there was no provision for pivoting to enhance stability. Next, an array for QR or LU factorization of a dense square matrix, in linear time, was given by Bojanczyk, Brent, and Kung [2]. A different array, better in that it could handle rectangular $m \times n$ matrices in time $O(m)$, was given by Gentleman and Kung [10]. The unit of time we use is the cycle time of a cell of the given array. In this time, every array accepts one set of inputs and produces one set of outputs. Finally, Heller and Ipsen developed a rectangular array for band matrix LU and QR factorization [11]. (See Figures 1 and 2.) Many practical details concerning the implementation of the Gentleman-Kung array were discussed by Schreiber and Kuekes; they also give an array for triangular systems with many right-hand sides [15].

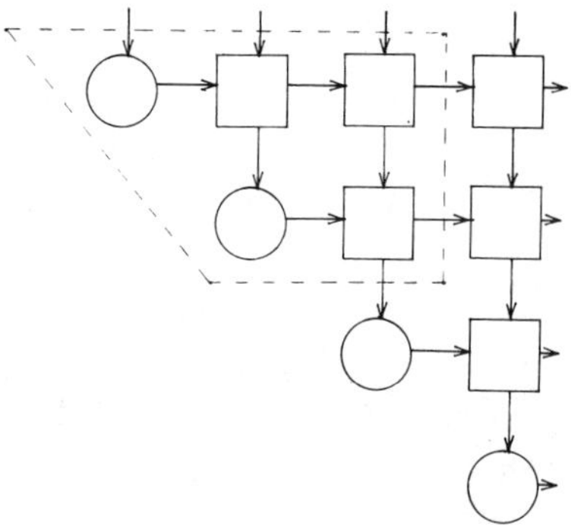

Figure 1.
The Gentleman-Kung Array

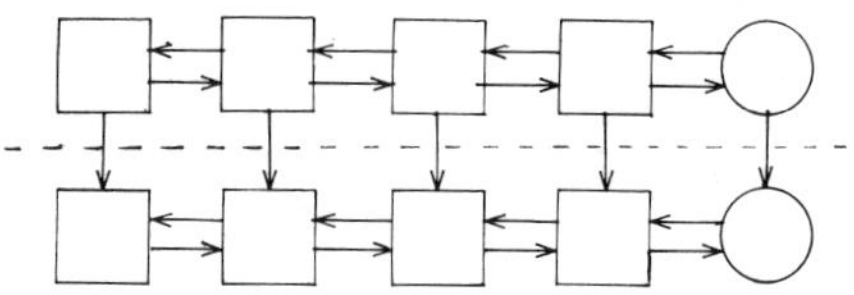

Figure 2.
The Heller-Ipsen Array

The QR factorization arrays employ Givens rather than Householder transformations. The LU arrays employ a stable variant of Gaussian elimination that uses "neighbor pivoting." When the matrix element a_{ij} is eliminated, the elimination is done by subtracting from row i a multiple of row i-1. These rows are first exchanged, if necessary, to make the multiplying factor smaller in modulus than one. Partial pivoting does not lend iteself to systolic implementation.

III. EIGENVALUE COMPUTATION

The obvious first approach was to attempt to implement the standard algorithm for the symmetric problem: reduction to tridiagonal form by orthogonal similarity transformation followed by a method, QR iteration for example, for the tridiagonal matrix. A linear array that requires $O(n^2)$ time is easy to find, but more parallelism has so far proved elusive. Schreiber showed, however, that reduction to a matrix with 2k+1 non-zero diagonals could be done in n^2/k cycle times using a $k \times n$ sub-array of the Gentleman-Kung array. The eigenvalues of the banded matrix can be found using QR iteration which was shown to be implementable by the Heller-Ipsen array in only a little more time than is needed for QR factorization [13]. If further speed is desired, then up to n/k QR iterations can be performed simultaneously by a pipeline of Heller-Ipsen arrays. Choosing $k = n^{1/2}$ yields a method using $O(n^{3/2})$ processors and $O(n^{3/2})$ time. (There is some difficulty concerning the choice of shifts in the QR iteration when iterations are pipelined.) These techniques are also applicable to nonsymmetric matrices [14].

The pursuit of a linear-time solution led Brent and Luk to consider Jacobi methods. They have found an implementation using an $n/2 \times n/2$ array, which implements a cyclic Jacobi sweep in n-1 cycle times [5]. Their experiments have shown that $O(\log n)$ sweeps are required for convergence. In practice, for $n \leq 1000$, no more than 10 sweeps would be required [5].

IV. SINGULAR VALUE DECOMPOSITION (SVD)

To both the approaches of Schreiber and Brent-Luk there are analogous approaches to the SVD. Schreiber's takes mn/k cycle times to reduce an $m \times n$ matrix to an upper triangular matrix with k+1 nonzero diagonals. QR iteration can then be carried out using Heller-Ipsen arrays [16]. Brent and Luk employ a linear array, and their method requires $O(mn \log n)$ time [4]. Brent, Luk, and Van Loan have just reported an improved method requiring almost linear time [6].

One of the uses of SVD is in solving ill-conditioned least-squares problems. An alternative method is to construct the generalized inverse of a rank-deficient matrix closest to the given matrix by an iterative method. An algorithm of Ben-Israel, improved by Schreiber, can be used. The generalized inverse is obtained in O(m log cond(A)) time using n^2 processors [18].

V. DECOMPOSING ARRAYS

A most important question from the practical point of view is this: can a physical array of fixed size (number of cells) be used to solve problems of arbitrary size? The question is important because, obviously, problems of widely differing sizes may be confronted in some applications. Not so obviously, in some applications the problems may have only one size, but they may occur infrequently enough that a full-sized array would not be kept busy for more than a fraction of the time. In these cases it would be more efficient to build a smaller array and have it simulate a full-sized array of the same kind. The question is whether this hardware/time tradeoff is possible.

For some arrays it is easy to see how to do this. Consider the Gentleman-Kung array (Figure 1). Suppose a sub-array of the size shown within the dashed outlines is available. The sub-array could be used to perform the work done by the outlined part of the whole array because all the data flowing into that part of the array is known at the outset. Data flowing out of the subarray has to be stored. The sub-array can next be moved to cover another part of the large array for which all inputs are now known. This continues until the whole array has been covered.

For some arrays, the only sub-array for which all inputs are known is the entire array. The Kung-Leisserson LU factorization array is an example. Such an array is not "decomposable" by this technique. We consider that this is a serious drawback.

The Heller-Ipsen array (Figure 2) is partially decomposable. If we cut it by a horizontal line (the dashed line in Figure 2) then, since all data flows through the cut from bottom to top, we can "run" the bottom part of the array, save the output, then run the top part using this saved data as input. But if a vertical cut is made, data flows across the cut in both directions. Moreover, these data cannot be known unless the computations performed by both the left and right halves of the array are done.

The Brent-Luk SVD array at first appears to be indecomposable. Despite this we have recently found a way to use this array to solve larger problems. The key is to use a p-processor Brent-Luk array as the basic "cell" in a q-processor "super-array" that works on matrices with 2pq columns. Columns move between these cells in groups of p. The idea also applies to the Brent-Luk eigenvalue array. We shall give the details in a later paper [19].

VI. MULTIGRID METHODS

To illustrate the applicability of systolic-array-like devices for elliptic problems, we consider implementation of multigrid algorithms. Full details are given by Chan and Schreiber [7]. Related work is reported by Brandt [3] and Gannon and Van Rosendale [9]. Consider a standard multigrid algorithm with n^d gridpoints on the finest grid and a coarse-to-fine mesh-length ratio of a, in which c coarse grid interations are done for every fine grid iteration. Suppose we build a processor grid with n^p processors for every point grid with n^d gridpoints, so that all computation on the fine grid take $O(n^{d-p})$ time. Then we can show that the time for an iteration T(n), satisfies

$$T(n) = \begin{cases} O(n^{d-p}) & \text{if } c < a^{d-p} \\ O(n^{d-p}\log n) & \text{if } c = a^{d-p} \\ O(n^{\log_a c}) & \text{if } c > a^{d-p} \end{cases}$$

Thus, we get a speedup proportional to the number of processors employed only in the first case. The loss of efficiency is not too bad $(O(1/\log n))$ in the second case, as for instance when there is one processor for every gridpoint $(p = d)$ and we take $c = 1$.

ACKNOWLEDGEMENT

This research was partially supported by the Office of Naval Research under contract N00014-83-K-0703. Part was done while the author was a guest researcher at the Royal Institute of Technology, Stockholm, and Uppsala University, Sweden.

REFERENCES

1. J. Blackmer, P. Kuekes, and G. Frank. A 200 MOPS systolic processor. SPIE Vol 298. SPIE, Bellingham, WA, 1981, pp. 10-18.

2. A. Bojanczyk, R.P. Brent, and H.T. Kung. Numerically stable solution of dense systems of linear equations using mesh-connected processors. SIAM J. Sci. and Stat. Comput., 1983.

3. A. Brandt. Multigrid solvers on parallel computers. ICASE report 80-23, ICASE, NASA Langley Res. Ctr., Hampton, VA, 1980.

4. R.P. Brent and F.T. Luk. A systolic architecture for the singular value decomposition. Tech Rep. TR-CS-82-09 Dept. of Computer Science, Australian National University, Canberra, August, 1982.

5. R.P. Brent and F.T. Luk. A systolic architecture for almost linear time solution of the symmetric eigenvalue problem. Tech. Rep. TR-CS-82-10, Comp. Sci., ANU, 1982.

6. R.P. Brent, F.T. Luk, and C. Van Loan. Almost linear time computation of the singular-value decomposition using mesh-connected processors. Rep. TR-82-528, Dept. of Comp. Sci., Cornell Uni., Ithaca, NY, November, 1982.

7. T. Chan and R. Schreiber. Highly parallel computing networks for multigrid algorithms. In preparation.

8. S.N. Cole. Real-time computation by n-dimensional iterative arrays of finite-state machines. IEEE Trans. Comp. C-18, pp. 349-365 (1969).

9. D. Gannon and J. Van Rosendale. Highly parallel multigrid solvers for elliptic PDEs: an experimental analysis. ICASE Report 82-36, ICASE, NASA Langley Research Center, Hampton, VA, 1982.

10. W.M. Gentleman and H.T. Kung. Matrix triangularization by systolic arrays. Proc. SPIE Vol 298, SPIE, Bellingham, WA, 1981, pp. 19-26.

11. D.E. Heller and I.C.F. Ipsen. Systolic networks for orthogonal equivalence transformations and their applications. Proc., Conf. on Advanced Research in VLSI, editor P. Penfield, Jr., Artech House, Dedham, MA, 1982, pp. 113-122.

12. H.T. Kung and C.E. Leiserson. Systolic arrays for VLSI, §8.3 of C.A. Mead and L.A. Conway, Introduction to VLSI Systems, Addison-Wesley, Reading, Mass., 1980.

13. R. Schreiber. Systolic arrays for eigenvalue computation. Proc. SPIE Vol 341. SPIE, Bellingham, Wash., 1982, pp. 27-34.

14. R. Schreiber. Systolic arrays for eigenvalues. Proc. Inter-American Workshop on Numerical Methods (Caracas, 1982), editors V. Pereyra and A. Reynosa, Springer-Verlag, to appear.

15. R. Schreiber and P. Kuekes. Systolic linear algebra machines in digital signal processing. VLSI and Modern Signal Processing, editors S.Y. Kung, H.J. Whitehouse and T. Kailath, Prentice-Hall, to appear.

16. R. Schreiber. A systolic architecture for singular-value decomposition. Proc. 1er Colloque International sur les Méthodes Vectorielles et Paraleles en Calcul Scientifique. Electicite de France Bulletin de la Direction des Études et Techerches, Série C, No 1-1983 (1 avenue du General de Gaulle, 92140 Clamart, France.), pp.

17. R. Schreiber. On the systolic arrays of Brent and Luk for the symmetric eigenvalue and singular value problems. In preparation.

18. R. Schreiber. An improved iterative method for computing the generalized inverse of a matrix. In preparation.

19. J.J. Symanski. Progress on a systolic processor implementation. Proc. SPIE Vol 341., SPIE, Bellingham, Wash., 1982, pp. 2-7.

III

Iterative Equation Solving

A SURVEY OF RECENT RESULTS ON ITERATIVE METHODS FOR SOLVING LARGE SPARSE LINEAR SYSTEMS

Richard S. Varga[1]

I. INTRODUCTION

The most spectacular advances, in my opinion, which have taken place in recent years in the iterative solution of large sparse systems of linear equations, have come in the following areas: i) preconditioned conjugate gradient methods; ii) multigrid methods for solving elliptic difference equations; iii) efficient adaptation of basic iterative methods to special computers, such as vector computers; and iv) robust software. It is not surprising that over fifty percent of the papers in this Proceedings deal with these same items.

Because these above items are well-represented in these Proceedings, I have chosen to concentrate here specifically on the recent development in algebraic tools, which, by virtue of their reoccurring use in these above items, furnishes a link between current technology and future advances. As an excellent example of an algebraic tool which opened new doors in the area of matrix iterative analysis, and which is still widely used today, it would be difficult to find one better than Young's famous functional equation (cf. [26] and [27, p. 142]):

$$(\lambda + \omega - 1)^2 = \lambda\omega^2\mu^2 , \qquad (1.1)$$

coupling the eigenvalues λ of the successive overrelaxation (SOR) iteration matrix with the eigenvalues μ of the associated Jacobi matrix, in the consistently ordered weakly-cyclic of index 2 case. It is interesting that conformal

[1]Institute for Computational Mathematics, Kent State University.

ISBN 0-12-100560-7

mapping ideas were basic both for the analysis of (1.1), as well as for the analysis of its extension

$$(\lambda + \omega - 1)^p = \lambda^{p-1}\omega^p\mu^p \tag{1.2}$$

in the consistently ordered weakly-cyclic of index p ($p \geq 2$) case (cf. [20] and [22, p. 106]). While it is true that (1.1) and (1.2) can be derived by purely algebraic methods, it is curious that all of the well-recognized textbooks covering this theory have suppressed such conformal mapping ideas in favor of a purely algebraic approach to this problem (cf. [2],[22],[27]).

On the other hand, the recent analyses for example of Niethammer and Varga [16] and Eiermann and Niethammer [5] make new use of ideas from summability theory and conformal mapping theory, to study optimized iterative methods. (The latter work also brings into play the well-known concepts of "maximal convergence" and "uniformly distributed nodes" from complex function theory and approximation theory, concepts which should prove to be invaluable as research tools in the future of the area.) Because these latter papers are long, and will appear before the publication of the Proceedings of this Conference, I will not survey this topic here.

What, then, are some common algebraic tools used today in research in iterative analysis? Obviously, diagonal dominance arguments abound in the literature, but it is perhaps not so well-known that such arguments have a basis in the theory of M- and H-matrices, as introduced by Ostrowski [17] in 1937. It is also true that concepts arising from the Perron-Frobenius theory of nonnegative matrices, such as weakly-cyclic matrices of index p (≥ 2), are in frequent use today. Because of this, I have chosen to focus on closely-related algebraic methods arising in the recent analysis of the symmetric successive overrelaxation (SSOR) iterative method, in the iterative solution of sparse least-squares problems, and in extensions in the theory of regular splittings.

II. A NEW IDENTITY FOR THE SSOR METHOD

Today, a popular preconditioning method used in conjunction with the conjugate gradient method, is one or more sweeps of the SSOR method. This new use of the SSOR iterative method has, interestingly enough, sparked recent interest into the general theory of this iterative method. The purpose of this section is to review some new developments in the basic theory for the SSOR method, in the weakly-cyclic case (of index $p \geq 2$), which parallel those for the SOR method.

To begin with, consider the matrix equation

$$A\underline{x} = \underline{k} \ , \tag{2.1}$$

where it is assumed that the $n \times n$ matrix has the partitioned form

$$A = \begin{bmatrix} A_{1,1} & A_{1,2} & 0 & \cdots & 0 & 0 \\ 0 & A_{2,2} & A_{2,3} & \cdots & 0 & 0 \\ \vdots & & & & & \vdots \\ 0 & 0 & 0 & \cdots & A_{p-1,p-1} & A_{p-1,p} \\ A_{p,1} & 0 & 0 & \cdots & 0 & A_{p,p} \end{bmatrix} . \tag{2.2}$$

Suppose also that $p \geq 2$ and that each diagonal submatrix $A_{i,i}$ is square and nonsingular. On setting $D := \operatorname{diag}[A_{1,1}; A_{2,2}; \ldots; A_{p,p}]$, the associated block-Jacobi matrix $J := I - D^{-1}A$ has the form

$$J = \begin{bmatrix} 0 & B_{1,2} & 0 & \cdots & 0 & 0 \\ 0 & 0 & B_{2,3} & \cdots & 0 & 0 \\ \vdots & & & & & \vdots \\ 0 & 0 & 0 & \cdots & 0 & B_{p-1,p} \\ B_{p,1} & 0 & 0 & \cdots & 0 & 0 \end{bmatrix} , \tag{2.3}$$

so that J is weakly-cyclic of index p (cf. [22, p. 39]). Thus, on expressing the matrix J as the sum of a strictly lower and strictly upper triangular matrix, i.e.,

$$J = L + U \ , \tag{2.4}$$

the associated SSOR iteration matrix S_ω is defined as usual (cf. [27, p. 461]) as

$$S_\omega := (I-\omega U)^{-1}[(1-\omega)I+\omega L]\cdot(I-\omega L)^{-1}[(1-\omega)I+\omega U] \ , \tag{2.5}$$

while the associated SOR (successive overrelaxation) iterative matrix $\mathcal{L}_\omega$ is similarly defined (cf. [27, p. 73]) as

$$\mathcal{L}_\omega := (I-\omega L)^{-1}[(1-\omega)I + \omega U] \ . \tag{2.6}$$

In both of the above matrices, ω is the <u>relaxation parameter</u>.

The new identity coupling the eigenvalues λ of the SSOR iteration matrix S_ω to the eigenvalues μ of the block-Jacobi matrix J of (2.3) is contained in

<u>Theorem 2.1</u> ([24]). <u>Given the matrix</u> A <u>of</u> (2.2) <u>with nonsingular square diagonal submatrices</u> $A_{i,i}$, $1 \le i \le p$, <u>let</u> J <u>of</u> (2.3) <u>be its associated weakly-cyclic of index</u> p <u>block Jacobi matrix, and assume that</u> $0 < \omega < 2$. <u>If</u> λ <u>is any eigenvalue of the associated SSOR iteration matrix</u> S_ω (<u>of</u> (2.5)) <u>for which</u> $\lambda \neq (1-\omega)^2$ <u>and if</u> μ <u>satisfies</u>

$$[\lambda-(1-\omega)^2]^p = \lambda[\lambda+1-\omega]^{p-2}(2-\omega)^2\omega^p\mu^p \ , \tag{2.7}$$

<u>then</u> μ <u>is an eigenvalue of the block-Jacobi matrix</u> J. <u>Conversely, if</u> μ <u>is any eigenvalue of</u> J <u>and if</u> $\hat{\lambda}$ <u>satisfies</u> (2.7) with $\hat{\lambda} \neq (1-\omega)^2$, <u>then</u> $\hat{\lambda}$ <u>is an eigenvalue of</u> S_ω.

It is interesting that the tools used to obtain these results stem from the Perron-Frobenius theory of nonnegative matrices, the canonical form of a weakly-cyclic of index p matrix, and the Frobenius-Romanovsky Theorem on the eigenvalues of a weakly-cyclic of index p matrix (cf. [22, p. 39, 40]).

We first remark that the functional equation (2.7), relating the eigenvalues μ of the block-Jacobi matrix J of (2.3) with the eigenvalues λ of the SSOR iteration matrix S_ω of (2.5), has very much the <u>flavor</u> of both Young's functional equation (1.1) as well as Varga's extension (1.2) of it (for $p > 2$) for the SOR iterative method. It is evident, from the tools used in the proof given in [24], that Theorem 2.1 <u>could have</u> been derived twenty years earlier. <u>Why wasn't it?</u> One interpretation for this late discovery of Theorem 2.1 may be the following. The <u>usual</u> textbook proof of (1.1)-(1.2) involve a preliminary result on a certain determinantal invariance (cf. [22, p. 102] and [27, p. 141]), a step which is not used in [24] in deriving Theorem 2.1. It is plausible that this lack of a similar determinantal invariance in the SSOR case may have been responsible for this delay in establishing Theorem 2.1.

It should also be mentioned that the special case p = 2 of (2.7) of Theorem 2.1 was derived earlier by D'Sylva and Miles [19]. In particular, on setting $\hat{\omega} := \omega(2-\omega)$, the special case p = 2 of (2.7) becomes

$$[\lambda + \hat{\omega} - 1]^2 = \lambda\hat{\omega}^2\mu^2 , \tag{2.8}$$

which reduces to Young's equation (1.1). From this, one easily recovers earlier results of Niethammer [14] and Lynn [9] concerning the SSOR method in the case when the Jacobi matrix is a weakly-cyclic of index 2 matrix.

In the case of applying the SOR iterative method to a block-Jacobi matrix which is weakly cyclic of index p, the effect of reorderings (or permutations) of the associated unknowns appears as a change in the exponent k of λ^k in the functional equation of (1.2):

$$(\lambda+\omega-1)^p = \lambda^k\omega^p\mu^p , \quad (k = 1,2,\ldots,p-1;\ p \geq 2), \tag{2.9}$$

(cf. [22, p. 109, Exercise 2], and Nickels and Fox [13]). It is an open question how such reorderings of unknowns similarly affect the functional equation of (2.7) of Theorem 2.1, for the SSOR iterative method.

III. CONVERGENCE AND DIVERGENCE DOMAINS FOR SSOR, APPLIED TO H-MATRICES

As mentioned in §II, the popularity of the SSOR iterative method as a preconditioner, in conjunction with the conjugate gradient method, has revived interest in the general theory of the SSOR iterative method. The purpose of this section is to review new developments in this theory, as it pertains to H-matrices. As we shall see, these new results parallel similar results for the SOR method.

To begin with, a certain amount of notation is necessary. Let $\mathbb{C}^{m,n}$ $(\mathbb{R}^{m,n})$ denote the set of all mxn matrices $A = [a_{i,j}]$ having complex (real) entries. For each $A = [a_{i,j}]$ in $\mathbb{C}^{n,n}$, the comparison matrix for A, $M(A) = [\alpha_{i,j}]$, has its entries $\alpha_{i,j}$ defined by

$$\alpha_{i,i} := |a_{i,i}|,\ 1 \le i \le n;\quad \alpha_{i,j} := -|a_{i,j}|,\ i \ne j,\quad 1 \le i,j \le n, \tag{3.1}$$

so that $M(A)$ is in $\mathbb{R}^{n,n}$. Further, for any $A = [a_{i,j}]$ in $\mathbb{C}^{n,n}$, set

$$\Omega(A) := \{B = [b_{i,j}] \in \mathbb{C}^{n,n} : |b_{i,j}| = a_{i,j},\ \text{for all } 1 \le i,j \le n\}. \tag{3.2}$$

The set $\Omega(A)$ is called the equimodular set of matrices associated with A. Note that both A and $M(A)$ are in $\Omega(A)$.

Next, let $\mathbb{C}_\pi^{n,n}$ denote the subset of matrices in $\mathbb{C}^{n,n}$ having all diagonal entries nonzero. For each $B = [b_{i,j}]$ in $\mathbb{C}_\pi^{n,n}$, we can express B as the sum

$$B = D(B)\{I - L(B) - U(B)\}, \tag{3.3}$$

where $D(B) := \operatorname{diag}[b_{1,1}, b_{2,2}, \ldots, b_{n,n}]$ is nonsingular, and where $L(B)$ and $U(B)$ are respectively strictly lower and strictly upper triangular matrices. Then,

$$J(B) = L(B) + U(B) \tag{3.4}$$

as before in (2.4), defines the associated (point) Jacobi matrix for B. From a classic paper of Ostrowski [17], we know that any real nonsingular matrix $A = [a_{i,j}]$, with $a_{i,j} \le 0$ for all $i \ne j$ $(1 \le i,j \le n)$ and with A^{-1} having only nonnegative entries, is a nonsingular M-matrix, and that any matrix A in $\mathbb{C}^{n,n}$, for which its comparison matrix $M(A)$ is a nonsingular M-matrix, is by definition a non-singular H-matrix. Finally, if $B = [b_{i,j}]$ is in $\mathbb{C}^{m,n}$, then the matrix $|B|$ is defined as usual as $|B| = [|b_{i,j}|]$, and $\rho(C) := \max\{|\lambda| : \det(\lambda I - C) = 0\}$ denotes the spectral radius of any nxn matrix C.

From the works of Alefeld and Varga [1] and Varga [23], we also know that, given any A in $\mathbb{C}_\pi^{n,n}$, $n \ge 2$, the following are equivalent:

i) A is a nonsingular H-matrix;

ii) for any $B \in \Omega(A)$, $\rho(J(B)) \le \rho(|J(B)|) = \rho(J(M(A))) < 1$;

iii) for each $B \in \Omega(A)$ and for each ω satisfying $0 < \omega < 2/[1+\rho(|J(B)|)]$, the associated SOR iteration matrix $L_\omega(B)$ for B of (2.6), satisfies

$$\rho(L_\omega(B)) \le |1-\omega| + \omega\rho(|J(B)|) < 1 \; ; \tag{3.5}$$

iv) for each $B \in \Omega(A)$ and for each ω satisfying $0 < \omega < 2/[1+\rho(|J(B)|)]$, the associated SSOR iteration matrix $S_\omega(B)$ for B of (2.5) satisfies

$$\rho(S_\omega(B)) < 1. \tag{3.6}$$

We remark that the inequality of (3.5) originates in the works of Kahan [7] and Kulisch [8].

To discuss the sharpness of the first inequality of (3.5), we set

$$H_\nu := \{A \in \mathbb{C}^{n,n},\ n \text{ arbitrary: } A \text{ is a nonsingular H-matrix with } \rho(|J(A)|) = \nu\}, \text{ where } 0 \le \nu < 1. \tag{3.7}$$

Thus, with the definition of (3.7), it directly follows from (3.5) that

$$\sup\{\rho(L_\omega(A)) : A \in H_\nu\} \le |1-\omega| + \omega\nu \quad (\text{for all } 0 < \omega < 2/(1+\nu)), \tag{3.8}$$

and it is natural to ask if equality holds in (3.8) for each ω with $0 < \omega < 2/(1+\nu)$. This was answered <u>affirmatively</u> in Neumann and Varga [12]:

$$\sup\{\rho(L_\omega(A)) : A \in H_\nu\} = |1-\omega| + \omega\nu \quad (\text{for all } 0 < \omega < 2/(1+\nu)). \tag{3.9}$$

We next ask in what sense iv) above is sharp for the SSOR iterative method. Since the SSOR iteration matrix $S_\omega(B)$ reduces to the identity matrix when $\omega = 0$ (cf. (2.5)), it is obvious that convergence of the SSOR matrix $S_\omega(B)$ cannot hold in iv) at the left endpoint of the interval $(0, 2/[1+\rho(|J(A)|)]$, for any B in $\Omega(A)$. As for the right endpoint of this interval in iv), Neumann [11] has recently shown that

$$\rho(S_{\omega'}(A)) < 1 \quad (\omega' := 2/[1+\rho(|J(A)|)]) \tag{3.10}$$

for each nonsingular H-matrix A, unless $\rho(|J(A)|) = 0$. This suggests that the upper bound, namely $2/[1+\rho(|J(A)|)]$, in iv) above might be <u>increased</u>.

That an improvement in iv) <u>can</u> be achieved has been recently announced in Neumaier and Varga [10], where a new SSOR-analog (cf. (3.11)) of the Kahan-Kulisch inequality (for SOR) of (3.5) is derived. Their result is

<u>Theorem 3.1</u> ([10]). <u>Let</u> A <u>be any nonsingular</u> H-<u>matrix</u>. <u>Then, the associated (point)</u> SSOR <u>matrix</u> $S_\omega(A)$ <u>for</u> A satisfies

$$\rho(S_\omega(A)) \leq (1-\omega)^2 + \omega(2-\omega)\{\omega+|1-\omega|\}\rho(|J(A)|) < 1, \tag{3.11}$$
$$\text{for all } 0 < \omega < \min\{2;(\rho(|J(A)|)+1)/2\rho(|J(A)|)\} .$$

It is interesting to remark that the proof of Theorem 3.1 depends on the following useful representation of the SSOR iteration matrix S_ω of (2.5) as

$$S_\omega = (1-\omega)^2 I + \omega(2-\omega)(I-\omega U)^{-1}(I-\omega L)^{-1}\{(1-\omega)J+\omega^2 LU\}, \tag{3.12}$$

on regular splittings of matrices (cf. Varga [21]), and on a little known result of Woźnicki [25] on comparing two regular splittings of a matrix.

As an immediate corollary of Theorem 3.1, it follows, for any nonsingular H-matrix A in H_ν, that (cf. iv) above) for each $B \in \Omega(A)$,

$$\rho(S_\omega(B)) \leq (1-\omega)^2 + \omega(2-\omega)\{\omega + |1-\omega|\}\nu < 1 , \tag{3.13}$$
$$\text{for all } 0 < \omega < \min\{2;(\nu+1)/2\nu\} .$$

In other words, the interval of convergence, $(0,2/[1+\nu])$, of iv) for the SSOR method has been <u>increased</u> to $(0,\min\{2;(\nu+1)/2\nu\})$ within the class H_ν.

It is not known if the above interval $(0,\min\{2;(\nu+1)/2\nu\})$, of convergence of the SSOR iterative method, is <u>sharp</u> for the class of H-matrices H_ν, where $0 \leq \nu < 1$. However, without settling this question of sharpness, new upper bounds for the domain of convergence of the SSOR method, applied to H-matrices, have been deduced in Varga, Niethammer, and Cai

[24]. Specifically, the idea is to consider special nxn point Jacobi matrices of the form

$$\begin{bmatrix} 0 & \alpha & & \\ & \ddots & \ddots & \\ & & \ddots & \alpha \\ \beta & & & 0 \end{bmatrix}, \tag{3.14}$$

where $-\alpha^{n-1}\beta = |\alpha^{n-1}\beta| = \nu^n$, with $0 \leq \nu < 1$. To such matrices, (2.7) of Theorem 2.1 can then be applied (with $p = n$). The result is

<u>Theorem 3.2</u> ([24]). <u>For each</u> ν <u>with</u> $1/2 < \nu < 1$, <u>set</u>

$$\hat{\omega}(\nu) := \frac{2}{1 + \sqrt{2\nu - 1}}. \tag{3.15}$$

<u>Then, for each</u> ω <u>satisfying</u> $\hat{\omega}(\nu) < \omega < 2$, <u>there holds</u>

$$\sup\{\rho(S_\omega(B)):\ B \in H_\nu\} > 1. \tag{3.16}$$

To illustrate the results of Theorems 3.1 and 3.2, we include Figure 1 below, indicating where convergence and divergence take place. It is an <u>open question</u> whether convergence (for all $B \in H_\nu$) or divergence (for some $B \in H_\nu$) holds in the unshaded region of Figure 1.

As suggested by Figure 1, we remark that the two curves, defined by the functions

$$\tilde{\omega}(\nu) = \frac{\nu+1}{2\nu}, \tag{3.17}$$

and $\hat{\omega}(\nu)$ of (3.15), can be shown to be <u>tangent</u> to one another at $\nu = 1$. This implies that the <u>interval of uncertainty</u> in ω, corresponding to the unshaded region in Figure 1, is <u>small</u> for ν close to unity. To illustrate this, for $\nu = 0.8$, we have

i) <u>convergence</u>, for <u>each</u> ω satisfying $0 < \omega < \tilde{\omega}(0.8) \doteq 1.111$ of the associated SSOR matrix corresponding to <u>any</u> matrix in $H_{0.8}$;

ii) <u>divergence</u>, for <u>each</u> ω satisfying $1.127 \doteq \hat{\omega}(0.8) < \omega < 2$, of an associated SSOR matrix corresponding to <u>some</u> matrix in $H_{0.8}$.

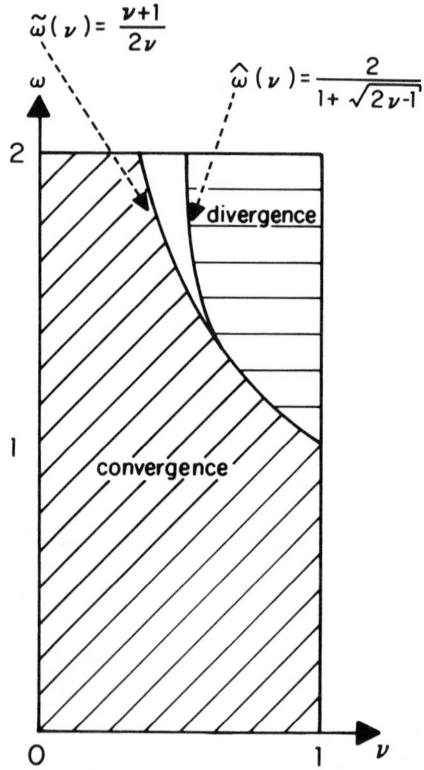

FIGURE 1.

IV. BLOCK ITERATIVE METHODS FOR SPARSE LEAST-SQUARES PROBLEMS

There has been much recent interest by numerical analysts in methods for accurately computing the least-squares solutions of very large sparse overdetermined linear systems of equations. In geodetical network problems, for example, such overdetermined systems take the form

$$A\underline{x} \doteq \underline{b} \, , \tag{4.1}$$

where A is a given real $m \times n$ matrix with $m \geq n$, and $\underline{b}$ is a given real vector with m components, (written $\underline{b} \in \mathbb{R}^m$). Usually, m is very much larger than n. Assuming that A has full column rank n, the least-squares solution of (4.1) is the unique vector $\underline{x}$ in $\mathbb{R}^n$ for which

$$||\underline{b} - A\underline{x}||_2 = \min_{\underline{y}\in\mathbb{R}^n} ||\underline{b} - A\underline{y}||_2 \quad (\text{where } ||\underline{u}||_2^2 := \underline{u}^*\underline{u}) \ . \tag{4.2}$$

For excellent surveys containing extensive bibliographies for such geodetic problems, see Golub and Plemmons [6], and Plemmons [18].

An equivalent formulation of the above least-squares problem is the following: determine (unique) vectors $\underline{x} \in \mathbb{R}^m$ and $\underline{r} \in \mathbb{R}^m$ such that

$$\underline{r} + A\underline{x} = \underline{b}, \quad \text{and} \quad A^T\underline{r} = \underline{0} \ . \tag{4.3}$$

Since A has full column rank n, we may assume, without loss of generality, that the rows of A have been permuted so that A has the block-partitioned form

$$A = \begin{bmatrix} A_1 \\ \\ A_2 \end{bmatrix} , \tag{4.4}$$

where A_1, an element of $\mathbb{R}^{n,n}$, is <u>nonsingular</u>. Partitioning the m-vectors $\underline{r}$ and $\underline{b}$ conformally with respect to the partitioning of A in (4.4), i.e.,

$$\underline{r} = \begin{bmatrix} \underline{v} \\ \\ \underline{w} \end{bmatrix} \quad \text{and} \quad \underline{b} = \begin{bmatrix} \underline{b}_1 \\ \\ \underline{b}_2 \end{bmatrix} , \ \text{where } \underline{v}, \underline{b}_1 \in \mathbb{R}^n ; \ \underline{w}, \underline{b}_2 \in \mathbb{R}^{m-n} , \tag{4.5}$$

the system of equations (4.3) can be equivalently expressed as the following system of m+n linear equations in m+n unknowns:

$$C\underline{z} = \underline{d} , \tag{4.6}$$

where

$$C := \begin{bmatrix} A_1 & 0 & I \\ A_2 & I & 0 \\ 0 & A_2^T & A_1^T \end{bmatrix} ; \quad \underline{z} := \begin{bmatrix} \underline{x} \\ \underline{w} \\ \underline{v} \end{bmatrix} ; \tag{4.7}$$

$$\underline{d} := \begin{bmatrix} \underline{b}_1 \\ \underline{b}_2 \\ \underline{0} \end{bmatrix} .$$

Because A_1 is nonsingular, the form of the matrix C in (4.7) shows that C is also nonsingular, from which it follows that the vector $\underline{z}$ in (4.6) is uniquely determined.

Our interest in the reformulation (4.6) of the least-squares problem (4.2) stems from the fact that the block-SOR iterative method has been recently suggested as a practical means of solving (4.6). The purpose in this section is to review the new theoretical results concerning such block-SOR applications.

To define this iterative method, set $D := \text{diag}(C) = \text{diag}(A_1, I, A_1^T)$, so that D is a nonsingular block-diagonal matrix. The associated block-Jacobi matrix J for the matrix C of (4.7) is then given, as before, by

$$J := I - D^{-1}C = \begin{bmatrix} 0 & 0 & -A_1^{-1} \\ -A_2 & 0 & 0 \\ 0 & -A_1^{-T}A_2^T & 0 \end{bmatrix} =: \begin{bmatrix} 0 & 0 & B_1 \\ B_2 & 0 & 0 \\ 0 & B_3 & 0 \end{bmatrix} . \tag{4.8}$$

Next, on writing the block-Jacobi matrix J of (4.8) as the sum $J = L + U$ where

$$L := \begin{bmatrix} 0 & 0 & 0 \\ B_2 & 0 & 0 \\ 0 & B_3 & 0 \end{bmatrix} ; \quad U := \begin{bmatrix} 0 & 0 & B_1 \\ 0 & 0 & 0 \\ 0 & 0 & 0 \end{bmatrix} , \tag{4.9}$$

then the block-SOR iterative method, applied to (4.6), can be expressed as

$$\underline{z}^{(m+1)} = \mathcal{L}_\omega \underline{z}^{(m)} + (I - \omega L)^{-1} D^{-1} \underline{d}, \quad (m = 0, 1, \ldots), \tag{4.10}$$

where the block-SOR iterative matrix $\mathcal{L}_\omega$ is defined as usual by

$$\mathcal{L}_\omega := (I - \omega L)^{-1}\{(1-\omega)I + \omega U\}. \tag{4.11}$$

For the convergence properties of the block-SOR iterative method (4.10), it is essential to observe, as in Chen [3] and

Plemmons [18], that the block-Jacobi matrix J of (4.8) is a *consistently ordered weakly-cyclic of index* 3 *matrix* (cf. [22],[27]). Moreover, from (4.8), it directly follows that

$$J^3 = \operatorname{diag}(-A_1^{-1}P^TPA_1; -PP^T; -P^TP) \quad (\text{where } P := A_2A_1^{-1})$$
$$= \operatorname{diag}(B_1B_3B_2;\ B_2B_1B_3;\ B_3B_2B_1)\ , \tag{4.12}$$

so that J^3 is similar to a real symmetric *negative semidefinite* matrix. As such, the eigenvalues of J^3 lie in the real interval

$$I_- := [-\rho^3(J), 0]\ , \tag{4.13}$$

where $\rho(J)$ denotes the spectral radius of J. Because J is a consistently ordered weakly-cyclic of index 3 matrix, it is known (cf. Varga [22, Theorem 4.3]; [20, Thm. 4]) that the functional equation, coupling the eigenvalues λ of L_ω of (4.11) with the eigenvalues μ of J of (4.8), is given by

$$(\lambda + \omega - 1)^3 = \lambda^2\omega^3\mu^3\ . \tag{4.14}$$

Using (4.14), the following precise domains for convergence and divergence of the block-SOR iterative, applied to (4.6), have been determined.

Theorem 4.1 ([15]). *The block-SOR iterative method of* (4.10), *applied to the least-squares matrix equation of* (4.6), *converges for*

$$0 < \omega < \omega_1(\beta) := \frac{2}{1+\beta},\ \text{when } 0 \le \beta \le 2,\ (\beta := \rho(J)), \tag{4.15}$$

converges for

$$\omega_2(\beta) := \frac{\beta-2}{\beta-1} < \omega < \omega_1(\beta),\ \text{when } 2 \le \beta < 3\ , \tag{4.16}$$

and diverges for all other real values of ω. *The optimal relaxation factor* $\omega_b = \omega_b(\beta)$ *is the unique positive root of*

$$4\beta^3\omega^3 + 27\omega - 27\omega = 0 \quad (0 \le \beta < 3)\ , \tag{4.17}$$

and ω_b *satisfies*

$$\frac{1}{2} < \omega_b \le 1 \quad \text{for all} \quad 0 \le \beta < 3\ . \tag{4.18}$$

Further, there holds

$$\rho(\mathcal{L}_{\omega_b}) = 2(1-\omega_b) \quad \text{for all} \quad 0 \leq \beta < 3 . \tag{4.19}$$

It is clear from Theorem 4.1 that one can find values of ω for which the block-SOR iteration matrix $\mathcal{L}_\omega$ is convergent, even when the block-Jacobi matrix J is divergent; for example, for each $\beta = \rho(J)$ satisfying $1 \leq \beta < 3$, there are intervals in ω (cf. (4.15) and (4.16)) for which $\rho(\mathcal{L}_\omega) < 1$. In this respect, Theorem 4.1 extends what is known theoretically in the literature for such block-SOR applications. More important, however, is the fact that Theorem 4.1 greatly increases the applicability of the block-SOR iterative method to least-squares problems which arise, as was previously mentioned, in geodetical network problems.

It is also important to note that Theorem 4.1 corrects results in the literature for such least-squares applications. Under the assumptions of Theorem 4.1, one finds in Plemmons [18, p. 166] the statement that, for any $\rho(J) < 1$, the associated block-SOR iterative method converges for all ω satisfying

$$0 < \omega < \frac{3}{2} \qquad (0 \leq \rho(J) < 1) , \tag{4.20}$$

whereas from (4.15) of Theorem 4.1, the correct statement is that the block-SOR iterative method converges in the subset of (4.20) consisting of all ω satisfying

$$0 < \omega < \frac{2}{1+\rho(J)} \qquad (0 \leq \rho(J) < 1) . \tag{4.21}$$

The same error occurs in Berman and Plemmons [2, p. 179].

As a counterpart of Theorem 4.1, the following new results on the convergence of the block-SOR iterative method have been determined when the eigenvalues of J^3 are nonnegative, i.e., the eigenvalues of J^3 lie in the interval

$$I_+ := [0, \rho^3(J)] . \tag{4.22}$$

Theorem 4.2 ([15]). *Let the block-Jacobi matrix* J *be a consistently ordered weakly-cyclic of index* 3 *matrix, such that the eigenvalues of* J^3 *are real and nonnegative. Then, the associated block-SOR iterative method converges for*

$$0 < \omega < \omega_3(\beta) := \frac{\beta+2}{\beta+1}, \quad \underline{\text{when}} \quad 0 \le \beta < 1 \ (\beta := \rho(J)), \tag{4.23}$$

and diverges for all other real values of ω. *The optimal relaxation factor* $\omega_b = \omega_b(\beta)$ *is the smallest positive root of*

$$4\beta^3\omega^3 - 27\omega + 27 = 0 \quad (0 \le \beta < 1), \tag{4.24}$$

and ω_b *satisfies*

$$1 \le \omega_b < \frac{3}{2} \quad \underline{\text{for all}} \quad 0 \le \beta < 1. \tag{4.25}$$

Further, there holds

$$\rho(L_{\omega_b}) = 2(\omega_b - 1). \tag{4.26}$$

Interestingly enough, while the proof of (4.24)-(4.26) is given in Varga [20] as the special case p = 3, the precise upper bound $\omega_3(\beta)$ of (4.23) for convergence is *new*. Previously, it had been shown in [20], under the assumptions of Theorem 4.2, that convergence of the associated block-SOR method holds in the *subset* of (4.23) defined by

$$0 < \omega < \frac{3}{2} \quad (0 \le \beta := \rho(J) < 1). \tag{4.27}$$

The results of Theorems 4.1 and 4.2 are graphically given in Figure 2. For convenience, we have introduced the variable $\tilde{\beta}$, where $\tilde{\beta} := \rho(J)$ if the eigenvalues of J^3 are nonnegative (cf. Theorem 4.2), and where $\tilde{\beta} = -\rho(J)$ if the eigenvalues of J^3 are nonpositive (cf. Theorem 4.1). Thus, an open bounded region Ω in the $\tilde{\beta}$-ω plane is obtained such that for each point of Ω (shown as the shaded region in Figure 2), the associated block-SOR matrix with relaxation factor ω (when applied to a system for which $\rho(J) = \tilde{\beta}$ or $\rho(J) = -\tilde{\beta}$) is *convergent*, and is *divergent* for all points in the complement of Ω. Also included in Figure 2 is the set of all optimum relaxation factors $\omega_b(\tilde{\beta})$, as a function of $\tilde{\beta}$; this appears as the dotted line in Figure 2.

V. REGULAR SPLITTINGS OF MATRICES

The theory of regular splittings of matrices has been a useful *algebraic tool* in the analysis of iterative methods for solving large systems of linear equations (cf. [2],[22], and

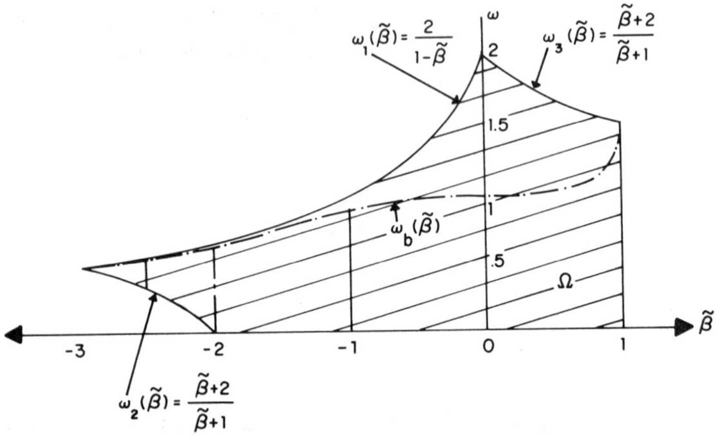

Figure 2

[27]). Our purpose in this section is to give some new comparison theorems for regular splittings of matrices, which generalize both the original results of Varga [21], and the subsequent unpublished thesis results of Woźnicki [25].

To begin, if $A \in \mathbb{C}^{n,n}$, then $A = M - N$ is said to be a regular splitting of A (cf. [21],[22]) if M is nonsingular, and if M^{-1} and N have all their entries nonnegative (written $M^{-1} \geq 0$ and $N \geq 0$). (We write $C \geq D$ and $C > D$ if $C - D \geq 0$ and if $C - D > 0$, respectively.) The following results of Theorem 5.1 and 5.2 are well-known.

Theorem 5.1 ([21]). *Let* $A = M - N$ *be a regular splitting of* A. *If* $A^{-1} \geq 0$, *then*

$$\rho(M^{-1}N) = \frac{\rho(A^{-1}N)}{1 + \rho(A^{-1}N)} < 1. \tag{5.1}$$

Conversely, if $\rho(M^{-1}N) < 1$, *then* $A^{-1} \geq 0$.

Theorem 5.2 ([21]). *Let* $A = M_1 - N_1 = M_2 - N_2$ *be two regular splittings of* A, *where* $A^{-1} \geq 0$. *If* $N_2 \geq N_1$, *then*

$$\rho(M_2^{-1}N_2) \geq \rho(M_1^{-1}N_1) . \tag{5.2}$$

In particular, if $N_2 \geq N_1$ with $N_2 \neq N_1$, *and if* $A^{-1} > 0$, then

$$\rho(M_2^{-1}N_2) > \rho(M_1^{-1}N_1) \quad . \tag{5.3}$$

Suppose we are given the system of n linear equations in n unknowns:

$$A\underline{x} = \underline{k} \ , \tag{5.4}$$

where $A \in \mathbb{C}^{n,n}$ is a given matrix, and where $\underline{x}$ and $\underline{k}$ are in $\mathbb{C}^n$, with $\underline{k}$ being given. If $A = M_1 - N_1$ is a regular splitting of A, we can write (5.4) as

$$\underline{x} = M_1^{-1}N_1\underline{x} + M_1^{-1}\underline{k} \ , \tag{5.5}$$

which induces the following iterative method:

$$\underline{x}^{(m+1)} = M_1^{-1}N_1\underline{x}^{(m)} + M_1^{-1}\underline{k}, \quad (m = 0,1,\ldots) \ , \tag{5.6}$$

where $\underline{x}^{(0)}$ is an arbitrary vector in $\mathbb{C}^n$. When $A^{-1} \geq 0$, the iterative method of (5.6) is necessarily <u>convergent</u>, by Theorem 5.1, to the unique solution to (5.4), independent of the choice of $\underline{x}^{(0)}$. If $A = M_2 - N_2$ is another regular splitting of A, one similarly obtains the convergent iterative method

$$\underline{x}^{(m+1)} = M_2^{-1}N_2\underline{x}^{(m)} + M_2^{-1}\underline{k} \ , \quad (m = 0,1,\ldots) \ . \tag{5.7}$$

If $N_2 \geq N_1$ with $N_2 \neq N_1$ and if $A^{-1} > 0$, the iterative method of (5.6) is, from Theorem 5.2, <u>asymptotically faster</u> than the iterative method of (5.7).

From a practical point of view, it is of interest to know other techniques for determining which of two regular splittings of a matrix induces the asymptotically faster iterative method. One such technique, which is less well-known but still very useful, is the following thesis result of Woźnicki [25].

Theorem 5.3 ([25]). *Let* $A = M_1 - N_1 = M_2 - N_2$ *be two regular splittings of* A, *where* $A^{-1} \geq 0$. *If* $M_1^{-1} \geq M_2^{-1}$, *then*

$$\rho(M_2^{-1}N_2) \geq \rho(M_1^{-1}N_1) \quad . \tag{5.8}$$

In particular, if $M_1^{-1} > M_2^{-1}$ *and if* $A^{-1} > 0$, *then*

$$\rho(M_2^{-1}N_2) > \rho(M_1^{-1}N_1) \quad . \tag{5.9}$$

As we shall see (cf. i) of Proposition 5.4), Woźnicki's Theorem 5.3 is more general than Theorem 5.2, and it is natural to ask if similar, more general results, based on the comparison of two regular splittings, can be deduced. To motivate our subsequent discussion, we state the result of

Proposition 5.4 ([4]). *Let* $A = M_1 - N_1 = M_2 - N_2$ *be two regular splittings of* A, *where* $A^{-1} \geq 0$. *Then,*

i) $N_2 \geq N_1$ *implies that* $M_1^{-1} \geq M_2^{-1}$;

ii) $M_1^{-1} \geq M_2^{-1}$ *implies that* $A^{-1}N_2A^{-1} \geq A^{-1}N_1A^{-1}$;

iii) *for each positive integer* j, $A^{-1}N_2A^{-1} \geq A^{-1}N_1A^{-1}$ *implies that* $(A^{-1}N_2)^jA^{-1} \geq (A^{-1}N_1)^jA^{-1}$.

Obviously, Proposition 5.4 gives progressively *weaker* hypotheses. As our new generalization of the first parts of Theorems 5.2 and 5.3, we give

Theorem 5.5 ([4]). *Let* $A = M_1 - N_1 = M_2 - N_2$ *be two regular splittings of* A, where $A^{-1} \geq 0$. *Assume that there exists a positive integer* j *for which*

$$(A^{-1}N_2)^jA^{-1} \geq (A^{-1}N_1)^jA^{-1} \; . \tag{5.10}$$

Then,

$$1 > \rho(M_2^{-1}N_2) \geq \rho(M_1^{-1}N_1) \; . \tag{5.11}$$

Continuing, let $A = M_1 - N_1 = M_2 - N_2$ be two regular splittings of A, where $A^{-1} \geq 0$. On setting

$$S := \{\text{positive integers } j: (A^{-1}N_2)^jA^{-1} \geq (A^{-1}N_1)^jA^{-1}\} \; , \tag{5.12}$$

we note that S not empty implies that Theorem 5.5 can be applied. It is easy to show that S is closed under addition, so that if 1 is in S, then S consists of all positive integers. Similarly, if S contains two relatively prime integers, then S necessarily contains all sufficiently large positive integers. Matrices A and two regular splittings can, however, be constructed (cf. [4]) such that the *smallest* positive integer in the associated set S can be made to be any *preassigned* positive integer!

For results which provide partial converses to Theorem 5.5, as suggested by the second parts of Theorems 5.2 and 5.3, we state the new result of

Theorem 5.6 ([4]). *Let* $A = M_1 - N_1 = M_2 - N_2$ *be two regular splittings of* A, where $A^{-1} > 0$. *If*

$$\rho(M_2^{-1}N_2) > \rho(M_1^{-1}N_1) , \tag{5.13}$$

there exists a positive integer j_o *for which*

$$(A^{-1}N_2)^j A^{-1} > (A^{-1}N_1)^j A^{-1}, \quad \text{for all} \quad j \geq j_o . \tag{5.14}$$

Thus, (5.13) *implies that the set* S *of* (5.12) *contains all sufficiently large positive integers. Conversely, if there is an integer* j *for which*

$$(A^{-1}N_2)^j A^{-1} > (A^{-1}N_1)^j A^{-1} , \tag{5.15}$$

then (5.13) *is valid*.

We finally remark that the case j = 1 of (5.15) generalizes the second parts of *both* Theorems 5.2 and 5.3, as shown in [4].

REFERENCES

1. G. Alefeld and R.S. Varga, "Zur Konvergenz des symmetrischen Relaxationsverfahrens," Numer. Math. 25 (1976), 291-295.

2. A. Berman and R.J. Plemmons, *Nonnegative Matrices in the Mathematical Sciences*, Academic Press, Inc., New York, 1979.

3. Y.T. Chen, "Iterative methods for linear least squares problems," Ph.D. Dissertation, Dept. of Computer Science, University of Waterloo, Waterloo, Ontario, Canada, 1975.

4. G. Csordas and R.S. Varga, "Comparisons of regular splittings of matrices," Numer. Math. (to appear).

5. M. Eiermann and W. Niethammer, "On the construction of semiiterative methods," SIAM J. Numer. Anal. (to appear).

6. G.H. Golub and R.J. Plemmons, "Large-scale geodetic least-squares adjustment by dissection and orthogonal decomposition," Linear Algebra and Appl. 34 (1980), 3-27.

7. W. Kahan, "Gauss-Seidel methods for solving large systems of linear equations," Doctoral Dissertation, University of Toronto, 1958.

8. U. Kulisch, "Über reguläre Zerlegungen von Matrizen und einige Anwendungen," Numer. Math. 11 (1968), 444-449.

9. M.S. Lynn, "On the equations of SOR, SSOR, and USSOR as applied to σ_1-ordered systems of linear equations," Comp. J. 7 (1964), 72-75.

10. A. Neumaier and R.S. Varga, "New upper bounds and convergence domains for the SSOR iterative method," (to appear).

11. M. Neumann, "On bounds for the convergence of the SSOR method for H-matrices," (to appear).

12. M. Neumann and R.S. Varga, "On the sharpness of some upper bounds for the spectral radii of S.O.R. iteration matrices," Numer. Math. 35 (1980), 425-433.

13. N.K. Nickels and L. Fox, "Generalized consistent ordering and optimum successive overrelaxation factor," Numer. Math. 13 (1969), 425-433.

14. W. Niethammer, "Relaxation bei komplexen Matrizen," Math. Zeitscrhift 86 (1964), 34-40.

15. W. Niethammer, J. de Pillis, and R.S. Varga, "Block iterative methods applied to sparse least-squares problems," Linear Algebra and Appl. (to appear).

16. W. Niethammer and R.S. Varga, "The analysis of k-step iterative methods for linear systems from summability theory," Numer. Math. 41 (1983), 177-206.

17. A.M. Ostrowski, "Über die Determinanten mit überwiegender Hauptdiagonale," Comment. Math. Helv. 10 (1937), 69-96.

18. R.J. Plemmons, "Adjustments by least squares in geodesy using block iterative methods for sparse matrices," Proceedings of the Annual U.S. Army Conference on Numerical Analysis and Computers, 1979, pp. 151-186.

19. E. D'Sylva and G.A. Miles, "The S.S.O.R. iterative scheme for equations with σ_1-ordering," Comp. J. 6 (1964), 366-367.

20. R.S. Varga, "p-cyclic matrices: a generalization of the Young-Frankel successive overrelaxation scheme," Pacific J. Math. 9 (1959), 617-628.

21. "Factorizations and normalized iterative methods," Boundary Problems in Differential Equations (R.E. Langer, ed.), pp. 121-142, University of Wisconsin Press, Madison, 1960.

22. ________, Matrix Iterative Analysis, Prentice-Hall, Inc., Englewood Cliffs, N.J., 1962.

23. "On recurring theorems on diagonal dominance," Linear Algebra and Appl. 13 (1976), 1-9.

24. R.S. Varga, W. Niethammer, and D.-Y, Cai, "p-Cyclic matrices and the symmetric successive overrelaxation method," Linear Algebra and Appl. (to appear).

25. Z. Woźnicki, "Two-sweep iterative methods for solving large linear systems and their applications to the numerical solution of multi-group multi-dimensional neutron diffusion equation," Doctoral Dissertation, Institute of Nuclear Research, Świerk k/Otwocka, Poland, 1973.

26. David M. Young, "Iterative methods for solving partial difference equations of elliptic type," Trans. Amer. Math. Soc. 76 (1954), 92-111.

27. _______, Iterative Solution of Large Linear Systems, Academic Press, New York, 1971.

NUMERICAL ALGORITHMS FOR INDEFINITE PROBLEMS

O. Axelsson*

I. INTRODUCTION

Mixed variable variational formulations for partial differential equations may arise by introducing a new dependent variable, typically a derivative, together with the original variable, to form a coupled system of partial differential equations of lower order than the equation for the primal variable. We illustrate this by an example,

Example 1. The diffusion problem. Here

$$-\nabla\cdot(\nu\nabla\hat{v}) = f \quad \text{in } \Omega, \quad \hat{v} = 0 \quad \text{on } \Gamma_D, \quad \nu\nabla\hat{v}\cdot\hat{n} = 0 \quad \text{on } \Gamma_N ,$$

where $\Gamma_N = \Gamma/\Gamma_D$, $\nu \geq \nu_0 > 0 \ \forall x \in \overline{\Omega}$ and $\Omega \in R^n$. (The diffusion tensor ν is a positive definite matrix.) Following [10], we let $\hat{u} = -\nu\nabla\hat{v}$ to get

$$\begin{cases} \nu^{-1}\hat{u} + \nabla\hat{v} = 0 \\ \nabla\cdot\hat{u} = f \end{cases}$$

with both $\hat{v} = 0$ on Γ_D and $\nu\hat{u}\cdot\hat{n} = 0$ on Γ_N esential boundary conditions. We may let $\hat{v} \in H^1(\Omega)$ and $\hat{u} \in (L^2(\Omega))^n$, when we use the variational formulation

$$\begin{cases} \int_\Omega \nu^{-1}\hat{u}\cdot u + \int_\Omega \nabla\hat{v}\cdot u = 0 \ \forall u \in (L^2(\Omega))^n, \ \nu u\cdot\hat{n} = 0 \quad \text{on} \quad \Gamma_N \\ \int_\Omega \hat{u}\cdot\nabla v = -\int_\Omega fv \ \ \forall \ v \in H^1(\Omega), \quad v = 0 \quad \text{on} \quad \Gamma_D. \end{cases}$$

*Department of Mathematics, Catholic University Nijmegen, The Netherlands

ISBN 0-12-100560-7

In the following constrained problem, the normal formulation is of mixed variable form.

Example 2. Stokes problem. We have

$$\begin{cases} -\nabla\cdot(\nu\nabla\hat{u}) + \nabla\hat{v} = f & \text{in } \Omega \\ \nabla\cdot\hat{u} = 0 & \text{in } \Omega \\ |\hat{u}| = 0 \text{ on } \Gamma_D, \quad \hat{u}\cdot\hat{n} = 0 & \text{on } \Gamma/\Gamma_D . \end{cases}$$

Its variational formulation is

$$\begin{cases} \int_\Omega \nu\nabla\hat{u}\cdot\nabla u - \int_\Omega \hat{v}\nabla\cdot u = \int_\Omega f\cdot u & \forall\ u \in U \\ \int_\Omega \nabla\cdot\hat{u}v = 0 & \forall\ v \in L^2(\Omega) , \end{cases}$$

where $U = \{u \in (H^1(\Omega))^n;\ |u| = 0 \text{ on } \Gamma_D,\ \hat{u}\cdot\hat{n} = 0 \text{ on } \Gamma/\Gamma_D\}$. Here $\hat{v}$, the pressure variable, acts as a Lagrangian multiplier and $\hat{u}$ is the velocity field. Note that in this method the finite element approximating functions do not have to satisfy the constraint $\nabla\cdot\hat{u} = 0$ exactly, which makes the choice of finite elements easier.

In general, mixed formulations lead to an operator of the form

$$\begin{bmatrix} M & B^* \\ B & 0 \end{bmatrix} \begin{bmatrix} \hat{u} \\ \hat{v} \end{bmatrix} = \begin{bmatrix} f_1 \\ f_2 \end{bmatrix} \tag{1.1}$$

which are the stationary equations for the constrained optimization problem

$$\inf_u \sup_v \{\tfrac{1}{2} u^*Mu - f_1^*u + v^*(Bu - f_2)\} ,$$

where v acts as a Lagrangian multiplier for the constraint $Bu = f_2$. In the following we shall assume that

(i) $M + \frac{1}{\varepsilon}B^*B$ is self adjoint and positive definite for $0 < \varepsilon \le \varepsilon_0$, where ε_0 is small enough.

This implies that $\hat{u}$ has a unique solution (see e.g., [5]). Note that, since in particular, (i) implies that M is positive definite on N(B), $\hat{v}$ is unique only modulo N(B*). In Example

1, $N(B^*) = \{0\}$ is the trivial null space if $\text{meas}(\Gamma_D) \neq 0$. If $\Gamma_N = \Gamma$, then as well as in Example 2, $N(B^*) = \{\text{constant functions}\}$. (We assume that the source function then satisfies the consistency relation $\int f\, d\Omega = 1$.) In this case the corresponding finite element matrix B will be rank deficient, which we may correct by simply deleting one of the rows in B (and the corresponding column of B^t). Equivalently, we may pick a value of the function v at any nodal point. Alternatively, we may consider a finite element subspace for the variable v such that its entries satisfy $\int_\Omega v \cdot d\Omega = 0$. In all cases, the solution on the corresponding subspaces is unique. Note however that in §3 we shall present a method (Method 2) for which a possible rank deficiency of B does not matter when we are mainly interested in the other variable $\hat{u}$.

As is well known, the operator in (1.1) is indefinite. The purpose of this paper is to discuss various methods for the reduction of (1.1) to a positive definite form, on which fast elliptic solvers, such as some multigrid method, or a preconditioned conjugate gradient method, may be applied (for details, see e.g., [1], [7]).

II. ITERATION METHODS FOR INDEFINITE MATRICES

A. *Indefinite Matrix Incomplete Factorization.*

We consider a discretized form of (1.1) by a finite element method. Let

$$A = \begin{bmatrix} M & B^t \\ B & 0 \end{bmatrix}$$

be the resulting matrix and let

$$C = \begin{bmatrix} L & 0 \\ BL^{-t} & U^t \end{bmatrix} \begin{bmatrix} L^t & L^{-1}B^t \\ 0 & -U \end{bmatrix} = \begin{bmatrix} LL^t & B^t \\ B & B(LL^t)^{-1}B^t - U^tU \end{bmatrix}$$

where L and U are lower and upper triangular, respectively, be a block matrix incomplete factorization of A. Here $M_0 = LL^t$ and U^tU are modified incomplete factorizations of M and of $BM_0^{-1}B^t$, respectively. Note that in the case M_0 is diagonal, then $B(LL^t)^{-1}B^t = BM_0^{-1}B^t$ may be calculated explicitly cheaply.

The iterative method for the solution of (1.1) takes the form

$$C\begin{bmatrix} u^{\ell+1}-u^{\ell} \\ v^{\ell+1}-v^{\ell} \end{bmatrix} = -\tau_{\ell}\left\{A\begin{bmatrix} u^{\ell} \\ v^{\ell} \end{bmatrix} - \begin{bmatrix} f_1 \\ f_2 \end{bmatrix}\right\}, \tag{2.1}$$

where $\{\tau_{\ell}\}$, $\ell = 0,1,\dots$ is a sequence of acceleration parameters, such as in a Chebyshev semiiterative method. A similar form arises for a conjugate gradient acceleration method (see, e.g., [1], [2], [7] and [12]). Note that at every step in (2.1) we solve first

$$\begin{bmatrix} L & 0 \\ BL^{-t} & U^t \end{bmatrix}\begin{bmatrix} \alpha^{\ell+1} \\ \beta^{\ell+1} \end{bmatrix} = \text{r.h.s. of } (2.1)$$

by forward substitution and then

$$\begin{bmatrix} L^t & L^{-1}B^t \\ 0 & -U \end{bmatrix}\begin{bmatrix} u^{\ell+1}-u^{\ell} \\ v^{\ell+1}-v^{\ell} \end{bmatrix} = \begin{bmatrix} \alpha^{\ell+1} \\ \beta^{\ell+1} \end{bmatrix}$$

by backward substitution. In these multiplication by L^{-t} and L^{-1} means solving systems with triangular matrices L^t and L, respectively, which is cheap since in our applications all matrices are also sparse.

For the aforementioned acceleration methods the rate of convergence is determined by the spectrum $\{\lambda_i\}$ of $C^{-1}A$. In particular, if the spectrum is real and nonnegative and if the quotient between the extreme positive eigenvalues is $H = \lambda_N/\lambda_1$, where $0 = \lambda_0 < \lambda_1 \le \lambda_2 \le \dots \le \lambda_N$, then an upper bound for the number of iterations needed is proportional to $H^{1/2}$. In order to find H we let $D_1 = M-M_0$, $D_2 = U^tU-BM_0^{-1}B^t$ be the defect matrices of the incomplete factorizations and consider the generalized eigenvalue problem

$$\lambda C\begin{bmatrix} x \\ y \end{bmatrix} = A\begin{bmatrix} x \\ y \end{bmatrix}, \tag{2.2}$$

which we write in the form

$$(\lambda-1)\begin{bmatrix} M & B^t \\ B & 0 \end{bmatrix}\begin{bmatrix} x \\ y \end{bmatrix} = \lambda\begin{bmatrix} D_1 & 0 \\ 0 & D_2 \end{bmatrix}\begin{bmatrix} x \\ y \end{bmatrix}$$

We note at first that by condition (i),

$$\lambda = 0 \leftrightarrow x = 0, \ y \in N(B^t).$$

Hence consider $y \in N(B^t)^{\perp}$. Then

$$(1 - \frac{1}{\lambda}) \begin{bmatrix} M & B^t \\ B & 0 \end{bmatrix} \begin{bmatrix} x \\ y \end{bmatrix} = \begin{bmatrix} D_1 & 0 \\ 0 & D_2 \end{bmatrix} \begin{bmatrix} x \\ y \end{bmatrix} .$$

If D_1, D_2 have the same signs (i.e., both are positive semidefinite or negative semidefinite) then the eigensolutions are real. If however, D_1, D_2 have opposite signs or if one of them is indefinite, then the eigensolutions of (2.2) may be complex. In that latter case, the estimates of the rate of convergence of (2.1) are not that promising. Unfortunately, in many applications of incomplete factorizations, D_2 may be indefinite. Hence, although this method has been applied and worked well for bi-harmonic problems [4], we don't always feel safe with it. Therefore, we shall look at the following alternative.

B. Coupled Inner-Outer Iteration Method.

For the outer iterations we use the iterative method

$$A_0 \begin{bmatrix} \gamma^{\ell} \\ \delta^{\ell} \end{bmatrix} = - \tau_{\ell} \begin{bmatrix} r^{\ell} \\ s^{\ell} \end{bmatrix} , \quad \ell = 0,1,... \tag{2.3}$$

where $\{\tau_{\ell}\}$ is a sequence of iteration parameters; the corrections are $\gamma^{\ell} = u^{\ell+1} - u^{\ell}$, $\delta^{\ell} = v^{\ell+1} - v^{\ell}$; the residuals satisfy

$$\begin{bmatrix} r^{\ell} \\ s^{\ell} \end{bmatrix} = A \begin{bmatrix} u^{\ell} \\ v^{\ell} \end{bmatrix} - \begin{bmatrix} f_1 \\ f_2 \end{bmatrix}$$

and where $A_0 = \begin{bmatrix} D & B^t \\ B & 0 \end{bmatrix}$. If we solve for the corrections exactly one finds that the rate of convergence of (2.3) is determined by the condition number $\{\tilde{\lambda}_N/\tilde{\lambda}_1\}^{1/2}$, where $\tilde{\lambda}_N$, $\tilde{\lambda}_1$ are the extreme eigenvalues of the set $1 \cup \{\lambda_i\}$ and λ_i are the eigenvalues of $Mx = \lambda Dx$. Note that in our applications both M and D are positive definite which means that the eigenvalues are positive.

In some applications such as for Example 1 where M is a mass-matrix, it may be possible to choose D as a simple matrix, sometimes even diagonal, but still have the spectral condition number λ_N/λ_1 not be large. In other applications it may not be possible to find such a matrix D, however.

In practice we do not solve exactly for the corrections in (2.3) and thus we have to consider a perturbed eigenvalue problem, for which the eigenvalues may be slightly complex. Then the acceleration parameters $\{\tau_\ell\}$ may be chosen accordingly (see e.g., [3] for details). For the calculation of the corrections from (2.3) we consider using (inner) iterations,

$$\begin{bmatrix} D & 0 \\ B & U^t \end{bmatrix} \begin{bmatrix} I & D^{-1}B^t \\ 0 & -U \end{bmatrix} \begin{bmatrix} \gamma^{\ell,j+1}-\gamma^{\ell,j} \\ \delta^{\ell,j+1}-\delta^{\ell,j} \end{bmatrix} =$$

$$- \rho_\ell \left\{ A_0 \begin{bmatrix} \gamma^{\ell,j} \\ \gamma^{\ell,j} \end{bmatrix} + \tau_\ell \begin{bmatrix} r^\ell \\ s^\ell \end{bmatrix} \right\}, \quad j = 0,1,2,\ldots,$$

$$\begin{bmatrix} \gamma^{\ell,0} \\ \delta^{\ell,0} \end{bmatrix} = \begin{bmatrix} 0 \\ 0 \end{bmatrix} .$$

where the $\{\rho_\ell\}$ are iteration parameters.

One finds (see [4]) that the rate of convergence of this method is determined by the eigenvalues of the generalized eigenvalue problem

$$\mu \begin{bmatrix} D & B^t \\ B & R \end{bmatrix} \begin{bmatrix} x \\ y \end{bmatrix} = \begin{bmatrix} D & B^t \\ B & 0 \end{bmatrix} \begin{bmatrix} x \\ y \end{bmatrix}$$

where $R = BD^{-1}B^t - U^tU$, for which the spectrum $\{1 \cup \{\mu_i\}\}$ is again real and positive, because μ_i satisfies $BD^{-1}B^ty = \mu_i U^tUy$. Here U^tU is an incomplete factorization of $BD^{-1}B^t$.

Hence the analysis of this method has been reduced to that of incomplete factorizations for definite matrix problems.

The total work of the above coupled outer-inner iteration method is proportional to $\{\lambda_N/\lambda_1 \cdot \mu_N/\mu_1\}^{1/2}$. Although this method is safer than the previous, it may hence be costly and furthermore, it may not always be possible to find a proper matrix D. We shall now consider two methods where the indefinite problem is reduced to positive definite ones in a more direct way and where the above mentioned difficulties do not arise.

III. DIRECT REDUCTIONS TO POSITIVE DEFINITE PROBLEMS

In this section, we analyze an iterative algorithm based on the calculation of the Lagrange multiplier (or update) $\{v^{(\ell)}\}$ alone.

We shall consider two methods for (1.1), i.e.,

$$\begin{bmatrix} M & B^* \\ B & 0 \end{bmatrix} \begin{bmatrix} \hat{u} \\ \hat{v} \end{bmatrix} = \begin{bmatrix} f_1 \\ f_2 \end{bmatrix} \tag{3.1}$$

<u>Method 1</u>. (M is positive definite). Since in particular, by the assumption for this method, M is nonsingular, we may eliminate u from (3.1) to get

$$\hat{u} = M^{-1}(-B^*\hat{v} + f_1) \tag{3.2}$$

and

$$BM^{-1}B^*\hat{v} = F = BM^{-1}f_1 - f_2 \tag{3.3}$$

Here $BM^{-1}B^*$ is positive semidefinite and (3.3) may be solved by a preconditioned conjugate gradient method. However, due to the rectangular form of B when discretized, it may not be obvious how to construct such a preconditioning. In some problems, such as Example 2 (Stokes problem), a preconditioning may however not be needed, because the condition number of $BM^{-1}B^*$ will not be large anyway. Heuristically this is seen because B, B* are the discretizations of the divergence and gradient operators, whereas M is the operator (- div grad). The condition number does depend on the exact type of boundary conditions however.

Note that in this method we need to solve several problems with operator M, namely for the calculation of $M^{-1}f_1$, for the calculation of $\hat{u}$ from (3.2) when $\hat{v}$ has been found and, most important, once at every (outer) iteration step, when we solve (3.3) by iteration. To solve these systems with M it may then pay off to calculate a good incomplete factorization of M and then solve them by (inner) iterations using a preconditioned iterative method.

For a bi-harmonic problem, a special case occurs because we may precondition $BM^{-1}B^*$ by the product of square matrices (see [11]).

We consider now an alternative method, where the task of constructing a preconditioner is made easier and where the number of outer iterations is made smaller.

<u>Method 2</u>. In case a) M is positive definite, in case b), M may be indefinite but then BB^* is assumed to be regular.

Let $0 < \varepsilon \leq \varepsilon_0$ (condition (i)). Then it follows from (3.1) that

$$\begin{cases} (M + \frac{1}{\varepsilon}B^*B)\hat{u} + B^*\hat{v} = f_1 + \frac{1}{\varepsilon}B^* f_2 \\ B\hat{u} = f_2 \end{cases} \tag{3.4}$$

which implies

$$\hat{u} = (\varepsilon M + B^*B)^{-1}[-\varepsilon B^*\hat{v} + \varepsilon f_1 + B^*f_2] , \tag{3.5}$$

and by $B\hat{u} = f_2$, that

$$B\hat{M}^{-1}B^*\hat{v} = B\hat{M}^{-1}(f_1 + \frac{1}{\varepsilon}B^* f_2) - \frac{1}{\varepsilon} f_2 \tag{3.6}$$

where

$$\hat{M} = \varepsilon M + B^*B . \tag{3.7}$$

We consider now the perturbed (or regularized) problem

$$\begin{bmatrix} M & B^* \\ B & -\varepsilon I \end{bmatrix} \begin{bmatrix} \hat{u}_\varepsilon \\ \hat{v}_\varepsilon \end{bmatrix} = \begin{bmatrix} f_1 \\ f_2 \end{bmatrix} .$$

Then we get

$$\begin{cases} (\varepsilon M + B^*B)\hat{u}_\varepsilon = B^*f_2 + \varepsilon f_1 \\ \hat{v}_\varepsilon = \frac{1}{\varepsilon}(B\hat{u}_\varepsilon - f_2) \end{cases}$$

or

$$\hat{v}_\varepsilon = B\hat{M}^{-1}[B^*(\frac{1}{\varepsilon} f_2) + f_1] - \frac{1}{\varepsilon} f_2 \tag{3.8}$$

Then (3.6) takes the form

$$B\hat{M}^{-1}B^*\hat{v} = \hat{v}_\varepsilon \tag{3.9}$$

which we solve by iteration:

$$\hat{v}^{(\ell+1)} = [I - B\hat{M}^{-1}B^*]\hat{v}^{(\ell)} + \hat{v}_\varepsilon, \quad \ell = 0,1,\ldots$$

or

$$\hat{v}^{(\ell+1)}-\hat{v}^{(\ell)} = [I-B\hat{M}^{-1}B^*](\hat{v}^{(\ell)}-\hat{v}^{(\ell-1)}), \; \ell = 0,1,\ldots \tag{3.10}$$

where $\hat{v}^{(-1)} = 0$, $\hat{v}^{(0)} = \hat{v}_\varepsilon$ as defined by (3.8). Naturally every multiplication by $\hat{M}^{-1}$ requires solving a linear system with operator $\hat{M}$.

Note first that if BB^* is singular (i.e., B is rank deficient in the matrix case), then $N(B^*)$ is nontrivial but every element in the sequence $\{v^{(\ell)}\}$ will have the <u>same</u> component along $N(B^*)$. Hence the rate of convergence of (3.10) is determined by spectral radius $\hat{\rho}$ of the spectrum of $I-B\hat{M}^{-1}B^*$ on $N(B^*)^\perp$, i.e.,

$$\hat{\rho} = \sup_{||x||=1, x\in N(B^*)^\perp} \{x^*(I-B\hat{M}^{-1}B^*)x\}$$

Having found $\hat{v}$ we evaluate $\hat{u}$ by use of (3.5) (or (3.2), if M is regular). Note that the (arbitrary) component of $\hat{v}$ along $N(B^*)$ does not influence $\hat{u}$.

We shall now prove that the rate of convergence of (3.10) is arbitrarily fast, because the spectral radius of the iteration matrix is $O(\varepsilon)$, i.e., arbitrarily small, if only ε is small enough.

<u>*Spectral radius of $(I-B\hat{M}^{-1}B^*)$*</u>. We first assume that M is positive definite:

<u>Lemma 3.1</u>. If M is positive definite and symmetric, then

a) $B(\varepsilon M+B^*B)^{-1} = (\varepsilon I+BM^{-1}B^*)^{-1}BM^{-1}$

b) $B(\varepsilon M+B^*B)^{-1}B^* = I - \varepsilon(\varepsilon I+BM^{-1}B^*)^{-1}$

Proof: For the proof we may use the Sherman-Morrison-Woodbury formula (see, e.g., [13]) or we may simply write

$$\begin{bmatrix} A & F \\ C & E \end{bmatrix} = \begin{bmatrix} M & B^* \\ B & -\varepsilon I \end{bmatrix}^{-1}$$

where we evaluate both F and C independently and then use $C = f^*$ to arrive at a). Part b) follows readily from a).

Corollary 3.1. The spectrum of $I-B\hat{M}^{-1}B^*$ on $N(B^*)$ is in the interval $(0,\varepsilon a)$, $0 < \varepsilon \leq \varepsilon_0$, where

$$a = \sup_{x\in N(B^*)^{\perp}} (x^*(\varepsilon I+BM^{-1}B^*)^{-1}x)/(x^*x)$$

i.e., a is bounded above by the inverse of the smallest positive eigenvalue of $BM^{-1}B^*$.

When M is indefinite, we use a different result.

Lemma 3.2. Let BB^* be regular. Then for $0 < \varepsilon \leq \varepsilon_0$, ε_0 small enough (assumption (i)),

$$B(\varepsilon M+B^*B)^{-1}B^* = I + \varepsilon E,$$

where

$$E = -(BB^*)^{-1}B[M-M(M+\frac{1}{\varepsilon}B^*B)^{-1}M]B^*(BB^*)^{-1} \, .$$

Here $||E||$ is bounded uniformly in ε.

Proof. Similar to the proof of the previous lemma, we have

$$\begin{bmatrix} M & B^* \\ B & -\varepsilon I \end{bmatrix}\begin{bmatrix} A & C^* \\ C & E \end{bmatrix} = \begin{bmatrix} I & 0 \\ 0 & I \end{bmatrix}$$

so

$$C = \frac{1}{\varepsilon}B\,A, \quad BB^*E = -BMC^* = -BM(\varepsilon M+B^*B)^{-1}B^*$$

and

$$B(\frac{1}{\varepsilon}A)B^* + I + \varepsilon E, \quad \frac{1}{\varepsilon}A = (\varepsilon M+B^*B)^{-1} .$$

The important point to make when comparing methods 1 and 2, is that in the latter (in algorithm (3.10)), the matrices B and B* appear only in matrix-vector multiplications. Hence we have reduced the indefinite problem (3.1) to a sequence of positive definite ones for the operator $\hat{M} = \varepsilon M+B^*B$. Again, it may pay off to explicitly calculate the entries of B^*B and $\hat{M}$ and to make an accurate incomplete factorization of the corresponding matrix and use a preconditioned conjugate gradient algorithm for the systems with $\hat{M}$.

<u>Remark 3.1</u>. Consider the augmented Lagrangian optimization problem

$$\inf \sup\{\frac{1}{2}u^*Mu-f_1^*u+v^*(Bu-f_2) + \frac{1}{2\varepsilon}(Bu-f_2)^*(Bu-f_2)\}$$

We see at once that the equations for the stationary points of this functional are precisely the equations in (3.4).

<u>Remark 3.2</u>. If we let $\varepsilon \to 0$ in (3.8) we readily find by use of Lemma 3.1, that, assuming both M and $BM^{-1}B^*$ are regular,

$$\hat{v}_\varepsilon \to (BM^{-1}B^*)^{-1}[BM^{-1}f_1 - f_2], \quad \varepsilon \to 0 .$$

Hence $\hat{v}_\varepsilon$ converges towards the solution of (3.3) as we might expect, since the perturbed problem goes over in the unperturbed as $\varepsilon \to 0$. This has been proven in a more general context in [5].

<u>Remark 3.3</u>. From (3.2), (3.3) follows that $\hat{u} = A_1f_1 + A_2f_2$ where $A_1 = [I-M^{-1}B^*(BM^{-1}B^*)^{-1}B]M^{-1}$ and $A_2 = M^{-1}B^*(BM^{-1}B^*)^{-1}$. It was observed in [14] and [9] that if $BM^{-1}B^*$ is regular, the solution $\hat{u}$ will be unchanged if we in this expression for $\hat{u}$, substitute $\hat{M}$ by $M = \varepsilon M+B^*B$ (see (3.7)). Hence even if M itself is indefinite, we may use Method 1, if we substitute M by $\hat{M}$, in order to get a positive definite operator $(0 < \varepsilon \leq \varepsilon_0)$. We would then first solve

$$B\hat{M}^{-1}B^*\tilde{v} = B\hat{M}^{-1}f_1 - f_2 \tag{3.11}$$

and then calculate $\hat{u}$ from

$$\hat{M}\hat{u} = f_1 - B^*\tilde{v} . \tag{3.12}$$

Note that the solution, $\tilde{v}$ for this method does not in general equal $\hat{v}$ in (3.1).

Anyhow if we are only interested in the varible $\hat{u}$, we may accordingly use the following modification of Methods 1 and 2.

Method 1'. This is (3.11), (3.12) where we may solve (3.11) by a preconditioned iterative method. The difficulties in choosing a preconditioner remains however.

Method 2'. Analogously to Method 2, we use an iterative method to solve (3.11), with ε small enough:

$$\tilde{v}^{(\ell+1)} = (I-B\hat{M}^{-1}B^*)\tilde{v}^{(\ell)} + \tilde{v}^{(0)}, \quad \ell = 0,1,\ldots$$

$$\tilde{v}^{(0)} = B\hat{M}^{-1}f_1 - f_2 .$$

When $\tilde{v}$ has been found to enough accuracy, we use (3.12) to evaluate $\hat{u}$. Since it is somewhat simpler to evaluate $\tilde{v}^{(0)}$ than $\hat{v}_\varepsilon$ in (3.8), Method 2' is a simplification of Method 2 in case we only want to calculate $\hat{u}$ and not $\hat{v}$.

Remark 3.4. We observe from Remark 3.3 that for any $\varepsilon > 0$, then $B\hat{A}_2 = I$, i.e., $\hat{A}_2 = B^+$, the generalized (right-sided) inverse of B, and $\hat{A}_1 = (I-\hat{A}_2B)\hat{M}^{-1}$. Hence

$$\hat{u} = (I-\hat{A}_2B)\hat{M}^{-1}f_1 + B^+f_2 ,$$

where B^+f_2 is the generalized solution of $B\hat{u} = f_2$. Hence we note that, given $B\hat{u} = f_2$, Method 2' may be used to calculate a generalized solution. We simply put $f_1 = 0$, i.e., consider

$$\begin{bmatrix} \hat{M} & B^* \\ B & 0 \end{bmatrix} \begin{bmatrix} \hat{u} \\ \hat{v} \end{bmatrix} = \begin{bmatrix} 0 \\ f_2 \end{bmatrix} ,$$

where M (symmetric and positive definite) may be chosen arbitrarily. To the author's best (but limited) knowledge in this field, this is a new iterative algorithm for the calculation of a generalized solution. The generalized solution will be the one which has a minimal norm in $||\cdot||M^{1/2}$. To get the solution with minimal Euclidean norm $||\cdot||$ we just consider

$$\begin{bmatrix} I & B^* \\ B & 0 \end{bmatrix} \begin{bmatrix} \hat{u} \\ \hat{v} \end{bmatrix} = \begin{bmatrix} 0 \\ f_2 \end{bmatrix}$$

and use Method 2.

Remark 3.5. In [15], a very general form of update formulas (diagonalized multiplier methods) are considered. The matrix used here belongs to a subclass of these. Note however that in [15], the corresponding sequences $\{u^{(\ell)}\}$, $\}v^{(\ell)}\}$ are calculated in parallel.

Remark 3.6. As has been noted, in Method 2, B may be rank deficient, but the algorithm will still converge. B being rank deficient corresponds to the so-called Babuška-Brezzi condition (see [8]),

$$\inf_u \sup_v (Bu)^*v/||u|| \; ||v|| \geq \gamma > 0$$

is not satisfied.

Acknowledgement. The author thanks Paul Boggs and Trond Steihaug for discussions on the relation between Method 2 and the diagonalized multiplier methods.

REFERENCES

1. Axelsson, O., "On Preconditioning and Convergence Acceleration in Sparse Matrix Problems", CERN 74-10, Geneva, 1974.

2. Axelsson, O., "A Class of Iterative Methods for Finite Element Equations", Comp. Meth. in Appl. Mech. and Eng. 9 (1976), 123-137.

3. Axelsson, O., "Notes on the Numerical Solution of the Biharmonic Equation", J. Inst. Maths. Applics. 11 (1973), 213-226.

4. Axelsson, I., and Gustafsson, I., "An Iterative Solver for a Mixed Variable Variational Formulation of the (First) Biharmonic Problem", Comp. Meth. in Appl. Mech. and Eng. 20 (1979), 9-16.

5. Axelsson, I., "Preconditioning of Indefinite Problems by Regularization", SIAM J. Numer. Anal. 16 (1979), 58-69.

6. Axelsson, I., "Conjugate Gradient Type Methods for Unsymmetric and Inconsistent Systems of Linear Equations", Linear Algebra and Its Applications 29 (1980), 1-16.

7. Axelsson, O., and Barker, A.V., Finite Element Solution of Boundary Value Problems, Theory and Computation. Academic Press, to appear.

8. Brezzi, F., "On the Existence, Uniqueness and Approximation of Saddle Point Problems Arising from Lagrangian Multipliers", R.A.I.R.O. R-2 (1974), 129-151.

9. Boggs, P.T., W. Tolle, J. and Wang, P., "On the Local Convergence of Quasi-Newton Methods for Constrained Optimization", SIAM J. Control and Optimization, 20 (1982), 161-171.

10. Ewing, R.E., and Wheeler, M.F., Computational Aspects of Mixed Finite Element Methods, Numerical Methods for Scientific Computing (R.S. Stepleman, editor), North Holland, Amsterdam, to appear.

11. Gustafsson, I., "A Preconditioned Iterative Method for the Numerical Solution of the Biharmonic Problem", 82.03R, Department of Computer Sciences, Chalmers University of Technology, Goteborg, Sweden.

12. Hageman, L.A., and Young, D.M., Applied Iterative Methods, Academic Press, New York, 1982.

13. Householder, A.S., The Theory of Matrices in Numerical Analysis, Blaisdell, New York, 1964.

14. Powell, M.J.D., "The Convergence of Variable Metric Methods for Nonlinearly Constrained Optimization Calculations", Nonlinear Programming (J.O. Mangasarian, R. Meyer and S. Robinson, eds.), Academic Press, New York, 1978, pp. 27-63.

15. Tapia, R.A., "Diagonalized Multiplier Methods and Quasi-Newton Methods for Constrained Optimization", J. Opt. Theory Appl., 22 (1977), 135-194.

PRECONDITIONED CONJUGATE GRADIENT METHODS FOR THE HELMHOLTZ EQUATION

Alvin Bayliss*, Charles I. Goldstein** and Eli Turkel***

I. INTRODUCTION

In this paper we investigate iterative methods for solving the two-dimensional Helmholtz equation

$$\Delta u + k^2 n(x,y) u = 0 \tag{1.1}$$

The function $n(x,y)$ in (1.1) is a piecewise smooth function. We consider the case that $n(x,y)$ is positive over a significant region. Equation (1.1) arises in scattering theory and underwater acoustics. Here, we will concentrate on the numerical difficulties, and methods for overcoming them, associated with solving the discretized approximation to (1.1) with appropriate boundary conditions.

The assumption $n > 0$ implies that the eigenvalues of the operator $\Delta + k^2 n$ will have both positive and negative real parts. The discretization of (1.1) results in a linear system of equations

$$Ax = b \tag{1.2}$$

Since the equation (1.1), with appropriate boundary conditions,

---

*Exxon Research & Engineering Co., Linden, New Jersey.

**Brookhaven National Laboratory, Upton, New York.

***University of Tel Aviv, Tel Aviv, Israel.

ISBN 0-12-100560-7

has an indefinite real part, the symmetric part of A will also be indefinite.

In applications, it is usually of interest to solve (1.1) in a region that is unbounded in at least one direction. To solve such a problem on a computer, one truncates the region and specifies some approximation to a radiation condition on an artificial boundary. This boundary condition can be either a global condition represented by an integral operator along the boundary [7] or a local differential operator [1], [2]. In this paper, we shall only consider local operators.

Local radiation boundary conditions are of the form

$$\alpha u_x + \beta u_y + \gamma u = 0 \tag{1.3a}$$

or,

$$\alpha u_{xx} + \beta u_{yy} + \gamma u_{xy} + \mu u_x + \nu u_y + \sigma u = 0 \ , \tag{1.3b}$$

where α, β, γ, μ, ν and σ are complex numbers. Due to the presence of these complex numbers, the differential equation (1.2) subject to boundary conditions of the form (1.3) (or a global boundary condition) is no longer self-adjoint; moreover its symmetric part is indefinite.

Finite element methods are convenient for treating boundary conditions of the type described above with regions of arbitrary shape; see [8] for a detailed description of some relevant techniques. The numerical results presented in this paper were obtained using piecewise linear functions and triangular elements.

A further difficulty with solving (1.1) is that solutions are oscillatory. A fine mesh is required to resolve the waves, even for moderate values of the wave number k. In summary, the system of equations (1.2) involves several difficulties:

(a) The symmetric part of A is indefinite due to the presence of $k^2 n$, $n > 0$.

(b) The matrix A is not self-adjoint because of the radiation boundary condition.

(c) The size of A is large since we wish to resolve an oscillatory wavelike solution.

For large problems, iterative methods are more efficient than methods based on Gaussian elimination both in terms of running time and even more in terms of memory requirements. Therefore the use of such direct methods can severely restrict the resolution that can be attained for any particular problem.

The methods that we consider are conjugate gradient methods applied to the normal equations with an appropriate preconditioning.[1] We first replace the linear system (1.2) by the equivalent system

$$By = f \quad \text{with} \quad B = Q_2 A Q_1, \quad Q_1 y = x, \quad f = Q_2 b \; . \tag{1.4}$$

This system is better conditioned than (1.2) if $M^{-1} = Q_2^{-1} Q_1^{-1}$ approximates A^{-1} in an appropriate sense. (1.4) is then converted to the normal equations

$$B^*By = B^*f \tag{1.5}$$

In order to construct M^{-1}, we wish to use properties of the Helmholtz operator (1.1). In particular, M^{-1} will be based on an approximation to (1.1) obtained by setting $k = 0$ in both the differential equation (1.1) and in the boundary conditions (1.3). The resulting matrix is both positive definite and self-adjoint. Furthermore, the preconditioned problem (1.4) now has a smaller condition number. In the next section, we will discuss the possible preconditionings in more detail. We will present numerical results in sections III and IV which demonstrate the effectiveness of the method.

II. ITERATION METHOD

The conjugate gradient method applied to (1.5) has a rate of convergence inversely proportional to $\kappa(B^* B)^{1/2}$ where κ is the condition number. Hence we wish to choose Q_1 and Q_2 so that $\kappa(B^* B)$ is reduced while $M^{-1} = Q_2^{-1} Q_1^{-1}$ is easy to evaluate. We will discuss three specific ways of splitting M between Q_1 and Q_2. These are

$$\text{(A)} \quad Q_1 = Q_2^*$$

---

[1] *A variety of iterative methods for solving other kinds of elliptic problems have been discussed Chandra [4] and Elman [6].*

$$\text{(B)} \quad Q_1 = M, \quad Q_2 = I \qquad \text{(C)} \quad Q_1 = I, \quad Q_2 = M \tag{2.1}$$

The preconditioning given by (A) is symmetric. Since we consider the normal equations, it is not necessary to use a symmetric preconditioning. It is our experience that preconditioning of type A is most effective when using a preconditioner based on SSOR (to be described below). The numerical results presented in section III were obtained with the symmetric preconditioning. For all these methods, the number of multiplications and divisions needed is 5N+2 plus two matrix vector products (Av and A^*v) and two preconditioning solves ($M^{-1}v$ and $(M^{-1})^*v$). The use of a preconditioner requires these two extra matrix inversions plus extra vectors for storage. One should choose Q_1 and Q_2 so that M is sparse and the computation of $M^{-1}v$ does not require excessive storage or computational effort.

As previously mentioned, we shall choose M to approximate (1.1) with $k = 0$ in both the differential equation and boundary condition. In this paper, we will compare preconditionings based on SSOR and ADI.

A. SSOR Preconditioning

Let the matrix A in (1.2) be written as $A = A_o + A_1$, where A_o is the matrix obtained by discretizing (1.1) with $k = 0$ in both the equation and the radiation boundary condition (1.3). Let D_o, L_o, U_o denote the diagonal, strictly lower and upper parts of A_o and let ω be a parameter with $1 \leq \omega \leq 2$. Then, the SOR iterative method is given by

$$X_{m+1} = (D_o - \omega L_o)^{-1}((1-\omega)D_o + \omega U_o)X_m + \omega(D_o - L_o)^{-1}b \tag{2.2}$$

We now define our preconditioning of type (A) by $Q_1^{-1}b = X_1$ with $X_o = 0$, and $M = Q_1 Q_1^*$.

B. ADI Preconditioning

Letting A_o be as above, ADI is given by

$$A_o = H + V \text{ and}$$

$$X_{m+1} = (r_{m+1}I + H)^{-1}(r_{m+1}I + V)^{-1} [(r_{m+1} - V)(r_{m+1} - H)X_m + 2\, r_{m+1}b] \tag{2.3}$$

where r_m are a sequence of positive constants. The preconditioning of type B is given by $M^{-1}b = X_m$ where m is the number of steps in the cycle and $X_o = 0$. (See [10].) In the usual ADI method, H and V are tridiagonal matrices corresponding to the x and y directions. When H and V commute, it can be proved that a judicious choice of the parameters $r_1, \ldots, r_m$ greatly accelerates the convergence rate. For the Poisson equation, computational experience indicates that the ADI iterative method converges rapidly much more generally (see e.g. [5]). This suggests that for the Helmholtz equation, where ADI is used as a preconditioning, increasing m can decrease the total computational effort particularly for fine meshes.

In sections III and IV, it will be dmonstrated that the preconditionings described above are effective even when $k^2n(x,y)$ is large. This acceleration is found even though the matrix is indefinite and M^{-1} is no longer an approximate inverse to the full matrix A. In a model problem, SSOR preconditioning and ADI preconditioning with one parameter have a convergence rate of O(h) while ADI with m parameters will have a convergence rate of $O(h^{1/m})$. The proofs will be presented elsewhere.

III. RESULTS FOR SSOR PRECONDITIONING

In this section, we evaluate the utility of SSOR preconditioning for a variety of problems.

We first consider plane waves scattered by a circular cylinder. We use cylindrical coordinates and ignore variations in the axial direction, z. The problem reduces to

$$\frac{1}{r}\frac{\partial}{\partial r}\left(r\frac{\partial u}{\partial r}\right) + \frac{1}{r^2}\frac{\partial^2 u}{\partial \theta^2} + k^2u = 0 \tag{3.1a}$$

$$u = f(\theta) \quad \text{at} \quad r = r_I , \tag{3.1b}$$

$$\text{radiation boundary condition at} \quad r = r_o \tag{3.1c}$$

We solve this problem by two methods. The first method is the conjugate gradient scheme preconditioned with SSOR as described above: (3.1a) is discretized using a finite element method and continuous piecewise linear elements. The second method is a modificiation of the following <u>limiting amplitude</u>

principle, used in [9] to solve a wide variety of problems.

Setting $v = ue^{ikt}$ in the wave equation

$$\frac{\partial^2 u}{\partial t^2} = \frac{1}{r}\frac{\partial}{\partial r}\left(r\,\frac{\partial u}{\partial r}\right) + \frac{1}{r^2}\frac{\partial^2 u}{\partial \theta^2} \tag{3.2}$$

we get

$$v_{tt} - 2ik\,v_t = \Delta v + k^2 v \tag{3.3}$$

The limiting amplitude principle states that, as $t \uparrow \infty$, v approaches a time-independent solution of (3.1).

The chief drawback to this method is that for a two-dimensional exterior region $v(x,t)$ decays to $u(x)$ only as $1/t$. Hence, convergence can be slow. An alternative approach is to replace (3.3) by

$$v_{tt} - 2ic(r)kv_t = d(r)\,[\Delta v + k^2 v] \tag{3.4}$$

with $c(r)$, $d(r)$ appropriately chosen. A detailed description of this method will appear elsewhere [3].

Two different solutions to (3.1) were computed using both methods. Exact solutions were generated by assuming a monopole source inside the cylinder; described in more detail in [1]. A relatively coarse convergence criteria was assumed. In both cases, the use of (3.4) required about 3 times as many iterations, but the total computer time was about the same. Thus, for this simple problem, the two methods were about equally efficient.

The second test case concerns scattering by a Helmholtz resonator. In this case, the circular cylinder has an opening so that a plane wave entering from infinity can enter into the interior of the cylinder. In a time dependent frame, one expects rays to bounce around in the interior of the resonator before exiting from the resonator and escaping to infinity. The smaller the aperture, the longer the rays take to escape [9]. To solve this problem, we decompose u into an incident and a scattered wave. Since the total field vanishes on the boundary, the scattered field satisfies a Dirichlet condition on the resonator. Hence, we wish to solve

$$\Delta u + k^2 u = 0 \quad \text{in all of space} \tag{3.5a}$$

$$u \quad \text{given for} \quad r = r_I = 2 \quad \text{and} \quad \theta > \theta_o \,. \tag{3.5b}$$

A radiation boundary condition is imposed along the artificial surface $r_o = 7.3$. We use symmetry about the x axis to reduce the computational effort.

For this problem, the modified equation (3.4) did not converge even for relatively large aperture openings. This problem was solved using the symmetric preconditioning (2.1(A)). For small aperture openings, the matrix A has a large nearly singular block since the problem approximates an interior Dirichlet problem in the interior of the resonator. Nevertheless, convergence required less than twice as many iterations as were required for the closed resonator problem (i.e. exterior to a cylinder). Using a grid of 81 radial points and 61 theta points and $K = 3.77$, an aperture opening of 14° required 502 iterations and an opening of 22° required 494 iterations. The cylinder required 268. In this and all subsequent problems, convergence was defined when the relative residual was less than 10^{-6}.

A more accurate computation of this problem would require a significantly finer mesh in the vicinity of the resonator, especially near the aperture opening. Nevertheless, the results indicate that the preconditioner is not significantly affected by complex wave interactions and nearly trapped modes. When a highly stretched radial mesh is used, the SSOR preconditioner should be replaced by a symmetric line SOR preconditioner ([4]).

As a third example, we consider the interaction of a plane wave with a cold overdense plasma column. This problem was studied in [9] using the limiting amplitude principle. The equation for the electric field u is given by

$$u + k^2 n(r) u = k^2 (n-1) e^{ikx} \,. \tag{3.6}$$

The index of refraction $n(r)$ can be modeled by

$$n(r) = \begin{cases} (r^2 - 4)/5 & r \le 3 \\ 1 & r \ge 3 \,. \end{cases} \tag{3.7}$$

Radiation boundary conditions are specified at $r = 8$. This problem was again solved by using SSOR preconditioning. Using a grid of 81 radial points and 61 angular points, convergence required 327 iterations for $k = 3.77$. Similar results were obtained for a variety of plasma models and frequencies.

IV. RESULTS FOR ADI PRECONDITIONING

In this section, we describe results obtained thus far using the ADI preconditioning. These results are preliminary and are confined to a simple model problem in a cartesian duct. The same problem was studied in [1] using SSOR as a preconditioning. The ADI preconditioner using one or two parameters will be compared to the SSOR preconditioner.

Consider the following problem

$$\Delta u + k^2 u = 0; \quad 0 \leq x \leq \pi, \quad 0 \leq y \leq \pi \tag{4.1a}$$

$$u_y(x,0) = u(x,\pi) = 0 \tag{4.1b}$$

$$u_x(0,y) = f(y) \tag{4.1c}$$

u satisfies an approximate radiation condition at $x = \pi$. (4.1d)

The boundary conditions (4.1b-c) are analogous to those occurring in a model of underwater wave propagation. As described in [1], we employ two approximate radiation boundary conditions in (4.1d). These are

$$B_1 u = \left(\frac{\partial}{\partial x} - i\sqrt{k^2 - 1/4}\right) u = 0 \tag{4.2a}$$

$$B_2 u = \left(\frac{\partial}{\partial x} - i\sqrt{k^2 - 9/4}\right)\left(\frac{\partial}{\partial x} - i\sqrt{k^2 - 1/4}\right) u = 0 \tag{4.2b}$$

The operator B_j, $(j = 1,2)$ annihilates the first j terms in a modal expansion of u for the continuation of (4.1) to the region $x \geq \pi$. Tables 1 and 2 list the number of iterations required for convergence for two different frequencies. The optimal parameters were determined experimentally. The Neumann data f in (4.1c) was chosen to make the exact solution $u = e^{i\sqrt{k^2 - 1/4}} \cos \frac{y}{2}$, but similar results were obtained for other data. A square $(N \times N)$ grid was used. In

TABLE 1. Results for ADI with One and Two Parameters, K = 2.77

N	Boundary Condition	Number of Parameters	r_1	r_2	Number of Iterations
61	(4.2a)	1	.043	-	290
91	(4.2a)	1	.031	-	451
61	(4.2b)	1	.027	-	245
91	(4.2b)	1	.018	-	359
61	(4.2a)	2	.063	.122	132
91	(4.2a)	2	.046	.050	176
61	(4.2b)	2	.021	.122	126
91	(4.2b)	2	.028	.050	174

TABLE 2. Results for ADI with One and Two Parameters, K = 4.16

N	Boundary Condition	Number of Parameters	r_1	r_2	Number of Iterations
61	(4.2a)	1	.043	-	290
91	(4.2a)	1	.028	-	458
61	(4.2b)	1	.027	-	249
91	(4.2b)	1	.018	-	342
61	(4.2a)	2	.063	.122	153
91	(4.2a)	2	.048	.050	203
61	(4.2b)	2	.028	.115	136
91	(4.2b)	2	.028	.045	186

the ADI preconditioning, the boundary terms are evaluated for $k = 0$ and included in the appropriate factor. The preconditioner appears to be very sensitive to the treatment of the boundary conditions. For comparison, the number of iterations required by using SSOR as a preconditioning is shown in Table 3.

As stated previously, ADI is implemented as a left preconditioner since the operators H and V do not commute with these boundary conditions. In this case, supplementary calculation is needed to compute the residual based on A, which is then monitored for convergence. SSOR is implemented as symmetric preconditioning. The number of iterations would be slightly increased if it were implemented as a left preconditioner.

TABLE 3. Results for SSOR Preconditioning

N	K	Boundary Condition	ω	Number of Iterations
61	2.77	(4.2a)	1.76	284
91	2.77	(4.2a)	1.77	469
61	2.77	(4.2b)	1.73	351
91	2.77	(4.2b)	1.77	567
61	4.16	(4.2a)	1.73	308
91	4.16	(4.2a)	1.77	506
61	4.16	(4.2b)	1.73	386
91	4.16	(4.2b)	1.83	621

We observe that ADI with one parameter requires slightly more work per iteration than SSOR, while ADI with two parameters requires slightly less than twice as many multiplications as one parameter. It can be concluded from these preliminary numerical experiments that the use of ADI can decrease the work in comparison with SSOR. It appears to be sensitive to the boundary condition, being more effective for the second order condition (4.2b).

The use of two parameters does seem to improve the convergence rate although we could not achieve a rate of $O(h^{1/2})$. The main improvement seems to be for very fine grids. We observe that r_1 does not scale in the expected manner for the second order boundary condition (4.2b); see [10]. In summary, the results to date indicate that ADI preconditioning can offer computational advantages; however, it is less robust than SSOR with respect to the parameters and more sensitive to the boundary conditions. It is also more difficult to implement in general.

REFERENCES

1. Bayliss, A., Goldstein, C. I., and Turkel, E., J. Comp. Phys., to appear.

2. Bayliss, A. and Turkel, E., J. Comp. Phys., 48 (1982), 182-199.

3. Bayliss, A. and Turkel, E., in preparation.

4. Chandra, R., Ph.D. Thesis, Department of Computer Science, Yale University (1978).

5. Doss, S. and Miller, K., SIAM J. Numer. Anal. 16 (1979), 837-856.

6. Elman, H. C., Ph.D. Thesis, Department of Computer Science, Yale University (1982).

7. Fix. G. J. and Marin, S. R., J. Comp. Phys., 28 (1978), 253-270.

8. Goldstein, C. I., in Elliptic Problem Solvers (M. Schultz, ed.), Academic Press, NY (1981), pp. 307-13.

9. Kriegsmann, G. A. and Morawetz, C. S., SIAM J. Sci. Stat. Comp. 1 (1980), 371-85.

10. Varga, R. S., Matrix Iterative Analysis, Prentice Hall, N.J., 1962.

SOLVING ELLIPTIC PROBLEMS ON REGIONS PARTITIONED INTO SUBSTRUCTURES

Petter E. Bjørstad* and Olof B. Widlund**

I. INTRODUCTION

In this paper, we will introduce and discuss some special iterative methods for the solution of finite element approximations for elliptic equations defined on regions regarded as unions of subregions. We are interested, in particular, in the design of methods for which the interaction between subproblems, i.e., discrete elliptic problems on the subregions, is computed by a conjugate gradient method while the subproblems are solved by a direct method. This approach differs from the standard one in industry in which a direct method is used throughout the entire solution process; see Section 2. As we will also see in that section, the whole problem can be reduced to finding accurate values of the solution on the interaces, i.e., the curves or surfaces which separate the subregions.

To simplify the discussion, we will only consider in detail the case of a plane region cut into two pieces. The region could equally well be subdivided into several subregions and the ideas can also be applied recursively. A fuller exposition will be given in Bjørstad and Widlund (1983), in

---

*Det Norske Veritas, FDIV, P.O. Box 300, 1322, Høvik, Norway.

**Courant Institute of Mathematical Sciences, New York University, 251 Mercer Street, New York, N.Y. 10012

This work was supported by the National Science Foundation under contract NSF-MCS-8203236 and by the U.S. Department of Energy under contract DE-AC02-76-ER03077-V at the Courant Mathematics and Computing Laboratory.

ISBN 0-12-100560-7

which the optimality of the so-called good and excellent methods; see Section 3, will be rigorously established for second order elliptic problems in the plane. This theory is quite complicated. Its development requires the use of elliptic regularity theory on Lipschitz regions; see Grisvard (1980) and some apparently new finite element results. We regard the extension of these results to more general elliptic problems as a significant open problem.

The partition of elliptic problems into subproblems is a very natural idea that is much used in practice. In finite element computations the modelling of entire mechanical systems can profitably be organized by discretizing the given partial differential equations using finite elements on subregions into which the region is partitioned or from which it was originally assembled. These tasks and the factorization of the resulting stiffness matrices for the subproblems can obviously be assigned to different engineering groups and computer systems or processors with coordination required only at the interfaces between the substructures to assure matching of the finite element triangulations. These ideas are particularly attractive if some of the substructures are identical or have previously been analyzed, as in the case when a simulation is repeated after the redesign or damage of one or a few of the substructures.

We know of only a few previous efforts to develop iterative methods similar to ours. In a paper on generalized conjugate gradient methods, Concus, Golub, and O'Leary (1975) reported on numerical experiments with Poisson's equation on a T-shaped region. No results for very large problems are reported in their work which makes assessing its potential difficult. In Section IV of this paper, we will report on results obtained by three other methods which are described in Section III. Methods inspired by the Schwarz alternating method and by control theory are considered in Glowinski, Periaux and Dihn (1982), and by Glowinski in this volume. After comparing their numerical results with ours, we have come to believe that our methods are more powerful.

The pioneer in the mathematical analysis of the kind which we consider is Maksymilian Dryja; see Dryja (1982). He has also contributed very important algorithmic ideas. We received his paper in March 1981 after our first series of numerical experiments. He there analyzes two of the methods for the five point formula for Laplace's equation on L-shaped regions. We also greatly benefited from his visit with the second author in October 1981. At that time, we were introduced to the so-called excellent method discussed in Sections III and IV.

We also note that both from algorithmic and theoretical points of view, our methods have much in common with capacitance matrix methods, see Bjørstad (1980), Dryja (1983), O'Leary and Widlund (1979) and Proskurowski and Widlund (1980), and the many references given in those papers.

II. SUBSTRUCTURED PROBLEMS AND BLOCK FORM OF THE STIFFNESS MATRICES

To simplify the discussion, we confine ourselves to Dirichlet problems defined on the union Ω of Ω_1, Ω_2 and Γ_3. Here Ω_1 and Ω_2 are plane, bounded, non-intersecting regions and Γ_3 the intersection of their closures. The boundaries of Ω_1 and Ω_2 are $\Gamma_1 \cup \Gamma_3$ and $\Gamma_2 \cup \Gamma_3$, respectively, and the boundary of Ω is $\Gamma_1 \cup \Gamma_2$. We assume that these boundaries are Lipschitz continuous; see Grisvard (1980). A linear, second order, positive definite, self adjoint elliptic operator is defined on Ω. Its symmetric bilinear form is denoted by $a_\Omega(u,v)$; see Ciarlet (1978) or Strang and Fix (1973). A simple example is given by the Laplace operator with homogeneous Dirichlet condition for which

$$a_\Omega(u,v) = \int_\Omega \nabla u \cdot \nabla v \, dx, \qquad u,v \in H_0^1(\Omega) .$$

Here $H_0^1(\Omega)$ is the subspace of elements, with zero boundary values, of the Sobolev space of square integrable functions with square integrable first derivatives. Triangulations of Ω_1 and Ω_2 are introduced in such a way that the nodes on Γ_3 coincide. We assume that each degree of freedom of the finite element subspaces is associated with a node and a basis function in the finite element space and that the support of any basis function coincides with the triangles to which its node belongs. An element of the stiffness matrix has the form

$a_\Omega(\phi_i,\phi_j)$, where ϕ_i and ϕ_j are basis functions. It vanishes unless the two nodes belong to a common triangle, To simplify, we will only consider Lagrangian finite elements in this discussion.

It is easy to see that

$$a_{\Omega_1}(u,v) + a_{\Omega_2}(u,v) = a_\Omega(u,v) \tag{2.1}$$

and that therefore the stiffness matrix of a problem on Ω can be constructed from those of Ω_1 and Ω_2. This fact is frequently used in practice to construct stiffness matrices from the stiffness matrices on the individual triangles; see Strang and Fix (1973).

We write the stiffness matrices corresponding to Ω_1 and Ω_2 respectively as

$$A^{(1)} = \begin{bmatrix} A_{11} & A_{13} \\ A_{13}^T & A_{33}^{(1)} \end{bmatrix} \text{ and } A^{(2)} = \begin{bmatrix} A_{22} & A_{23} \\ A_{23}^T & A_{33}^{(2)} \end{bmatrix}, \tag{2.2}$$

where A_{11}, represents couplings between the pairs of nodes in Ω_1, A_{13} couplings between the pairs belonging to Ω_1 and Γ_3 respectively and $A_{33}^{(1)}$ couplings between nodes on Γ_3, etc. The stiffness matrix of the entire problem is of the form,

$$A = \begin{bmatrix} A_{11} & 0 & A_{13} \\ 0 & A_{22} & A_{23} \\ A_{13}^T & A_{23}^T & A_{33} \end{bmatrix}$$

where, by a simple computation,

$$A_{33} = A_{33}^{(1)} + A_{33}^{(2)}. \tag{2.3}$$

We note that the degrees of freedom have been partitioned into three sets of which the third, the separator set, corresponds to the nodes on Γ_3. From the point of view of graph theory, the undirected graph of A becomes disconnected into two components if the nodes of the separator set and their incident edges are removed. If conforming finite elements are used, then it also follows from the assumptions on $a_\Omega(u,v)$, that A is positive definite and symmetric, and as a consequence so are A_{11}, A_{22} and A_{33}; see Ciarlet (1978) or Strang and Fix (1973). In other cases we assume that A has these properties.

We consider the linear system of algebraic equations of the form

$$Ax = \begin{bmatrix} A_{11} & 0 & A_{13} \\ 0 & A_{22} & A_{23} \\ A_{13}^T & A_{23}^T & A_{33} \end{bmatrix} \begin{bmatrix} x_1 \\ x_2 \\ x_3 \end{bmatrix} = \begin{bmatrix} b_1 \\ b_2 \\ b_3 \end{bmatrix} \tag{2.4}$$

By using block-Gaussian elimination, we can reduce this system to the positive definite, symmetric system,

$$\begin{aligned} Sx_3 &= (A_{33} - A_{13}^T A_{11}^{-1} A_{13} - A_{23}^T A_{22}^{-1} A_{23})x_3 \\ &= b_3 - A_{13}^T A_{11}^{-1} b_1 - A_{23}^T A_{22}^{-1} b_2 = \tilde{b}_3 . \end{aligned} \tag{2.5}$$

It is common practice to complete the process by soling (2.5) by a direct method.

The right hand side $\tilde{b}_3$ can be obtained at the expense of solving the two subproblems on Ω_1 and Ω_2, multiplying the resulting vectors by the sparse matrices A_{13}^T and A_{23}^T respectively and by subtracting the resulting vectors from b_3. The matrix S, which is a so-called Schur complement (see Cottle (1974)) can be expensive to compute and store. However, we notice that Sy can be computed for a given vector y at the expense of solving the two subproblems with the sparse right hand sides $A_{13}y$ and $A_{23}y$, respectively, and certain sparse matrix and vector operations. In the next section, we will develop iterative methods which only require S in terms of such matrix-vector products.

The cost of computing Sy depends primarily on the efficiency of the solvers for the subproblems. It should also be noted that if a Gaussian elimination method is used, advantage can be taken of the sparsity of the vectors $A_{13}y$ and $A_{23}y$. Thus when the lower triangular systems of equations are solved, the computation can begin with the first equation which has a nonzero component in the right hand side. Similarly, the solution of the upper triangular systems can be stopped as soon as all the components of $A_{11}^{-1} A_{13}y$ and $A_{22}^{-1} A_{23}y$ necessary for computing $A_{13}^T(A_{11}^{-1} A_{13}y)$ and $A_{23}^T(A_{22}^{-1} A_{23}y)$ have been found. This effectively reduces the size of the triangular systems necessary to carry out the iteration steps. It is particularly advantageous if all the variables at nodes adjacent to Γ_3 are ordered late. It should be noted, however, that such a constraint may be hard to impose on existing software or may lead to an increase in the time required to factor A_{11} and A_{22} into their triangular factors.

In the next section, we will need the Schur complements with respect to the matrices $A^{(1)}$ and $A^{(2)}$ defined in (2.2). They are,

$$S^{(1)} = A_{33}^{(1)} - A_{13}^T A_{11}^{-1} A_{13} \text{ and } S^{(2)} = A_{33}^{(2)} - A_{23}^T A_{33}^{-1} A_{23}, \qquad (2.6)$$

Using equations (2.3) and (2.6), we find that

$$S = S^{(1)} + S^{(2)} . \qquad (2.7)$$

III. CONJUGATE GRADIENT ALGORITHMS FOR SUBSTRUCTURED PROBLEMS AND AN INFORMAL THEORY

The general theory of conjugate methods is quite well known and will therefore be discussed only very briefly; see Concus, Golub and O'Leary (1975), Hestenes (1956) or Luenberger (1973).

Let $Ax = b$ be a linear system of algebraic equations with a positive definite, symmetric matrix A. Let $x^{(0)}$ be an initial guess and $r^{(0)} = b-Ax^{(0)}$ the initial residual. The k-th iterate in the standard conjugate gradient method, $x^{(k)}$, can then be characterized as the minimizing element for the variational problem

$$\min[(\tfrac{1}{2})y^TAy - y^Tb]$$

where $y-x^{(0)}$ varies in the linear space spanned by $r^{(0)}$, $Ar^{(0)},\ldots,A^{k-1}r^{(0)}$. By expanding in eigenvectors of A, it can be established that

$$(x^{(k)}-x)^T A(x^{(k)}-x)/(x^{(0)}-x)^T A(x^{(0)}-x)$$

is bounded from above by,

$$\min_{p \in P_{k-1}} \max_{\lambda \in \sigma(A)} (1 + \lambda p(\lambda))^2 , \tag{3.1}$$

see Luenberger (1973). Here x is the exact solution, P_{k-1} the space of all polynomial of degree k-1 and $\sigma(A)$ the spectrum of A. This bound can be used to establish that the convergence is rapid if A is well conditioned and that the rate of convergence can be bounded uniformly for entire families of operators if all the eigenvalues fall in a fixed interval.

Preconditioned conjugate gradient methods have been studied extensively in recent years. The idea goes back to the mid-fifties; see Hestenes (1956). Let A_0 be another positive definite, symmetric operator for which it is feasible to solve auxiliary systems of the form $A_0 y = c$ repeatedly for different right hand sides. In one of the versions of the method, the original problem $Ax = b$ is transformed into $AA_0^{-1} y = b$. The iterate $y^{(k)}$ is sought as the sum of an initial guess $y^{(0)}$ and a linear combination of $r^{(0)}$, $AA_0^{-1} r^{(0)}$, $\ldots, (AA_0^{-1})^{k-1} r^0$. If an appropriate inner product is used, a convenient recursion formula results, see e.g., Proskurowski and Widlund (1980). Each step of this algorithm requires the solution of an auxiliary linear system. It is important to note that the estimate (3.1) still holds, but that now the eigenvalues of AA_0^{-1}, i.e., those of the generalized eigenvalue problem $A\phi = \lambda A_0 \phi$, are of relevance rather than those of A. It is also worth noting that the estimate (3.1) can be used to show particularly rapid convergence if the eigenvalues are clustered.

For the problems at hand, we first consider the solution of equation (2.5) without preconditioning. The vector x_3 represents nodal values on Γ_3 and provide part of the Dirichlet data for two discrete subproblems. The restriction of Ax to the same set is computed when evaluating the residual We will now give an informal argument which shows that this mapping involves a loss of one derivative in $L^2(\Gamma_3)$. For a rigorous argument; see Bjørstad and Widlund (1983).

Let us recall that by (2.7), $S = S^{(1)} + S^{(2)}$ and show that the positive definite, symmetric operators $S^{(1)}$ and $S^{(2)}$ each have the property mentioned above. A discrete Dirichlet problem for the region Ω_1 can be written as

$$\begin{bmatrix} A_{11} & A_{13} \\ 0 & I \end{bmatrix} \begin{bmatrix} x_1 \\ x_3 \end{bmatrix} = \begin{bmatrix} 0 \\ g_D \end{bmatrix}$$

if the right hand side f and the Dirichlet data on Γ_1 vanish. Similarly, a Neumann problem takes the form,

$$\begin{bmatrix} A_{11} & A_{13} \\ A_{13}^T & A_{33}^{(1)} \end{bmatrix} \begin{bmatrix} x_1 \\ x_3 \end{bmatrix} = \begin{bmatrix} 0 \\ g_N \end{bmatrix}$$

It is straightforward to show that

$$\begin{bmatrix} A_{11} & A_{13} \\ A_{13}^T & A_{33}^{(1)} \end{bmatrix} \begin{bmatrix} A_{11} & A_{13} \\ 0 & I \end{bmatrix}^{-1} \begin{bmatrix} 0 \\ g_D \end{bmatrix} = \begin{bmatrix} 0 \\ S^{(1)} g_D \end{bmatrix}$$

Thus we can regard $S^{(1)}$ as a discretization of a mapping which maps the boundary values of a solution of a homogeneous elliptic problem onto its normal derivate on the boundary. Such a mapping will have a condition number proportional to the number of nodes on Γ_3. We call this the mediocre method.

In view of what we just learned, it is natural to try to find a preconditioner which also involves the loss of a derivative in $L_2(\Gamma_3)$. A natural choice would be a tangential derivative but that is not a symmetric operator. Instead we use the square root of a discretization of the Laplacian on Γ_3. Such a method is practical for problems in the plane and has been tested; see Section IV. We call this the good method.

An even better method, referred to as the excellent method, involves the solution of a block lower triangular system,

$$\begin{bmatrix} A_{11} & A_{13} & 0 \\ A_{13}^T & A_{33}^{(1)} & 0 \\ 0 & A_{23} & A_{22} \end{bmatrix} \begin{bmatrix} y_1 \\ y_3 \\ y_2 \end{bmatrix} = \begin{bmatrix} 0 \\ f_3 \\ 0 \end{bmatrix}$$

We obtain, straightforwardly, that

$$\begin{bmatrix} A_{11} & A_{13} & 0 \\ A_{13}^T & A_{33} & A_{23}^T \\ 0 & A_{23} & A_{22} \end{bmatrix} \begin{bmatrix} y_1 \\ y_3 \\ y_2 \end{bmatrix} = \begin{bmatrix} 0 \\ SS^{(1)^{-1}} f_3 \\ 0 \end{bmatrix}$$

It can be shown that $SS^{(1)^{-1}}$ has a very good spectrum. We note that this mapping involves solving a Neumann problem on Ω_1 and then a Dirichlet problem on Ω_2, using the values on Γ_3 just computed. The last step to find $SS^{(1)^{-1}} f_3$ is the computation of the residual on Γ_3.

IV. NUMERICAL EXPERIMENTS

We will only give a brief report on a few of the many experiments which we have carried out for the five point approximation of Poisson's equation on regions which are unions of two rectangles. The choice of such regions greatly simplifies the experiments and makes it feasible to conduct many experiments with very many degrees of freedom since fast Poisson solvers can be used to solve the subproblems. We note that this simple finite difference approximation can be viewed as a conforming finite element approximation on a mesh of right triangles using piecewise linear functions; see Strang and Fix (1973).

In the experiments considered here, we consider the union of two rectangles with corners at the points (0,0), (1,0), (1,1/2), (0,1/2) and (1/8,1/2), (5/8,1/2), (5/8,1), (1/8,1) respectively. The relevant parameters are the number of mesh points q on Γ_3, the interval between (1/8,1/2) and (5/8,1/2), and the total number of degrees of freedom. We have used data which are consistent with an exact solution $u(x,y) = x^2+y^2-xe^x \cos y$. We have found no real difference between the performance of our method for this and other cases.

We first show, in Table 1, how the number of iterations grow as a function of q using the excellent method. We stopped the iterations at the level of the truncation error. The initial guess was the zero function. We note that the overall number of degrees of freedom increases quadratically with q and equals 48641 for q = 127.

TABLE 1

q	Number of iterations	Maximum error on Ω
3	2	3.66×10^{-4}
7	3	9.59×10^{-5}
15	3	2.45×10^{-5}
31	4	6.09×10^{-6}
63	4	1.49×10^{-6}
127	5	3.02×10^{-7}

In Table 2, we compare the maximum error for the excellent and good methods for the case of $q = 127$. Ten to twelve iterations are required to decrease the maximum error by a factor of 10 if the mediocre method is used.

TABLE 2

Number of iterations	Excellent Method	Good Method
0	3.79×10^{-1}	3.79×10^{-1}
1	1.25×10^{-2}	3.22×10^{-2}
2	7.48×10^{-4}	4.01×10^{-3}
3	2.56×10^{-5}	5.26×10^{-4}
4	4.42×10^{-7}	8.74×10^{-5}
5	3.02×10^{-7}	1.05×10^{-5}
6	3.02×10^{-7}	1.33×10^{-6}
7	3.02×10^{-7}	3.08×10^{-7}
8	3.02×10^{-7}	3.03×10^{-7}

Finally, we give some information on the spectrum of $SS^{(1)^{-1}}$ for $q = 63$, i.e., for the excellent method. The smallest eigenvalues were found to be $\lambda_1 = 1.713$, $\lambda_2 = 1.777$, $\ldots, \lambda_5 = 1.992$ with the others less than or equal to 2.000. The corresponding spectrum for the good method shows a somewhat less pronounced cluster with $\lambda_1 = 1.733$, $\lambda_2 = 1.804$, $\ldots$ $\lambda_5 = 2.008$, $\ldots$, $\lambda_{63} = 2.828$.

REFERENCES

Bjørstad, P.E., "Numerical Solution of the Biharmonic Equation", Ph.D. Thesis, Stanford 1980. See also SIAM J. Numer. Anal. 20, 1983, pp. 59-71.

Bjørstad, P.E. and Widlund, O.B., to appear 1983.

Ciarlet, P.G., "The Finite Element Method for Elliptic Problems", North Holland, 1978.

Concus, P., Golub, G. and O'Leary, D., Proc. Symp. Sparse Matrix Computations, (Argonne National Lab., 1975), Academic Press, 1975.

Cottle, R., Lin. Alg. Appl. 8, 1974, pp. 189-211.

Dryja, M., "A Capacitance Matrix Method for Dirichlet's Problem on a Polygonal Region", Numer. Math. 39, 1982, pp. 51-64.

Dryja, M., "A Finite Element-Capacitance Matrix Method for Elliptic Problems", to appear in SIAM J. Numer. Anal., (1983).

Glowinski, R., Periaux, J. and Dihn, Q.V., "Domain Decomposition Methods for Nonlinear Problems in Fluid Dynamics", INRIA report #147, July 1982.

Grisvard, P., "Boundary Value Problems in Non-Smooth Domains", Dept. of Math., Univ. of Maryland, Lecture notes 19, 1980.

Hestenes, M., Proc. Symp. Appl. Math., VI, McGraw-Hill, 1956, pp. 83-102.

Luenberger, D.G., "Introduction to Linear and Nonlinear Programming", Addison-Wesley, 1973.

O'Leary, D. and Widlund, O., Math. Comp. 33, (1979), pp. 849-879.

Proskurowski, W. and Widlund, O., SIAM J. Sci. Stat. Comput., 1, (1980), pp. 410-425.

Strang, G. and Fix. G.J., "An Analysis of the Finite Element Method", Prentice-Hall, 1973.

THE AD-HOC SOR METHOD: A LOCAL RELAXATION SCHEME

L.W. Ehrlich[1]

I. INTRODUCTION

In solving a system of five-point difference equations by the successive overrelaxation (SOR) method, one needs an estimate of the optimal relaxation parameter. This parameter can be determined from the eigenvalues of the Jacobi matrix. However, the eigenvalues of the Jacobi matrix are not easily estimated in general, although methods have been developed to approximate the relaxation factor dynamically [5].

The ad-hoc SOR method proposed in [4] attempts to circumvent the need to estimate a relaxation factor by computing, for each equation, its own relaxation factor based on the difference equation coefficients of that equation and the boundary conditions and region shape. In [4], the method is fully described and is compared with the SOR method by noting the number of iterations necessary for convergence for a set of sample problems.

In this paper, the comparison is extended. For certain linear problems, the spectral radii of appropriate matrices are compared. The application to complex systems is briefly considered. Also, the effectiveness of the method on nonlinear systems is discussed.

Recently, a paper has appeared by Botta and Veldman [1] advocating a technique that is essentially equivalent to the ad-hoc SOR method. Somewhat more appropriately, they refer to

[1]Milton S. Eisenhower Research Center, Applied Physics Laboratory, The Johns Hopkins University, Laurel, Maryland.

ISBN 0-12-100560-7

the method as a <u>local relaxation technique</u>. We shall, from time to time, refer to their work.

II. THE BASIC PROCEDURE

For convenient reference, we first summarize the ad-hoc SOR technique. Consider the five-point difference equation

$$
\begin{aligned}
& a_{i+1,j}x_{i+1,j} + a_{i,j+1}x_{i,j+1} + a_{i-1,j}x_{i-1,j} \\
& + a_{i,j-1}x_{i,j-1} - a_{ij}x_{ij} = b_{ij}
\end{aligned}
\tag{2.1}
$$

where the a_{ij} are, in general, functions of position over some grid. for notational convenience, we use the numbering system indicated below and rewrite the above in stencil notation as

$$
\text{a)}\quad
\begin{array}{ccccccccc}
 & & | & & | & & | & & \\
i+1 & - & 7 & - & 2 & - & 6 & - & \\
 & & | & & | & & | & & \\
i & - & 3 & - & 0 & - & 1 & - & \\
 & & | & & | & & | & & \\
i-1 & - & 8 & - & 4 & - & 5 & - & \\
 & & | & & | & & | & & \\
 & & j+1 & & j & & j+1 & &
\end{array}
\qquad
\text{b)}\quad
\begin{bmatrix} & a_2 & \\ a_3 & -a_0 & a_1 \\ & a_4 & \end{bmatrix} x = b.
\tag{2.2}
$$

At each point, we assume that the difference equation has the coefficients of that point constant throughout a rectangular region and that we have Dirichlet boundary conditions. Under these conditions, the Jacobi matrix, J, of (2.1) is separable leading to the eigenvalues

$$
\mu_{p,q} = \frac{2}{a_0}\left[\sqrt{a_1 a_3}\cos\left(\frac{\pi p}{N+1}\right) + \sqrt{a_2 a_4}\cos\left(\frac{\pi q}{M+1}\right)\right]
\tag{2.3}
$$

with $N \times M$ unknowns in the region.

From the structure of the matrix, it is known that if all the eigenvalues of J are real, the optimal SOR relaxation factor, ω, is determined by

$$
\omega_{ij} = \frac{2}{1 + \sqrt{1 - \bar{\mu}^2}}
\tag{2.4}
$$

where $\bar{\mu}$ is the spectral radius of J, i.e., the eigenvalue of maximum modulus. On the other hand, if the eigenvalues are pure imaginary, then

$$\omega_{ij} = \frac{2}{1 + \sqrt{1 + \bar{\mu}^2}} \tag{2.5}$$

However, if the eigenvalue of largest magnitude is complex, we can use Rigal's [8] single-eigenvalue solution of the optimization of the relaxation factor. This formula is quite complicated. Botta and Veldman [1], however, present an approximation of this equation which is quite accurate in general, being only two percent off at worst when the imaginary part is small. Their approximation is

$$\omega_{ij} = 2/(1 + \sqrt{1 - \mu_r^2 + \mu_i^2(1 - \mu_r^{2/3})^{-1}}\ ). \tag{2.6}$$

III. BOUNDARY CONDITION AND REGION SHAPE MODIFICATIONS

The eigenvalues (2.3) were determined assuming Dirichlet boundary conditions. In [4], we discussed modifications to the eigenvalues due to other boundary conditions such as normal derivatives on one side or both, or periodic conditions. In effect, the eigenvalues are similar to (2.3) except, one might also be required to solve a transcendental equation. Approximations have been suggested. We refer the interested reader to [4] for more details.

The N and M used in (2.3) are obvious if our region is a rectangle. For nonrectangular regions, however, empirical results indicate the following: at a given point, the values of M and N that should be used are the number of unknowns between the nearest boundaries along a horizontal and vertical line, respectively, through the point in question. This is illustrated in Figure 1 of [4]. Brazier [2], who proposed a variable relaxation technique, also suggested this. (His method is compared to ours in [4].)

IV. REAL LINEAR SYSTEMS

For real difference equations with constant coefficients over a rectangle with Dirichlet boundary conditions, the SOR and the ad-hoc SOR are identical. This follows since the eigenvalues of the Jacobi matrix lie in a rectangle in the complex plane as per (2.3). (See also [6].) For nonconstant coefficients, however, such analysis is not possible.

In order to study the methods in general, we have chosen as our model stencil

$$\begin{bmatrix} & 1 - Rb_{ij} & \\ 1 + Ra_{ij} & -4 & 1 - Ra_{ij} \\ & 1 + Rb_{ij} & \end{bmatrix} \tag{4.1}$$

with R varying from 0 to ∞. Matrices of this form have the real part of their Jacobi eigenvalues with magnitude less than one, a necessary condition for the SOR method to converge. Also, matrices of this type occur in fluid flow problems where R is the Reynolds number. Indeed, the a_{ij} and b_{ij} were chosen, among others, from several fluid flow problems the author had occasion to solve at various times.

The eigenvalues of the appropriate matrices were determined using EISPACK. From the eigenvalues of the Jacobi matrix, the optimal relaxation factor for SOR was determined, followed by the evaluation of the spectral radii of the SOR method and the ad-hoc SOR method. To keep computer costs reasonable, our region was limited to 10×10 unknowns, or 100×100 matrices. The results are summarized in Graph I.

As R varied from 0 upward, the eigenvalues of the corresponding Jacobi matrix varied from all real, to general complex, to all purely imaginary. This afforded a convenient separation of the runs into three distinguishable cases.

i) When the eigenvalues of the Jacobi matrix are all real, the two methods have comparable rates of convergence. Even though the SOR method has a slightly smaller spectral radius, it is usually not possible to guess which method will produce the least iterations.

ii) As the eigenvalues of the Jacobi matrix begin to become complex, several things happen. The SOR method deteriorates rapidly at the optimal relaxation factor. Also this factor becomes increasingly more difficult to find without knowing virtually the entire spectrum of the Jacobi matrix. On the other hand, the ad-hoc method shows surprising improvement.

iii) As the eigenvalues of the Jacobi matrix become purely imaginary, the two methods again approach each others' rate of convergence. However, the optimal relaxation factor for SOR is generally quite small, along with the range of convergent factors. Thus, for example, we might have a case where the optimal factor is 0.095, but convergence occurs only in the range 0 to 0.0955, making estimations virtually impossible. This is graphically illustrated in Figures 3 and 4 of [8].

On the other hand, the ad-hoc SOR method has no problem computing the local relaxation factors. However, there are situations where the ad-hoc method can diverge. These situations are generally physically unrealistic.

It is possible to improve the convergence of the ad-hoc SOR method by perturbing the relaxation factors upward in Region i) and downward in Region iii), even to the point where divergent cases converge. However, this aspect needs more investigation.

Summaries of some representative results along with details appear in Tables I to V of [4]. Botta and Veldman ran a series of problems with results consistent to Graph I.

GRAPH I: SPECTRAL RADIUS VERSUS R

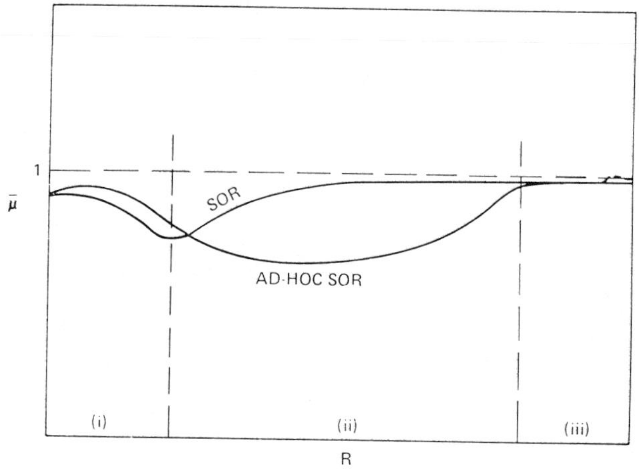

It might be noted that each iteration of ad-hoc SOR takes about three and one-half to four percent longer than each iteration of SOR.

V. COMPLEX LINEAR SYSTEMS

For complex linear systems, the situation is more involved. Indeed, one may have a constant Jacobi matrix over a rectangle with Dirichlet boundary conditions where the SOR method will converge, but the ad-hoc method will not. Thus, consider the constant stencil

$$\begin{bmatrix} & 1+iR & \\ 1-iR & -4 & 1-iR \\ & 1+iR & \end{bmatrix}. \tag{5.1}$$

The eigenvalues of the corresponding Jacobi matrix are

$$\mu_{p,q} = \frac{1}{2}[\cos(\frac{\pi p}{N+1}) + \cos(\frac{\pi q}{M+1}) + iR[\cos(\frac{\pi q}{M+1}) - \cos(\frac{\pi p}{N+1})]] , \tag{5.2}$$

and these lie in a diamond-shaped region in the complex plane. For a mesh size of 5×5, when $R = 1$, the SOR method is convergent, but the ad-hoc SOR method is not; i.e., the spectral radius of J does not determine the optimal relaxation parameter in this case. (See Table I.)

TABLE I

$M = N = 5$

R	SOR ω	SOR $\lvert\lambda\rvert$	ADHOC ω	ADHOC $\lvert\lambda\rvert$
0	1.333	.333	1.333	.333
1	0.995	.753	1.333	1.94
1.5	0.773	.841	0.758	.846
5	0.477	.919	0.477	.919

The only example of a complex system we have studied to date is

$$(\nabla^2 - ik)\psi = 1 \tag{5.3}$$

on a quarter circle with Dirichlet boundary conditions. Table II contains some results. It is seen that the behavior is

TABLE II

 $(\Delta - ik)\psi = 1$

h^{-1}	k	adhoc λ	adhoc ω	adhoc Iter	SOR λ	ω_b	SOR Iter
10	1	.591	1.57-1.23	28	.597	1.56	25
10	10	.665	1.56-1.23	32	.678	1.52	31
10	100	.844	1.3 - 1.06	63	.853	1.09	64
15	1	–	1.67-1.25	41	–	1.64	34
15	10	–	1.67-1.25	46	–	1.64	44
15	100	–	1.43-1.17	126	–	1.21	130
31	1	–	1.82-1.07	77	–	1.80	66
31	10	–	1.82-1.07	86	–	1.80	83
31	100	–	1.61-1.07	310	–	1.42	331

similar to that of Region i) in the real linear case. We might note if the region were a rectangle, the eigenvalues of this equation would lie on a straight line in the complex plane through the origin and for this distribution, the spectral radius does determine the optimal factor [6]. The eigenvalues for the quarter circle lie almost on a straight line. This may account for the effectiveness of the ad-hoc method with this system.

VI. NONLINEAR SYSTEMS

For nonlinear systems, the ad-hoc SOR is, in general, surprisingly effective. Unfortunately, since the systems are nonlinear, the ad-hoc relaxation factors must be computed for each point for each iteration. Frequently, this causes the effectiveness of the method to be somewhat reduced; however, there are instances where some local relaxation technique must be used.

Suppose we have the nonlinear system

$$f_{ij}(x) = f_i(\cdots,x_j,\cdots) = 0, \quad i,j = 1,\ldots,n. \tag{6.1}$$

In order to use the ad-hoc SOR method, we rewrite (6.1) as

$$\begin{aligned} f_0(x) &= a_1x_1 + a_2x_2 + a_3x_3 + a_4x_4 - g_0(x)x_0 \\ &\quad - f_0 = 0 \end{aligned} \tag{6.2}$$

using the stencil (2.1), where the a_p may be functions of the solution. We shall compare nonlinear ad-hoc SOR with nonlinear SOR. Using the notation of (6.1), the nonlinear SOR method can be written as

$$x_0^{(k+1)} = x_0^{(k)} - \omega \frac{f_0(\tilde{x})}{\left(\frac{\partial f_0}{\partial x_0}(\tilde{x})\right)} \tag{6.3}$$

where the $\tilde{x}$ indicates that the new values of x_i are to be used as obtained. On the other hand, using the notation of (6.2), the nonlinear ad-hoc SOR method can be written as

$$x_0^{(k+1)} = x_0^{(k)} - \omega \frac{f_0(\tilde{x})x_0}{g_0(\tilde{x})} \tag{6.4}$$

Comparing (6.3) with (6.4) gives the relative complexities of the two methods except for the relaxation factor computation.

Several test problems were tried. The first was

$$\nabla^2 x = x^3. \tag{6.5}$$

Results are shown in Table III. The reason ad hoc took three times longer than SOR was because it took 0.00002 seconds per

TABLE III

$$\Delta x = x^3$$

$$f(x_0) = x_1 + x_2 + x_3 + x_4 - 4x_0 - h^2 x_0^3 = 0$$

SOR:

$$x^{k+1} = x^k + \omega \frac{f(\tilde{x}_0^{(k)})}{4 + 3h^2 x_0^{(k)2}}$$

$$x^{(0)} \equiv 50$$

AD-HOC:

$$x^{k+1} = x^{(k)} + \omega_0 \frac{f(\tilde{x}_0^{(k)})}{4 + h^2 x_0^{(k)2}}$$

h^{-1}	ADHOC N	SOR N	ω	TIME/IT ADHOC	TIME/IT SOR
19	65	66	1.73	.0119	.0042
39	130	129	1.855	.0495	.0167

point to compute the relaxation factors and half that time to evaluate the equation because of its stark simplicity. Similar results were obtained using e^x on the right side of (6.5).

A class of nonlinear problems that has been, and is being, solved by local relaxation techniques is partial differential systems over arbitrary regions that have been mapped into rectangles. This approach has been made popular by J.F. Thompson and associates, for fluid flow problems. (See [10] and references cited there.) For example, using a Y-shaped region, the following equations are applied

$$\begin{aligned} &\alpha y_{uu} - 2\beta y_{uv} + \gamma y_{vv} = 0 \\ &\alpha x_{uu} - 2\beta x_{uv} + \gamma x_{vv} = 0 \qquad (6.7) \\ &\alpha = y_v^2 + x_v^2;\ \beta = y_u y_v + x_u x_v;\ \gamma = y_u^2 + x_u^2 \end{aligned}$$

and the region in (x,y) coordinates is transformed into a rectangle with a slit in (u,v) coordinates. This transformation is usually computed using SOR with a relaxation factor culled from experience, frequently 1.75. This approach is effective, but the ad-hoc SOR method is also quite applicable. See Table IV.

TABLE IV

NUMERICAL MAP

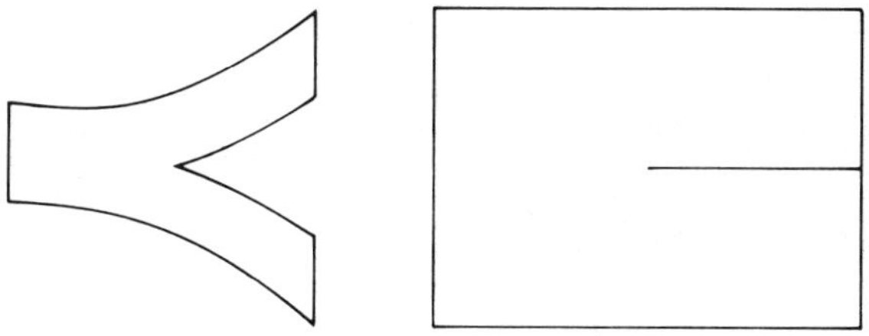

	SOR	ADHOC
M	105	69
ω	1.75	1.69–1.83
time-iter	.441	.503

If the original system of equations is the Navier-Stokes equations for stream-function-vorticity formulation on rectangular coordinates, the resulting equations may look like

$$\begin{aligned} &\psi_{xx} + \psi_{yy} = - \Omega \\ &\Omega_t + \mathrm{Re}(\psi_y \Omega_x - \Omega_x \psi_y) = \Omega_{xx} + \Omega_{yy}. \end{aligned} \tag{6.8}$$

After the transformation (6.7), they become

$$\begin{aligned} &\alpha\psi_{uu} - 2\beta\psi_{uv} + \gamma\psi_{vv} = -J^2\Omega \\ &\Omega_t + J\,\mathrm{Re}(\psi_v\Omega_u - \psi_u\Omega_v) = \alpha\Omega_{uu} - 2\beta\Omega_{uv} + \gamma\Omega_{vv} \\ &J = x_u y_v - x_v y_u \end{aligned} \tag{6.9}$$

in the u-v coordinate system. Note that the cross derivative term will produce a nine-point stencil whereas the original equations (6.8) needed only five-point stencils. Since most (if not all) local relaxation techniques use only five points, what is usually done is to ignore the coefficients due to the cross derivative term in computing the relaxation factor, but include these terms in the relaxation. Much literature exists on the solution of (6.9) but little is usually said about the local relaxation techniques used. One exception is [7] where it is reported that if (2.3) is real, then (2.4) is used. Otherwise, the absolute value of (2.3) is used in (2.5).

Finally, a rather interesting nonlinear equation that we tried was the equation of minimal surface investigated by Concus [3]. The equation is

$$\begin{aligned} &(1 + u_y^2)u_{xx} - 2u_x u_y u_{xy} + (1 + u_x^2)u_{yy} = 0 \\ &u(0,y) = u(2,y) = u(x,1) = 0;\; u(x,0) = K \sin [\tfrac{\pi x}{2}]. \end{aligned} \tag{6.10}$$

Concus considered a vectored form of this equation to derive a nonlinear difference equation with a symmetric matrix, which can be found in [3]. This, in turn, can be written in the ad-hoc formulation as

$$f_0 = A_0u_0 - A_1u_1 - A_2u_2 - A_3u_3 - A_4u_4 = 0$$

$$A_1 = \gamma_1 + \gamma_6;\; A_2 = \gamma_2 + \gamma_6;\; A_3 = \gamma_0 + \gamma_2;$$

$$A_4 = \gamma_0 + \gamma_1$$

$$A_0 = A_1 + A_2 + A_3 + A_4 \tag{6.11}$$

$$|\Delta_0|^2 = \frac{1}{2K^2}[(u_o - u_4)^2 + (u_0 + u_3)^2 + (u_3 - u_8)^2 + (u_4 - u_8)^2]$$

$$\gamma_0 = 1/\sqrt{1 + |\Delta_0|^2} \ .$$

Details on the computational aspects of the problem can be found in [3]. Table V contains a summary of results where we refer to (6.11) as Ad-hoc I. As can be seen, even the initial

TABLE V

MINIMAL SURFACE

INITIAL VALUES

a) $u(x,y) = K \dfrac{\sin(\pi x/2)\sinh \pi(1-y)/2}{\sinh(\pi/2)}$

b) $u(x,y) = 0$

c) $u(x,y) = K$

10 x 20:

	K	SOR	ω_b	ADHOC I	ADHOC II
a)	1	37	1.66	97	47
	5	103	1.615*	327	71
b)	1	43	1.65	112	51
	5	59	1.73**	304	56
c)	1	116	1.45†	125	58
	5	277	1.33††	389	81

BLOW UP AT* - 1.62, ** - 1.8, † - 1.46, †† - 1.34
TIME/ITERATION: SOR – .014 sec.; AD-HOC I – .019 sec.; AD-HOC II – .012 sec.

starting values affected the behavior of the iterative techniques. Also, a look at the optimal ω's and the convergent ranges indicates a substantial amount of work went into finding those ω's. Application of the ad-hoc method for comparison was somewhat disappointing.

However, equation (6.11) was derived to accommodate the SOR method. Consider, instead, the differential form (6.10). Replacing this with standard finite differences, we have

$$\begin{aligned} & B_1u_1 + B_2u_2 + B_3u_3 + B_4u_4 - B_0u_0 \\ & \qquad + B_5(u_6 + u_8 - u_7 - u_5) = 0 \\ & B_1 = B_3 = \left[1 + \left[\frac{u_2 - u_4}{2h}\right]^2\right]; \\ & B_2 = B_4 = \left[1 + \left[\frac{u_1 - u_3}{2h}\right]^2\right] \\ & B_0 = B_1 + B_2 + B_3 + B_4; \quad B_5 = -\frac{(u_1 - u_3)(u_2 - u_4)}{2h^2}. \end{aligned} \tag{6.12}$$

This set of equations was solved by the ad-hoc method quite efficiently, as indicated in Table V, referring to Ad-hoc II.

VII. CONCLUSIONS

Local relaxation factors are computed in a variety of ways. Botta and Veldman [1] compared ad-hoc SOR and SOR with four other approaches. In [4], we considered some of these and yet another one. In effect, ad-hoc SOR seems to be the technique for all seasons. It was the only one that performed consistently well.

In and of itself, the ad-hoc SOR method is a surprisingly effective local relaxation technique. Generally, the SOR method might be recommended if the optimal relaxation factor is known, and if the problem is not likely to fall into Region ii) of Graph I and is to be solved many times. However, with no knowledge of the optimal SOR factor available, ad-hoc SOR is not a poor strategy to use. Certainly, if the problem is to be solved only once or a very few times, ad-hoc SOR can be cost effective. What is needed now is some analytic results on when we can formally expect it to be effective.

BIBLIOGRAPHY

1. Botta, E.F.F., and Veldman, A.E.P., J. of Comp. Phys. 48, 127-149 (1982).

2. Brazier, P.H., Comput. Meth. Appl. Mech. Engin. 3, 335-347 (1974).

3. Concus, P., M. O. C. 21, 340-350 (1967).

4. Ehrlich, L.W., J. of Comp. Phys. 44, 31-45 (1981).

5. Hageman, L.A., and Young, D.M., Applied Iterative Methods, Academic Press, New York, (1981).

6. Kredell, B., Bit 2, 143-152 (1962).

7. Moitra, A., Ph.D. Dissertation, Mississippi State, Mississippi (1982).

8. Rigal, A., J. of Comp. Phys. 32, 10-23 (1979).

9. Schultz, M.H., Elliptic Problem Solvers, Academic Press, New York, (1981).

10. Thompson, J.F., and Warsic, J.U.A., J. of Comp. Phys. 47, 1-108 (1982).

11. Young, D.M., and Eidson, H.D., Report CNA-1, Center for Numerical Analysis, University of Texas at Austin (1970).

ITERATIVE METHODS FOR NON-SELF-ADJOINT ELLIPTIC PROBLEMS

Howard C. Elman[1]

1. INTRODUCTION

The numerical solution of non-self-adjoint elliptic boundary value problems requires the solution of systems of linear equations of the form

$$Ax = b, \tag{1}$$

where A is a large sparse nonsingular nonsymmetric matrix of order N. In most cases, direct methods for solving (1) entail an LU-decomposition of A, which can be prohibitively expensive for large problems. Iterative methods have low storage requirements, but must converge rapidly to be effective. Very effective iterative methods exist for solving the symmetric positive-definite linear systems arising from the discretization of self-adjoint elliptic problems (see e.g. [12]), but little of the convergence theory carries over to the nonsymmetric case. In this paper, we discuss recent progress in the development of iterative methods for solving nonsymmetric systems.

Progress has been made along two directions:

1. *basic iterative methods*, i.e. techniques for computing the i'th approximate solution x_i from previously computed quantities; and
2. *preconditioning techniques*, in which an auxiliary operator Q is used to generate an equivalent linear system such as

[1]Computer Science Department, Yale University, New Haven, CT

This work was supported by the U.S. Office of Naval Research under grant N00014-82-K-0184.

ISBN 0-12-100560-7

$$AQ^{-1}\tilde{x} = b, \quad x = Q^{-1}\tilde{x}$$

that is in some sense easier to solve.

In general, these two approaches are independent: any basic iterative method can be combined with any preconditioning technique.

Most of the effort in basic iterative methods has been directed toward Krylov subspace methods, in which x_i is chosen from the translated Krylov space $x_0 + S_i$, where

$$S_i = \text{span}\{r_0, Ar_0, \ldots, A^{i-1}r_0\},$$

and $r_0 = b - Ax_0$. One approach has been to generalize the conjugate gradient method (CG) [13], producing a method that needs no a priori parameter estimates. One such method, which depends explicitly on preconditioning by the symmetric part of A, was proposed by Concus and Golub [2] and Widlund [24]. More flexible methods include a collection of CG-like methods based on a truncated or restarted Gram-Schmidt-like computation, developed by several authors [1, 5, 6, 18, 23, 25, 26]; and the biconjugate gradient method, originally proposed by Fletcher [10] for symmetric indefinite systems (see [6, 12] for other references). A second important approach is Manteuffel's adaptive Chebyshev method [14, 15], which inherits certain asymptotic properties from the Chebyshev polynomials but requires estimates for the eigenvalues of A.

Most of the preconditionings used for nonsymmetric systems are straightforward generalizations of techniques for symmetric systems. For general matrices, they include incomplete factorizations, such as those developed by Dupont, Kendall and Rachford [4], Gustafsson [11], and Meijerink and van der Vorst [16]. Special techniques applicable to discretized elliptic problems include partial elimination (construction of a "reduced system;" see [12]) and fast direct methods [3, 6].

In this paper, we describe what we regard as the most salient features of the basic iterative methods, present some recent enhancements, and briefly discuss some preconditioning techniques. In Section 2, we describe representative examples and properties of the Krylov subspace methods. In Section 3, we discuss a new hybrid Chebyshev/CG-like method recently developed by Elman, Saad and Saylor [7]. In Section 4, we discuss the incomplete factorization and reduced system preconditioning techniques. In Section 5, we present some numerical experiments in which the methods discussed are used to solve a two-dimensional non-self-adjoint elliptic boundary value problem, and in Section 6, we present some open questions.

2. KRYLOV SUBSPACE METHODS

In this section, we consider three Krylov subspace methods: two different generalizations of the conjugate gradient method, and the adaptive Chebyshev method.

For symmetric positive-definite systems, the conjugate gradient method computes the unique point in $x_0 + S_i$ whose A-norm $(x-x_i,A(x-x_i))^{1/2}$ is smallest. We refer to this as a "global" property of CG: at step i, it solves an i-dimensional minimization problem. CG accomplishes this with a small amount of work per step (5N multiplications plus a matrix-vector product) and without a priori parameter (e.g. eigenvalue) estimates.

These properties of rapid convergence, low cost per step, and independence of parameter estimates have led to efforts to generalize CG. One type of generalization sacrifices global properties to retain low cost per step. By way of introduction, consider the following method [5, 6]:

Algorithm 1: The generalized conjugate residual method (GCR).

Choose x_0. Compute $r_0 = b\text{-}Ax_0$. Set $p_0 = r_0$.

FOR i=0 STEP 1 UNTIL Convergence DO

$$\alpha_i = (r_i,Ap_i)/(Ap_i,Ap_i),$$
$$x_{i+1} = x_i + \alpha_i p_i,$$
$$r_{i+1} = r_i - \alpha_i Ap_i,$$
$$b_j^{(i)} = -(Ar_{i+1},Ap_j)/(Ap_j,Ap_j),$$
$$p_{i+1} = r_{i+1} + \sum_{j=0}^{i} b_j^{(i)} p_j.$$

This method is convergent if the symmetric part $(A+A^T)/2$ is positive-definite. The directions $\{p_i\}$ are constructed to be A^TA-orthogonal, and as a result, the choice of the steplength α_i is such that x_{i+1} is the point in $x_0 + S_{i+1}$ with smallest residual norm $\|r_{i+1}\|_2$. Thus, GCR is optimal with respect to the residual norm. Unlike CG, however, GCR achieves global optimality at a high cost: the recurrence for p_{i+1} requires O((i+1)N) operations and storage. Though optimal, GCR is not practical if more than a few (say, twenty) iterations are required.

Indeed, as Faber and Manteuffel [9] have recently shown, except in special cases there are no optimal CG-like methods with short recurrences for nonsymmetric matrices. However, nonoptimal but effective techniques that do not require parameter estimates have been developed by forcibly limiting the cost per step. For example, the direction update can be truncated so that at most $k \ll N$ previous directions are used after iteration k:

$$p_{i+1} = r_{i+1} + \sum_{j=i-k+1}^{i} b_j^{(i)} p_j.$$

In this case, the iterate x_{i+1} is a *local* minimum, the point in

$$x_{i-k} + \text{span}\{p_{i-k},\ldots,p_i\}$$

whose residual norm $\|r_{i+1}\|_2$ is minimized. Alternatively, with a maximum of k directions saved, GCR can be restarted every k+1 steps. We refer to the truncated and restarted versions of GCR as ORTHOMIN(k) and GCR(k) respectively. One matrix-vector product is required at each step (see Table 1). Let M denote the symmetric part of A, and R the skew-symmetric part $(A-A^T)/2$. When M is positive-definite, the residuals generated by both methods satisfy [5, 6]

$$\|r_i\|_2 \le \left[1 - \frac{\lambda_{min}(M)^2}{\lambda_{min}(M)\lambda_{max}(M)+\rho(R)^2}\right]^{i/2} \|r_0\|_2, \tag{2}$$

where $\lambda_{min}(M)$ and $\lambda_{max}(M)$ denote the smallest and largest eigenvalues of M, respectively, and $\rho(R)$ denotes the spectral radius of R. Hence, both methods are convergent. In practice, we have found this bound to be pessimistic and these methods to be effective solution techniques (see [6] and Section 5).

A second generalization of CG is the biconjugate gradient method (BCG) [10], which attempts to impose a global condition with short recurrences.

Algorithm 2: The biconjugate gradient method.

Choose x_0. Compute $r_0 = b-Ax_0$. Set $p_0=r_0$, $\bar{r}_0=r_0$, $\bar{p}_0=p_0$.
FOR i=0 STEP 1 UNTIL Convergence DO

$$\alpha_i = (\bar{r}_i,r_i)/(\bar{p}_i,Ap_i),$$
$$x_{i+1} = x_i + \alpha_i p_i,$$
$$r_{i+1} = r_i - \alpha_i A p_i, \qquad \bar{r}_{i+1} = \bar{r}_i - \alpha_i A^T \bar{p}_i,$$
$$\beta_{i+1} = (\bar{r}_{i+1},r_{i+1})/(\bar{r}_i,r_i),$$
$$p_{i+1} = r_{i+1} + \beta_i p_i, \qquad \bar{p}_{i+1} = \bar{r}_{i+1} + \beta_i \bar{p}_i.$$

The introduction of the auxiliary vectors $\{\bar{r}_i\}$ and $\{\bar{p}_i\}$ results in the construction of the auxiliary Krylov space

$$\overline{S}_i = \text{span}\{r_0, A^T r_0, \ldots, (A^T)^{i-1} r_0\}.$$

The choice of the scalars $\{\alpha_i\}$ and $\{\beta_i\}$ establish the global Galerkin conditions

$$(r_i, v) = 0, \quad v \in \overline{S}_i, \qquad (\bar{r}_i, v) = 0, \quad v \in S_i.$$

However, these relations are imposed at the cost of an extra matrix-vector product, by A^T, at each step (see Table 1). Morever, although BCG can in principle be applied to any nonsingular linear system, little is known about convergence. We know of no error norm that is monotonically decreasing, and although the method will terminate after at most N iterations, it may break down with $\alpha_i = 0$ before the solution is found. (See [12, 19].)

A contrasting approach is Manteuffel's adaptive Chebyshev method [14, 15], which displays rapid convergence at low cost per step but is highly dependent on parameter estimates. This method generates a sequence of iterates $\{x_i\}$ whose residuals satisfy

$$r_i = C_i(A) r_0, \tag{3}$$

where C_i is a scaled, translated i'th-degree Chebyshev polynomial:

$$C_i(z) = T_i\left(\frac{d-z}{c}\right) \Big/ T_i\left(\frac{d}{c}\right). \tag{4}$$

The scaling and translation depend on locating an ellipse E that is small in a certain sense which encloses the spectrum of A but whose interior does contain the origin; d is the center of E, and $d \pm c$ are its foci. Rapid convergence follows from the asymptotic optimality in L_∞ of the Chebyshev polynomials on ellipses in the complex plane [15]. Moreover, the three-term recurrence for Chebyshev polynomials induces a three-term recurrence

$$x_{i+1} = x_i + \alpha_i r_i + \beta_i (x_i - x_{i-1}), \tag{5}$$

in which the scalars α_i, β_i are computed with scalar arithmetic, so that the work per step is lower than for the CG-like methods. (See Table 1.)

The drawback is the explicit dependence on the ellipse parameters c and d, which are computed from the estimated extreme eigenvalues of A. Starting with (possibly arbitrary) initial values for c, d, the adaptive Chebyshev method estimates these eigenvalues dynamically for systems whose symmetric part is positive-definite [14]. With given values for c, d, the generated residuals satisfy

$$r_i \approx S(A)^i r_0, \quad i \to \infty, \tag{6}$$

where S(A) is the linear operator induced by

$$S(z) = S_{c,d}(z) = \frac{d\text{-}z + [(d\text{-}z)^2 - c^2]^{1/2}}{d + [d^2 - c^2]^{1/2}}.$$

That is, the residuals resemble the vectors generated by the power method for S(A). If some eigenvalue of S(A) has modulus greater than one and r_0 has a component in the corresponding eigenvector, then the residuals will diverge but will eventually become rich in that eigenvector. In this case, the residuals can be used to compute estimates for the eigenvalues of S(A). These are then used to compute eigenvalue estimates for A, which in turn are used to compute new ellipse parameters. The Chebyshev iteration is then restarted with the new ellipse parameters. Note that since the eigenvalue computation is based on a variant of the power method, this computation will actually be facilitated if the residuals are allowed to diverge. Numerical experiments [6] indicate that the norms of the residuals generated by the adaptive Chebyshev method may increase by several orders of magnitude before good parameters are obtained, but that convergence is rapid afterwards (see Section 5).

	ORTHOMIN(k)	GCR(k)	BCG	Chebyshev[2]
Work/Loop	(3k+4)N, Av	((3/2)k+4)N, Av	7N, Av, $A^T v$	2N, Av
Storage	(2k+4)N	(2k+4)N	6N	8N

Table 1: Work and storage costs.

Table 1 contains the work (number of floating point multiplications and divisions) and storage costs (not including storage for A, b) of the methods of this section. Matrix-vector products are counted separately: Av denotes a matrix multiplication by A and $A^T v$ denotes a matrix multiplication by A^T.

3. A HYBRID METHOD

A method that attempts to combine two of the approaches of the previous section is the hybrid Chebyshev/gradient method recently devised by Elman, Saad and Saylor [7]. It uses the basic Chebyshev iteration (5) but replaces the Manteuffel adaptive procedure with a Lanczos-type method that can be used to simul-

[2]The overhead of 14N operations for the adaptive step is not included.

taneously estimate eigenvalues and improve the solution iterate. Again, it is applicable to systems with positive-definite symmetric part.

In the hybrid method, the eigenvalue estimation is performed by Arnoldi's method [20], which estimates the eigenvalues of A by reducing it to upper-Hessenberg form. Starting with an initial vector v_1 with $||v_1||_2=1$, this method generates an orthonormal basis for the Krylov space span$\{v_1,Av_1,...,A^{k-1}v_1\}$ by an iteration of the form

$$h_{i+1,i}v_{i+1} = Av_i - \sum_{j=1}^{i} h_{ji}v_j.$$

The orthonormal matrix $V_k = [v_1,...,v_k]$ and the k'th order upper-Hessenberg matrix $H_k = [h_{ij}]$ satisfy $V_k^T A V_k = H_k$. Taking this relation as an approximate similarity transformation, Arnoldi's method uses the eigenvalues of H_k as estimates for the eigenvalues of A.

Information obtained from Arnoldi's method can be used as the basis of an iterative method for solving (1). Given some initial guess, x_0, let v_1 denote the normalized residual $r_0/||r_0||_2$. Then the k Arnoldi vectors $\{v_i\}_{i=1}^k$ can be used to compute the point

$$\tilde{x} \in x_0 + \text{span}\{r_0,Ar_0,...,A^{k-1}r_0\}$$

whose residual norm $||b-A\tilde{x}||_2$ is minimized. Let $\overline{H}_k$ denote the (k+1)xk matrix obtained by appending to H_k a row with single nonzero entry $h_{k+1,k}$ in column k. Then $\tilde{x}$ is given by

$$\tilde{x} = \sum_{i=1}^{k} \alpha_i v_i,$$

where the scalars $\{\alpha_i\}_{i=1}^k$ solve the least squares problem

$$\min || \overline{H}_k\alpha - ||r_0||_2 e_1 ||_2.$$

This is the generalized minimum residual (GMRES) method of Saad and Schultz [21]. It is mathematically equivalent to GCR when the symmetric part of A is positive-definite, and it is also applicable to arbitrary nonsingular matrices.

The hybrid method combines the Chebyshev method with the CG-like method GMRES, using Arnoldi's method for eigenvalue estimates. In the following simple form, the adaptive step is invoked if the residual norm generated by the Chebyshev iteration exceeds a specified tolerance τ relative to the smallest residual encountered, or after at most n Chebyshev steps. It is also used to generate initial eigenvalue estimates.[3]

Algorithm 3: The hybrid method.

Choose x_0. Compute $r_0 = b - Ax_0$.

UNTIL Convergence DO

Adaptive Step: Set v_1 = the current normalized residual, perform k Arnoldi/GMRES steps, and use the new eigenvalue estimates to update (or initialize) the ellipse parameters.

Chebyshev Step: Set $i_{max}=i+n$.

WHILE $(\|r_i\|_2/\|r_{min}\|_2 \leq \tau$ and $i \leq i_{max})$

Compute x_i by the Chebyshev iteration (5).

The adaptive step uses $(k^2+4k+1)N$ multiplications and k matrix-vector products Av, plus kN storage.

The hybrid method has two advantages over Manteuffel's original adaptive Chebyshev method. First, by not using the approximate powers of S(A) (6), the hybrid method is not as dependent on divergence of the Chebyshev iteration for eigenvalue estimates. The use of Arnoldi's method allows much more stringent monitoring of the residuals [7]. Second, in addition to providing new eigenvalue estimates, the adaptive step improves the quality of the approximate solution provided to the next Chebyshev run. Indeed, the GMRES steps approximately annihilate the components in the residual of the eigenvectors corresponding to the computed eigenvalue estimates [7, 21].

4. SOME PRECONDITIONING TECHNIQUES

In this section, we briefly describe some choices for the preconditioning operator Q. See [6] for more details and other preconditionings.

One effective class of preconditionings is based on approximate factorization of A. Q is the product of a lower-triangular matrix L and an upper-triangular matrix U that resemble the factors obtained from Gaussian elimination but are forced to be sparse, so that the preconditioned matrix-vector product $AQ^{-1}v$ is not too ex-

[3]Manteuffel's method for computing ellipse parameters from eigenvalue estimates is still used [15].

pensive. Suppose $\mathbf{Z} \subseteq \{(i,j)|1\leq i,j\leq N\}$ is given. In the incomplete LU-factorization (ILU) [16], L_{ij} and U_{ij} are defined to be zero for $(i,j) \in \mathbf{Z}$, and $[LU]_{ij} = A_{ij}$ for $(i,j) \notin \mathbf{Z}$. In the modified incomplete factorization, (MILU) [4, 11], an attempt is made to compensate for the error caused by restricting the fill-in of the factors. $[LU]_{ij} = A_{ij}$ for $(i,j) \notin \mathbf{Z}$ and $i\neq j$, but the diagonal entry L_{ii} is modified so that $\sum_{j=1}^{N} (A_{ij} - [LU]_{ij}) = 0$. Note that these preconditionings can be defined for any matrix. For example, the zero pattern $\mathbf{Z}$ can consist of the set of indices for which $A_{ij}=0$. Locations of additional fill-in for matrices arising from the finite-difference discretization of elliptic problems have been considered; see e.g. [11, 16].

When five-point (in two dimensions) or seven-point (in three dimensions) finite difference schemes are used to discretize elliptic problems, the resulting matrices are two-cyclic, i.e., the rows and columns can be permuted symmetrically so that (1) has the form

$$Ax = \begin{bmatrix} D_1 & C_1 \\ C_2 & D_2 \end{bmatrix} \begin{bmatrix} x^{(1)} \\ x^{(2)} \end{bmatrix} = \begin{bmatrix} b^{(1)} \\ b^{(2)} \end{bmatrix},$$

where D_1 and D_2 are diagonal matrices. If

$$Q = \begin{bmatrix} D_1^{-1} & 0 \\ -C_2D_1^{-1} & I \end{bmatrix},$$

then the left-preconditioned system has the form

$$Q^{-1}Ax = \begin{bmatrix} I & D_1^{-1}C_1 \\ 0 & D_2-C_2D_1^{-1}C_1 \end{bmatrix} \begin{bmatrix} x^{(1)} \\ x^{(2)} \end{bmatrix} = \begin{bmatrix} D_1^{-1}b^{(1)} \\ b^{(2)}-C_2D_1^{-1}b^{(1)} \end{bmatrix},$$

in which the block unknowns $x^{(1)}$ and $x^{(2)}$ are decoupled. Block $x^{(2)}$ can be computed by solving the reduced system

$$[D_2-C_2D_1^{-1}C_1]\, x^{(2)} = b^{(2)}-C_2D_1^{-1}b^{(1)} \tag{7}$$

whose order is typically approximately N/2. Block x_1 is then given by $D_1^{-1}[b^{(1)}-C_1x^{(2)}]$.

Note that with this reduced system preconditioning, the coefficient matrix and right hand side of (7) are formed explicitly. This enables (7) to also be preconditioned by, for example, an incomplete factorization. If no second preconditioning is used, then the reduced system need not be formed.

5. NUMERICAL EXPERIMENTS

In this section, we describe some numerical experiments for solving the two-dimensional elliptic partial differential equation

$$-(e^{-xy}u_x)_x - (e^{xy}u_y)_y + \gamma[(x+y)u_y + ((x+y)u)_y] + [1/(1+x+y)] = G, \qquad (8)$$

with the right hand side G chosen so that the solution is

$$u(x,y) = x\, e^{xy} \sin(\pi x) \sin(\pi y).$$

We pose (8) on the unit square $\{0 \leq x,y \leq 1\}$ with homogeneous Dirichlet boundary conditions and discretize using centered finite differences on a uniform 47x47 grid, producing a linear system of order N = 2209.

We present two sets of experiments, corresponding to the values $\gamma=5$ and $\gamma=50$ for the scalar γ in (8). In each case, we solve (8) using ORTHOMIN(1), BCG, adaptive Chebyshev, and the hybrid method with four Arnoldi vectors, together with the ILU preconditioning. In addition, we solve the reduced system using ORTHOMIN(1) and the ILU preconditioning. As a benchmark, we show the performance of the conjugate gradient method applied to the normal equations of the ILU preconditioned system.[4] Figures 2 and 3 graph work vs. residual norm (on a logarithmic scale) for $\gamma=5$ and $\gamma=50$, respectively. All overhead for computing preconditionings and adapting is included in the operation counts. The tests were run on a VAX11-780 in double precision (55 bit mantissa). More extensive experiments and details on implementation are given in [6, 7].

6. OPEN QUESTIONS

Two open questions are:

1. how well do these methods generalize for indefinite problems;
2. can strong error bounds be found.

Most of the iterative methods that we have discussed are rigorously applicable only when the symmetric part of the coefficient matrix is positive-definite. There are two ways to handle indefinite systems. One is to choose an accurate enough preconditioning so that the preconditioned matrix AQ^{-1} has positive-definite sym-

---

[4]This method is equivalent to Paige and Saunders' LSQR method [17], which is more stable but requires more work per step. We also compared these two methods in these tests and found the residual norms to be nearly identical at each step.

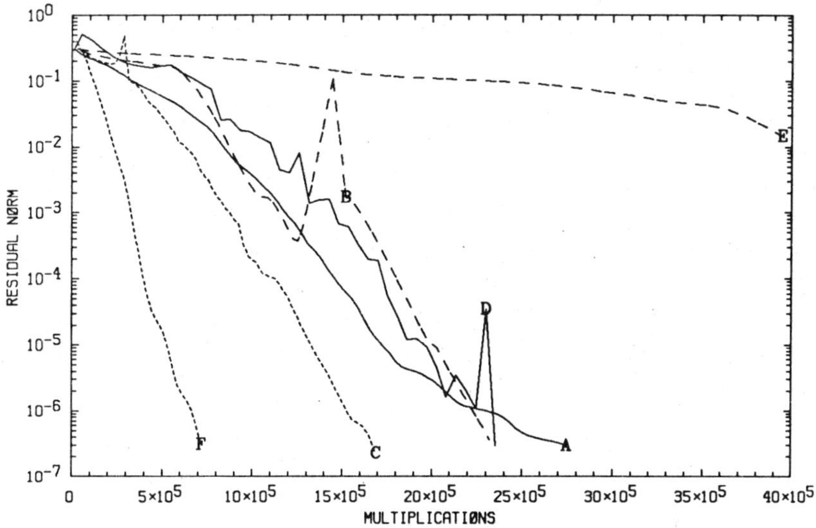

A: ORTHOMIN(1)
B: CHEBYSHEV
C: HYBRID, 4 ARNOLDI VECTORS
D: BICONJUGATE GRADIENT
E: CG / NORMAL EQUATIONS
F: REDUCED SYSTEM + ORTHOMIN(1)/ILU

Figure 1: Residual norm vs. multiplications, γ=5, ILU preconditioning.

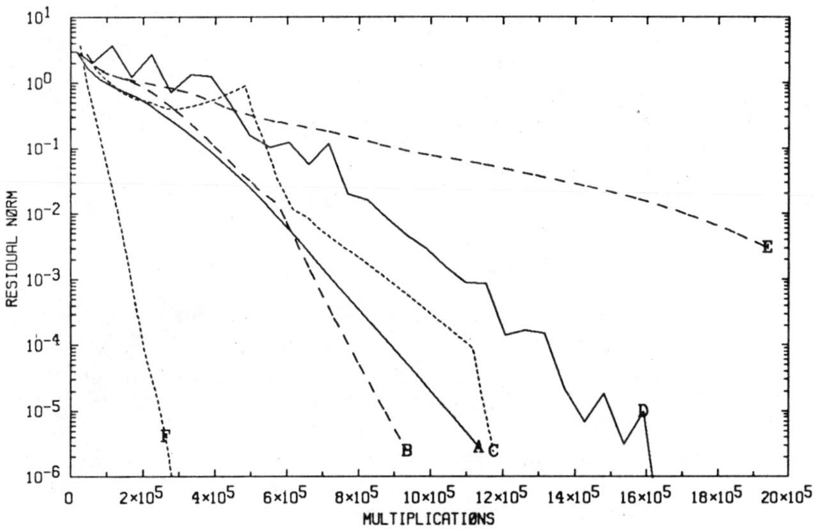

A: ORTHOMIN(1)
B: CHEBYSHEV
C: HYBRID, 4 ARNOLDI VECTORS
D: BICONJUGATE GRADIENT
E: CG / NORMAL EQUATIONS
F: REDUCED SYSTEM + ORTHOMIN(1)/ILU

Figure 2: Residual norm vs. multiplications, γ=50, ILU preconditioning.

metric part. We have, however, encountered situations where apparently effective preconditionings lead to indefinite preconditioned systems [6]. An alternative is for the basic iterative method to handle indefiniteness. Several techniques may be applicable in this domain: the GMRES method mentioned in Section 3; the incomplete orthogonalization method of Saad [18]; ORTHODIR, proposed by Young and Jea [12, 26]; the biconjugate gradient method; a version of Broyden's method developed for sparse systems by Engleman, Strang and Bathe [8]; and a new adaptive method devised by Smolarski and Saylor [22]. None of these ideas have been stringently tested, though, and little is known about convergence rates.

Finally, the analysis of the CG-like methods [5, 6] and of the Chebyshev method [15] prove that these techniques converge, but it is weaker than the analysis of methods for symmetric positive-definite problems. We do not know whether existing error bounds, such as (2), are tight or if there are stronger ones.

References

[1] O. Axelsson. Conjugate gradient type methods for unsymmetric and inconsistent systems of linear equations. *Linear Algebra and Appl.* 29:1-16, 1980.

[2] P. Concus and G. H. Golub. A generalized conjugate gradient method for nonsymmetric systems of linear equations. In R. Glowinski and J. L. Lions, Editors, *Lecture Notes in Economics and Mathematical Systems, Vol. 134*, Springer-Verlag, Berlin, 1976, pp. 56-65.

[3] P. Concus and G. H. Golub. Use of fast direct methods for the efficient numerical solution of nonseparable elliptic equations. *SIAM J. Numer. Anal.* 10:1103-1120, 1973.

[4] T. Dupont, R. P. Kendall and H. H. Rachford Jr. An approximate factorization procedure for solving self-adjoint elliptic difference equations. *SIAM J. Numer. Anal.* 5:559-573, 1968.

[5] S. C. Eisenstat, H. C. Elman and M. H. Schultz. Variational iterative methods for nonsymmetric systems of linear equations. *SIAM J. Numer. Anal.* 20:345-357, 1983.

[6] H. C. Elman. *Iterative Methods for Large, Sparse, Nonsymmetric Systems of Linear Equations.* Ph.D. Thesis, Department of Computer Science, Yale University, 1982. Also available as Technical Report 229.

[7] H. C. Elman, Y. Saad and P. E. Saylor. *A Hybrid Chebyshev Krylov-Subspace Method for Nonsymmetric Systems of Linear Equations.* 1983. In preparation.

[8] M. S. Engleman, G. Strang and K.-J. Bathe. The application of quasi-Newton methods in fluid mechanics. *Int. J. Numer. Meth. Eng.* 17:707-718, 1981.

[9] V. Faber and T. Manteuffel. *Conjugate Gradient Methods.* Technical Report LA-UR-82-1967, Los Alamos National Laboratory, 1982.

[10] R. Fletcher. Conjugate gradient methods for indefinite systems. In G. A. Watson, Editor, *Numerical Analysis Dundee 1975*, Springer-Verlag, New York, 1976, pp. 73-89.

[11] I. Gustafsson. A class of first order factorizations. *BIT* 18:142-156, 1978.

[12] L. A. Hageman and D. M. Young. *Applied Iterative Methods.* Academic Press, New York, 1981.

[13] M. R. Hestenes and E. Stiefel. Methods of conjugate gradients for solving linear systems. *J. Research Nat. Bur. Standards* 49:409-435, 1952.

[14] T. A. Manteuffel. Adaptive procedure for estimation of parameters for the nonsymmetric Tchebychev iteration. *Numer. Math.* 31:187-208, 1978.

[15] T. A. Manteuffel. The Tchebychev iteration for nonsymmetric linear systems. *Numer. Math.* 28:307-327, 1977.

[16] J. A. Meijerink and H. A. van der Vorst. An iterative solution method for linear systems of which the coefficient matrix is a symmetric M-matrix. *Math. Comp.* 31:148-162, 1977.

[17] C. C. Paige and M. A. Saunders. LSQR: An algorithm for sparse linear equations and sparse least squares. *ACM Trans. Mathematical Software* 8:43-71, 1982.

[18] Y. Saad. Krylov subspace methods for solving large unsymmetric linear systems. *Math. Comp.* 37:105-126, 1981.

[19] Y. Saad. The Lanczos biorthogonalization algorithm and other oblique projection methods for solving large unsymmetric systems. *SIAM J. Numer. Anal.* 19:485-506, 1982.

[20] Y. Saad. Variations on Arnoldi's method for computing eigenelements of large unsymmetric matrices. *Linear Algebra and Appl.* 34:269-295, 1980.

[21] Y. Saad and M. H. Schultz. *GMRES: A Generalized Minimal Residual Algorithm for Solving Nonsymmetric Linear Systems.* Technical Report 254, Yale University Department of Computer Science, 1983.

[22] D. C. Smolarski and P. E. Saylor. *Optimum Parameters for the Solution of Linear Equations by Richardson's Iteration.* May 1982. Unpublished manuscript.

[23] P. K. W. Vinsome. Orthomin, an iterative method for solving sparse sets of simultaneous linear equations. In *Proceedings of the Fourth Symposium on Reservoir Simulation*, Society of Petroleum Engineers of AIME, 1976, pp. 149-159.

[24] O. Widlund. A Lanczos method for a class of non-symmetric systems of linear equations. *SIAM J. Numer. Anal.* 15:801-812, 1978.

[25] D. M. Young. *Accelerating non-symmetrizable iterative methods.* 1983. To appear in Proceedings of the Elliptic Problem Solvers Conference held at The Naval Postgraduate School, Monterey CA, Jan. 1983.

[26] D. M. Young and K. C. Jea. Generalized conjugate gradient acceleration of nonsymmetrizable iterative methods. *Linear Algebra and Appl.* 34:159-194, 1980.

IMPLICIT BLOCK EXPLICIT OVERRELAXATION SCHEMES

David J. Evans*

I. INTRODUCTION

Point iterative methods in which, at any one time only a single equation of the linear system is treated are well known (Varga, 1962), Young (1971).

The motivation in a group (or block) iterative method is to group a certain number of individual equations (mesh points) and treat this group implicitly, i.e., similar to the way a single point is treated in the point iterative methods. The solution of the points within each such group is invariably carried out directly by a specially designed efficient algorithm.

The main concern of this paper is to construct new groupings of the mesh points of the network into small order groups of 4 and 9 points and to investigate their advantages when used explicitly in iterative methods.

We consider the model problem of the Laplace Equation in the unit square (Figure 1) with Dirichlet boundary conditions, (Varga, 1962).

For a large class of 2-dimensional linear elliptic differential equations in which a 5-point approximation scheme on a uniform mesh (Figure 2) is used, it is a simple procedure to approximate the partial derivatives of the PDE by suitable finite difference expressions using Taylor series and quadrature techniques. Thus, the finite difference equation at the point P has the form

*Department of Computer Studies, Loughborough University of Technology, Loughborough, Leicestershire, U.K.

ISBN 0-12-100560-7

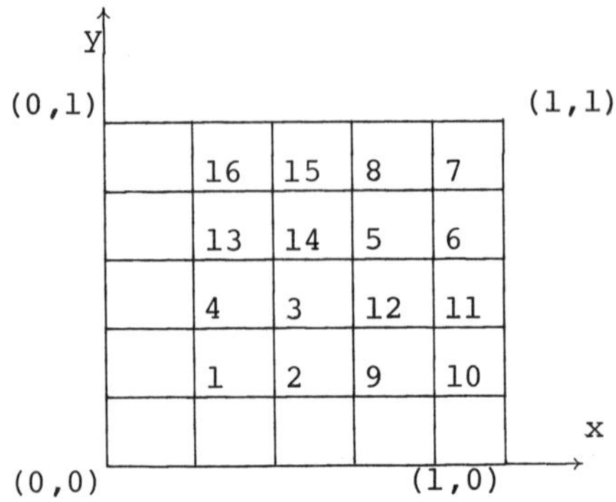

FIGURE 1

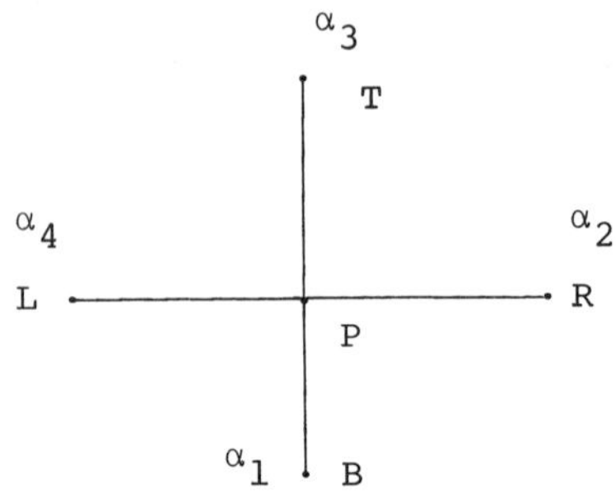

FIGURE 2

$$\alpha_4 u_{L,P} + \alpha_1 u_{B,P} + u_P + \alpha_2 u_{R,P} + \alpha_3 u_{T,P} = b_P \tag{1.1}$$

for all points within a coordinate square region as in Fig. 1. Here we have denoted the points on the network to the Left, Top, Right and Bottom of the representative point P in Fig. 2.

For such equations, the coefficients α_1, α_2, α_3 and α_4 satisfy the relationship

$$1 \geq \alpha_1 + \alpha_2 + \alpha_3 + \alpha_4 , \tag{1.2}$$

and when grouped together in matrix form reduce the problem to one of solving a sparse system of linear difference equations $A\underline{u} = \underline{b}$ where the coefficient matrix A is of large order, non-singular, positive definite and symmetric.

II. GROUP ITERATIVE METHODS

Consider a system of linear equations,

$$\sum_{j=1}^{n} a_{i,j} u_j = b_i , \quad i = 1,2,\ldots,n , \tag{2.1}$$

which can be written in matrix form as

$$A\underline{u} = \underline{b} . \tag{2.2}$$

In group iterative methods, the equations in (2.1) are first assigned into different groups or subsets numbered 1,2,...,N such that each of the above equations belongs to one and only one group. Then, the corresponding unknowns u_i of each group are solved in which the other unknowns belonging to the remaining groups are treated as known quantities.

Definition 2.1:

An ordered grouping π of T, the set of the first n integers, is a sub-division of T into disjoint subsets $G_1, G_2, \ldots, G_N$ such that $G_1 \cup G_2 \cup G_3, \ldots, \cup G_N = T$.

We now define an ordered grouping π of the equations in (2.1). The submatrices $A_{k,\ell}$ for $k,\ell = 1,2,\ldots,N$ are such that $A_{k,\ell}$ is formed from the matrix A by deleting all rows except those corresponding to G_k and all columns except those corresponding to G_ℓ. A similar grouping is chosen for the vectors $\underline{u}$ and $\underline{b}$.

We now rewrite the system as

$$\sum_{\ell=1}^{N} A_{k,\ell} u_\ell = b_k , \quad k = 1,2,\ldots,N . \tag{2.3}$$

If we now assume that all submatrices $A_{k,k}$ are non-singular then we define the following group SOR iterative method

$$\underline{u}^{(n+1)} = \mathscr{L}_\omega^{(\pi)} \underline{u}^{(n)} + (I-\omega L^{(\pi)})^{-1} \omega \underline{c}^{(\pi)} , \tag{2.4}$$

where

$$\mathscr{L}_\omega^{(\pi)} = (I-\omega L^{(\pi)})^{-1} (\omega U^{(\pi)} + (1-\omega) I) ,$$

with

$$L^{(\pi)} = (D^{(\pi)})^{-1} C_L^{(\pi)} ,$$

and

$$U^{(\pi)} = (D^{(\pi)})^{-1} C_U^{(\pi)} , \quad \text{with} \quad A = D^{(\pi)} - C_L^{(\pi)} - C_U^{(\pi)} .$$

It is of interest to know which partition π will yield the fastest convergence. Indeed Varga (1962, p. 87) shows that for regular splittings of A, the more non-zero elements $D^{(\pi)}$ contains, the faster the convergence. In the limiting case, when $D^{(\pi)} = A$, we would obtain the solution

$\underline{u} = (D^{(\pi)})^{-1}\underline{b}$ in one iteration and the method then becomes a direct one.

It is clear that a faster rate of convergence can always be obtained by selecting a partitioning which results in blocks which will have more non-zero elements. However, the amount of work involved in each iteration will also sharply increase.

III. BLOCK EXPLICIT ITERATIVE METHODS

In this section we reconsider the approach by Shortley & Weller (1938) of using a small block of fixed size, i.e., 4 points and explicitly inverting it within the iteration. Then the loss of sparseness which will inevitably occur when an explicit form of block SOR is used will be small and independent of the size of the given problem.

Thus, for the simple case of the unit square and $\Delta x = \Delta y = 1/5$ as illustrated in Figure 1, a red/black ordering of the 4-point groups will result in a coefficient matrix which has the block structure

$$A = \left[\begin{array}{c|c} D_1 & C_1 \\ \hline C_2 & D_2 \end{array}\right], \tag{3.1}$$

where $D_1 = D_2 = R_0 I_8$, with $I_8 \equiv$ the (8×8) identity matrix

and

$$C_1 = \left[\begin{array}{c|c} R_2 & R_3 \\ \hline R_1 & R_4 \end{array}\right], \quad C_2 = \left[\begin{array}{c|c} R_4 & R_3 \\ \hline R_1 & R_2 \end{array}\right], \tag{3.2}$$

with

$$R_0 = \begin{bmatrix} 1 & \alpha_2 & 0 & \alpha_3 \\ \alpha_4 & 1 & \alpha_3 & 0 \\ 0 & \alpha_1 & 1 & \alpha_4 \\ \alpha_1 & 0 & \alpha_2 & 1 \end{bmatrix}, \quad R_1 = \begin{bmatrix} 0 & 0 & 0 & \alpha_1 \\ 0 & 0 & \alpha_1 & 0 \\ 0 & 0 & 0 & 0 \\ 0 & 0 & 0 & 0 \end{bmatrix} \tag{3.3}$$

$$R_2 = \begin{bmatrix} 0 & 0 & 0 & 0 \\ \alpha_2 & 0 & 0 & 0 \\ 0 & 0 & 0 & \alpha_2 \\ 0 & 0 & 0 & 0 \end{bmatrix}, \quad R_3 = \begin{bmatrix} 0 & 0 & 0 & 0 \\ 0 & 0 & 0 & 0 \\ 0 & \alpha_3 & 0 & 0 \\ \alpha_3 & 0 & 0 & 0 \end{bmatrix}$$

and
$$R_4 = \begin{bmatrix} 0 & \alpha_4 & 0 & 0 \\ 0 & 0 & 0 & 0 \\ 0 & 0 & 0 & 0 \\ 0 & 0 & \alpha_4 & 0 \end{bmatrix}$$

Further for larger meshes, it is a simple matter to deduce that when the 4-point blocks are taken in natural row ordering and because the block equations only refer to adjacent blocks then the coefficient matrix has the familiar structure, i.e.,

$$A \equiv \begin{bmatrix} B_1 & C_1 & & & \\ A_1 & B_2 & C_2 & & 0 \\ & \ddots & \ddots & \ddots & \\ & & \ddots & \ddots & C_{k-1} \\ 0 & & & A_k & B_k \end{bmatrix}, \tag{3.4}$$

where $C_i \equiv R_3 I$, $A_i \equiv R_1 I$, and

$$B_i \equiv \begin{bmatrix} R_0 & R_2 & & & \\ R_4 & \ddots & \ddots & & 0 \\ & \ddots & \ddots & \ddots & \\ & & \ddots & \ddots & R_2 \\ 0 & & & R_4 & R_0 \end{bmatrix}, \quad i = 1,2,\ldots,k .$$

Now an explicit block SOR method can be considered as a point SOR method applied to a transformed matrix $A^E = [\text{diag}\{R_0\}]^{-1}A$ and vector $\underline{b}^E = [\text{diag}\{R_0\}]^{-1}\underline{b}$ and because R_0 is a small order matrix which can be easily inverted, then the matrix A^E and vector $\underline{b}^E$ can be evaluated explicitly. For the linear system $A^E\underline{u} = \underline{b}^E$, with $A^E = I-L^E-U^E$, the method can be written compactly as,

$$\underline{u}^{(n+1)} = \underline{u}^{(n)} + \omega[L^E\underline{u}^{(n+1)} + U^E\underline{u}^{(n)} + \underline{b}^E - \underline{u}^{(n)}] \ . \qquad (3.5)$$

The matrix $[\mathrm{diag}\{R_0\}^{-1}]$ is simply diag $\{R_0^{-1}\}$ and the inverse of R_0 given by

$$R_0^{-1} = (d^{-1}) \begin{bmatrix} \alpha_5 & -\alpha_2\alpha_6 & 2\alpha_2\alpha_3 & -\alpha_3\alpha_7 \\ -\alpha_4\alpha_6 & \alpha_5 & -\alpha_3\alpha_7 & 2\alpha_3\alpha_4 \\ 2\alpha_1\alpha_4 & -\alpha_1\alpha_7 & \alpha_5 & -\alpha_4\alpha_6 \\ -\alpha_1\alpha_7 & 2\alpha_1\alpha_2 & -\alpha_2\alpha_6 & \alpha_5 \end{bmatrix} \qquad (3.6)$$

where $d = (\alpha_1\alpha_3-\alpha_2\alpha_4)^2 - 2(\alpha_1\alpha_3+\alpha_2\alpha_4)+1$ and $\alpha_5 = 1-\alpha_1\alpha_3-\alpha_2\alpha_4$, $\alpha_6 = 1+\alpha_1\alpha_3-\alpha_2\alpha_4$, $\alpha_7 = 1+\alpha_2\alpha_4-\alpha_1\alpha_3$.

The block structure of A^E is the same as that of A (i.e., eqn. 3.4) with the submatrices R_0 replaced by identity matrices, and the submatrices R_i, replaced by $R_0^{-1}R_i$, i = 1,2, 3,4. In addition, where R_i, i = 1,2,3,4 has a column or row of zeros so does $R_0^{-1}R_i$, and where an element α_i occurs as the (p,q)th element of R_i, the qth column of $R_0^{-1}R_i$ is the pth column of R_0^{-1}, multiplied by α_i. So, for example,

$$R_0^{-1}R_1 = (d^{-1}) \begin{bmatrix} 0 & 0 & -\alpha_1\alpha_2\alpha_6 & \alpha_1\alpha_5 \\ 0 & 0 & \alpha_1\alpha_5 & -\alpha_1\alpha_4\alpha_6 \\ 0 & 0 & -\alpha_1^2\alpha_7 & 2\alpha_1^2\alpha_4 \\ 0 & 0 & 2\alpha_1^2\alpha_2 & -\alpha_1^2\alpha_7 \end{bmatrix} . \qquad (3.7)$$

For the model problem and a square grid $\alpha_1 = \alpha_2 = \alpha_3 = \alpha_4 = -1/4$ so that,

$$R_0^{-1} = \frac{1}{6}\begin{bmatrix} 7 & 2 & 1 & 2 \\ 2 & 7 & 2 & 1 \\ 1 & 2 & 7 & 2 \\ 2 & 1 & 2 & 7 \end{bmatrix} \quad \text{and } R_0^{-1}R_1 = -\frac{1}{24}\begin{bmatrix} 0 & 0 & 2 & 7 \\ 0 & 0 & 7 & 2 \\ 0 & 0 & 2 & 1 \\ 0 & 0 & 1 & 2 \end{bmatrix}$$

from which we can establish the computational stencil at the point P to be

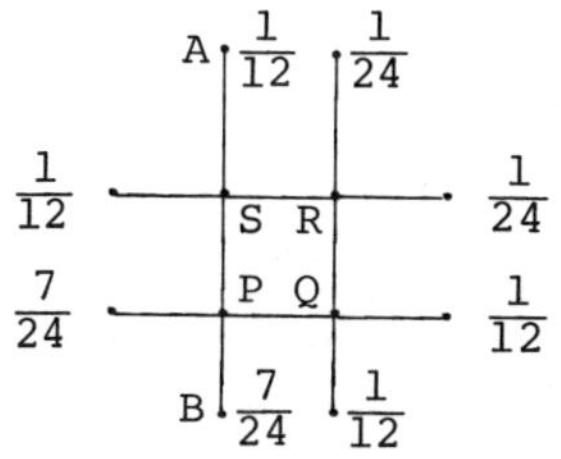

FIGURE 3

Further analyses will yield analogous relationships for the remaining points Q,R and S of the 4 point block.

The use of this computational stencil to derive the solution of self adjoint P.D.E.'s is given in Evans & Biggins (1982).

Finally, the transformed matrix A^E has the block form, given by eqn. (3.4), where

$$B_i = \begin{bmatrix} I & R_0^{-1}R_2 & & & 0 \\ R_0^{-1}R_4 & I & R_0^{-1}R_2 & & \\ & \ddots & \ddots & \ddots & \\ & & & & R_0^{-1}R_2 \\ 0 & & & R_0^{-1}R_4 & I \end{bmatrix}, \qquad (3.8)$$

$$C_i = R_0^{-1}R_3, \quad A_i = R_0^{-1}R_i , \quad i = 1,2,\ldots,k .$$

and the implicit block explicit SOR method for solving (3.8) is obtained by solving

$$B_r^{-1}A_r\underline{u}_{r-1}^{(n+1)} + \underline{u}_r^{*(n+1)} + B_r^{-1}C_r\underline{u}_{r+1}^{(n)} = B_r^{-1}\underline{b}_r^E, \quad 1 \le r \le k ,$$

or

$$\underline{u}_r^{*(n+1)} = B_r^{-1}[\underline{b}_r^E - C_r\underline{u}_{r+1}^{(n)} - A_r\underline{u}_{r-1}^{(n+1)}] , \qquad (3.9a)$$

and then extrapolating by

$$\underline{u}_r^{(n+1)} = \underline{u}_r^{(n)} + [\underline{u}_r^{*(n+1)} - \underline{u}_r^{(n)}] , \qquad (3.9b)$$

to obtain the final solution.

IV. THE SOLUTION OF BOUNDARY VALUE PROBLEMS BY BLOCK EXPLICIT ITERATIVE METHODS

Consider now the solution of the model problem in the region depicted by Figure 4.

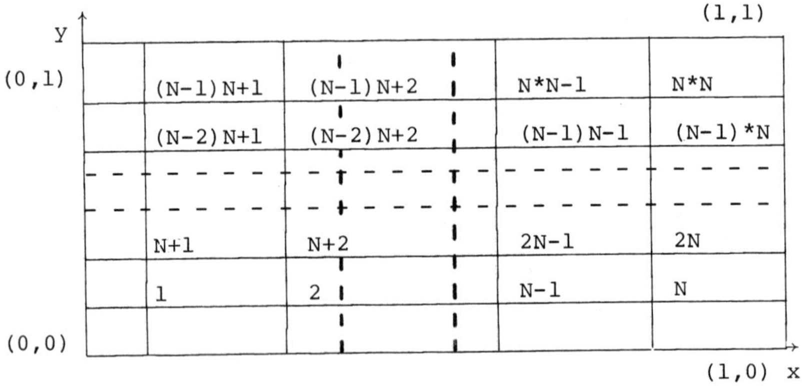

FIGURE 4

We construct two new groupings of mesh points in the following manner:

1) Place the 4 mesh points as shown in Figure 5 in each group where $k = (\ell N+1)(\ell+1)N-1$ and $\ell = 0(2)N-2$. In this case, where N is an even number we have the subset G_M of Definition 2.1, $M = 1,2,\ldots,N^2/4$, such that each G_M consists of the four elements $\{k,k+1,N+k,N+k+1\}$ while the matrix $A_{M,M}$ is of order 4, i.e.,

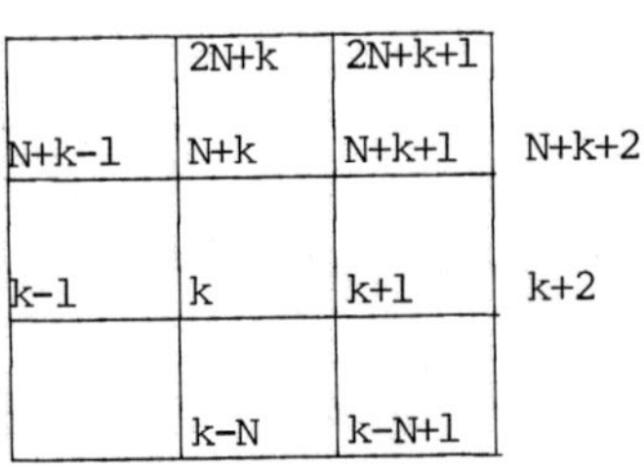

FIGURE 5

$$A_{M,M} = \begin{bmatrix} 4 & -1 & -1 & 0 \\ -1 & 4 & 0 & -1 \\ -1 & 0 & 4 & -1 \\ 0 & -1 & -1 & 4 \end{bmatrix} \qquad (4.1)$$

$$M = 1,2,\ldots,N^2/4 .$$

Thus, for the model problem, we can state the implicit group Jacobi method to be

$$
\begin{aligned}
4u_k^{(n+1)} - (u_{k+1}^{(n+1)} + u_{N+k}^{(n+1)}) &= u_{k-N}^{(n)} + u_{k-1}^{(n)} + f_k \\
4u_{k+1}^{(n+1)} - (u_k^{(n+1)} + u_{N+k+1}^{(n+1)}) &= u_{k+2}^{(n)} + u_{k-N+1}^{(n)} + f_{k+1} \\
4u_{N+k}^{(n+1)} - (u_k^{(n+1)} + u_{N+k+1}^{(n+1)}) &= u_{N+k-1}^{(n)} + u_{2N+k}^{(n)} + f_{N+k} \\
4u_{N+k+1}^{(n+1)} - (u_{k+1}^{(n+1)} + u_{N+k}^{(n+1)}) &= u_{2N+k+1}^{(n)} + u_{N+k+2}^{(n)} + f_{N+k+1}
\end{aligned}
\tag{4.2}
$$

such that, each iteration involves the solution of $N^2/4$ systems of order 4, with the matrix $A_{M,M}$ given in (4.1). However, we now propose that group systems of the type (4.2) are solved explicitly, i.e., where the system of equations (4.2) may be written as,

$$
\begin{aligned}
u_k^{(n+1)} = \frac{1}{24}[7(u_{k-N}^{(n)} + u_{k-1}^{(n)}) + 2(u_{k-N+1}^{(n)} + u_{k+2}^{(n)} + u_{2N+k}^{(n)} \\
+ u_{N+k-1}^{(n)}) + u_{N+k+2}^{(n)} + u_{2N+k+1}^{(n)}] ,
\end{aligned}
\tag{4.3}
$$

with analogous equations for $u_{k+1}^{(n+1)}$, $u_{N+k+1}^{(n+1)}$ and $u_{N+k}^{(n+1)}$ obtained by rotating the stencil in Figure 3 anti-clockwise and $k = (\ell N+1)(2)(\ell+1)N-1$ and $\ell = 0(2)N-2$. Note that, if any of $u_M^{(n)}$ in the right hand sides of (4.3) is a point of the boundary $\partial\Omega$ then we compute the appropriate value of u at this point. Obviously similar equations can be written down for the 4-point explicit JOR and SOR methods. Clearly, equations (4.3) can be derived immediately from the computational stencil (Figure 3) applied to the given network of points.

2) The second grouping of the mesh points proposed is when each group is formed from 9 points of the net region Figure 4 in accordance with Figure 6, where $k = (\ell N+1)(3)(\ell+1)N-2$ and $\ell = 0(3)N-3$. For this grouping of the mesh points, N must be divisible by 3. In this case, each subset G_M, $M = 1,2,\ldots,N^2/9$, of Definition (2.1) consists of 9 elements and the submatrices $A_{M,M}$, $M = 1,2,\ldots,N^2/9$, of the original matrix A are of order 9, i.e.,

	3N+k	3N+k+1	3N+k+2	
2N+k-1	2N+k	2N+k+1	2N+k+2	2N+k+3
N+k-1	N+k	N+k+1	N+k+2	N+k+3
k-1	k	k+1	k+2	k+3
	k-N	k-N+1	k-N+2	

FIGURE 6

$$A_{M,M}$$

$$\begin{bmatrix} 4 & -1 & 0 & -1 & . & . & . & . & . \\ -1 & 4 & -1 & 0 & -1 & . & . & . & . \\ 0 & -1 & 4 & 0 & 0 & -1 & . & . & . \\ -1 & 0 & 0 & 4 & -1 & 0 & -1 & . & . \\ . & -1 & 0 & -1 & 4 & -1 & 0 & -1 & . \\ . & . & -1 & 0 & -1 & 4 & 0 & 0 & -1 \\ . & . & . & -1 & 0 & 0 & 4 & -1 & . \\ . & . & . & . & -1 & 0 & -1 & 4 & -1 \\ . & . & . & . & . & -1 & 0 & -1 & 4 \end{bmatrix} \quad (4.4)$$

$$M = 1,2,\ldots,N^2/9 .$$

The matrix $A_{M,M}$ again, can be easily inverted, and for the solution of the model problem the explicit group Jacobi method is given by the following formulae.

$$\begin{aligned} u_k^{(n+1)} = \frac{1}{224}[& 67(u_{k-N}^{(n)} + u_{k-1}^{(n)}) + 22(u_{k-N+1}^{(n)} + u_{N+k-1}^{(n)}) \\ & + 7(u_{k-N+2}^{(n)} + u_{k+3}^{(n)} + u_{3N+k}^{(n)} + u_{2N+k-1}^{(n)}) + 6(u_{N+k+3}^{(n)} \\ & + u_{3N+k+1}^{(n)}) + 3(u_{2N+k+3}^{(n)} + u_{3N+k+2}^{(n)})] , \end{aligned} \quad (4.5a)$$

with analogous equations for $u_{k+2}^{(n+1)}$, $u_{2N+k}^{(n+1)}$ and $u_{2N+k+2}^{(n+1)}$ chosen as before,

$$u_{k+1}^{(n+1)} = \frac{1}{112}[37u_{k-N+1}^{(n)} + 11(u_{k-1}^{(n)} + u_{k-N}^{(n)} + u_{k-N+2}^{(n)} + u_{k+3}^{(n)})$$

$$+ 7(u_{N+k+3}^{(n)} + u_{N+k-1}^{(n)}) + 5u_{3N+k+1}^{(n)} + 3(u_{2N+k+3}^{(n)}$$

$$+ u_{3N+k+2}^{(n)} + u_{3N+k}^{(n)} + u_{2N+k-1}^{(n)})] \ , \qquad (4.5b)$$

with similar equations for $u_{N+k}^{(n+1)}$, $u_{N+k+2}^{(n+1)}$ and $u_{2N+k+1}^{(n+1)}$, and finally,

$$u_{N+k+1}^{(n+1)} = \frac{1}{16}[2(u_{k-N+1}^{(n)} + u_{N+k+3}^{(n)} + u_{3N+k+1}^{(n)} + u_{N+k+1}^{(n)})$$

$$+ u_{k-1}^{(n)} + u_{k-N+2}^{(n)} + u_{k-N}^{(n)} + u_{k+3}^{(n)} + u_{2N+k+3}^{(n)}$$

$$+ u_{3N+k+2}^{(n)} + u_{3N+k}^{(n)} + u_{2N+k-1}^{(n)}] \ , \qquad (4.5c)$$

where $k = (\ell N+1)(3)(\ell+1)N-2$ and $\ell = 0(3)N-3$. Again, similar sets of equations can be written down for the 9-point explicit Group JOR and SOR methods.

Similarly, equations (4.5) can be derived by applying the computational molecule (Figure 7) where $\beta_1 = 67/224$, $\beta_2 = 22/224$, $\beta_3 = 7/224$, $\beta_4 = 6/224$ and $\beta_5 = 3/224$ to the given mesh.

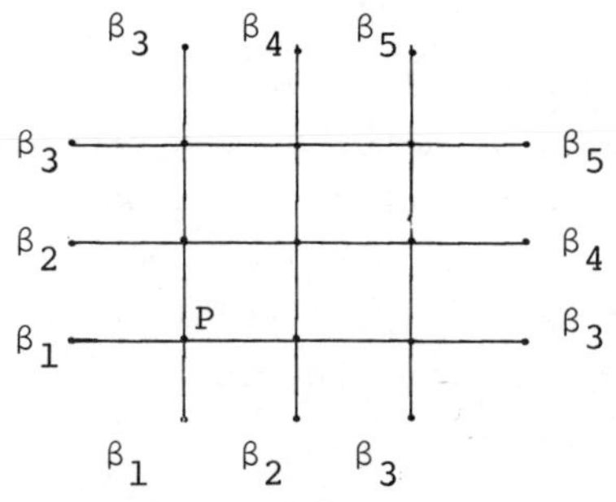

FIGURE 7

The advantages of these groupings of the mesh points obviously lie in the improved convergence rates obtained in the solution of the problem by group iterative methods. Also, if we consider each group of 4 mesh points as shown in Figure 5 'lumped' as a single point P_M, then, we have different methods of labelling the 'composite' points P_M. For example, if we choose the 4-point groups of mesh points corresponding to the first two rows from left to right and then similar groups corresponding to the third and fourth rows etc. Then the labelling points P_M are known as the "natural" ordering and it can be seen that the matrix $Z = (z_{k,\ell})$ of Definition (2.1) in Young (1971) has property (A) and is consistently ordered. Thus, the matrix A is a consistently ordered, weakly cyclic matrix of index 2. By this consideration we have different choices of labelling the points P_M which could result in the matrix A having property $(A^{(\pi)})$ and be π-consistently ordered, such as "red-black" ordering or "diagonal" ordering. Therefore, in each case the theory of group successive over-relaxation would be valid.

V. NUMERICAL RESULTS FOR EXPLICIT BLOCK SOR METHODS

We now describe some numerical experiments we made to compare the new block methods with existing methods.

Problem 1: The solution of Laplace's equation

$$\frac{\partial^2 u}{\partial x^2} + \frac{\partial^2 u}{\partial y^2} = 0 \quad \text{in} \quad \Omega \tag{5.1}$$

and

$$u = 0 \quad \text{on} \quad \partial\Omega , \tag{5.2}$$

where Ω is a unit square with boundary $\partial\Omega$.

We use a red-black ordering of points of the mesh (Figure 4) and also of blocks of 4-points and 9-points where each point on the mesh can represent a single point, a block of four points or a block of nine points depending on the size of the grouping scheme. Therefore, the resulting matrices are π-consistently ordered, and thus the optimum ω, ω_b (say) are calculated by the formula

$$\omega_b = 2/(1 + \{1-\rho(G)\}^{\frac{1}{2}}) , \tag{5.3}$$

where $\rho(G)$ is the spectral radius of the Gauss-Seidel iteration matrix and is estimated by use of the Power method from

$$\rho(G) = \lim_{n\to\infty} \frac{||\underline{d}^{(n)}||}{||\underline{d}^{(n-1)}||} \tag{5.4}$$

where $\underline{d}^{(n)} = \underline{u}^{(n)} - \underline{u}^{(n-1)}$ and the norm $||\underline{d}^{(n)}||$ can be defined as

$$||\underline{d}^{(n)}|| = [\sum_{i=1}^{N} (u_i^{(n)} - u_i^{(n-1)})^2]^{\frac{1}{2}} . \tag{5.5}$$

An experimental optimum value of ω is also found by solving the problem using different values of ω. The condition for convergence of the SOR method is

$$|u_{i,j}^{(n+1)}| < \varepsilon , \quad \text{for all } i,j , \tag{5.6}$$

where $\varepsilon = 5*10^{-6}$, and similarly for the Power method where the difference between successive estimates of $\rho(G)$ is chosen to be less than the same value of ε. Different mesh sizes were used which produced (20×20), (32×32), (44×44), (56×56) and (62×62) networks. The results obtained from these experiments are recorded in Table 1 where n is the minimum number of iterations required for convergence. In addition, results for SLOR and S2LOR, which represent the successive block overrelaxation method for 1-line and 2-line groupings of the mesh points, are included for comparison.

Method / h^{-1} mesh spacing	4-point explicit block SOR			9-point explicit block SOR			SOR	SLOR	S2LOR
	$\rho(G)$	ω	n	$\rho(G)$	ω	n	n	n	n
19	0.9469	1.635	30	0.9220	1.573	24	40	30	21
31	0.9796	1.755	46	0.9698	1.709	38	64	47	33
43	0.9894	1.813	63	0.9841	1.781	52	88	63	46
55	0.9934	1.853	80	0.9902	1.825	67	111	81	57
61	0.9946	1.864	88	0.9920	1.839	73	123	90	64

TABLE 1

To compare the efficiency of each of the iterative methods it is necessary to consider the computational complexity for each iteration as well as the number of iterations required for convergence. One way of estimating this computational complexity is to count the number of arithmetic operations performed per iteration. For m^2 internal points on the mesh these are given in Table 2, when, in Table 2, SEBOR indicates the explicit block SOR method.

The results shown in Table 1 when combined with the number of operations per iteration required by each method given in Table 2 give the total number of arithmetic operations required by each of the point and block SOR methods together with the actual times in computer mill units to obtain the solution using an optimal value of ω for $h^{-1} = 61$.

Method	Computational Complexity		Execution Time/ Solution for $h^{-1} = 61$
	Adds.	Mults.	Problem 1
SOR	$5m^2$	$2m^2$	370
SLOR	$5m^2$	$m(3m-1)$	367
S2LOR	$m(6m-1)$	$m(10m-5)/2$	272
4 pt. SEBOR	$11m^2/2$	$5m^2/2$	303
9 pt. SEBOR	$53m^2/9$	$34m^2/9$	202

TABLE 2

These results confirm the statement that if $\rho(kL)$ and $\rho(kB)$ denote the dominant eigenvalues of the k-line Jacobi scheme and $(k \times k)$ block Jacobi scheme respectively, then $\rho(kL) \simeq 1-k\pi^2h^2$ and $\rho(kB) \simeq 1-k\pi^2h^2/2$ (Parter, 1981).

It can be seen from Table 2 that comparing the number of iterations, the amount of arithmetic operations per iteration and the total time required to obtain the solution shows that the 9-point block appears to be the more efficient strategy.

Indeed from further experiments on more general self-adjoint equations (Evans, 1983a) it appears that the implicit 2-line SOR method may be challenged as the most efficient iterative method by the smaller size block methods, i.e., the 9-point and 4-point explicit group SOR schemes which not only give improved solution times under many circumstances, but appear to be more flexible in the way the blocks can be selected.

VI. IMPLICIT BLOCK EXPLICIT OVERRELAXATION SCHEMES

The emergence of the new explicit 4-point group computational stencil (Figure 3) for use in iterative methods obviously replaces the use of the more usual 5-point molecule

Method	SOR	SLOR	S2LOR	4 pt. exp. group SOR	4 pt. imp./exp. group SOR
$h^{-1} = 25$	52 iter.	40 iter.	27 iter.	37 iter.	23 iter.

TABLE 3

(Figure 2). Indeed, we can now assemble the group explicit equations (4.3) at each mesh point immediately and form a new matrix system $A^E\underline{u} = \underline{b}^E$ where $A^E = I - L^E - U^E$ on which all known iterative methods, i.e., Jacobi, JOR, Gauss-Seidel and SOR schemes together with their associated Preconditioning strategies (Evans, 1983b) can be applied to give improved results.

Finally, as briefly indicated in Section III the new explicit equations can be grouped together again in an implicit iterative method. From a columnwise ordering of the points on the network the points A, S, P and B in the new 4-point group stencil Figure 3, etc., can be grouped along the columns of the grid similar to implicit line overrelaxation as in SLOR. The resulting implicit block explicit overrelaxation scheme can then be written as

$$D^E_r \underline{u}^{*(n+1)}_r - L^E_r \underline{u}^{(n+1)}_r - U^E_r \underline{u}^{(n)}_r = \underline{b}^E_r , \quad 1 \leq r \leq k \tag{6.1}$$

and

$$\underline{u}^{(n+1)}_r = \underline{u}^{(n)}_r + \omega(\underline{u}^{*(n+1)}_r - \underline{u}^{(n)}_r) , \tag{6.2}$$

where D^E_r is now a quindiagonal matrix which is to be solved along each column of the net.

Preliminary results for this new strategy for problem 1 are given in Table 3. They are encouraging and confirm its superior convergence properties.

VII. REFERENCES

Evans, D.J., (1983a), in pp. 789-98 of "3rd Int. Symp. on Num. Meth. in Eng.", Vol. 2, (P. Lascaux ed.) Pluralis Press.

Evans, D.J., (1983b), "Preconditioning Methods: Analysis and Applications", Gordon & Breach, New York.

Evans, D.J. and Biggins, M.J., (1982), Int. Jour. Comp. Math. 10, 269.

Parter, S.V., (1981), in pp. 375-82, "Elliptic Problem Solvers", (M.H. Schultz, ed.), Academic Press, New York.

Shortley, G.H. and Weller, R., (1938), J. Appl. Phys. 9, 334.

Varga, R.S., (1962), "Matrix Iterative Analysis", Prentice Hall Inc., Englewood Cliffs, N.J.

Young, D.M., (1971), "Iterative Solution of Large Linear Systems", Academic Press, New York.

DYNAMIC ACCELERATION OF NONLINEAR ITERATIONS

James M. Hyman,* Thomas A. Manteuffel**

I. INTRODUCTION

Many large and complicated production codes require iteration to the solution of a system of nonlinear equations. For example, the solution to a time-dependent partial differential equation may require the solution of a nonlinear system at each time step. Unfortunately, these codes often implement suboptimal methods and convergence is very slow.

Due to unstructured programming and often political considerations, it is difficult to implement major changes in an existing applications computer code. In this paper we describe algorithms for accelerating the convergence of iteration methods with a minimal impact on the original code. In most cases a single subroutine call and additional work space are sufficient.

---

* Center for Nonlinear Studies, MS-B284, Los Alamos National Laboratory, Los Alamos, NM 87545
** Computing Division, MS-B265, Los Alamos National Laboratory, Los Alamos, NM 87545

ISBN 0-12-100560-7

Suppose we monitor an iteration by examining a sequence of iterates,

$$\ldots\ \underline{v}_{i-1},\ \underline{v}_i,\ \underline{v}_{i+1}\ \ldots\ .$$

If each iteration is a function of the previous iterate, then

$$\underline{v}_{i+1} = F(\underline{v}_i) \tag{1}$$

for some function F. The iteration has converged when a $\underline{v}$ is found such that

$$\underline{v} = F(\underline{v}) \quad . \tag{2}$$

The basis for this work is the assumption that the suboptimally implemented algorithm yields a sequence of iterates that can be well approximated by an iteration based on a linear system of equations when the convergence is very slow; that is,

$$\underline{v}_{i+1} = G_i\ \underline{v}_i + \underline{b}_i \tag{3}$$

where G_i, $\underline{b}_i$ vary slowly. This model allows us to obtain eigenvalue estimates of G at each step from a sequence of previous iterates. These eigenvalue estimates can be used to find parameters for one-step, two-step, or other acceleration schemes.

Since the underlying iteration is nonlinear, the algorithm must be able to determine when acceleration will be helpful. The eigenvalue estimates can be used to predict how viable the assumption of linearity is. Also, trust regions are used to avoid catastrophic moves.

We will discuss one-step and two-step acceleration schemes and show how eigenvalue estimates may be found. These estimates can be used to choose the acceleration parameters. We will then briefly discuss the implications of nonlinearity. Finally we will present numerical results.

In this work we adapt the methods for linear systems derived in Manteuffel [4,5] to the context of nonlinear systems. Similar methods for a system with a real spectrum can be found in Hageman and Young [1]. We introduce the concept of a changing convex hull of approximate eigenvalues of the Jacobian. Also, we consider the implications of trust region strategies on the choice of parameters. Moreover, we are primarily interested in applications where the Jacobian matrix is not available. This black box function approach makes the algorithms more robust, but necessarily limits the range of possible acceleration techniques. We must determine when the iteration is behaving in a manner amenable to this type of acceleration and then determine acceleration parameters.

II. ONE-STEP METHOD

If G_i and b_i are independent of i, then the convergence of the iteration

$$\underline{v}_{i+1} = G\underline{v}_i + \underline{b} \tag{4}$$

is determined by the spectral radius of G, $\sigma(G)$. Let

$$\underline{v} = G\underline{v} + \underline{b} \tag{5}$$

be the exact solution, and $\underline{e}_i = \underline{v} - \underline{v}_i$ be the error vector. Subtracting (4) from (5) yields

$$\underline{e}_{i+1} = G\,\underline{e}_i = G^{i+1}\underline{e}_0 \quad . \tag{6}$$

If the spectral radius, $\sigma(G) < 1$, then (4) will converge. One-step methods attempt to effectively reduce the spectral radius of G. Given $\underline{v}_i$, let

$$\underline{r}_i = (G\underline{v}_i + \underline{b}) - \underline{v}_i = \underline{v}_{i+1} - \underline{v}_i \quad . \tag{7}$$

We see that (4) can be written as

$$\underline{v}_{i+1} = \underline{v}_i + \underline{r}_i \quad . \tag{8}$$

If we let

$$\underline{v}_{i+1} = \underline{v}_i + \alpha\,\underline{r}_i \tag{9}$$

then (6) becomes

$$\underline{e}_{i+1} = ((1-\alpha)I + \alpha G)\,\underline{e}_i = ((1-\alpha)I + \alpha G)^{i+1}\underline{e}_0 \quad . \tag{10}$$

If

$$\sigma[(1-\alpha)I + \alpha\, G] < \sigma(G) \quad ,$$

then the new iteration will converge faster. The relation between the eigenvalues of G and of $(1-\alpha)I + \alpha\, G$ is simple, so the spectral radii are easily compared. Also note that the new iteration is consistent in that if $\underline{v}_i = \underline{v}$, then $\underline{v}_{i+1} = \underline{v}$.

The iteration (9) is called a stationary one-step method. We can also implement a nonstationary one-step method by letting α also depend upon i:

$$\underline{v}_{i+1} = \underline{v}_i + \alpha_i \underline{r}_i \quad . \tag{11}$$

The error equation becomes

$$e_{i+1} = ((1-\alpha_i)I + \alpha_i G)\, \underline{e}_i = \prod_{j=0}^{i} ((1-\alpha_j)I + \alpha_j G)\, \underline{e}_0 \quad . \tag{12}$$

It is clear that there is some choice for $\{\alpha_j\}_{j=0}^{i}$ such that

$$\sigma \prod_{j=0}^{i} ((1-\alpha_j)I + \alpha_j G) \leq \sigma(G^{i+1}) \quad , \tag{13}$$

because $\alpha_j = 1$ yields equality.

To correctly choose α or $\{\alpha_i\}$ requires a knowledge of the spectrum of G. Consider the sequence $\{\underline{v}_i\}$ as generated by (8). The corresponding $\{\underline{r}_i\}$ has the property that

$$\underline{r}_i = G\, \underline{r}_{i-1} \quad . \tag{14}$$

For sufficiently large i, the vector $\underline{r}_i$ will be dominated by a linear combination of the eigenvectors of G associated with the eigenvalues of largest modulus. Let $p_s(\lambda)$ be the orthogonal polynomial of degree s with respect to the Krylov sequence based upon G and $\underline{r}_i$ (see Householder [2,p. 24] or Manteuffel [4, p. 190]). The roots of $p_s(\lambda)$ are approximations to the eigenvalues of

G. The polynomial

$$p_s(\lambda) = \lambda^s + \rho_{s-1}^{(s)} \lambda^{s-1} + \ldots + \rho_0^{(s)}$$

satisfies the least squares problem

$$\begin{pmatrix} | & | & & | \\ \underline{r}_{i+s} & \underline{r}_{i+s-1} & \cdots & \underline{r}_i \\ | & | & & | \end{pmatrix} \begin{pmatrix} 1 \\ \rho_{s-1}^{(s)} \\ \cdot \\ \cdot \\ \cdot \\ \rho_0^{(s)} \end{pmatrix} \cong \begin{pmatrix} 0 \\ 0 \\ \cdot \\ \cdot \\ \cdot \\ 0 \end{pmatrix} \tag{15}$$

The residuals associated with the stationary one-step method also form a Krylov sequence. From Eq. (10) we see that

$$\underline{r}_{i+1} = ((1-\alpha)I + \alpha\, G)\, \underline{r}_i \quad . \tag{16}$$

Thus, eigenvalue estimates of the eigenvalues of $[(1-\alpha)I + \alpha G]$ with largest modulus can be found. From these, eigenvalues of G can be estimated.

The residuals formed by a nonstationary one-step method can also be used to find the orthogonal polynomials with respect to the Krylov sequence formed by G and $\underline{r}_i$. As in (12) we have

$$\underline{r}_{i+1} = ((1-\alpha_i)I + \alpha_i\, G)\, \underline{r}_i \quad . \tag{17}$$

Thus,

$$\underline{r}_{i+s} = \prod_{j=0}^{s-1} ((1-\alpha_{i+j})I + \alpha_{i+j}\, G)\, \underline{r}_i \quad . \tag{18}$$

This polynomial relationship can be used to construct the Krylov sequence. More importantly, the orthogonal polynomial, $p_s(\lambda)$, can be constructed directly from the sequence of residuals.

Eigenvalue estimates can be used to choose new parameters. The convex hull of the eigenvalue estimates can be used to find the optimal stationary one-step parameter. If the residual is almost a linear combination of a few eigenvectors, then the parameters may be chosen so that the polynomial in (18) has roots at the corresponding eigenvalues. We will refer to this as annihilation.

III. TWO-STEP METHOD

The one-step method only uses information from the previous step. The two-step method uses the two previous iterates. Instead of (9) we have

$$\underline{v}_{i+1} = \underline{v}_i + \alpha\, \underline{r}_i + \beta(\underline{v}_i - \underline{v}_{i-1}) \quad . \tag{19}$$

The error equation becomes

$$\underline{e}_{i+1} = ((1+\beta-\alpha)I + \alpha G)\underline{e}_i - \beta\, \underline{e}_{i-1} \quad . \tag{20}$$

This error equation can also be written as a system of dimension 2N,

$$\begin{pmatrix} \underline{e}_{i+1} \\ \underline{e}_i \end{pmatrix} = \begin{pmatrix} (1+\beta-\alpha)I + \alpha\, G & -\beta I \\ I & 0 \end{pmatrix} \begin{pmatrix} \underline{e}_i \\ \underline{e}_{i-1} \end{pmatrix} \quad . \tag{21}$$

To simplify the notation we rewrite this as

$$\tilde{\underline{e}}_{i+1} = \tilde{G}\,\tilde{\underline{e}}_i = \tilde{G}^{i+1}\,\tilde{\underline{e}}_0 \quad . \tag{22}$$

We would like to choose α,β to make $\sigma(\tilde{G})$ as small as possible. The solution to this can be well approximated by fitting the best ellipse around the spectrum of G and choosing the parameters to minimize the iteration polynomial over this ellipse. We use the algorithm derived in Manteuffel [4].

Notice that if both sides of (20) are multiplied by I-G the $\underline{e}$'s are replaced by $\underline{r}$'s. Thus, as in (22), we have

$$\tilde{\underline{r}}_{i+1} = \tilde{G}\,\tilde{\underline{r}}_i \quad . \tag{23}$$

The residuals may be used to form a Krylov sequence in the operator $\tilde{G}$. Using the techniques described above we can find eigenvalue estimates of the operator $\tilde{G}$. Examination of (21) shows that the eigenvalues of $\tilde{G}$, say μ, are related to the eigenvalues of G, say λ, by the formula

$$\mu^2 - ((1+\beta-\alpha) + \alpha\lambda)\mu + \beta = 0 \quad . \tag{24}$$

One may also consider a nonstationary two-step method. One such method is the Chebychev acceleration described in Manteuffel [4]. For real spectrum, Chebychev acceleration is the optimal two-step method measured in the max norm. In Manteuffel [5] it is shown that the Chebychev iteration is asymptotically equivalent to a stationary two-step iteration. Moreover, there is a simple relationship between the optimal parameters for a Chebychev iteration and the optimal parameters for a two-step method. The stationary two-step iteration has the advantage that eigenvalue estimates do not depend upon an asymptotic relationship and can be formed from any window of consecutive residual vectors.

IV. DETERMINING LINEARITY

Since the analysis above is based upon a linear model, it is important to determine when this model is viable. If convergence is swift, then this algorithm is superfluous. It is when each iterate differs little from its predecessor that some improvement may be found. From (7) we see that a small change from $\underline{v}_i$ to $\underline{v}_{i+1}$ implies that $\underline{r}_i$ is relatively small. This, of course, does not imply that $\underline{e}_i$ is small. When the change in $\underline{r}_i$ is linear, then the acceleration methods in Secs. II and III may be used to enhance the convergence.

Linearity can be ascertained by monitoring a moving window of past residuals. Consider the matrix in (15). If the previous residuals are linearly dependent then this matrix has a null space. If this null space differs little from step to step we may assume that the iteration is behaving linearly and the error is in the direction of only a few eigenvectors of G. A sequence of acceleration parameters may be chosen to annihilate these errors.

Even if these previous residuals are not linearly dependent, linear behavior may be detected by monitoring the eigenvalue estimates obtained as described above. If the eigenvalue estimates differ little from step to step we may assume that the iteration is behaving linearly and use the eigenvalue estimates to choose stationary one- or two-step parameters. This is done by accumulating a pile of previous eigenvalue estimates and forming the convex hull of the pile. Because the underlying iteration is nonlinear, these estimates must be aged and discarded after an appropriate number of steps.

Note that the eigenvalue estimates can be used to determine the relationship between $\underline{r}_i$ and $\underline{e}_i$. Thus, the convergence criteria can be greatly improved.

V. TRUST REGIONS

It is assumed that the underlying nonlinear iteration will converge, albeit slowly, if left alone. From (8) we see that $\underline{r}_i$ is a safe step. From (9) and (19) it is clear how this safe step is being altered. If the new step is shorter, then no problems should arise. However, if the step is lengthened substantially then care must be taken. The eigenvalue estimates from which the parameters are chosen also yield a predicted reduction in the residual. If the actual reduction is substantially different then the step length will have to be shortened. For more on the theory of trust regions see for example Moré [3].

VI. NUMERICAL RESULTS

To test the effectiveness of the acceleration methods, we studied the convergence of a second-order Runge-Kutta method to the approximate steady-state solution of the scalar three-dimensional Burger's equation

$$u_t + uu_x + uu_y + uu_z = \varepsilon\Delta u, \tag{25a}$$

on the unit cube. The small parameter $\varepsilon = 0.02$ and Dirichlet boundary conditions

$$\begin{aligned} u(0,y,z) &= u(x,0,z) = u(x,y,0) = 0 \\ u(1,y,z) &= u(x,1,z) = u(x,y,1) = 1 \end{aligned} \tag{25b}$$

were chosen so there was a thin boundary layer. The equation was differenced in divergence form with second order centered finite differences on a uniform grid with N grid points. The time step was chosen to be $\Delta t = \frac{1}{2}\Delta x/||u||_{max} = \frac{1}{2}\Delta x$.

The results for N = 8,000 are shown in Fig. 1. Similar results, not described here, are obtained for N = 27,000 and N = 64,000. We compared three methods:

Method 1 - No acceleration.

Method 2 - After four time steps, eigenvalue estimates of G are obtained and used to generate stationary two-step acceleration parameters. From then on, up to four eigenvalue estimates are obtained on every iteration using the past five residuals. If the residual increases on an iteration, as happened on iterations 34,

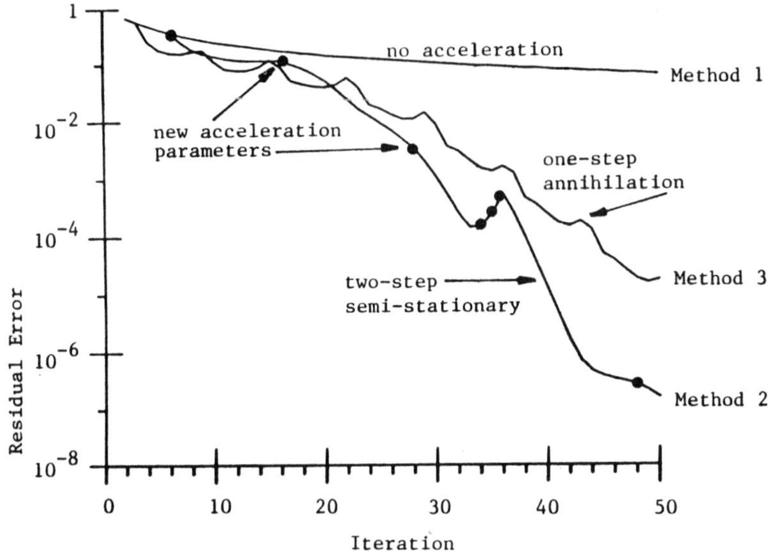

Fig. 1. The residual errors in solving the steady-state solution to the 3-D scalar Burgers equation (25).

35, and 36, or if twelve iterations have been made since new acceleration parameters were calculated, as happened on iterations 16, 28, and 48, then new parameters are calculated. These are marked by dots in Fig. 1. The new parameters are generated based on the best ellipse enclosing the past twenty five eigenvalue estimates. The maximum accelerated step-length in this run was limited to less than five times the original step-length.

Method 3 - This method is included because of its simplicity and effectiveness. If a single real eigenvalue estimate is obtained on every time step using only the previous two residuals,

$$\lambda_i = \lambda_{i-1} + (1-\lambda_{i-1}) \langle \underline{r}^i, \underline{r}^{i-1} \rangle / \langle \underline{r}^{i-1}, \underline{r}^{i-1} \rangle , \quad (26)$$

then the annihilation one-step acceleration parameter is

$$\alpha_i = 1/(1-\lambda_i) \quad . \tag{27}$$

Again we limited the maximum step size to less than five times the original step size ($\alpha_i \leq 5$).

Its clear that both acceleration methods substantially improve the convergence to the steady-state solution. Method 3 works extremely well on this problem, but we mention that it is much more prone to instabilities, especially when the real part of the some of the complex eigenvalues of G are close to one.

Without the trust region model the iterations soon diverge. If the underlying equation is linear (as in some examples not presented here including a linear Burger's equation and an SOR iterative solution of Laplace's equation) the trust region is less critical and methods can recover.

REFERENCES

1. Hageman, L. A. and Young, D. M., Applied Iterative Methods, Academic Press, London, 1981.

2. Householder, A. S., The Theory of Matrices in Numerical Analysis, Blaisdell Publishing Co., New York, 1964.

3. Moré, J. J., "Levenberg Marquant Algorithm: Implementation and Theory," Numerical Analysis, Dundee 1977, G. A. Watson, ed., Lecture Notes in Mathematics 630, Springer-Verlag.

4. Manteuffel, T. A., "Adaptive Procedure for Estimations Parameters for the Nonsymmetric Tchebychev Iteration," Numer. Math. 31 (1978), pp. 183-208.

5. Manteuffel, T. A., "Optimal Iteration Parameters for Linear Second-Degree Stationary Iterative Methods," SIAM J. Numer. Anal. 19, (1982), pp. 833-839.

NAVIER-STOKES PRESSURE EQUATION ITERATION

Eugene L. Wachspress[1]

I. INTRODUCTION

Solution on a CYBER-205 of the discrete Poisson-type pressure equations that arise from mass conservation constraints on fluid velocities in the Navier-Stokes equations [1] will be considered. Analysis and numerical studies for these equations reveals that the solution of what might appear at first glance to be a standard problem to be solved with packaged software [2] has many interesting features.

There are several efficient algorithms for solving Poisson's equation over a rectangle [3]. Any of these procedures may be used as a part of a "compound iteration" such as preconditioned conjugate gradients [4]. Column rebalancing [5] of a basic Chebyshev iteration will be examined here. The main thrust of this study is to illustrate some of the considerations in designing an efficient solver for a particular vector computer.

The TEACH method [6] of constraining the velocities to satisfy mass continuity constraints is used. The pressure equations derived from the mass conservation constraint form an elliptic system with a symmetric positive semidefinite coefficient matrix. Matrix elements vary in an unpredictable manner governed by local fluid flow patterns [7,8]. For example, turbulence dependent eddy viscosities used in the momentum equations affect the pressure equations.

In the TEACH method for numerical solution of the Navier-Stokes equations there is an iteration to consistency of

[1]Department of Mathematics, University of Tennessee, Knoxville, Tennessee.

ISBN 0-12-100560-7

velocities and pressures. The number of iterations depends on how accurately the momentum equations are solved with given pressures to update the velocities and on how accurately pressures are found in order to modify the velocities to satisfy mass conservation. An underrelaxation parameter introduced in the momentum equations [6,8] through which the velocity is updated leads to strong diagonal dominance in these equations so that simple point-Jacobi velocity "inner iteration" converges quite rapidly. The pressure inner iteration, however, requires greater effort for solution sufficiently accurate to not hamper the "outer iteration" to consistency.

II. ANALYSIS OF A MODEL PROBLEM

Some important characteristics of the pressure equations are revealed by considering the five-point difference equation for solving Poisson's equation in a rectangle with a uniform rectangular grid. The difference equation is as usual

$$
\begin{aligned}
&(2/k^2 + 2/h^2)x(i,j) - \{x(i-1,j) + x(i+1,j)\}/h^2 \\
&\qquad - \{x(i,j-1) + x(i,j+1)\}/k^2 = g(i,j) \ .
\end{aligned} \tag{1}
$$

The case where the aspect ratio, $s = k/h$, is much less than unity is of particular interest for many applications. Neumann boundary conditions are applied on the entire boundary. The coefficient matrix of the discrete equations is thus singular, with its zero eigenvalue belonging to an eigenvector whose components are all unity. Just prior to the pressure calculation, exit velocities are normalized so that exit flow is exactly equal to inlet flow. The source term $g(i,j)$ in the discrete pressure equation at node i,j is equal to the outflow from a box around the node. Hence, the sum of the source components is zero. The singular system is therefore consistent, and there is a unique solution up to an additive normalization constant.

The eigenvectors of the coefficient-matrix are $e(m,n)$ with component at node i,j of

$$e(m,n,i,j) = \cos(m\pi i/I)\cos(n\pi j/J)$$

with $m,i = 0,1,\ldots,I$ and $n,j = 0,1,\ldots,J$. The eigenvalues of the simple point iteration matrix are:

$$\lambda(m,n) = [\cos(n\pi/J) + s\cos(m\pi/I)]/(1+s^2) \qquad (2)$$

The spectral radius (s.r.) of the iteration matrix is unity. This value is attained when $m = n = 0$. No error is introduced by the corresponding eigenvector, all of whose components are unity, since the mean pressure is normalized to an arbitrary level. The rate of error decay is governed by the next largest eigenvalue. This is the larger of $\lambda(0,1)$ and $\lambda(1,0)$.

The number of iterations for prescribed error reduction with Chebyshev extrapolation of the error vectors generated by simple iteration varies as

$$t(\text{pt.-Cheby}) = \{1 - \text{s.r.}(\text{pt})\}^{-1/2} \qquad (3a)$$

which is proportional to I/s when $s \ll 1$. The number of iterations with simultaneous improvement of all nodes on a column (line-Chebyshev) varies as

$$t(\text{line-Cheby}) = \{1 - \text{s.r.}(\text{line})\}^{-1/2} \qquad (3b)$$

which is roughly proportional to I.

For equal spacing ($s = 1$) the ratio of t in Eq. (3a) to t in Eq. (3b) is about $\sqrt{2}$. This may no longer be true on a vector computer. The CYBER 205 computes with vectors of length greater than a thousand about twice as fast as it computes with vectors of length one hundred. Point iteration is performed with vectors of length IJ while line iteration is performed with vectors of length I. Each line iteration takes twice as long as one point iteration if I is of order 100 and J is greater than ten. Now point iteration is faster than line iteration by the $\sqrt{2}$ factor.

Considerable improvement is achieved with line relaxation when $s \ll 1$. For example, let $I = 50$, $J = 10$, and $s = 0.1$. Then, the spectral radius for point-Jacobi iteration and the average rate of error decay (R) for point-Chebyshev iteration are

$$\text{s.r.}(\text{pt-Jacobi}) = 1 - e(\text{pt}), \text{ where } e(\text{pt}) = 2 \times 10^{-5}, \text{ and}$$

$$R(\text{pt-Cheby}) = \exp\{-\operatorname{arccosh}(1/(1-e))\} = 0.9937.$$

For error reduction by a factor of 100, $t > -2/\log .9937$, and

729 iterations are required. The corresponding values for line iteration are:

$$\text{s.r.(line-Jacobi)} = 1 - e(\text{line}), \text{ where } e(\text{line}) = .002, \text{ and}$$

$$R(\text{line-Cheby}) = \exp\{-\text{arccosh}(1/.998)\} = 0.93866.$$

Now only 73 iterations are required for error reduction by a factor of 100. The ratio of line to point iterations is approximately s.

When using point relaxation, $n = 0$ in Eq. (2) leads to the large spectral radius when $s \ll 1$. The error component associated with this mode is a function of i only. This error may be removed by an additive column renormalization [5] that is a simple form of two-level iteration [9] which will now be described.

III. ADDITIVE COLUMN RENORMALIZATION

Sum the equations over each column but the last (on which the arbitrary pressure normalization allows one to set the column correction to zero) to obtain a three-term linear system

$$C = \text{tridiagonal } \{-c_w(i), c_w(i) + c_e(i), -c_e(i)\} \tag{4}$$

with first row of

$$\{c_e(1), -c_e(1), 0, \ldots, 0\} \ , \tag{5}$$

and last row of

$$\{0, \ldots, 0, -c_w(I-1), c_w(I-1) + c_e(I-1)\} \ . \tag{6}$$

The coefficients in these expressions are

$$c_e(i) = \sum_j c_e(i,j) \quad \text{and} \quad c_w(i) = \sum_j c_w(i,j) \tag{7}$$

The system to be solved is of order I. The right-hand side of the system has component s(i) for column i of the grid, where s(i) is the sum over column i of the residuals at nodes i,j on this column. Let n range over the neighbors of node i,j. Then

$$s(i) = \sum_j [\sum_{n=1}^{4} c_n(i,j)\{p_n(i,j) - p(i,j)\} + g(i,j)] \ . \qquad (8)$$

Let the components of $\underline{q}$ be the additive corrections to the column pressures. The three-term linear system may be written as

$$C\underline{q} = \underline{s} \qquad (9)$$

The corrected pressures are then

$$p(i,j) = p(i,j) \text{ before normalization} + q(i) \ . \qquad (10)$$

On one test problem with s = 0.1, the number of outer iterations for a prescribed accuracy when the pressures were solved exactly (except for roundoff) at each iteration was 75. About 50 successive-line-overrelaxation iterations led to convergence in 77 outer iterations. This was achieved with about 300 point Chebyshev inner iterations in the absence of the additive normalization. Convergence was achieved in 75 outer iterations when the normalization was added after only 20 Chebyshev inner iterations.

A subtle point arises in applying Chebyshev extrapolation to these pressure equations. Eigenvectors of point-Jacobi iteration with eigenvalues of 1 and -1 coallesce in SOR or SLOR to a single eigenvector, all of whose components are unity, with an eigenvalue of unity. No error is associated with this eigenvector which is absorbed in the arbitrary pressure normalization. Hence, the extrapolation parameter may be chosen as a function of the next larger eigenvalue, say a:

$$w = 2/(1 + \sqrt{1 - a^2}) \ . \qquad (11)$$

No such eigenvector degeneracy occurs when Chebyshev extrapolation is applied. If one were to act on an assumed spectrum of (-a,a), the error associated with the eigenvector belonging to the -1 eigenvalue would persist. The number of iterations for damping an eror over an assumed spectrum (-1,a) differs little from that required for (-a,a). To retain 2-cyclic Chebyshev iteration, which requires spectrum symmetry about zero, one could eliminate the error associated with the -1 eigenvalue by an initial full-mesh sweep and underrelaxation with 1/2. Although this analysis is of a model problem, the same considerations apply to more general problems arising in practice. More general fluid flow problems are modelled with nonorthogonal grid lines. The resulting equations are nine-point rather than five-point. This precludes use of 2-cyclic point-Chebyshev extrapolation. One need not compute a sophisticated estimate of the spectrum bound, a, for the Chebyshev iteration. One may estimate a low value and append a few conjugate-gradient (CG) steps. Each CG iteration requires evaluation of two inner products in addition to the arithmetic common to both Chebyshev and CG iteration. The combination of Chebyshev and CG can be quite efficient [4d]. The additive column normalization may be performed between the Chebyshev and the CG iterations.

Numerical studies revealed that it was better to cycle through the Chebyshev and renormalization steps a few times than it was to follow one long Chebyshev cycle with renormalization. When this strategy was used with a reasonable estimate of the eigenvalue interval of the basic iteration matrix, little was gained by appending CG iteration.

The pressure equations are symmetric and positive semidefinite. When the grid is not rectangular, the off-diagonal elements are not uniformly signed. The upper bound on the iteration matrix spectrum is still unity. However, it can be shown that the lower bound decreases to -2. An initial spectrum bound estimate of (-2,a) may be updated during the iteration by conventional techniques [10].

IV. SUMMARY

There are classes of problems for which considerable gains in efficiency may be realized by clever combination of techniques, some of which may be prefabricated general purpose routines. Characteristics of the specific computer on which a program is implemented must be considered in design of a solution algorithm. For example, point-Chebyshev iteration is particularly well-suited for the CYBER-205. When one solves the singular pressure equations arising in the TEACH method, one must exercise care in estimating eigenvalue bounds. Periodic interruption of point-Chebyshev iteration with an additive column renormalization enhances convergence considerably, especially for grids with large aspect ratios.

V. REFERENCES

1. Roache, P.J. (1976). Computational Fluid Dynamics, Hermosa Publishers, Albuquerque, N.M.

2. Rice, J.R. (1977). "ELLPACK: A research tool for elliptic partial differential equations software," in Mathematical Software III edited by J.R. Rice, Academic Press, pp. 319-341.

3. Dorr, F.W. (1970). "The direct solution of the discrete Poisson equation on a rectangle," SIAM Rev. 12, pp. 248-263.

4. Preconditioned conjugate gradients:

   a. Wachspress, E.L. (1962). "Extended application of ADI model problem theory," J. SIAM 11, pp. 994-1016 (alternative-direction-implicit iteration preconditioner).

   b. Bartels, R. and Daniel, J.W. (1973). "A conjugate gradient approach to nonlinear elliptic boundary problems in irregular regions," Rep. CNA 63, Center for Numerical Analysis, Univ. of Texas at Austin (fast Poisson solver preconditioner).

   c. Meijerink, J.A. and Van der Vorst, H.A. (1977). "An iterative method for linear systems of which the coefficient matrix is a symmetric M-matrix," Math Comp 31, pp. 148-162 (incomplete Cholesky preconditioner).

   d. Johnson, O.G., Micchelli, C.A. and Paul, G. (1983). "Polynomial preconditioners for conjugate gradient calculations," SIAM J. Numer. Anal. 20, pp. 362-376 (Chebyshev polynomial preconditioner).

5. Watts, J.W. (1971). "An iterative matrix inversion method suitable for anisotropic problems," J. Soc. Pet. Eng. 11, pp. 47-51.

6. Gosman, A.D. and Ideriah, F.J.K. "TEACH-T: A general computer program for 2-D, turbulent recirculating flows" Fluid Section Report, Mechanical Engineering Department, Imperial College, London, 1976.

7. Wachspress, E.L. (1981). "Discrete pressure equations in incompressible flow problems," KAPL-4139 (Available from T.I.S., U.S. Dept. of Commerce, Springfield, VA. 22151).

8. ________. (1982). "Mathematical foundations for numerical solution of Navier-Stokes Equations," KAPL-4152 (In ISBN 0-89448-111-8, Advances in Reactor Computations, proceedings of a Topical Meeting of the ANS Math & Comp Division held in Salt Lake City, March 1983).

9. ________. (1976). "Two-level finite element computation," in Formulations and Computational Algorithms in Finite Element Analysis (edited by Bathe, Oden and Wunderlich), MIT Press.

10. Hageman, L.A. and Young, D.M. (1981). Applied Iterative Methods, Academic Press.

ACCELERATING NONSYMMETRIZABLE ITERATIVE METHODS[1]

David M. Young[2], Kang C. Jea[3] and David R. Kincaid[2]

I. INTRODUCTION

In this paper we consider some iterative procedures for solving the linear system

$$Au = b \tag{1.1}$$

where A is a given real nonsingular $N \times N$ matrix and b is a given real $N \times 1$ vector. Occasionally, we will assume that A is symmetric and positive definite (SPD).

The algorithms which we will consider involve the use of a simple one-step, or "basic," iterative method usually combined with an acceleration procedure such as Chebyshev acceleration or conjugate gradient (CG) acceleration. For a certain class of basic iterative methods, referred to as "symmetrizable methods," these acceleration procedures are very effective. However, if the matrix A of the system (1.1) is not SPD, or is not at least <u>similar</u> to an SPD matrix, the basic iterative method may not be symmetrizable. Moreover, some iterative methods are not symmetrizable even if the matrix A is SPD. Acceleration procedures which are efficient

---

[1]This work was supported, in part, by the National Science Foundation under Grant MCS-7919829 and by the Department of Energy under Grant DE-AS05-81ER10954 with The University of Texas at Austin.

[2]Center for Numerical Analysis, The University of Texas at Austin, Austin, Texas.

[3]Present address: Department of Mathematics, Fu Jen University, Hsinchuang, 242 Tapei, Taiwan.

ISBN 0-12-100560-7

and effective for the symmetrizable case may not be appropriate for the nonsymmetrizable case.

The object of this paper is to describe some recent work which has been done at The University of Texas at Austin and elsewhere on the development of procedures for accelerating nonsymmetrizable iterative methods. Because of space limitations our discussion cannot be in any sense complete, and a number of important contributions will not be discussed. Approaches which we do discuss include the use of the GCW method of Concus and Golub [1976] and of Widlund [1978], the use of the normal equations and related equations, Chebyshev acceleration, and some generalized CG acceleration procedures. Normally, the generalized CG acceleration procedures, in their idealized form, are too expensive in terms of computer time and storage for practical use. In the symmetrizable case the formulas simplify since many of the terms vanish. In some nonsymmetrizable cases, it is possible to achieve similar simplifications. We describe work on such simplifications done by Jea [1982] and Jea and Young [1983] as well as work of Faber and Manteuffel [1983]. We also show that the simplifications of Jea and Young [1983] can be used to derive three alternative forms of the Lanczos method (see Lanczos [1950, 1952]). A theoretical basis is given for choosing one of the forms in preference to the other two more frequently used forms.

The companion paper in this volume by Kincaid and Young [1983] describes the ITPACK project. Currently the ITPACK package is designed to handle several symmetrizable iterative methods. However, the package is being expanded to handle a greater variety of basic methods and acceleration procedures with special emphasis on the nonsymmetrizable case. It is anticipated that the expanded version of ITPACK will provide a tool for extensively testing a number of methods including those described in this paper.

II. BASIC ITERATIVE METHODS

A basic iterative method for solving the system (1.1) is a method of the form

$$u^{(n+1)} = Gu^{(n)} + k \tag{2.1}$$

It can be shown, see e.g., Young [1971], that the method (2.1) is "completely consistent" if and only if for some nonsingular matrix Q we have

$$G = I - Q^{-1}A, \quad k = Q^{-1}b \tag{2.2}$$

The matrix Q is known as the "splitting matrix."

We list below the splitting matrices corresponding to some of the more familiar basic iterative methods. We represent the matrix A in the form

$$A = D - C_L - C_U$$

where D is diagonal, C_L is strictly lower triangular, and C_U is strictly upper triangular.

Method	Splitting Matrix (Q)
Richardson	I
Jacobi	D
Gauss-Seidel	$D - C_L$
SOR	$\omega^{-1}D - C_L$ (ω = relaxation factor)
SSOR	$(\omega/(2-\omega))(\omega^{-1}D - C_L)D^{-1}(\omega^{-1}D - C_U)$
Approximate Factorization	LU

For the approximate factorization methods, L and U are lower and upper triangular matrices, respectively, such that LU approximates A in some sense. For one class of such methods, the (i,j)-elements of A and LU agree for any pair (i,j) such that $a_{i,j} \neq 0$.

The basic iterative method (2.1) is <u>symmetrizable</u> if $I - G$ ($= Q^{-1}A$) is similar to an SPD matrix. Equivalently, the method is symmetrizable if $Z(I - G)$ is SPD for some SPD matrix Z. By (2.2) it follows that if A and Q are SPD, then the method (2.1) is symmetrizable and Z can be chosen as A or Q. Evidently, if A is SPD, then the Richardson method, the Jacobi method, and the SSOR method are symmetrizable. If the method (2.1) is symmetrizable, the eigenvalues of G are real and less than unity. We let m(G) and M(G) denote, respectively, the smallest and largest eigenvalues of G.

III. ACCELERATION OF SYMMETRIZABLE METHODS

We first consider Chebyshev acceleration. To apply Chebyshev acceleration one must have available good estimates for m(G) and M(G). The formulas are

$$u^{(n+1)} = \rho_{n+1}(u^{(n)} + \gamma_{n+1}\delta^{(n)}) + (1 - \rho_{n+1})u^{(n-1)} \tag{3.1}$$

$$\gamma_{n+1} = \gamma = 2(2 - M(G) - m(G))^{-1} = 2(m(\hat{A}) + M(\hat{A}))^{-1} \tag{3.2}$$

$$\rho_{n+1} = \begin{cases} 1 & , \text{ if } n = 0 \\ (1 - \frac{1}{2}\sigma^2)^{-1} & , \text{ if } n = 1 \\ (1 - \frac{1}{4}\sigma^2\rho_n)^{-1} & , \text{ if } n \geq 2 \end{cases} \tag{3.3}$$

$$\sigma = \frac{M(G) - m(G)}{2 - M(G) - m(G)} = \frac{M(\hat{A}) - m(\hat{A})}{M(\hat{A}) + m(\hat{A})} \tag{3.4}$$

Here $\hat{A} = Q^{-1}A$ and $\delta^{(n)} = k - \hat{A}u^{(n)}$ is the pseudo-residual corresponding to $u^{(n)}$. The _optimum extrapolated procedure_ based on (2.1) is defined by (3.1) with $\rho_1 = \rho_2 = \ldots = 1$. It can be shown that the number of iterations required to achieve a fixed level of convergence is asymptotically proportional to $K(\hat{A})$ for optimum extrapolation and to $\sqrt{K(\hat{A})}$ for Chebyshev acceleration. Here $K(\hat{A}) = M(\hat{A})/m(\hat{A})$ is the _condition_ of $\hat{A}$. The improvement is substantial if $K(\hat{A})$ is large. To achieve this large increase in the convergence rate one must have accurate estimates of m(G) and M(G). Adaptive procedures for automatically obtaining good estimates of these quantities are described in the book of Hageman and Young [1981] and are incorporated in the ITPACK package described in Kincaid _et al._ [1982].

Conjugate gradient (CG) acceleration provides an alternative to Chebyshev acceleration which yields faster convergence (measured in an appropriate norm) and which does not require estimates for m(G) and M(G). To apply CG acceleration one must first choose an auxiliary matrix Z such that $Z\hat{A}$ and Z are SPD. We choose the form of CG acceleration which is used in ITPACK; however, we will consider other forms later in this paper. The procedure is defined by (3.1) where $\rho_1 = 1$ and

$$\gamma_{n+1} = (Z\delta^{(n)},\delta^{(n)})/(Z\hat{A}\delta^{(n)},\delta^{(n)}) \tag{3.5}$$

$$\begin{aligned}\rho_{n+1} &= \left[1 - \frac{\gamma_{n+1}}{\gamma_n}\,\frac{(Z\delta^{(n)},\delta^{(n)})}{(Z\delta^{(n-1)},\delta^{(n-1)})}\,\frac{1}{\rho_n}\right]^{-1} \\ &= \left[1 + \gamma_{n+1}\frac{(Z\hat{A}\delta^{(n)},\delta^{(n-1)})}{(Z\delta^{(n-1)},\delta^{(n-1)})}\right]^{-1}, \quad n = 1,2,\ldots\end{aligned} \tag{3.6}$$

It can be shown that the error obtained with CG acceleration does not exceed that obtained with Chebyshev acceleration when measured in the $(Z\hat{A})^{\frac{1}{2}}$-norm defined by $||v||_{(Z\hat{A})^{\frac{1}{2}}} = (v,Z\hat{A}v)^{\frac{1}{2}}$. From this it follows that the number of iterations required with CG acceleration is asymptotically proportional to $\sqrt{K(\hat{A})}$. We remark that in many problems CG acceleration is _much_ faster than Chebyshev acceleration. This occurs, for example, if there are only a few scattered eigenvalues of G outside of a small subinterval of the interval $[m(G),M(G)]$. In this case, the convergence is almost as rapid as it would have been if the eigenvalues outside the subinterval were absent.

IV. NONSYMMETRIZABLE METHODS: THE GCW METHOD

The nonsymmetrizable case is much more difficult than the symmetrizable case. In the first place, in the nonsymmetrizable case the eigenvalues of G need not be real and less than unity, as in the symmetrizable case. Quite the contrary, they may be complex and they may even have real parts greater than unity. In the second place, the Jordan canonical form of G need not be diagonal. All of these complications must be considered in choosing an effective acceleration procedure.

If the matrix A of the system (1.1) is not SPD, or at least similar to an SPD matrix, it is almost certain that any feasible iterative method will be nonsymmetrizable. Let us consider a case where A is nonsymmetric but where we assume A is positive real (PR) in the sense that $A + A^T$ is SPD. In some cases it is feasible to consider an iterative method based on the (SPD) splitting matrix

$$Q = \frac{1}{2}(A + A^T) \tag{4.1}$$

This method, which we refer to as the _GCW method_, was

considered by Concus and Golub [1976] and by Widlund [1978]. In order that the GCW method be feasible it is important that one be able to solve conveniently systems of the form $Qx = y$, for x, for any given y. This can be done for some problems arising from partial differential equations using fast direct methods.

A generalized CG acceleration procedure for the GCW method was derived by Concus and Golub [1976] and by Widlund [1978]. We refer to this procedure as the GCW-GCG method. Formulas similar to those of (3.1)-(3.3) were obtained. Numerical experiments showed that the GCW-GCG method was very effective in many cases, especially in cases where the matrix A was nearly symmetric. Hageman, Luk, and Young [1980] showed that the GCW-GCG method is equivalent to standard CG acceleration applied to the "double GCW method" in which each iteration consists of two iterations of the regular GCW method. Further details will be given later in the paper.

V. NORMAL EQUATIONS

Let us now consider the case where the matrix A of (1.1) is not SPD and is not similar to an SPD matrix. In this case the Richardson method is not symmetrizable. Nevertheless, we can obtain a new system where Richardson's method is symmetrizable by constructing the <u>normal equations</u>

$$A^TAu = A^Tb \tag{5.1}$$

Evidently A^TA is SPD since A is nonsingular.

While CG acceleration can be applied to Richardson's method for solving (5.1), the resulting procedure may not be effective since the condition of A^TA may be much worse than that of A. (If A were SPD we would have $K(A^TA) = K(A)^2$ and the use of (5.1) would not help.)

As an alternative to using the normal equations for (1.1) one could use the normal equations for the <u>preconditioned system</u>

$$\hat{A}u = Q^{-1}Au = \hat{b} \tag{5.2}$$

where $\hat{b} = Q^{-1}b = k$. If the condition of $Q^{-1}A$ were less than that of A, one might expect that the condition of $\hat{A}^T\hat{A}$ would be much less than that of A^TA. The use of the normal equations for (5.2), as well as related systems, has been considered by Dongarra et al. [1981], Elman [1982], Kershaw [1980], as well as others. Good results have been reported in many cases.

We now describe a heuristic scheme for constructing systems related to the modified normal equations (5.2). Starting with the preconditioned system (5.2) we choose a matrix Σ and construct the modified preconditioned system

$$\Sigma Q^{-1}Au = \Sigma Q^{-1}b$$

Next we construct the normal equations for this system obtaining

$$A^TQ^{-T}\Sigma^T\Sigma Q^{-1}Au = A^TQ^{-T}\Sigma^T\Sigma Q^{-1}b$$

Now we multiply both sides by an SPD matrix V obtaining

$$VA^TQ^{-T}\Sigma^T\Sigma Q^{-1}Au = VA^TQ^{-T}\Sigma^T\Sigma Q^{-1}b \tag{5.3}$$

Evidently Richardson's method for (5.1) is symmetrizable (with $Z = V^{-1}$) so that we can apply CG acceleration. We try to choose V and Σ so that the condition of the matrix of (5.3) is not much larger than that of $Q^{-1}A$.

To illustrate the above procedure, let us first consider the SOR method, where $Q = \omega^{-1}D - C_L$, for the case where A is SPD. We let $\Sigma = D^{\frac{1}{2}}$ and $V = A^{-T} = A^{-1}$. Since $C_L^T = C_U$, we obtain

$$Q^{-T}DQ^{-1}Au = Q^{-T}DQ^{-1}b$$

On the other hand, the preconditioned equation for the SSOR method is

$$\frac{2-\omega}{\omega}\, Q^{-T}DQ^{-1}Au = \frac{2-\omega}{\omega}\, Q^{-T}DQ^{-1}b$$

Moreover, it follows from results of Young [1971, 1977] that in some cases the condition of $I - S_\omega$, where S_ω is the matrix of the SSOR method, is not much larger than that of $Q^{-1}A$. Thus, in such cases the normal equations as modified do not have as bad a condition as one might normally expect.

The above procedure can be extended as follows. Let A be SPD and consider two iterative methods

$$u^{(n+1)} = Gu^{(n)} + k, \quad G = I - Q^{-1}A, \quad k = Q^{-1}b$$

$$u^{(n+1)} = G'u^{(n)} + k', \quad G' = I - Q^{-T}A, \quad k' = Q^{-T}b$$

The "product method" is defined by

$$u^{(n+1)} = G'Gu^{(n)} + G'k + k' \tag{5.4}$$

where

$$G'G = I - Q^{-T}EQ^{-1}A \tag{5.5}$$

and

$$E = Q + Q^T - A$$

If E is SPD, and if we can apply (5.3) with $\Sigma^T\Sigma = E$ and with $V = A^{-T}$, we obtain

$$Q^{-T}EQ^{-1}Au = (I - G'G)u = Q^{-T}EQ^{-1}b \tag{5.6}$$

Thus by (5.5) and (5.6) the modified normal equations correspond to the product method (5.4).

As another example, let us consider the GCW method for the case where A is PR. The splitting matrix Q is given by (4.1). We let

$$A = Q - R$$

If we let $\Sigma = Q^{\frac{1}{2}}$ and $V = Q^{-1}$, then from (5.3) we get

$$Q^{-1}A^TQ^{-1}Au = Q^{-1}A^TQ^{-1}b \tag{5.7}$$

Evidently Richardson's method for (5.7) is symmetrizable with $Z = Q$. We now relate this to the double GCW method defined by

$$u^{(n+1)} = G(Gu^{(n)} + k) + k = G^2u^{(n)} + (I + G)k$$

where $G = Q^{-1}R$ and $k = Q^{-1}b$. The corresponding preconditioned system is $(I - G^2)u = (I + G)Q^{-1}b$. But since $(I - G^2) = (I+G)(I-G)$ $= Q^{-1}A^TQ^{-1}A$, since $R^T = -R$, we get (5.7).

It was shown by Hageman, Luk, and Young [1980] that the standard CG acceleration method applied to the double CGW method, with $Z = (I+G^T)^{-1}Q(I+G)^{-1}$, is in a sense equivalent to the GCW-GCG method. Thus here again we have a situation where the use of the normal equations as modified does not lead to a substantially larger condition than that of the preconditioned system (5.2).

VI. CHEBYSHEV ACCELERATION IN THE NONSYMMETRIZABLE CASE

From now on we shall consider the acceleration of Richardson's method for solving the linear system (1.1). Our discussion also applies to the preconditioned system (5.2) corresponding to the basic iterative method (2.1). Thus instead of A and b we would have $\hat{A}$ and $\hat{b}$, respectively. Also, the residual vector $r^{(n)}$ would be replaced by the pseudo-residual $\hat{r}^{(n)} = \hat{b} - \hat{A}u^{(n)} = \delta^{(n)}$.

Let us now assume that the eigenvalues of A have positive real parts or, equivalently, that the eigenvalues of the iteration matrix G have real parts less than unity. We define Chebyshev acceleration as follows. We choose the real numbers d (< 1) and σ^2 (σ^2 may be negative) and consider the procedure defined by (3.1) and (3.3) but with $\gamma_1 = \gamma_2 = \ldots = (1-d)^{-1}$. It can be shown that the error vector $\varepsilon^{(n)} = u^{(n)} - \bar{u}$, where $\bar{u}$ is the true solution $A^{-1}b$ of (1.1), satisfies the equation $\varepsilon^{(n)} = P_n(G)\varepsilon^{(0)}$ where $P_n(G)$ is related to the Chebyshev polynomial of degree n. The choice of d and σ^2 is optimum, in some sense, if

$$S = (a+b)[1 - d + \sqrt{(1-d)^2 + c^2}]^{-1}$$

is minimized, where $a^2 - b^2 = c^2 = \sigma^2(1-d)^2$ and all eigenvalues μ of G lie in the closed ellipse

$$[(\mathrm{Re}(\mu)-d)/a]^2 + [\mathrm{Im}(\mu)/b]^2 = 1$$

Manteuffel [1977] has given a procedure for finding the optimum values of d and σ^2 given the eigenvalues of G. He has also developed [1978] an adaptive procedure for determining d and σ^2 for cases when the eigenvalues of G are not known.

The gain attainable by using Chebyshev acceleration depends on the eigenvalues of G. If all of the eigenvalues of G lie in a circle with center at the origin and if they are densely distributed on the boundary of the circle, then the best (or near best) choices of d and σ^2 are $d = \sigma^2 = 0$, which corresponds to the unaccelerated method. On the other hand, if all of the eigenvalues of G are real, or have small imaginary parts, then Chebyshev acceleration is very effective.

VII. GENERALIZED CG ACCELERATION

In this section we describe some generalizations of CG acceleration which can be applied to nonsymmetrizable iterative methods. The objective of the development of such methods was to obtain convergence as rapid as with Chebyshev acceleration but without requiring parameter estimates or a knowledge of the eigenvalues of the iteration matrix G. It was not expected that the generalized CG acceleration procedures would be effective in cases, such as those described in the previous section, where Chebyshev acceleration is not effective.

Young and Jea [1980] considered a generalized CG acceleration procedure, called the <u>idealized generalized CG acceleration method</u> (IGCG acceleration). To apply IGCG acceleration, one chooses a nonsingular auxiliary matrix Z and determines $u^{(1)}, u^{(2)}, \ldots,$ corresponding to the initial vector $u^{(0)}$, by

$$u^{(n)} - u^{(0)} \in K_n(r^{(0)})$$

$$(Zr^{(n)}, v) = 0, \quad \text{for all} \quad v \in K_n(r^{(0)}). \tag{7.1}$$

Here $K_n(r^{(0)})$ is the Krylov space spanned by the vectors $r^{(0)}$, $Ar^{(0)}, \ldots, A^{n-1}r^{(0)}$. If ZA is SPD, the condition (7.1) can be replaced by the minimization condition

$$||u^{(n)} - \bar{u}||_{(ZA)^{\frac{1}{2}}} \leq ||v - \bar{u}||_{(ZA)^{\frac{1}{2}}}$$

for all v such that $v - u^{(0)} \in K_n(r^{(0)})$. Here $\bar{u} = A^{-1}b$ is the true solution of (1.1).

Young and Jea [1980] derived three alternative forms of IGCG acceleration. These three forms, which are referred to as ORTHODIR, ORTHOMIN, and ORTHORES, are given in Table VII.1. In the symmetrizable case the terms inside the brackets [·] vanish. In this case ORTHOMIN reduces to the classical CG procedure considered by Hestenes and Stiefel [1952], and ORTHORES reduces to the procedure defined by (3.1), (3.5),

TABLE VII.1. Formulas for the IGCG Method

(In each case $u^{(0)}$ is arbitrary)

ORTHODIR

$$u^{(n+1)} = u^{(n)} + \hat{\lambda}_n q^{(n)}$$

$$\hat{\lambda}_n = (Zr^{(n)},q^{(n)})/(ZAq^{(n)},q^{(n)})$$

$$q^{(n)} = Aq^{(n-1)} + \beta_{n,n-1}q^{(n-1)} + \beta_{n,n-2}q^{(n-2)} + [\beta_{n,n-3}q^{(n-3)} + \cdots + \beta_{n,0}q^{(0)}],$$

$$n = 1,2,\ldots,;\ (q^{(0)} = r^{(0)})$$

$$\beta_{n,i} = -\frac{(ZA^2q^{(n-1)},q^{(i)}) + [\sum_{j=0}^{i-1} \beta_{n,j}(ZAq^{(j)},q^{(i)})]}{(ZAq^{(i)},q^{(i)})},$$

$$i = 0,1,\ldots,n-1, \quad n = 1,2,\ldots$$

ORTHOMIN

$$u^{(n+1)} = u^{(n)} + \lambda_n p^{(n)}$$

$$\lambda_n = (Zr^{(n)},r^{(n)})/(ZAp^{(n)},p^{(n)})$$

$$p^{(n)} = r^{(n)} + \alpha_{n,n-1}p^{(n-1)} + [\alpha_{n,n-2}p^{(n-2)} + \cdots + \alpha_{n,0}p^{(0)}],$$

$$n = 1,2,\ldots,(p^{(0)} = r^{(0)})$$

$$\alpha_{n,i} = -\frac{(ZAr^{(n)},p^{(i)}) + [\sum_{j=0}^{i-1} \alpha_{n,j}(ZAp^{(j)},p^{(i)})]}{(ZAp^{(i)},p^{(i)})},$$

$$i = 0,1,\ldots,n-1, \quad n = 1,2,\ldots$$

ORTHORES

$$u^{(n+1)} = \lambda_n r^{(n)} + f_{n+1,n}u^{(n)} + f_{n+1,n-1}u^{(n-1)} + [f_{n+1,n-2}u^{(n-2)} + \cdots + f_{n+1,0}u^{(0)}]$$

$$\sigma_{n+1,i} = \frac{(ZAr^{(n)},r^{(i)}) - [\sum_{j=0}^{i-1} \sigma_{n+1,j}(Zr^{(j)},r^{(i)})]}{(Zr^{(i)},r^{(i)})}, \quad i = 0,1,\ldots,n$$

$$\lambda_n^{-1} = \sigma_{n+1,n} + \sigma_{n+1,n-1} + [\sigma_{n+1,n-2} + \cdots + \sigma_{n+1,0}]$$

$$f_{n+1,i} = \lambda_n \sigma_{n+1,i}$$

and (3.6). Thus, if we let $\rho_{n+1} = f_{n+1,n}$ and $\gamma_{n+1} = \lambda_n/\rho_{n+1}$, we get (3.1), (3.5), and the second form of ρ_{n+1} given by (3.6). The procedure thus defined was considered by Engeli et al. [1959]. Concus, Golub, and O'Leary [1976] also considered this procedure and used the second form of ρ_{n+1} given by (3.6).

We remark that if ZA is SPD then the terms in the brackets $[\cdot]$ vanish for $\beta_{n,i}$ for ORTHODIR and for $\alpha_{n,i}$ for ORTHOMIN. Similarly, if Z is SPD, the terms in the brackets $[\cdot]$ vanish for $\sigma_{n+1,i}$ for ORTHORES. We also note that if Z and ZA are both SPD (such a Z always exists in the symmetrizable case) the formulas "truncate" in the sense that $\beta_{n,i} = 0$ for $i < n-2$ for ORTHODIR, $\alpha_{n,i} = 0$ for $i < n-1$ for ORTHOMIN, and $\sigma_{n+1,i} = 0$ for $i < n-1$ for ORTHORES.

It follows from the results of Jea [1982] that for each of the three methods there exists $t \leq N$ such that if the method does not break down for t iterations then $u^{(t+1)} = \bar{u}$, i.e., we have convergence. It is shown that ORTHOMIN converges if and only if ORTHORES converges, and if both converge then ORTHODIR converges. Young and Jea [1980] showed that the condition that ZA is PR is sufficient for the convergence of ORTHODIR while the conditions that ZA is PR and that Z is PR are sufficient for the convergence of ORTHOMIN and ORTHORES. We remark that, given A, one can always find Z such that ZA is PR; indeed, we can let $Z = A^T Y$ for any PR matrix Y. We can even make ZA be SPD by letting $Z = A^T K$ for any SPD matrix K.

In general, the three forms of the IGCG method often require an excessive amount of computer time and storage. This is because in general the computation of $u^{(n+1)}$ involves at least one vector from each of the n previous iterations. An exception is the symmetrizable case where the computation of $u^{(n+1)}$ only involves one vector from, at most, each of the two previous iterations. The situation in the symmetrizable case suggests the use of truncation where one arbitrarily discards all but a few terms in the formulas for the IGCG method. Thus, given an integer $s \geq 1$, we require that $\beta_{n,i} = 0$ if $i < n-s$ for ORTHODIR. The resulting procedure is referred to as "ORTHODIR(s)." Similarly, we obtain ORTHOMIN(s) by requiring that $\alpha_{n,i} = 0$ for $i < n-s$. For ORTHORES(s) we require that

$\sigma_{n+1,i} = 0$ for $i < n-s$. Evidently, in the symmetrizable case, if Z and ZA are SPD, then ORTHODIR, ORTHOMIN, and ORTHORES reduce to ORTHODIR(2), ORTHOMIN(1), and ORTHORES(1), respectively.

The convergence properties of the IGCG method no longer hold, in general, for the truncated procedures. However, Elman [1982] has shown that if A is PR, then ORTHOMIN(s) will always converge. He also gives bounds on the rate of convergence. Numerical experiments using the truncated schemes are described by Elman [1982], by Young and Jea [1981], and by Jea [1982]. The linear systems considered were derived from the convection-diffusion equation $u_{xx} + u_{yy} + Du_x = f(x,y)$ in a rectangle. ORTHOMIN(s) and ORTHORES(s) were found to work quite well in most cases for small s $(2 \leq s \leq 4)$ while ORTHODIR(s) was generally not effective. A great deal more thoeretical and experimental work is needed to accurately assess the effectiveness of the truncated methods.

The idea of truncation appears to have been first used by Vinsome [1976]. Vinsome applied truncation to ORTHOMIN and, in fact, introduced the term "ORTHOMIN."

VIII. SIMPLIFICATIONS OF GENERALIZED CG ACCELERATION

For the remainder of this paper we will concentrate on cases where certain truncated versions of the IGCG method are equivalent to the idealized versions. We show that it is always possible in theory, though it may not be feasible in practice, to choose the auxiliary matrix Z so that the formulas simplify for the generalized CG acceleration procedures. This, of course, is true in the symmetrizable case where we choose the auxiliary matrix Z such that Z and ZA are SPD.

Jea and Young [1983] showed that a similar simplification can be obtained if one can find a matrix H satisfying the following condition:

Definition. The nonsingular matrix H satisfies Condition I with respect to the matrix A if $HA = A^T H$.

Such a matrix H always exists. However, as shown by Jea and Young [1983], the determination of H requires finding a matrix P such that $P^{-1}AP = J_A$, the Jordan canonical form of A.

In other words, one needs all eigenvectors and principal vectors of A. Clearly, this is seldom practical.

As shown by Jea and Young [1983], if the auxiliary matrix Z satisfies Condition I, then ORTHODIR, ORTHOMIN, and ORTHORES reduce to ORTHODIR(2), ORTHOMIN(1), and ORTHORES(1), respectively. The formulas are the same as in the symmetrizable case except that there is one term in the brackets [·] for $\beta_{n,i}$ for ORTHODIR(2) unless ZA is SPD.

In the <u>symmetric indefinite case</u> $A = A^T$ (but A is not SPD). If an SPD matrix H exists which satisfies Condition I, then A is <u>similar</u> to a symmetric matrix. In this case, if we let $Z = A^T H$, then Z satisfies Condition I and $ZA = A^T HA$ is SPD. In this case, as stated in Section VII, we are guaranteed that ORTHODIR converges. Moreover, ORTHODIR is equivalent to ORTHODIR(2). The procedure thus obtained was considered by Chandra, Eisenstat, and Schultz [1977] and by Chandra [1978]; it is called the "modified conjugate residual method."

Without going into detail, we give an indication of the proof of the simplification process for ORTHODIR. The vectors $q^{(0)}, q^{(1)}, q^{(2)}, \ldots$ are determined so that $(ZAq^{(i)}, q^{(j)}) = 0$ for $j < i$. If Z satisfies Condition I, then the first term in the numerator for $\beta_{n,i}$ is

$$\begin{aligned}(ZA^2 q^{(n-1)}, q^{(i)}) &= (A^T ZAq^{(n-1)}, q^{(i)}) = (ZAq^{(n-1)}, Aq^{(i)}) \\ &= (ZAq^{(n-1)}, q^{(i+1)} - \beta_{n+1,i} q^{(i)} \\ &\quad - \cdots - \beta_{n+1,0} q^{(0)})\end{aligned} \tag{8.1}$$

which vanishes if $i < n-2$. From this it can be shown that $\beta_{n,i} = 0$ if $i < n-2$. Also, there is at most one nonzero term in the brackets [·] of the numerator of $\beta_{n,i}$.

Faber and Manteuffel [1982] considered the simplification of a class of generalized CG acceleration methods. Motivated by their work we now consider the following generalization of Condition I.

<u>Definition</u>. Given the integer $s \geq 1$, the nonsingular matrix H satisfies <u>Condition I(s) with respect to the matrix A</u> if $HA = P_s(A^T)H$ where $P_s(A^T)$ is a polynomial of degree s.

Evidently if H satisfies Condition I then it also satisfies Condition I(1). We remark that an SPD matrix H exists which satisfies Condition I(s) if and only if

$$CAC^{-1} = P_s((CAC^{-1})^T) \tag{8.2}$$

where C is any matrix satisfying $C^TC = H$. This implies that CAC^{-1} is normal; thus, A is similar to a normal matrix. Conversely, it can be shown that if A is similar to a normal matrix, then (8.2) holds for some polynomial $P_s((CAC^{-1})^T)$ of degree s and for some matrix C. For a proof see Drazin [1951] or Faber and Manteuffel [1982].

Using an argument similar to that given in (8.1) one can show that if Z satisfies Condition I(s), then ORTHODIR, ORTHOMIN, and ORTHORES reduce to ORTHODIR(s+1), ORTHOMIN(s), and ORTHORES(s), respectively.

As an example, we consider the GCW method. Instead of the system (1.1) we consider the preconditioned system

$$\hat{A}u = \hat{b}$$

where $\hat{A} = Q^{-1}A$, $\hat{b} = Q^{-1}b$, and where Q is given by (4.1). We show that Q satisfies Condition I(1) with respect to $\hat{A}$. Thus we show that (8.2) holds with $C = Q^{\frac{1}{2}}$. Since $A = Q - R$, where $R^T = -R$ we have

$$C\hat{A}C^{-1} = I - Q^{-\frac{1}{2}}RQ^{-\frac{1}{2}}$$

$$(C\hat{A}C^{-1})^T = I - Q^{-\frac{1}{2}}R^TQ^{-\frac{1}{2}} = I + Q^{-\frac{1}{2}}RQ^{-\frac{1}{2}}$$

Therefore $C\hat{A}C^{-1} = 2I - (C\hat{A}C^{-1})^T = P_1((C\hat{A}C^{-1})^T)$. Since $C^TC = Q$ it follows that Q satisfies Condition I(1). Evidently, then, ORTHORES, with $Z = Q$, is equivalent to ORTHORES (1). The method thus obtained is equivalent to the GCW-GCG method.

The convergence of ORTHORES, and hence of the GCW-GCG method, follows from the fact that $Z = Q$ is SPD and that $Z\hat{A}$ is PR; see Section VII. Thus, $Z\hat{A} = Q(Q^{-1}A) = A$ which is PR.

We remark that if H satisfies Condition I(s) then so does A^TH. If A is similar to a normal matrix we can take H to be SPD. In that case, if $Z = A^TH$, then ZA is SPD and the convergence of ORTHODIR, which is equivalent to ORTHODIR(2), is guaranteed. (Thus for the GCW method we could use ORTHODIR(2) with $Z = \hat{A}^T Q$.) If $Z = A^T H$, then ORTHOMIN is equivalent to ORTHOMIN(1), but convergence is not guaranteed unless Z is PR. Similarly, if we let Z = H, then ORTHORES is equivalent to ORTHORES(1). In this case convergence is not guaranteed unless, as in the case of the GCW method, $Z\hat{A}$ is PR.

Faber and Manteuffel [1983] considered necessary as well as sufficient conditions for the simplification of certain generalized CG methods. They made certain assumptions which are equivalent to assuming ZA to be SPD. Their results imply that, except for certain special cases, ORTHODIR reduces to ORTHODIR(s+1) only if Condition I(s) holds for some SPD matrix H.

IX. THE LANCZOS METHOD

The method of Lanczos [1950,1952] provides an alternative to generalized CG acceleration for speeding up the convergence of a nonsymmetrizable iterative method. In this section we show how three forms of the Lanczos method can be derived from a simplification of CG acceleration.

Let us introduce a fictitious system

$$A^T \tilde{u} = \tilde{b} \tag{9.1}$$

For the present we will not specify $\tilde{b}$. Evidently, for any $\tilde{b}$, the vector Ⓤ $= [u\ \tilde{u}]^T$ satisfies the (double) system

$$\text{Ⓐ}\text{Ⓤ} = \text{Ⓑ} \tag{9.2}$$

where

$$\text{Ⓐ} = \begin{bmatrix} A & 0 \\ 0 & A^T \end{bmatrix}, \quad \text{Ⓤ} = \begin{bmatrix} u \\ \tilde{u} \end{bmatrix}, \quad \text{Ⓑ} = \begin{bmatrix} b \\ \tilde{b} \end{bmatrix}$$

if and only if u satisfies (1.1) and $\tilde{u}$ satisfies (9.1).

We now consider the application of generalized CG acceleration to Richardson's method for solving (9.2). We choose the auxiliary matrix Ⓩ = Ⓗ (where we let Ⓗ $= \begin{bmatrix} 0 & I \\ I & 0 \end{bmatrix}$). If Ⓩ = Ⓗ, then both Ⓩ and Ⓩ Ⓐ are symmetric. Hence ORTHODIR, ORTHOMIN, and ORTHORES reduce to ORTHODIR(2) (with all terms in the brackets zero), ORTHOMIN(1), and ORTHORES(1), respectively. In Table IX.1 we give formulas for the three methods but with formulas involving the fictitious vectors $\tilde{u}^{(n+1)}, \tilde{u}^{(n)}, \ldots$ omitted. We refer to the resulting procedures as Lanczos/ORTHODIR, Lanczos/ORTHOMIN, and Lanczos/ORTHORES. We note that in the symmetrizable case, if Z and ZA are SPD and if one lets $\tilde{r}^{(0)} = Zr^{(0)}$ then the three procedures reduce to the corresponding CG procedures.

TABLE IX.1. The Lanczos Method for the Nonsymmetrizable Case (In each case $u^{(0)}$ and $\tilde{r}^{(0)}$ are arbitrary)

Lanczos/ORTHODIR

$$u^{(n+1)} = u^{(n)} + \hat{\lambda}_n q^{(n)},$$

$$\hat{\lambda}_n = [(\tilde{r}^{(0)}, q^{(n)}) + (r^{(n)}, \tilde{q}^{(n)})]/2(Aq^{(n)}, \tilde{q}^{(n)})$$

$$q^{(n)} = Aq^{(n-1)} - a_n q^{(n-1)} - b_n q^{(n-2)}, \quad n = 1,2,\ldots; \quad (q^{(0)} = r^{(0)})$$

$$\tilde{q}^{(n)} = A^T\tilde{q}^{(n-1)} - a_n \tilde{q}^{(n-1)} - b_n \tilde{q}^{(n-2)}, \quad n = 1,2,\ldots; \quad (\tilde{q}^{(0)} = \tilde{r}^{(0)})$$

$$a_n = (Aq^{(n-1)}, A^T\tilde{q}^{(n-1)})/(Aq^{(n-1)}, \tilde{q}^{(n-1)}), \quad n = 1,2,\ldots;$$

$$b_n = [(Aq^{(n-2)}, A^T\tilde{q}^{(n-1)}) + (Aq^{(n-1)}, A^T\tilde{q}^{(n-2)})]/2(Aq^{(n-2)}, \tilde{q}^{(n-2)}), \quad n = 2,3,\ldots; \quad (b_1 = 0)$$

$$r^{(n+1)} = r^{(n)} - \hat{\lambda}_n Aq^{(n)}; \quad \tilde{r}^{(n+1)} = \tilde{r}^{(n)} - \hat{\lambda}_n A^T\tilde{q}^{(n)} .$$

Lanczos/ORTHOMIN

$$u^{(n+1)} = u^{(n)} + \lambda_n p^{(n)}$$

$$\lambda_n = (r^{(n)}, \tilde{r}^{(n)})/(Ap^{(n)}, \tilde{p}^{(n)})$$

$$p^{(n)} = r^{(n)} + \alpha_n p^{(n-1)}, \quad n = 1,2,\ldots; \quad (p^{(0)} = r^{(0)})$$

$$\tilde{p}^{(n)} = \tilde{r}^{(n)} + \alpha_n \tilde{p}^{(n-1)}, \quad n = 1,2,\ldots; \quad (\tilde{p}^{(0)} = r^{(0)})$$

$$\alpha_n = (r^{(n)}, \tilde{r}^{(n)})/(r^{(n-1)}, \tilde{r}^{(n-1)}), \quad n = 1,2,\ldots ;$$

$$r^{(n+1)} = r^{(n)} - \lambda_n Ap^{(n)}, \quad \tilde{r}^{(n+1)} = \tilde{r}^{(n)} - \lambda_n A^T\tilde{p}^{(n)}$$

Lanczos/ORTHORES

$$u^{(n+1)} = \rho_{n+1}(u^{(n)} + \gamma_{n+1} r^{(n)}) + (1 - \rho_{n+1}) u^{(n-1)}$$

$$\gamma_{n+1} = (r^{(n)}, \tilde{r}^{(n)})/(Ar^{(n)}, \tilde{r}^{(n)})$$

TABLE IX.1. (CONT'D)

$\rho_{n+1} = \left[1 - \frac{\gamma_{n+1}}{\gamma_n} \frac{(r^{(n)}, \tilde{r}^{(n)})}{(r^{(n-1)}, \tilde{r}^{(n-1)})} \frac{1}{\rho_n}\right]^{-1}$,	$n = 1,2,\ldots;$ $(\rho_1 = 1)$
$r^{(n+1)} = \rho_{n+1}(r^{(n)} - \gamma_{n+1}Ar^{(n)}) + (1 - \rho_{n+1})r^{(n-1)}$	
$\tilde{r}^{(n+1)} = \rho_{n+1}(\tilde{r}^{(n)} - \gamma_{n+1}A^T\tilde{r}^{(n)}) + (1 - \rho_{n+1})\tilde{r}^{(n-1)}$	

We remark that the Lanczos/ORTHOMIN procedure is equivalent to the "biconjugate gradient method" considered by Fletcher [1975,1976]. Numerical experiments based on this procedure are described by Wong [1979].

Since Ⓗ Ⓐ is not necessarily PR, there is no guarantee that any of the three methods will not break down. However, the results of Section VII imply that for any one of the three methods there exists $t \leq N$ such that if the method does not break down within t iterations, then we have convergence; in fact, $u^{(t+1)} = \bar{u}$. Moreover, the Lanczos/ORTHOMIN procedure converges if and only if the Lanczos/ORTHORES procedure converges, and if both converge then the Lanczos/ORTHODIR procedure converges. These results suggest that the Lanczos/ORTHODIR method is safest. Numerical experiments are needed to see which of the three methods is actually better in practice.

REFERENCES

Chandra, R. (1978). "Conjugate gradient methods for partial differential equations," Res. Rep. 129, Dept. Comp. Sci., Yale Univ., New Haven, CT.

Chandra, R., Eisenstat, S.C., and Schultz, M.H. (1977). "The modified conjugate residual method for partial differential equations," in Advances in Computer Methods for Partial Differential Equations (R. Vichnevetsky, ed.), p. 13. Vol. II, IMACS.

Concus, P., and Golub, G.H. (1976). "A generalized conjugate gradient method for nonsymmetric systems of linear equations," Report STAN-CS-76-646, Comp. Sci. Dept., Stanford Univ., CA.

Concus, P., Golub, G.H., and O'Leary, D.P. (1976). "A generalized conjugate gradient method for the numerical solution of elliptic partial differential equations," in Sparse Matrix Computation (J.R. Bunch and D.J. Rose, eds.), p. 309. Academic Press, New York.

Dongarra, J.J., Leaf, G.K., and Minkoff, M. (1981). "A preconditioned conjugate gradient method for solving a class of non-symmetric linear systems," Argonne Nat. Lab. Rep. ANL-81-71.

Drazin, M. (1951). "On diagonable and normal matrices," Q.J. Math. (Oxford) 2, 189.

Elman, H.C. (1982). "Iterative methods for large, sparse, nonsymmetric systems of linear equations," Res. Rep. 229, Dept. Comp. Sci., Yale Univ., New Haven, CT.

Engeli, M., Ginsburg, T., Rutishauser, H., and Stiefel, E. (1959). "Refined iterative methods for computation of the solution and the eigenvalues of self-adjoint boundary value problems," Mitt. Inst. Angew. Math. ETH, Zurich, Nr. 8, Basel-Stuttgart.

Faber, V., and Manteuffel, T. (1982). "Conjugate gradient methods," unpublished manuscript.

Fletcher, R. (1975). "Conjugate gradient methods for indefinite systems," in Proc. Dundee Biennial Conf. on Numer. Anal. (G.A. Watson, ed.), p. 73. Springer-Verlag, Berlin and New York.

Fletcher, R. (1976). "Conjugate gradient methods for indefinite systems," In Lecture Notes in Mathematics 506, Springer-Verlag, Berlin and New York.

Hageman, L.A., Luk, F.T., and Young, D.M. (1980). "On the equivalence of certain iterative acceleration methods," SIAM J. Numer. Anal. 17, 852.

Hageman, L.A., and Young, D.M. (1981). Applied Iterative Methods, Academic Press, New York.

Hestenes, M.R., and Stiefel, E.L. (1952). "Methods of conjugate gradients for solving linear systems," J. Res. Nat. Bur. Standards 49, 409.

Jea, K. (1982). "Generalized conjugate gradient acceleration of iterative methods," Ph.D. thesis, Dept. Math., Univ. of Texas at Austin; also CNA-176, Center for Numer. Anal., Univ. of Texas at Austin.

Jea, K.C., and Young, D.M. (1983). "On the simplification of generalized conjugate gradient methods for nonsymmetrizable linear systems," J. of L. A. A., to appear.

Kershaw, D. (1980). "On the problem of unstable pivots in the incomplete LU-conjugate gradient method," J. Comp. Phy. 38, 114.

Kincaid, D., Respess, J., Young, D.M., and Grimes, R. (1982). "ITPACK 2C: A FORTRAN package for solving large sparse linear systems by adaptive accelerated iterative methods," T.O.M.S. 8, 302.

Kincaid, D.R., and Young, D.M. (1983). "The ITPACK project: past, present, future," in Elliptic Problem Solvers (A. Schoenstadt, ed.), Academic Press, N.Y.

Lanczos, C. (1950). "An iterative method for the solution of the eigenvalue problem of linear differential and integral operators," J. Res. Nat. Bur. Standards 45, 255.

Lanczos, C. (1952). "Solution of systems of linear equations by minimized iterations," J. Res. Nat. Bur. Standards 49, 33.

Manteuffel, T.A. (1977). "The Tchebychev iteration for nonsymmetric linear systems," Numer. Math. 28, 307.

Manteuffel, T.A. (1978). "Adaptive procedure for estimating parameters for the nonsymmetric Tchebyshev iteration," Numer. Math. 31, 183.

Vinsome, P.K.W. (1976). "ORTHOMIN, an iterative method for solving sparse sets of simultaneous linear equations," in Proc. 4th Symp. of Numer. Simulation of Reservoir Performance of the Soc. of Petr. Engrs. of the AIME, Los Angeles, CA.

Widlund, O. (1978). "A Lanczos method for a class of nonsymmetric systems of linear equations," SIAM J. Numer. Anal. 15, 801.

Wong, Y.S. (1979). "Conjugate gradient type methods for unsymmetric matrix problems," Tech. Rep. 79-36, The Inst. of Appl. Math. and Stat., Univ. of Brit. Columbia, Canada.

Young, D.M. (1971). Iterative Solution of Large Linear Systems. Academic Press, New York.

Young, D.M. (1977). "On the accelerated SSOR method for solving large linear systems," Adv. in Mathematics 23, 215.

Young, D.M., and Jea, K.C. (1980). "Generalized conjugate-gradient acceleration of nonsymmetrizable iterative methods," J. of L. A. A. 34, 159.

Young, D.M., and Jea, K.C. (1981). "Generalized conjugate gradient acceleration of iterative methods. Part II: The nonsymmetrizable case," Rep. CNA-163, Center for Numerical Analysis, Univ. of Texas at Austin.

IV
Finite Element and Multigrid Methods

ADAPTIVE FINITE ELEMENT PROCESSES IN STRUCTURAL MECHANICS

Ivo Babuška[1] and Werner C. Rheinboldt[1]

I. INTRODUCTION

In general, engineering computations serve to analyze specific physical phenomena with the aim of reaching decisions about proposed designs; that is, for instance, the expected behavior of a construction, its safety features, and similar concerns. For a general assessment of the structure, qualitative results about the displacements, stress-distributions, etc., are needed with an engineering accuracy of, say 10-20%. On the other hand, decisions are usually based on relatively few items known with higher accuracy, say in the range of 2-5%, such as the displacements or stresses at certain known critical points, stress-intensity factors, etc.

In order to achieve these diverse objectives, computational procedures should include the following features among others:

(i) Facilities for efficient and reliable estimation of the errors of the computed results. These estimates should correspond to the specific aims of the particular computation.

(ii) Adaptively controlled computations, in order to achieve the desired error tolerances with a minimal cost and to reduce the preparatory work required of the user.

[1]University of Maryland, College Park, Maryland and University of Pittsburgh, Pittsburgh, Pennsylvania, respectively. This work was in part supported under ONR contracts N0014-77-C-0623 and N0014-80-C-0455 and NSF grant MCS-78-05299.

ISBN 0-12-100560-7

(iii) Post-processing facilities which are designed to allow for the extraction of the high-accuracy data needed for the crucial decisions.

For the solution of classes of linear problems by finite element approximations, algorithms for all three features are now beginning to be well understood. On the other hand, the nonlinear case is as yet not well developed, and much still needs to be done before satisfactory algorithms including all three features can be constructed.

This paper is not intended to present an overview of the work in this area. Instead, it addresses certain specific questions relating to the theoretical and computational aspects of the three features. In particular, the design of effective adaptive procedures requires a clear understanding of the objectives that are to be achieved.

Accordingly, in Section II we discuss some new results about feedback and adaptivity. In line with the summary nature of this paper, these results are only illustrated for a one-dimensional, linear model problem. Then Section III addresses their extension to the more general two-dimensional, linear, elliptic problems which can be solved by the adaptive solver FEARS. In this generality, some results can as yet be tested only experimentally. Section IV presents some comments about the h-version, p-version, and the combined h-p-version of the finite element method. In Section V we then turn to some of the post-processing approaches for linear problems which may meet the demands for the third of the desirable features indicated above.

All the material in these sections concerns linear problems. As noted before, for nonlinear problems, the corresponding theory is not nearly so well developed. In part, this is due to the fact that the problem formulations differ considerably between the two cases. In particular, in practical nonlinear problems we must usually consider parametrized systems of equations, whose solutions form manifolds in the space of the state and parameter variables. The problem then becomes the computational determination of the characteristic features of such solution manifolds. This is the topic of Section VI.

For these parametrized equations it is not always readily apparent how to define the discretization errors needed for the first of our three features. In Section VII we present some recent results about a priori estimates of such discretization errors which in turn lend themselves to the development of the desired a posteriori estimates. Finally, in Section VIII, we outline a numerical method for the computation of paths on the solution manifolds and its combination with mesh-refinement strategies. This may provide the basis for procedures which include the second of the cited facilities.

II. FEEDBACK AND ADAPTIVITY

The term "adaptive" is widely, but generally very loosely used today in scientific computing. We will give to it here the technical meaning proposed in [24]. In artificial intelligence a related definition was used in [20], and for discussions along the same line in control theory we refer to [33].

Without entering into the details of [24], we explain in this section the needed concepts of feedback and adaptivity and summarize some of our theoretical conclusions for a simple model problem. The next section then addresses the extension of these results to more general problems.

Consider the two-point boundary value problem

$$L[u] \equiv -\frac{d}{dx}[a(x)\frac{du}{dx}] + b(x)u = f(x), \quad x \in I = (0,1), \qquad (2.1)$$

$$u(0) = u(1) = 0 \qquad (2.2)$$

where

$$0 \leq \underline{\alpha} \leq a(x) \leq \overline{\alpha}, \quad 0 \leq b(x) \leq \beta, \quad x \in \overline{I} \qquad (2.3)$$

As usual, we define on $\mathring{H}^1(I) \times \mathring{H}^1(I)$ the bilinear form

$$B(u,v) = \int_I (au'v' + buv)dx \qquad (2.4)$$

and note that the norm on $H^1(I)$ is equivalent with the energy norm $||u||_E = (B(u,u))^{1/2}$. The weak solution $u_o \in \mathring{H}^1(I)$ of (2.1)-(2.2) is defined by

$$B(u_o,v) = F(v) \equiv \int_I fv\,dx, \forall\, v \in H^1(I). \tag{2.5}$$

Let F be a family of finite-dimensional subspaces $S \subset \overset{\circ}{H}{}^1(I)$. For any $S \in F$ the approximate solution $u(S) \in S$ of (2.1)-(2.2) is specified by

$$B(u(S),v) = F(v), \ \forall\, v \in S \tag{2.6}$$

and hence the optimal error achievable with subspaces of F of a fixed dimension $d > 0$ is

$$\phi(u_o,d) = \inf\{||u_o-u(S)||_E;\ S \in F,\ \dim s = d\}. \tag{2.7}$$

For example, we obtain a set $F = F_P$ by considering all meshes

$$\Delta:\ 0 = x_o^\Delta < x_1^\Delta < \ldots < x_{m(x)}^\Delta = 1 \tag{2.8}$$

and specifying $S \in F_P$ as the space of all C^o-functions on I which are polynomials of degree at most P on each subinterval $I_i = [x_{i-1}^\Delta, x_i^\Delta]$, $i = 1,\ldots,m(\Delta)$, of Δ. This selection leads to the classical h-version of the finite element method for (2.1)-(2.2) with elements of order P. If we use instead a fixed mesh but vary the order of the polynomial pieces separately on each subinterval, then we arrive at the p-version of the method (see [4], [10], [11]).

A <u>feedback system</u> for (2.1)-(2.2) on some given set F may be characterized as a process which produces sequences (or <u>trajectories</u>) of subspaces $\{S_k\}_1^\infty \subset F$. This is accomplished by some transition function τ which produces for any index (time) k the successor $S_{k+1} \in F$ on the basis of information about the sequence up to k.

For a precise definition of this concept we refer to [18]. Here we present only an example of a simple transition function $\tau = \tau_P$ for the case $F = F_P$. For any $S = S(\Delta) \in F_P$ consider for each i, $1 \le i \le m(\Delta)$, the exact (weak) solutions z_i of (2.1) on $I_i^\Delta = [x_{i-1}^\Delta, x_i^\Delta]$ subject to the boundary conditions $z_i(x_{i-1}^\Delta) = u(S)(x_{i-1}^\Delta)$, $z_i(x_i^\Delta) = u(S)(x_i^\Delta)$. Then the error indicators

$$\eta_i^\Delta = ||z_i - u(S(\Delta))||_{E,I_i^\Delta}, \quad i = 1,\ldots,m(\Delta), \tag{2.9}$$

can be computed approximately or estimated from above and below in terms of the residuals $r_i^\Delta = L[u(S)] - f$, on I_i^Δ. Moreover

$$\varepsilon(\Delta) = [\sum_{i=1}^{m(\Delta)} (\eta_i^\Delta)^2]^{1/2} \tag{2.10}$$

represents an estimator for the error $||u_o - u(S)||_E$, (see [6], [7], [8]). Now a simple mesh-refinement strategy constructs from Δ a new mesh Δ' by halving the subinterval I_i^Δ for which η_i^Δ is maximal. This corresponds to the transition function $\tau = \tau_P$ which produces as the successor $\tau(S(\Delta))$ of the current space $S(\Delta) \in F_P$ the space $S' = S(\Delta') \in F_P$.

Now any feedback system on a set F will be called an adaptive system if it is optimal with respect to some <u>performance</u> measure, or, more precisely, if under this measure its performance is not worse than that of any other feedback system.

Generally, a performance measure is a function μ which associates with any trajectory $\{S_k\} \in F$ produced by a feedback system an element $\mu\{S_k\}$ of a given partially order set M. In the simplest case we might use the set $M = \{0,1\}$ which allows us to distinguish only between good (= 0) or bad (= 1) performance. Two typical performance measures of this type may be defined as follows:

$$\mu\{S_k\} = \begin{cases} 0 \text{ if } \lim_{k\to\infty} ||u_o - u(S_k)||_E = 0 \\ 1 \text{ otherwise} \end{cases} \tag{2.11}$$

$$\mu\{S_k\} = \begin{cases} 0 \text{ if } ||u_o - u(S_k)||_E \le C\,\phi(u_o, \dim(S_k)),\ \forall\, k \ge 0 \\ 1 \text{ otherwise} \end{cases} \tag{2.12}$$

where in (2.12) the constant C depends on the problem and the solution u_o, but not on k.

These and other measures are discussed in [1], [2], [12]. We mention here only a few partial results for our model problem phrased not necessarily in their most general form:

<u>Theorem 2.1</u>: For the problem (2.1)-(2.3) and any $u_o \in \overset{\circ}{H}{}^1(I)$ consider the simple feedback system on $F = F_P$ defined by the transition function τ_P. Then the resulting trajectories $\{S_k\}$ satisfy $\lim_{k\to\infty} ||u_o - u(S_k)||_E = 0$ and hence the system is adaptive with respect to the performance measure (2.11).

For the performance measure (2.12) more restrictive conditions are required. In essence, we compare here the convergence rate for the computed sequence of meshes with an "ideal" rate. For this we need to know--irrespective of any feedback system--the existence of a sequence of meshes for which some comparison with the ideal rate applies. We call this a comparison sequence; its existence is a condition on the solution $u_o \in \overset{\circ}{H}{}^1(I)$ which we wish to approximate. In essence, a theorem of the following form can then be proved (see [12]).

<u>Theorem 2.2</u>: Suppose that the coefficients a,b of (2.1-2.3) are sufficiently smooth and that for the given $u_o \in \overset{\circ}{H}{}^1(I)$ there exists a comparison sequence of meshes with certain properties. Then, for the feedback system on F_P defined by τ_P, the resulting trajectories satisfy

$$||u_o - u(S_k)||_E \le C\,\phi(u_o, \dim S(u_k)), \quad k \ge 0 \tag{2.13}$$

with a constant that depends on the problem and the comparison sequence, but not on k. Hence, the feedback system is adaptive under the performance measure (2.12).

The result implies that under the particular conditions the rate of convergence is independent of the smoothness of the solution. The conditions of the theorem are not overly stringent. For example, they hold for $u_o = x^\mu - x \in \overset{\circ}{H}{}^1(I)$, $\mu > \frac{1}{2}$, in which case the rate of convergence is of the order of $(\dim S(\Delta_k))^{-P}$ independent of μ, (see [12]). We refer also to [7] for related results.

The simple transition function τ_P used in both theorems is computationally costly since it modifies the meshes very little from step to step. Moreover, in practice we can work only with a finite segment $S_1,\ldots,S_n$ of any trajectory. In that case, it is reasonable to demand that the computation of the final approximate solution $u(S_n)$ is at most, say, twice or thrice the cumulative cost of the computation up to that point.

For this we have to introduce an appropriate stopping criterion which, obviously, needs to be based on an error estimator such as (2.10). The construction of such a stopping rule is suggested by the following two results.

Theorem 2.3: For $u_o \in \mathring{H}^1(I)$ and under the condition (2.3) we have

$$C_2\ \varepsilon(\Delta) \leq ||u_o - u(S(\Delta))||_E \leq C_1\ \varepsilon(\Delta) \tag{2.14}$$

where the constants $C_1, C_2 > 0$ are independent of $S(\Delta)$.

Theorem 2.4: Under the conditions of Theorem 2.2 we have

$$\lim_{k\to\infty} \frac{\varepsilon(\Delta_k)}{||u_o - u(S(\Delta_k))||_E} = 1. \tag{2.15}$$

For some results about the adaptivity of feedback systems with transition functions incorporating such stopping criteria see [1], [12]. It should be noted also that the results indicated here remain valid for finite sets of right-hand sides f in (2.1).

III. THE FEARS SYSTEM; EFFECTIVITY INDICES

The results sketched in the previous section allow extensions to more general problems in two space dimensions. In particular, results presented in [2] address the family of problems and feedback approaches upon which the FEARS-program [Finite Element Adaptive Research Solver] is based.

For a detailed description of FEARS we refer to [22] and [34]. In brief, FEARS solves certain systems of elliptic equations in two space dimensions and permits also combinations with one-dimensional problems, as, for instance, those needed for modelling stiffeners in elasticity problems. Curved quadrilateral elements of first order are used. As in (2.9) error indicators are associated with each element and from them an estimator (2.10) of the solution-error is obtained. For details about these a posteriori estimates see [2].

FEARS utilizes various modes of mesh-refinement. In particular, a so-called long pass involves the recomputation of the solution on the new mesh, while in a short pass no such recomputation is performed but instead decisions about further refinements are based on an extrapolation of the refinement patterns observed in previous meshes. Thus, by combining these short and long passes in various ways different feedback mechanisms are obtained.

Theorem 2.1 carries over, in full generality, to trajectories generated by long passes in FEARS, and the same is true for Theorem 2.3. On the other hand, the analogue of Theorem 2.2 has not been proved in sufficient generality. Yet, all numerical experiments performed with FEARS suggest that such a theorem should be valid also in this case. Theorem 2.4 does extend, however, under certain additional assumptions about the meshes (see [2]). By appropriate use of the short passes it is possible to ensure that the cost of the computation of the solution on the final meshes of the refinement sequence is proportional to the cumulative cost of the computation up to that point. In fact, experience shows that the proportionality factor is of the order of 1.5 to 3.0.

As an illustration we present here a typical example of the performance of FEARS. More specifically we consider the problem $\Delta u = 0$ on the domain shown in Figure 1 with the indicated boundary conditions. The figure shows also the

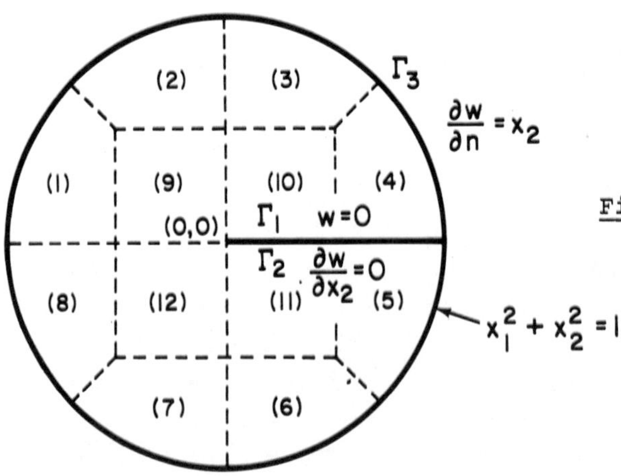

Figure 1

dotted partition of the domain used in the mesh-construction of FEARS (see [22]). The solution has a strong singularity at the origin of the type $r^{1/4}$; that is, the solution belongs to Besov space $B^{5/4}_{2,\infty}$. Hence for sequences of uniform meshes the rate of convergence is of the order of $N^{-1/8}$ with respect to the degrees of freedom N. This is certainly a slow rate.

In contrast FEARS constructs meshes with refinement leading to the rate of convergence $O(N^{-1/2})$. Figure 2 shows one such mesh in the subregions 9 through 12.

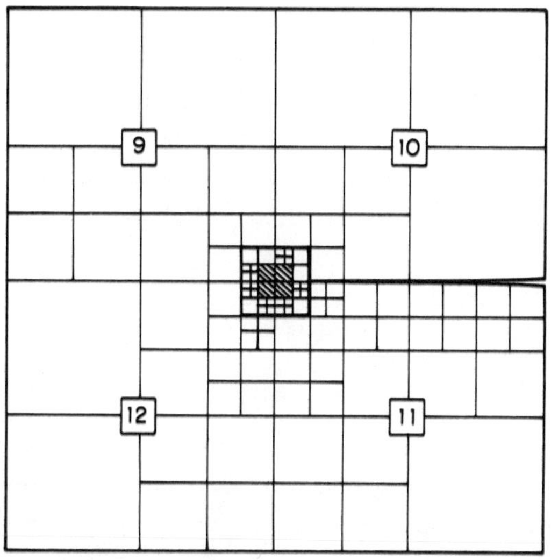

(i) Mesh on subregions 9, 10, 11 and 12.

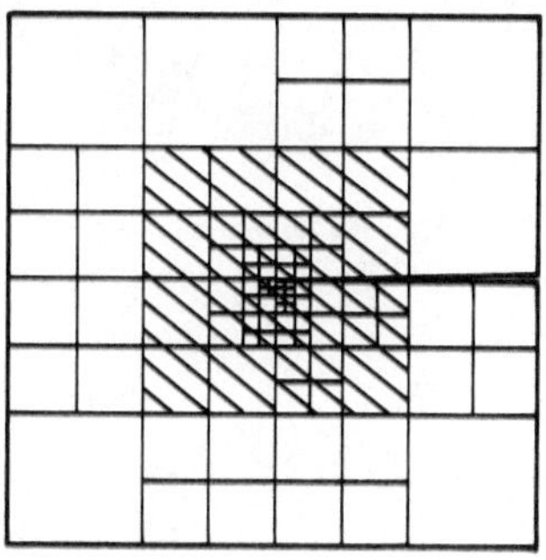

Figure 2

(ii) Details of mesh in shaded region of (i).

Figure 3 depicts the achieved accuracy--measured in the energy norm--as a function of the number of elements N* (~N). More specifically, the true errors are given for a sequence (a) of uniform meshes, (b) of adpatively constructed meshes based on the combination of long (L) and short (S) passes LSLSSLSSSLSSSSLSSSL, and (c) of adaptively constructed meshes using long passes only. In addition, the computed error estimates are shown. All data are in percent of the square root of the exact energy

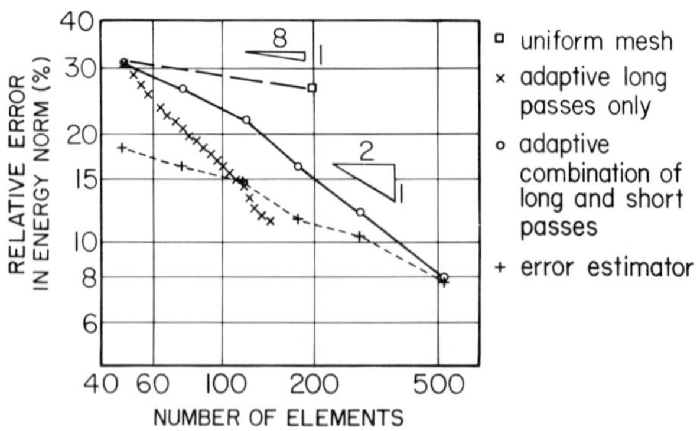

Figure 3

The figure indicates that the procedure is convergent in accordance with the two-dimensional analogue of Theorem 2.1. The rate of convergence for the adaptively constructed meshes has the maximum rate $N^{-1/2}$, which confirms experimentally the analogue of Theorem 2.2. Actually we see a slightly better rate because the higher errors are here principally confined to a neighborhood of the origin. On the other hand, the rate of convergence for the uniform meshes is very close to $N^{-1/8}$. In fact, the accuracy of 8.5% obtained by the adaptive mesh of 537 elements would require a uniform mesh of more than 10^6 elements.

The adaptive meshes based on long passes only give better results than those involving a combination of passes. In terms of dependence on N this is certainly expected. The situation is different in Figure 4 where the errors are shown in relation to the total machine time. Here the effectiveness of the combination of long and short passes shows up very clearly.

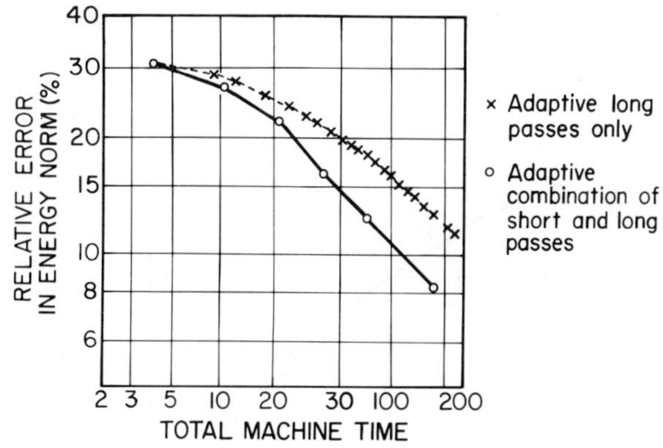

Figure 4

Figure 5 gives the earlier mentioned factor of the cost of the final long pass to the total cost for the sequence of long and short passes. Evidently the factor is of the order of two.

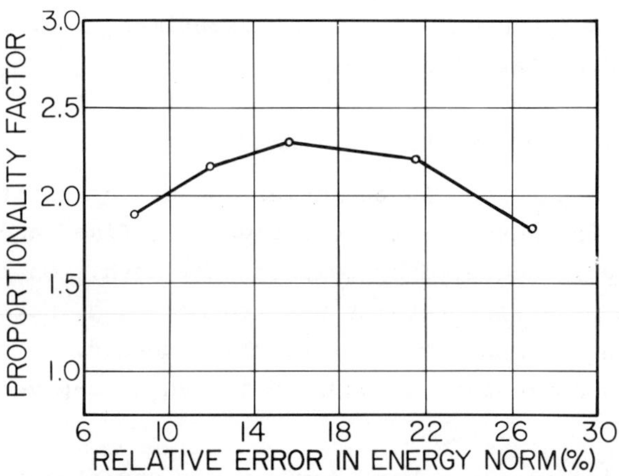

Figure 5

Finally, Figure 6 shows the effectivity index θ = est. error/true error as a function of the achieved accuracy. Experience shows that often, but not always, the estimated errors are smaller than the true errors. But for accuracies of 10% or better the effectivity index usually exceeds 0.9. In line with this, experience has shown that the stopping criterion based on the estimators is very reliable. Generally, the effectivity index is better for smoother solutions; it is also better for the procedure based only on long passes.

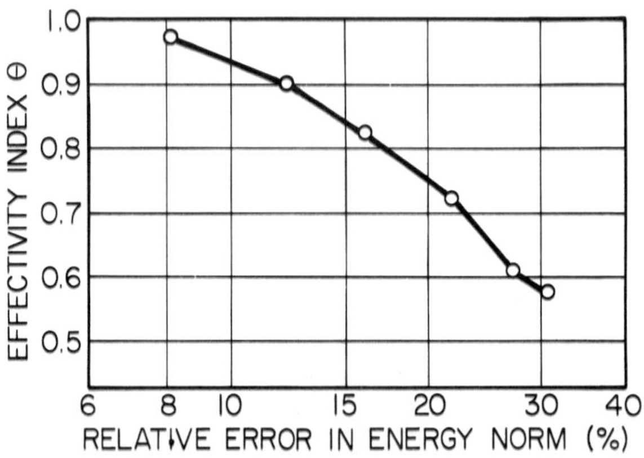

Figure 6

The results discussed here for this single example are consistent with and typical of all the experimental evidence we have obtained so far with FEARS.

IV. OTHER VERSIONS OF THE FINITE ELEMENT METHOD

As mentioned earlier, in the h-version of the finite-element method the degree p of the elements is fixed and the desired accuracy is achieved by refining the mesh. On the other hand, the p-version retains a fixed mesh and increases the degree of the elements to improve the accuracy. A combination of both approaches is usually called an h-p-version of the method.

The h-version has long been dominant in engineering practice. In contrast, for structural problems, a first research code COMET-X based on the p-version was developed less than a decade ago at Washington University [13], and the first commercial code, FIESTA [17], appeared only in the Fall of 1982. Correspondingly, the h-version has been intensively studied while the principal results about the p-version and h-p-version were only published relatively recently [10], [11], [14].

These three versions of the finite element method differ considerably in their program architecture. We will not enter into details. For some aspects of the hierarchic element types and the input controls used in the p-version see [5], [13], [17], [27]. Here we summarize only some of the principal differences between the three versions.

a) In the energy norm the rate of convergence of the p-version with respect to the number of degrees of freedom cannot be worse than that of the h-version for uniform meshes, and, in fact, it can be much better (see [10], [11]).

b) The convergence rate of the p-version is only influenced by the smoothness of the solution, and hence behaves similar to that of the spectral methods.

c) The p-version has twice the rate of convergence of the h-version (for uniform meshes) when a singularity of the solution, such as that caused by a corner or edge of the domain, is located in a nodal point of the mesh or on an edge, ([4], [10], [11], [14]).

d) The p-version is insensitive to incompressibility of the material when applied to elasticity problems; that is, to the case of a Poisson ratio σ near 1/2. Accordingly, it avoids the so-called locking problem of the h-version ([10], [29], [30]).

e) The h-p-version, with properly selected meshes and distribution of the element-degrees, exhibits arbitrarily high algebraic rates of convergence with respect to the degrees of freedom, or even exponential rates if the data are piecewise C^∞ or analytic, [4].

These properties suggest the need for appropriately designed adaptive approaches based on the p and h-p-version. But as yet such techniques are only insufficiently developed; see [28], [32], [35], [36], for example.

As an illustration of some of the above points we end this section with a numerical example involving a plane strain, elasticity problem on the unit-square with an imposed shear displacement. The material is assumed to be homogeneous and isotropic with $\sigma = 0.3$. The scheme of the problem is shown in Figure 7. The solution has singularities in the corners of the domain and belongs to the Besov space $B^{\mu}_{2,\infty}$ where for our value of σ we have $\mu = 1.76$. In view of the obvious symmetries only the upper quarter needs to be considered. Figure 8 gives the

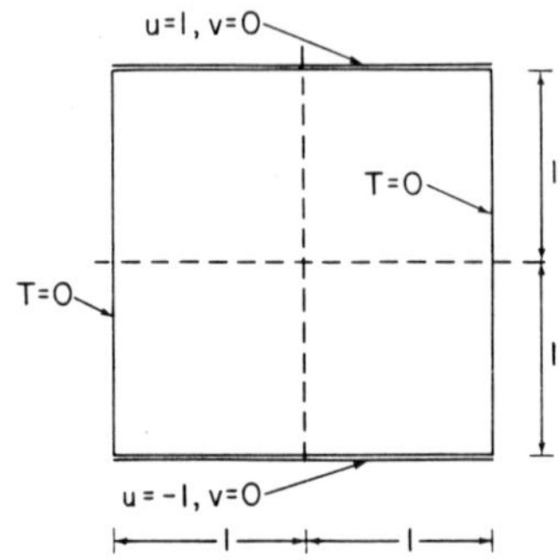

Figure 7

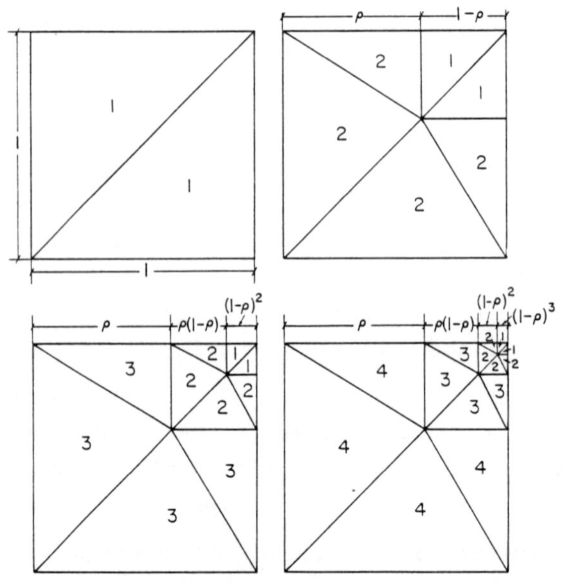

Figure 8

mesh-refinement as a function of a parameter ρ, with the degree p written in each element. The 18-element mesh with $p_{max} = 5$ was not included. Then Figure 9 shows the errors of the h-p-version for two choices of ρ and that of the p-version based on only two elements. Evidently, the rate of convergence of the h-p-version is better than algebraic and may be expected to be of order $e^{-c\sqrt{N}}$.

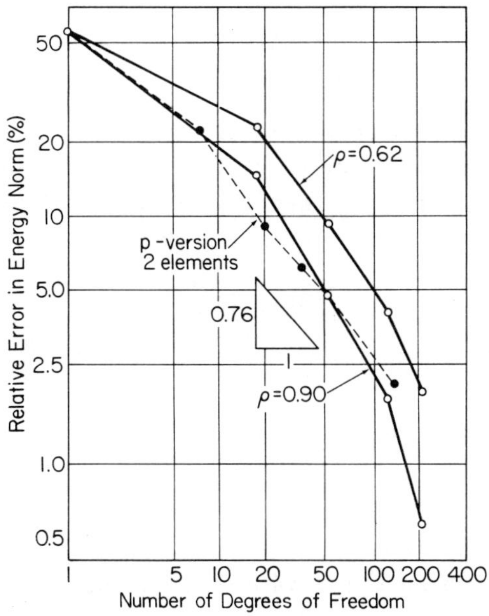

Figure 9

V. POST-PROCESSING

As mentioned in the Introduction, in practical engineering computations important decisions are usually based on a relatively few, but highly accurate data items. A very effective approach to the computation of such data is the application of post-processing techniques to the computer solutions. For a detailed study of this approach we refer to [3]. Here we sketch only the principal ideas for the simple model problem

$$-\Delta u = 1, \quad \text{on} \quad \Omega = (0,1) \times (0,1) \tag{5.1}$$

$$u = 0, \quad \text{on} \quad \partial\Omega. \tag{5.2}$$

Suppose that we are interested in the values $\phi_1 = u(0,0)$ and $\phi_2 = \frac{\partial u}{\partial x}(1,0)$. The main idea is to write

$$\phi_i = \int_\Omega u_o \xi_i \, dx + R_i, \quad i = 1,2 \tag{5.3}$$

where u_o again denotes the exact solution of (5.1)-(5.2) and the R_i are integrals that may be computed from the input data. For instance, for ϕ_1 we may use

$$\psi(x) = \frac{1}{2\pi}(\ell n||x||_2 - \frac{1}{2}\ \ell n[\frac{1}{2}(1+x_1^2)(1+x_2^2)]).$$

The verification of these formulas is based essentially on the use of integration by parts.

Once (5.3) is available, an approximate value of $\tilde{\phi}_i$ can be computed by replacing in (5.3) the exact solution u_o with the computed solution u(S) (see the notation of Section II). Then the error can be written in the form

$$\tilde{\phi}_i - \phi_i = B(u(S)-u_o, \psi_i(S)-\psi_{oi})$$

where

$$B(u,v) = \int_\Omega \left[\frac{\partial u}{\partial x_1}\frac{\partial v}{\partial x_1} + \frac{\partial u}{\partial x_2}\frac{\partial v}{\partial x_2}\right]dx$$

and ψ_{oi} is the solution of $-\Delta\psi = \xi_i$ subject to the boundary conditions (4.2) while $\psi_i(S)$ is the corresponding approximate solution. The computation of $\psi_i(S)$ can be done simultaneously with the computation of u(S) at no great additional expense.

Table 1 shows a comparison of the results for ϕ_1 and ϕ_2 obtained both from the solution computed by FEARS and the post-processing approach.

TABLE 1

No. of elements in quarter domain	4	16	64
Energy norm error of u(S)	30.1%	15.2%	7.6%
Error in ϕ_1 computed from u(S)	5.4%	1.3%	.31%
Error in ϕ_1 computed by post-processing	2.5%	.63%	.16%
Error in ϕ_2 computed from u(S)	29%	16%	8.7%
Error in ϕ_2 computed from post-processing	1.3%	.32%	.076%

Table 2 gives the effectivity index for the error estimates of the values of ϕ_1 and ϕ_2 computed by the post-processing approach; that is, the quotient of the estimated error divided by the exact error.

TABLE 2

No. of elements in quarter domain	4	16	64
Effectivity index for ϕ_1	1.07	.98	.99
Effectivity index for ϕ_2	.93	1.02	1.06

VI. PROBLEM FORMULATIONS IN THE NONLINEAR CASE

By far the majority of today's finite-element computations involve linear problems; but, in recent years, the nonlinear analysis of structural systems has become increasingly important in engineering applications. Not unexpectedly, there are many fundamental differences between linear and nonlinear problems, and, in fact, the problem formulations for the latter show many special features not present in the linear case.

First and foremost, it has to be recognized that the theoretical approaches and computational techniques depend strongly on the type and properties of the nonlinearities. In structural problems we may distinguish four principal sources of nonlinearities, namely (see [16])

(i) *geometric nonlinearities*, due to nonlinear strain-displacement relations

(ii) *material nonlinearities*, due to nonlinear constitutive equations

(iii) *force nonlinearities*, due to nonlinear stress boundary conditions

(iv) *kinematic constraint nonlinearities*, due to nonlinear displacement boundary conditions.

The source of nonlinearity affects the form of the resulting nonlinear equations and hence the solution methods. The corresponding numerical problems are as yet not fully understood. This is particularly true for the nonlinearities of type (ii) to (iv).

As indicated in the Introduction, engineering computations serve to predict the behavior of a physical system under study so as to reach decisions about proposed designs. In the case of structural problems, the term "behavior" refers, say, to deformations in response to the action of external influence quantities, such as loads, changes in material properties and geometrical data, etc. This means that the equations of the system also depend on a set of influence parameters and the objective is to assess the behavior of the solutions under variations of these parameters. In the linear case, this variation is also typically assumed to be linear and we need only compute a few specific solutions to achieve our objective. In nonlinear problems, however, the computation of a few specific solutions provides little or no insight into the behavior of the system and we are led to consider the set of all solutions depending on all the parameters in a specific range.

Mathematically, this means that we are faced with parametrized equations of the form

$$F(y,\lambda) = b \tag{6.1}$$

where y varies in a space Y of state (behavior) variables, λ in a parameter space Λ, F is a given mapping from $Y \times \Lambda$ into Y, and $b \in Y$ a prescribed element in the range of F. In general, the set of all solutions (y,λ) of (6.1) forms a manifold in $Y \times \Lambda$. In structural mechanics, this set is now usually called the equilibrium manifold of the structure. The objective then becomes a computational analysis of the form and characteristic features of this manifold.

Clearly, the specific features to be determined for a particular problem depend on the objectives of the computation. Generally, interest centers on computing solutions on the manifold along specified paths which represent, for instance, particular loading regimes, etc. Then, the boundaries of the stability regions on the manifold are to be explored, and the form of the manifold near specific singularities is to be determined.

Besides these general objectives, there are various specific solution data that are needed for certain decisions about the structure under consideration. A typical example of this is the so-called question about conservative input data. In general, the external input data influencing a structure are only known within certain ranges; for instance, this is certainly true for the loads to which the structure will be subjected during its lifetime. Thus, for the design it is essential to know the values of the parameters which represent worst possible cases. These are called conservative input data. Their determination represents a search problem on the solution manifold.

As an example, consider a clamped, thin, shallow circular arch which has been used as a test case by various authors (see e.g., [21], [31]). Let U and W be the radial and axial displacements, R the arch-radius, A its cross-sectional area, h the thickness, and E Young's modulus. With the dimensionless displacements $u = U/h$, $w = W/h$, the total potential energy--non-dimensionalized by dividing by $EAR(h/R)^2$--is given by

$$\frac{1}{2}\int_{-\theta_o}^{\theta_o}\left[\left(\frac{dw}{d\theta}-u\right)+\frac{1}{2}\,\frac{h}{R}\left(\frac{du}{d\theta}\right)^2\right]^2 d\theta+\frac{\alpha_1}{2}\int_{-\theta_o}^{\theta_o}\left(\frac{d^2u}{d\theta^2}\right)^2 d\theta-\alpha_2\int_{-\theta_o}^{+\theta_o} pu\,d\theta \ .$$

Here $p = p(\theta)$ is the dimensionless radial load and α_1, α_2 are dimensionless constants. Each end is assumed to be pinned; that is, we have the boundary conditions

$$u(\pm\theta_o) = 0, \quad w(\pm\theta_o) = 0, \quad \frac{d^2u}{d\theta^2}(\pm\theta_o) = 0 \ .$$

The finite element approximations introduced in [31] were applied. More specifically, we used a uniform mesh with eight elements, $\theta_o = 15°$, and the constants $\alpha_1 = 3.8716 \times 10^{-6}$, $\alpha_2 = 8.2752 \times 10^{-2}$ corresponding to data in [21].

Suppose that the following two-parameter family of loads is to be considered

$$p(u,\nu)(\theta) = \begin{cases} \left(1-\frac{4}{\theta_o}(\nu-\theta)\right)\mu, & \text{for } \max(-\theta_o, \nu-\frac{\theta_o}{4}) \le \theta \le \nu \\ \left(1-\frac{4}{\theta_o}(\theta-\nu)\right)\mu, & \text{for } \nu \le \theta \le \min(\theta_o, \nu+\frac{\theta_o}{4}) \\ 0 & \text{, otherwise.} \end{cases}$$

Thus the load is a piecewise-linear hat function which has the value μ at $\theta = \nu$ and is zero outside the interval centered at ν of width $\theta_o/2$. For any fixed position ν the path on the manifold determined by varying μ encounters a first limit point which represents a point where the arch is expected to buckle. Figure 10 shows the variation of this buckling load with the value of ν. We see that the most dangerous loads occur about for $\nu = \pm 0.16\theta_o$; hence, these values represent the desired conservative data. The computations were performed with the limit-point facilities of the continuation package PITCON (see [25]). We postpone to Section VIII below a discussion of this package and hence of the way Figure 10 was computed.

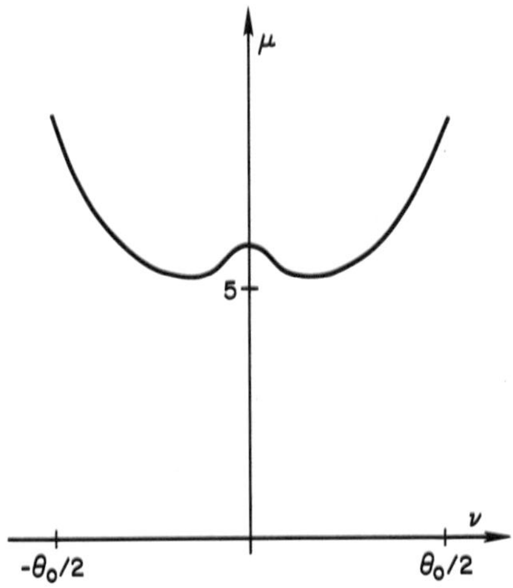

Figure 10

VII. ERROR ESTIMATES

As in the linear case, a desirable feature for nonlinear structural computations is the estimation of the errors of the computed results. But, not unexpectedly, for nonlinear problems the situation is considerably more difficult and much remains to be done. In fact, for the solutions of the parametrized equations of the previous section, it is not always readily apparent how to define the error. In fact, if, say, M denotes the equilibrium manifold of a given infinite-dimensional equation (6.1) and $\tilde{M}$ is the solution manifold of a finite-dimensional approximation of this equation, then it is not obvious how to relate points of M and $\tilde{M}$. Frequently, the parameter space Λ of (6.1) is finite-dimensional and hence there is no need to discretize $\lambda \in \Lambda$. In that case, we might wish to compare points of M and $\tilde{M}$ corresponding to the same values of λ. But already simple examples show that for specific λ there may exist a point only on one of the two manifolds.

In order to resolve this problem it is necessary to consider first the equation of a priori estimates for parametrized equations. For mildly nonlinear elliptic boundary value problems and one-dimensional parameter spaces such estimates were first developed in [15]. A different approach was used in [18] to provide similar results for more general boundary value problems. In [19] this approach was extended to the case of any finite-dimensional parameter space. We present here only a brief summary of the results in [18], [19].

Let X,Y be Banach-spaces and $F: S \subset X \to Y$ a mapping of class $C^r(S)$, $r \geq 1$, on an open connected subset S of X. Suppose that F is a Fredholm operator of index $m \geq 1$ on S and that the regularity set $R(F) = \{x \in S;\ \mathrm{rge}\ DF(x) = Y\}$ is non-empty. Then $R(F)$ is an open subset of X and for any regular value $b \in F(R(F))$ the regular solution set

$$M(b) = \{x \in R(F);\ F(x) = b\}$$

is an m-dimensional C^r-manifold in X. For proofs we refer to [18].

For any theoretical and computational study of $M(b)$ we need to specify which local parametrizations of the manifold we will use. At a point $x_o \in M(b)$ a local parametrization of $M(b)$ shall be a triple $\{V,A,T\}$ consisting of

(i) a closed subspace V of X such that $V \cap \ker DF(x_o) = \{0\}$,

(ii) an isomorphism $A \in L(Y,V)$ from Y onto V, and

(iii) an m-dimensional subspace T of X such that $X = V \oplus T$.

The meaning of this concept is contained in the following result (see [18]):

Theorem 7.1: Under the stated conditions for F and $M(b)$, let $\{V,A,T\}$ be a local parametrization of the manifold at $x_o \in M(b)$. Then there exists an open ball $B \subset T$ with $0 \in B$, an open neighborhood $U \subset X$ of x_o, and a unique C^r-function $\eta: B \to Y$ such that

$$M(b) \cap U = \{x \in X;\ x = x_o + t + A\eta(t),\ \ \forall t \in B\} .$$

The geometric interpretation is sketched in the composite Figure 11. In many applications certain parameters are identified explicitly; that is, one has a natural splitting $X = W \oplus Z$, $\dim Z = m$, where W is isomorphic with the range space Y. The question then arises at what points $x_o \in M(b)$ this natural splitting gives rise to a local parametrization; that is, when the given parameters can be used to parametrize the manifold near x_o. Clearly, the condition here is $W \cap \ker DF(x_o) = \{0\}$ and this holds exactly if $\text{rge}\, D_W F(x_o) = Y$. Then $\{W, D_W F(x_o)^{-1}, Z\}$ indeed is a local parametrization of $M(b)$ at x_o. Such points are called nonsingular points of $M(b)$ with respect to the natural parameters.

At points $x_o \in M(b)$ where $W \cap \ker DF(x_o) \neq \{0\}$ the natural parameters cannot be used to parametrize $M(b)$. Such points can be characterized by the dimension of this intersection. The simplest case arises when $\dim(W \cap \ker DF(x_o)) = 1$; these are limit points with respect to the direction of the intersection. For more details we refer to [19].

We turn now to a discretization of our problem. Once again, it is useful to assume the existence of a natural splitting $X = W \oplus Z$, $\dim Z = m$, where W is isomorphic with Y. More specifically, we assume that there is an operator Q such that

$$Q \in L(X,Y), \qquad Z = \ker Q, \qquad \text{rge}\, Q|W = Y.$$

Often we have $F:\ S \subset Y \times R^m \to Y$ in which case Q can be taken as the natural projection from $Y \times R^m$ onto Y.

The natural splitting has to be considered since in such applications only the state space W is discretized and the parameters are left untouched. Suppose now that the discretization is defined by a family of linear maps $P_h \in L(Y)$ of finite rank indexed by a positive $h > 0$ such that $\lim_{h\to 0} P_h y = y$ for all $y \in Y$. Then the finite-dimensional subspaces

$$Y_h = P_h Y \subset Y, \quad W_h = (Q|W)^{-1} Y_h \subset W, \quad X_h = W_h \oplus Z \subset X$$

are well-defined and we may consider the approximating problem

$$F_h(x) = y_{oh}, \quad x \in R(F_h), \quad y_{oh} = P_h y_o$$
$$\tag{7.1}$$
$$F_h: S_h \subset X_h \to Y_h, \quad F_h(x) = P_h F(x), \quad x \in S_h = S \cap X_h.$$

In order to compare the solution manifolds of these discretized problems with $M(b)$ we need to assume that the approximation is sufficiently close to ensure that the same local parametrization can be used on these manifolds, (see Figure 11). As usual, a stability condition is required to ensure the convergence. Without elaborating upon the specific details of this condition, we sketch here only the general form of the resulting theorems (see [18] and [19]):

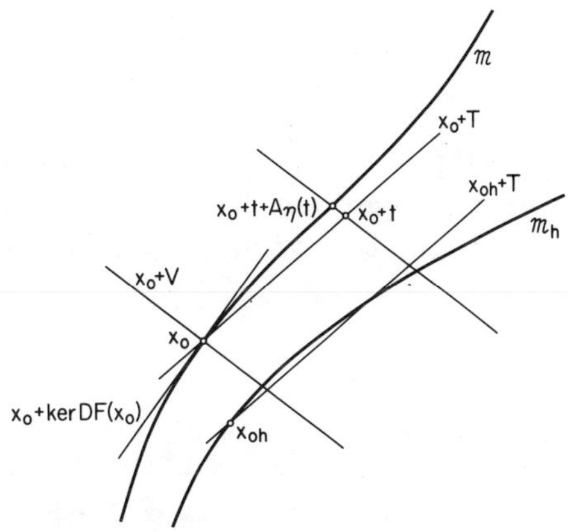

Figure 11

Theorem 7.2: Under the stated conditions about F and $M(b)$, let $\{V,A,T\}$ be a local parametrization of $M(b)$ at $x_o \in M(b)$ and suppose that the indicated stability condition holds. Then, for all sufficiently small $h > 0$, the approximate problems (7.1) possess solutions $x_{oh} \in R(F_h)$ for which $\lim_{h\to 0} x_{oh} = x_o$.

Theorem 7.3: In addition to the conditions of Theorem 7.2, let DF be Lipschitz-continuous on bounded subsets of $S \subset X$. Denote by $x: B \subset T \to S$ the representation of $M(b)$ near x_o defined on the open ball $B \subset T$, $0 \in B$. Then there exists a closed ball $B_o \in B$, $0 \in B_o$ such that for all sufficiently small $h > 0$ the approximate problems have solutions $x_h: B_o \to S_h$ for which

$$||x(t)-x_h(t)|| \leq C||(I-P_h)Qx(t)||, \quad t \in B_o \tag{7.2}$$

where the constant C is independent of h and t.

As a typical example consider the mildly nonlinear problem in the weak form

$$a(u,w) + \int_\Omega g(u,\lambda)w\, d\xi = 0, \quad w \in \overset{\circ}{H}{}^1(\Omega), \quad \lambda \in R^m,$$

where Ω is a suitable bounded domain in R^n, and the bilinear form

$$a(u,w) = \int_\Omega \sum_{i,j=1}^{n} a_{ij}(\xi) \frac{\partial u}{\partial \xi_j} \frac{\partial w}{\partial \xi_i} d\xi, \quad u,w \in \overset{\circ}{H}{}^1(\Omega)$$

has sufficiently smooth coefficients and is strongly elliptic. Then the operator $K: L^2(\Omega) \to H^1(\Omega)$ defined by $a(Ku,w) = (u,w)_o$, $u \in L^2(\Omega)$, $w \in \overset{\circ}{H}{}^1(\Omega)$, is compact and the problem is equivalent with

$$u + KG(u,\lambda) = y_o$$

where $G: \overset{\circ}{H}{}^1(\Omega) \times R^m \to L^2(\Omega)$ is defined by the C^r-map g. It turns out that this is a Fredholm map of index m and that, with Q as the natural projection from $\overset{\circ}{H}{}^1(\Omega) \times R^m$ onto $\overset{\circ}{H}{}^1(\Omega)$, the mentioned stability condition is satisfied and Theorems 7.2 and 7.3 apply.

On the basis of a priori estimates of the form (7.2) we can now develop a posteriori error estimates along the same lines as for the linear problems. Some examples for this are presented in [9] and, as in the linear case, the effectivity of these estimates turns out to be consistently high. However, the theory of these a posteriori estimates for nonlinear problems is not, as yet, very well developed and, in particular, such results as Theorem 2.3 are still lacking. In [9] it was shown also that a posteriori estimates can be constructed for buckling loads and the location of limit points.

VIII. CONTINUATION PROCESSES

For the computational analysis of the solution manifolds considered here the basic methods are general continuation processes for the trace of specified paths on the manifold. If the problem formulation involves natural parameters, then these paths are defined usually by combinations of the parameter values with one degree of freedom. But interest may center also on paths which are specified implicitly, for example, as paths in the critical boundary of the problem (see e.g., [23]).

The literature on continuation methods is rather large. For a survey relating to structural mechanics we refer only to [26] where also further references may be found. No attempt shall be made to give here an overview of these methods. Instead we sketch only the outline of the continuation package PITCON [25] which is essentially based on the theory of Section VII above.

The code supposes that the desired path on the manifold is defined as the solution set of an equation

$$F(x) = 0 \tag{8.1}$$

with a user-specified mapping $F: R^n \to R^{n-1}$ of class C^1 for which the derivative DF is locally Lipschitzian on R^n. In the notation of Section VII, the aim of the program is to approximate a non-empty connected component M^* of the regular solution set $M = \{x \in R(F),\ F(x) = 0\}$. From differential-geometry we know that M^* is diffeomorphic either to the circle or to some interval of R^1. Hence M^* is uniquely determined by any of its points x^o and we denote this by writing $M^* = M^*(F,x^o)$.

For the discussion it is useful to consider a parametrization by arclength of $M^*(F,x^o)$; that is, a function $x: J \to M^*(F,x^o)$ which maps some interval $J \subset R^1$ diffeomorphically onto some open subset of $M^*(F,x^o)$ such that $||\dot{x}(s)||_2 = 1$ for $s \in J$. We may also assume that $x(0) = x^o$,

$0 \in J$. Starting from x^o the program produces a sequence of approximations $x^k \doteq x(s_k)$, $k = 0,1,\ldots$, corresponding to some sequence $0 = s_o < s_1 < s_2 \ldots$ of arclength values. Note, however, that, in general, the values $s_1, s_2, \ldots$ are only approximately computable and are of limited interest in most applications. They are not computed by PITCON.

In PITCON the principal steps performed during one continuation step are as follows:

1. Initialization of the step.
2. Check for possible target point at this step. If desired then compute this point and return.
3. Calculate the tanget vector (8.2) at the current point.
4. Define a new local parametrization of the path near this point.
5. Check for a possible limit point at this step. If desired compute this point and return.
6. Determine a steplength.
7. Use the steplength to compute a predicted point further along the path.
8. Start a corrector iteration from the predicted point. If it diverges then reduce the steplength and go to 7. If it converges then use the final point as the next approximate point on the path.
9. Store all relevant data and return.

The sequencing of these steps is dictated by the data flow. We sketch here only some of the details. For this we postpone the discussion of Steps 2. and 5. and address first the basic steps 3., 4., 6., 7., and 8.

Under the stated assumptions about F the tangent-map

$$T: R(F) \to R^n, \quad DF(x)Tx = 0, \quad ||Tx||_2 = 1, \quad \det\begin{pmatrix} DF(x) \\ (Tx)^T \end{pmatrix} > 0 \tag{8.2}$$

is uniquely determined and locally Lipschitzian on $R(F)$. Let $e^1,\ldots,e^n$ be the natural basis vectors of R^n. Then for any index i, $1 \leq i \leq n$, such that $(e^i)^T x \neq 0$ the solution $v \in R^n$ of the linear system

$$\begin{pmatrix} DF(x) \\ (e^i)^T \end{pmatrix} v = e^n \tag{8.3}$$

is well-defined, and with it, the tangent vector (8.2) can be computed as

$$Tx = \sigma \frac{v}{||v||_2}, \quad \sigma = \text{sign}(v^T e^i) \text{ sign det} \begin{pmatrix} DF(x) \\ (e^i)^T \end{pmatrix}. \tag{8.4}$$

The orientation procedure used here is satisfactory as long as the process does not jump over a (simple) bifurcation point as happens frequently in applications. Accordingly, PITCON augments (8.4) with a more elaborate, adaptive orientation procedure based on the directions of prior tangents and secants.

The choice of the local parametrization follows from Theorem 7.1. More specifically, PITCON uses as the space $T \subset R^n$ the one-dimensional subspace spanned by one of the basis vectors as e^{i_k} and defines V as the (n-1) dimensional space orthogonal to T. The term $A\eta(t)$ in Theorem 7.1 then represents the action of the corrector process. The choice of the index i_k is controlled by an adaptive procedure which takes into account anticipated changes of the curvature of the path as well as the observed behavior of the corrector process. The same index i_k is also used in the system (8.3) needed for the tangent calculation at the next point. For details we refer to [25].

A further adaptive procedure is now used to determine the steplength $h_k > 0$ and the predicted point $\hat{x}^k = x^k + h_k Tx^k$. As before, anticipated curvature-changes are used here. Moreover, the corrector distance between $\hat{x}^k$ and the desired next point x^{k+1} is estimated, and, on the basis of an expected convergence behavior of the corrector iteration, h_k is chosen so as to ensure a relatively fixed number of corrector steps. Once again, we refer to [25] for details.

In line with the orthogonality between the spaces V and T used in the parametrization, the corrector system is taken as

$$\hat{F}(x) = \begin{pmatrix} F(x) \\ (e^{i_k})^T(x-\hat{x}^k) \end{pmatrix} = 0 \qquad (8.5)$$

Either the full Newton method or the chord-Newton method is applied to (8.5), with $\hat{x}^k$ as starting point, to compute the desired next point. The process incorporates adaptive features to abort the iteration as quickly as possible when divergence is suspected and, in the case of convergence, to stop it with a good approximation of a point on $M^*(F,x^o)$.

As the schematic algorithm indicates, the determination of a target or limit point represents an interruption of the normal flow of the program. For the target point calculation an index i, $1 \leq i \leq n$, and associated value $\bar{x}_i$ is given and it is desired to compute a point $x = x(s)$ on $M^*(F,x^o)$ for which $x_i = \bar{x}_i$. The adaptive target algorithm checks whether $\bar{x}_i$ lies between $(e^i)^T x^{k-1}$ and $(e^i)^T x^k$ and then sets up a suitable corrector iteration to compute the desired point.

For the computation of a limit point with respect to the direction e^i we have to determine points $x = x(s)$ of $M^*(F,x^o)$ where $(e^i)^T Tx = 0$. The adaptive procedure incorporated in PITCON for this purpose checks for the occurrence of $\text{sign}\,(e^i)^T Tx^{k-1} \neq \text{sign}\,(e^i)^T Tx^k$ indicating a simple limit point. Then a special local parametrization ϕ: $[0,1] \to M^*(F,x^o)$ is constructed between x^{k-1} and x^k and a one-dimensional zero-finder is applied to the solution of $(e^i)^T Tx(\phi(t)) = 0$. For the details of the adaptive control of the overall procedure see again [25].

In the case of the problem discussed at the end of Section VI, PITCON was used to compute a number of paths on the manifold defined by different, fixed values of ν. On each path the first limit point was determined and Figure 10 shows the values of μ corresponding to these points. A different procedure, well-suited to this problem of conservative data, is to start with one such limit point and to trace directly a path of limit points on the manifold. Two adaptive procedures for that purpose are presented in [23].

As this entire discussion shows the various parts of PITCON use extensive feedback as the basis of the various decisions. Moreover, in general, the feedback mechanism turns out to be rather complex in nature. This raises the question whether the overall process can be shown to be adaptive in the sense of Section II. Up to now, no such results have been formulated, although, there appears to be the possibility of proving at least some partial results. There is certainly a need for more studies of this area.

On the basis of the estimates of the previous section it turns out that the discretization errors along any continuation path often vary considerably. On the other hand, our objectives usually require that these errors be maintained reasonably well within a tolerance interval. This suggests again the use of mesh-refinement (and de-refinements) in conjunction with the continuation process.

In its simplest form such a combined process begins with a mesh that meets a prescribed error-tolerance requirement. Then, during the continuation process, the error estimators are monitored and, at any computed point where these estimators exceed the upper tolerance a mesh-refinement strategy is applied to meet the tolerance requirement. Similarly, we may work also with a lower error tolerance and de-refine the mesh whenever the estimates fall below that lower value.

Figure 12 indicates the effect of such a strategy for a problem involving a nonlinear rod. We shall not present here the details of the problem-formulation but refer, for that, to [9]. In brief, the indicated values of N refer to the number of quadratic C^0-elements that were used during the corresponding segments of the path. The error tolerance was 2%. Since the continuation process takes relatively large steps along the path, there is no guarantee that between these points the error tolerance does not exceed the required tolerance. A more sophisticated control-strategy may alleviate the situation somewhat. At the same time, problems of this type cause difficulties in developing theoretical results of the type of Theorems 2.1 and 2.2 for the combined continuation and mesh-refinement processes. The level of expected difficulty of such results for nonlinear problems suggests that at first some emphasis should be placed upon carefully designed experiments which confirm or refute various conjectures about the

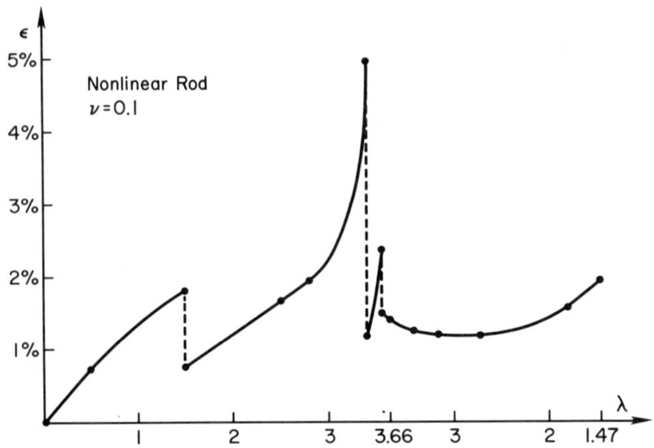

Figure 12

adaptivity of the feedback mechanisms that are used in these settings.

IX. REFERENCES

1. Babuška, I., Miller, A., and Vogelius, M., Adaptive Methods and Error Estimation for Elliptic Problems of Structural Mechanics, Univ. of Maryland, Inst. for Phys. Sci. and Tech., Tech. Note BN-1009,1983.

2. Babuška, I. and Miller, A., A Posteriori Error Estimates and Adaptive Techniques for the Finite Element Method, Univ. of Maryland, Inst. for Phys. Sci. and Tech., Tech. Note BN-968, 1981.

3. Babuška, I. and Miller, A., The Post-Processing Approach in the Finite Element Method, Univ. of Maryland, Inst. for Phys. Science and Techn.

   Part I: Technical Note BN-992, Dec. 1982
   Part II: Technical Note BN-993, Dec. 1982
   Part III; Technical Note, BN-1007, June 1983
   To appear in Int. J. for Num. Meth. in Eng.

4. Babuška, I. and Dorr, M.R., Error Estimates for Combined h and p Versions of the Finite Element Method, Num. Math. 37, 1981, 257-277.

5. Babuška, I., Katz, I.N. and Szabo, B.A., Hierarchic Families for the p-Version of the Finite Element Method, in "Advances in Computer Methods for Partial Differential Equations III" ed. by R. Vichnevetski and R.S. Stepleman, IMACS 1979, 278-286.

6. Babuška, I. and Rheinboldt, W., Error Estimates for Adaptive Finite Element Computations, SIAM J. Num. Analysis 15, 1978, 736-754.

7. Babuška, I. and Rheinboldt, W., Analysis of Optimal Finite Element Meshes in R^1, Math. of Comp. 33, 1979, 435-463.

8. Babuška, I. and Rheinboldt, W., A Posteriori Error Analysis of Finite Element Solutions for One-Dimensional Problems, SIAM J. Num. Anal. 18, 1981, 565-589.

9. Babuška, I. and Rheinboldt, W., Comp. Error Estimates and Adaptive Processes for Some Nonlinear Structural Problems, Comp. Meth. in Appl. Mech. and Eng. 34, 1982, 895-937.

10. Babuška, I. and Szabo, B.A., On the Rates of Convergence of the Finite Element Method, Int. J. for Num. Meth. in Eng. 18, 1981, 323-341.

11. Babuška, I. and Szabo, B.A. and Katz, I.N., The p-Version of the Finite Element Method, SIAM J. Num. Anal. 18, 1981, 515-545.

12. Babuška, I. and Vogelius, M., Feedback and Adaptive Finite Element Solution of One-Dimensional Boundary Value Problems, Univ. of Maryland, Inst. for Phys. Sci. and Tech., Tech. Note BN-1006, June 1983.

13. Basu, P.K., Rossow, M.P. and Szabo, B.A., Theoretical Manual and Users Guide for COMET-X, Center for Computational Mechanics, Washington Univ., St. Louis, MO, 1977.

14. Dorr, M.R., Piecewise Polynomial Approximation in the p-Version of the Finite Element Method, Ph.D. Dissertation in Mathematics, Univ. of Maryland, 1982.

15. Brezzi, F., Rappaz, J. and Raviart, P.A., Finite Dimensional Approximation of Nonlinear Problems

Part I: Branches of Nonsingular Solutions, Num. Math. 36, 1980, 1-25
Part II: Limit Points, Num. Math. 37, 1981, 1-28
Part III: Simple Bifurcation Points, Num. Math. 38, 1981, 1-30.

16. Felippa, C.A., Procedures for Computer Analysis of Large Nonlinear Systems, in "Large Engineering Systems", ed. by A. Wexler, Pergamon Press, Oxford, 1977.

17. FIESTA--Software System for Static Analysis of Solid Structures Based on the p-version of the Finite-Element Method, Users Manual Version 2, ISMES Bergamo Italy, WUTA St. Louis, MO, September 1982.

18. Fink, J.P. and Rheinboldt, W.C., On the Discretization Error of Parametrized Nonlinear Equations, SIAM J. Num. Anal. 1983, in press.

19. Fink, J.P. and Rheinboldt, W.C., On the Error Between the Solution Manifolds of Parametrized Equations and Their Discretizations; Univ. of Pittsburgh, Inst. f. Comp. Math. and Appl., Technical Report, 1983, in preparation.

20. Holland, J.H., Adaptation in Natural and Artificial Systems, The University of Michigan Press, Ann Arbor, MI, 1975.

21. Mau, S.T. and Gallagher, R.H., A Finite Element Procedure for Nonlinear Prebuckling and Initial Postbuckling Analysis, NASA Contractors Rept. NASA-CR-1936, Jan. 1972.

22. Mesztenyi, C. and Szymczak, W., FEARS User's Manual for the Univac-1100, Univ. of Maryland, Inst. for Phys. Sci. and Tech., Tech. Note BN-991, Oct. 1982.

23. Rheinboldt, W.C., Computation of Critical Boundaries on Equilibrium Manifolds, SIAM J. Num. Anal. 19, 1982, 653-669.

24. Rheinboldt, W.C., Feedback Systems and Adaptivity for Numerical Computations; in "Proceedings ARO Workshop on Adaptive Methods for Partial Diff. Equations 1983", to appear.

25. Rheinboldt, W.C. and Burkardt, J.V.,

I: A locally parameterized continuation process, ACM-TOMS 9, 1983, 215-235

II: Algorithm 596, A program for a locally parameterized continuation process, ACM-TOMS 9, 1983, 236-241.

26. Rheinboldt, W.C. and Riks, E., A Survey of Solution Techniques for Nonlinear Finite Element Equations; in "State of the Art Surveys in Finite Element Technology" ed. by A.K. Noor, W.D. Pilkey, Am. Society of Mech. Engineering Publications, in press.

27. Szabo, B.A., Some Recent Developments in Finite Element Analysis, Comp. Math. and Appl. 5, 1979, 99-115.

28. Szabo, B.A., Basu, P.K. and Rossow, M.P., Adaptive Finite Element Analysis Based on p Convergence. Presented at the Symposium on Future Trends in Computerized Structural Analysis and Synthesis, Washington, D.C., Nov. 1978, NASA Conference Publication 2059, 1978, 43-50.

29. Vogelius, M., An Analysis of the p-Version of the Finite Element Method for Nearly Incompressible Materials. Uniformly, Optimal Error Estimates, Num. Math. 41, 1983, 19-38.

30. Vogelius, M., A Right-Inverse for the Divergence Operator in Spaces of Piecewise Polynomials. Applications to the p-Version of the Finite Element Method, Num. Math. 41, 1983, 39-55.

31. Walker, A.C., A Nonlinear Finite Element Analysis of Shallow Circular Arches, Int. J. Solids Structures 5, 1969, 97-107.

32. Weiser, A., Local Mesh, Local Order Adaptive Finite Element Method with A-Posteriori Error Estimates for Elliptic Partial Differential Equations, Yale University, Comp. Science Dept., Tech. Rept. 213, December 1981.

33. Zadeh, L.A., On the Definition of Adaptivity, Proc. I.R.E. 51, 1963, 3-4.

34. Zave, P. and Rheinboldt, W., Design of an Adpative Parallel Finite Element System, ACM Transactions on Math. Software 5, 1979, 1-17.

35. Zienkiewicz, O.C., Kelly, D.W., Gago, J. and Babuška, I., Hierarchic Finite Element Approaches, Error Estimates and Adaptive Refinement; in "The Mathematics of Finite Elements and Applications IV" ed. by J. Whiteman, Acad. Press, London 1982, 313-347.

36. Zienkiewicz, O.C., Adaptive Methods for Structural Mechanics, in "Proceedings ARO Workshop on Adaptive Methods for Partial Differential Equations" to appear.

FINITE ELEMENT APPROXIMATIONS TO COMPRESSIBLE FLOW PROBLEMS

George J. Fix*

I. INTRODUCTION

This paper contains a survey of results on the finite element approximation to flow problems that have been obtained by the author and colleagues at Carnegie-Mellon University. Attention has been equally divided between compressible and incompressible flows, however in this paper we shall be primarily concerned with the former, and in particular, the special case of steady potential flows.

The second section deals with intrinsic instabilities that have long impaired finite element formulations. Recent work surveyed in this section shows how they can be avoided.

The third section deals with finite element methods based on least squares formulations. These do not have all of the problems cited in Section II, and they have the special advantage of working equally well with subsonic and supersonic flows. As a result they are quickly becoming the most popular finite element approach in the compressible case. They do have some important defects, and these are discussed in Section III along with methods for how to avoid them.

The final section deals with a poorly understood phenomenon the author has called the "phoenix effect". This is illustrated for a problem of modelling high frequency underwater acoustic waves. Because of the high frequencies very

*Department of Mathematics, Carnegie-Mellon University, Pittsburgh, Pennsylvania. This work was supported in part by ONR under contract No. N00014-81-K-0306.

ISBN 0-12-100560-7

fine grids are needed for resolution. On coarser grids very ragged solutions are obtained with for example, very large phase errors. However, when one takes certain key functionals of the solution, extraordinarily accurate approximations emerge. In this case it is the attenuation of the wave with radial distance that is distinguished.

II. INTRINSIC INSTABILITIES IN GALERKIN FORMULATIONS

We consider a steady potential flow in a bounded region Ω having Γ as its boundary and ν as its outer normal. We let ϕ denote the potential, v the velocity (so $\underset{\sim}{v} = \nabla\phi$), ρ the density, and $\underset{\sim}{u} = \rho\underset{\sim}{v}$ the mass flow. The boundary value problem takes the form

$$\frac{1}{\rho}\underset{\sim}{u} - \nabla\phi = 0 \quad \text{in} \quad \Omega \tag{2.1}$$

$$\text{div}\,\underset{\sim}{u} = 0 \quad \text{in} \quad \Omega \tag{2.2}$$

where

$$u_\nu = \underset{\sim}{u}\cdot\underset{\sim}{\nu} = \text{given on} \quad \Gamma. \tag{2.3}$$

The equation (2.2) represents conservation of mass, and the momentum balance is reflected through Bernoulli's equation, which gives the functional relationship

$$\rho = \rho(u) \ , \qquad u = |\underset{\sim}{u}| \ , \tag{2.4}$$

between density and velocity.

For (2.1)-(2.3) to have a solution it is necessary that the compatibility condition

$$0 = \int_\Omega \text{div}\,\underset{\sim}{u} = \int_\Gamma \underset{\sim}{u}\cdot\underset{\sim}{\nu} = \int_\Gamma u_\nu \tag{2.5}$$

be satisfied. We shall assume this, and reduce to the case of inhomogeneous boundary data by letting

$$\underset{\sim}{u} = \nabla\xi + \hat{\underset{\sim}{u}} \tag{2.6}$$

where

$$\nabla^2\xi = 0 \quad \text{in} \quad \Omega \tag{2.7}$$

$$\frac{\partial \xi}{\partial \nu} = u_\nu \quad \text{on} \quad \Gamma . \tag{2.8}$$

This introduces a nonlinear source term f into (2.1) and replaces (2.3) with $\hat{\underset{\sim}{u}} \cdot \underset{\sim}{\nu} = 0$ on Γ.

Observe that the resulting system can be written

$$A(u) - G\phi = f \tag{2.9}$$

$$Du \quad = 0 , \tag{2.10}$$

which is often called the Brezzi canonical form [1]. Systems with this form are uniquitous in continuum mechanics, a second example being the steady Navier-Stokes equations for incompressible flows where

$$A(u) = \nu \nabla^2 \underset{\sim}{u} - (\underset{\sim}{u} \cdot \nabla) \underset{\sim}{u} . \tag{2.11}$$

In the latter, $\underset{\sim}{u}$ denotes the velocity, ϕ the pressure, and $\nu > 0$ the viscosity. The operators G; D remain $\vec{\text{grad}}$, div, respectively, as in the compressible case. Boundary conditions are normally specified on all components of the velocity $\underset{\sim}{u}$.

In the general setting first studied by Brezzi [1] we have Hilbert Spaces V, S with dual spaces V', S'. G is the dual of $-D$ and

$$A: \quad V \to V' \tag{2.12}$$

$$G = -D': \quad S \to S' \tag{2.13}$$

are continuous (with A being possibly nonlinear). In the compressible case these spaces are the following:

$$S = L_0^2(\Omega) = \{\phi \in L_2(\Omega): \int_\Omega \phi = 0\} \tag{2.14}$$

$$V = \vec{H}_0(\Omega, \text{div}) = \{\underset{\sim}{u} \in \vec{L}^2(\Omega): \nabla \cdot \underset{\sim}{u} \in L^2(\Omega), \underset{\sim}{u} \cdot \underset{\sim}{\nu} = 0 \quad \text{on} \quad \Gamma\} \tag{2.15}$$

(See [3]). In the compressible case V is changed to

$$V = \vec{H}_0^1(\Omega) = \{\underset{\sim}{u} \in \vec{H}^1(\Omega): \underset{\sim}{u} = \underset{\sim}{0} \quad \text{on} \quad \Gamma\}^1 \tag{2.16}$$

---

[1] Throughout this paper $H^r(\Omega)$ denotes the r-th order Sobolev space with $||\cdot||_r$ denoting its norm [3].

Galerkin approximations go as follows. Let

$$\mathcal{V}_h \subset \mathcal{V} \qquad S_h \subset S \tag{2.17}$$

be finite-dimensional spaces. We seek

$$u_h \in \mathcal{V}_h \qquad \phi_h \in S_h \tag{2.18}$$

such that

$$(A(u_h), v^h) + (Dv^h, \phi_h) = f(v^h) \quad \text{all} \quad v^h \in \mathcal{V}_h \tag{2.19}$$

$$(Du_h, \psi^h) = g(\psi^h) \quad \text{all} \quad \psi^h \in S_h \ . \tag{2.20}$$

Once a basis is selected for $\mathcal{V}_h$ and S_h, (2.19)-(2.20) reduces to a set of nonlinear equations [2].

A fundamental feature of the Brezzi form that is common throughout continuum mechanics is that the approximate system (2.19)-(2.20) is unstable for most choices of $\mathcal{V}_h$ and S_h. In particular, this is true for both the compressible and incompressible cases cited here. These instabilities have seriously impaired many attempts to use finite element methods in fluid mechanics (see for example [10]).

As a consequence many workers have sought out pairs $\mathcal{V}_h$, S_h which do not have these instabilities. The incompressible case is now regarded as being understood. Work of the French school is summarized in [3]. Papers which summarize the activity at Carnegie-Mellon are [4]-[8], and important results from the Scandinavian school can be found in [9].

All of these results (on incompressible flows) center around the following angle condition:

$$\inf_{\psi \in S_h} \sup_{v \in \mathcal{V}_h} \frac{\int_\Omega \psi \operatorname{div} v}{||\psi||_0 ||v||_1} \geq \beta > 0 \ , \tag{2.21}$$

where β is a number independent of the dimensions of S_h and $\mathcal{V}_h$. In effect, (2.21) states that the angle between the spaces S_h and $\operatorname{div}(\mathcal{V}_h)$ should be uniformly bounded away from $\pi/2$, as $h \downarrow 0$. This condition is both necessary and sufficient for stability and optimal order of convergence (see [3] and [5]).

The most popular element which satisfies this condition is the so-called box element. Here S_h consists of piecewise constant functions in boxes while ν_h consists of piecewise bilinear functions in the same boxes, subdivided once:

Observe that this discretization is very close to staggered grid. This type of grid has been very popular in finite difference methods. In addition there is a very close connection between the two approaches (see for example [18]).

The combination where $h_S = h_\nu$ (i.e., pressures and velocities are defined on the same grid) is known to be unstable. Moreover, the trend suggested by the stable pair, i.e.,

$$h_\nu/\nu_S < c \tag{2.22}$$

is true in general. That is, given any pair of finite element spaces ν_h, S_h, (2.21) will hold if the mesh spacings satisfy (2.22) for sufficiently small c ([11]).

Another popular combination which satisfies (2.21) is the so-called bubble element. Here ν_h consists of continuous piecewise quadratic functions on a triangulation augmented by continuous piecewise trilinear (bubble) functions. S_h, on the other hand, consists of all piecewise linear functions (continuous and discontinuous) on the given grid (see [3]). Finally, Boland and Nicolaides (see [8]) have shown that higher order elements generally avoid the stability problems encountered with lower order elements.

The situation for compressible flows is a good deal more subtle. In this case there is also an angle condition which takes the following form:

$$\inf_{\psi \in S_h} \sup_{v \in \nu_h} \frac{(\psi, \operatorname{div} v)_{-1}}{||\psi||_{-1} ||v||_0} \geq \beta > 0 \ . \tag{2.23}$$

However, for compressible flows the angle condition is necessary but not sufficient. For example, all of the elements cited above satisfy (2.23), and yet none are really satisfactory for compressible flows. Defects range from severe instabilities to suboptimal rates of convergence ([7],[13]).

Necessary and sufficient conditions are not known and remain a major problem of both theoretical and practical interest. Sufficient conditions, on the other hand, are known. One is the inclusion condition

$$S_h = \operatorname{div} \nu_h \tag{2.24}$$

(along with (2.23)). This condition is mathematically equivalent to exact mass conservation, i.e., that the finite element mass flow $u_h \in \nu_h$ satisfies

$$\operatorname{div} u_h = 0 \ . \tag{2.25}$$

None of the elements cited above has this property. In fact, only two very special elements are known to satisfy (2.24). One of these is the so-called crisscross element which is shown below:

ν_h = continuous piecewise linear in triangles S_h = piecewise constant functions in triangles

See [7] for details. This element gives second order accuracy in the mass flow u and the velocity u/ρ. Another second order element is shown below:

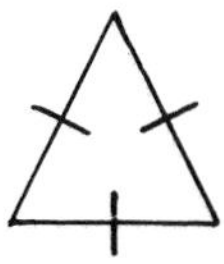

V_h = piecewise linear v with v and div v in L_2 (v is continuous at the midpoint of each edge)

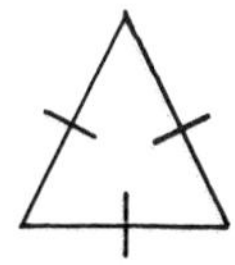

S_h = piecewise constants

Generalization of the elements to three dimensional problems will be treated in forthcoming publications.

III. LEAST SQUARES METHODS

Finite element approximations based on least squares have proven to be very popular alternatives to Galerkin-based schemes (see [14]-[17]). For the potential compressible flow equations they take the form

$$\int_\Omega \left\{ \left|\frac{u}{\rho} - \nabla\phi\right|^2 + |\text{div } u|^2 \right\} = \min \tag{3.1}$$

or equivalently,

$$\int_\Omega \left\{ \left|\text{curl } \frac{v}{\rho}\right|^2 + |\text{div } u|^2 \right\} = \min . \tag{3.2}$$

Their most striking property is their insensitivity to type; i.e., they work equally well for subsonic flows as well as supersonic flows. This has not been established rigorously, but the numerical evidence is very impressive. (See for example [19].)

These least squares approximations do have one major defect, namely, they require far more smoothness in solution than other types of methods. This was first seen in applications to fracture mechanics [20]. In the presence of corner singularities very poor approximations were obtained. Moreover, neither special singular elements nor mesh refinement seemed to improve the approximation, quite contrary to what could be obtained with Galerkin-based finite element schemes.

The linear elliptic case was treated by Fix and Stephan [21]. It was shown that for problems with corner singularities the use of weighted formulations of the form

$$\int_\Omega \{|\frac{u}{\rho} - \nabla\phi|^2 + r^\alpha|\text{div } u|^2\} = \min \tag{3.3}$$

significantly improved the accuracy of the approximation. In (3.3), r denotes the distance to the corner singularity and the exponent α depends on the nature of the singularity. The numerical results reported in [20] confirmed this analysis.

Similar considerations arise when shocks are present. In such cases many of the flow variables are not square integrable, hence a formulation based on (3.1) or (3.2) is clearly inappropriate. Indeed, the entire thrust of the modern theory of shock waves have emphasized spaces such as BV (functions with bounded variation) and $L^1(\Omega)$. Preliminary numerical evidence, on the other hand, suggest that appropriately weighted least squares can produce the correct behavior. This approach does require that the shock location be identified, or accurately approximated, hence it may not be practical for problems having a large number of shocks.

IV. THE PHOENIX EFFECT

There are many situations in computational fluid dynamics where the number of degrees of freedom used in approximations is far too small to completely resolve the flow field. In such cases traditional measures of error such as L_2 norms are large. Nevertheless, such approximations are on occasion not altogether useless. We illustrate this effect with a problem from acoustic wave propagation. To fix ideas we consider an axially symmetric acoustic potential ϕ which satisfies the Helmholtz equation

$$\nabla^2\phi + \omega^2\phi = 0 \tag{4.1}$$

in Ω subject to the boundary conditions

$$\begin{aligned} &\phi = g \quad \text{on} \quad r = r_0 , \\ &\phi = 0 \quad \text{on} \quad z = 0 , \\ &\frac{\partial\phi}{\partial n} = 0 \quad \text{on} \quad z = H(r) \end{aligned} \tag{4.2}$$

(see Figure 1) where z denotes the vertical axis and r denotes the radial coordinate.

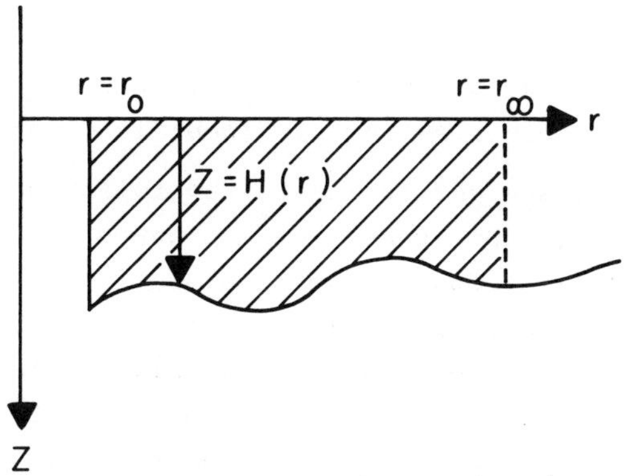

Figure 1. The flow region ω

Following the ideas introduced in [22], one can work on the truncated region

$$\hat{\Omega} = \{(r,z) \in \Omega: \quad r_0 < r < r_\infty\} \ , \tag{4.3}$$

by applying nonreflecting generalized radiation conditions of the form

$$\frac{\partial \phi}{\partial n} = T(\phi) \tag{4.4}$$

on the outer boundary $r = r_\infty$.

The boundary problem (4.1)-(4.2) and (4.4) admits a variational formulation in which the normal derivative conditions in (4.2) and (4.4) become natural boundary conditions. Indeed, this formulation was developed in [22] and takes the following form:

Find $\phi \in H^1$ satisfying

$$\phi = g \quad \text{on} \quad r = r_0, \qquad \phi = 0 \quad \text{on} \quad z = 0 \tag{4.5}$$

such that

$$\int_{\hat{\Omega}} [-\nabla\phi \cdot \nabla\psi + \omega^2 \phi\psi] + \int_{r=r_\infty} T(\phi)\psi = 0 \tag{4.6}$$

for all $\psi \in H^1$ satisfying

$$\psi = 0 \quad \text{on} \quad r_0 \quad \text{and} \quad z = 0. \tag{4.7}$$

By introducing finite dimensional finite element spaces one can convert this into a system of algebraic equations.

The issue we are addressing is the following. For a fixed frequency ω this scheme will produce quadratically converging approximations as the mesh spacing h goes to zero. However, as ω increases h must become small before "reasonable" approximations are obtained. We illustrate this for a flat bottom ($H(r) = 1$) and $r_\infty = 2$. In Figure 2, the value of ψ at $z = 1$ is plotted as a function of r for the case $\omega = 3$. The solid line is the exact solution given in [25]. The dotted lines are the approximations obtained with

$$\omega h = 1, \frac{1}{2}, \frac{1}{4} \tag{4.8}$$

The case $\omega h = \frac{1}{4}$ falls on top of the solid line. Observe that the classical criteria

$$\omega h \leq \frac{1}{2} \tag{4.9}$$

for resolving waves is evident in this figure. Similar results are obtained for $\omega = 6$ and are pictured in Figure 3.

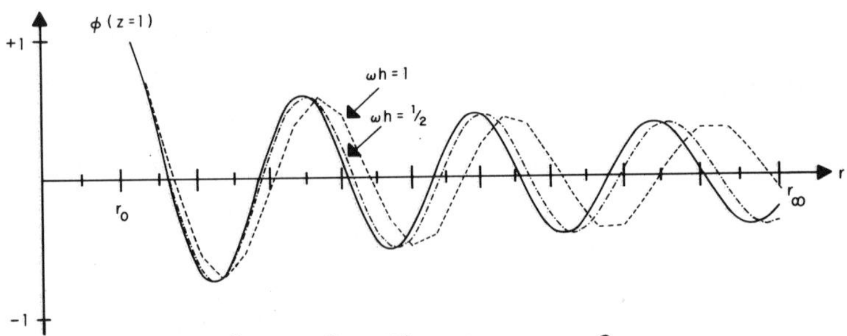

Figure 2. The case $\omega = 3$

Observe that the errors in the case $\omega h = 1$ are quite large due to the very large phase errors. Unfortunately, in many applications in underwater acoustics ω can be so large

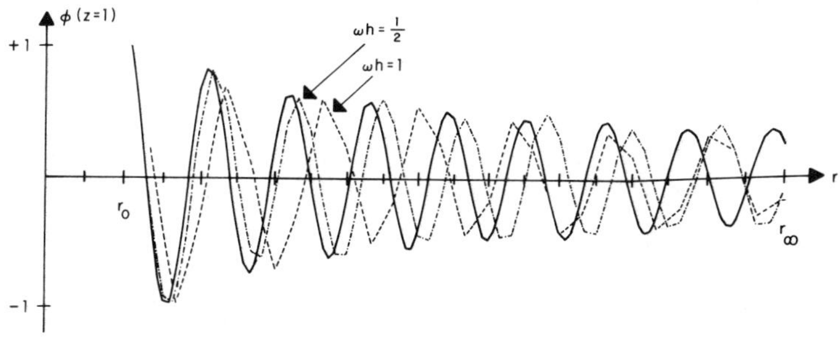

Figure 3. The case $\omega = 6$

that criteria like (4.9) would be impossible to satisfy even with modern supercomputers. The results in Figures 2 and 3 would seem to indicate that the numerical approximations are of limited interest. This is indeed the case if traditional measures of errors are used. However, in applications, having an accurate potential ϕ is not very important. Of far greater significance is the strength or attenuation of the wave as a function of the radial variable r. Using this as the criteria we see that all of the approximations--even for large values of ωh--are remarkably accurate.

This then is the phoenix effect, out of the ashes of poorly resolved numerical approximations can arise interesting and useful functionals of the solution. This raises the question of when such effects will occur. In this model mass conservation plays a crucial role (as in Section II). Indeed, (4.6) exactly conserves mass and this can be shown to give the correct asymptotic behavior for large r.

To illustrate this we compare the solution obtained from (4.6) (solid line in Figure 4) with that from a nonconservative scheme (dotted line in Figure 4). The latter was obtained by using biased or windward differencing for the first order system

$$u - \text{grad}\ \phi = 0$$

$$\text{div}\ u + \omega^2 \phi = 0.$$

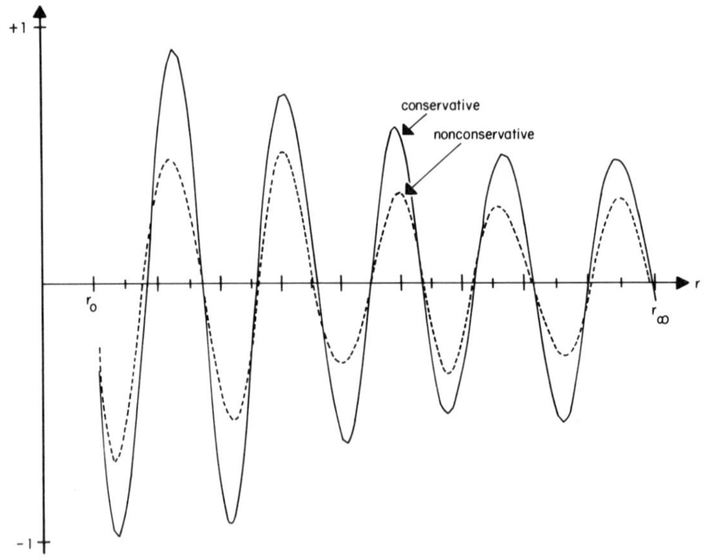

Figure 4. The case $\omega = t$, $\omega h = 1$

This type of situation is not special to acoustic waves, and for example can also be seen in geophysical fluid mechanics where again the question of conservation laws plays a fundamental role (see [26]). Similar effects occur in combustion problems ([27]-[29]).

V. CONCLUSIONS

The performance of finite element methods in fluid problems is quite different from that normally seen in solid mechanics. In the latter, the method typically yields best approximations (suitably defined), and thus effective computation requires only the selection of finite element spaces (i.e., grids and local polynomial behavior) that can accurately represent the solution. This is an important issue in fluid applications, but in addition, there are many finite element spaces which also have damaging instabilities when applied to flow problems.

The incompressible case has been well studied, and a number of practical elements have been identified which do not suffer from these instabilities. Unfortunately, most of these, when applied to compressible flows, do develop new types of instabilities. The problems appear to be related to a lack of mass conservation. In this paper we introduce new elements which do not have these defects.

In recent years there has been a growing interest in finite element schemes based on least squares principles. These typically do not have the instabilities cited above, however they do require a priori knowledge of the structure of the solution to get best results. This includes the nature of singular behavior as well as the presence of shocks. By formulating the least squares method in suitably weighted norms, these can be effectively modelled in the approximation.

The above considerations are based on traditional notions of stability and accuracy, typically L_2, or more generally, in Sobolev spaces. In many applications these may not be entirely relevant. In this paper examples are given of approximations which have very large L_2 errors, but nevertheless where certain key functionals of the solution are remarkably accurate. This is commonly called the phoenix effect, and it deserves further analysis.

REFERENCES

1. Brezzi, F., R.A.I.R.O. 8, 129-150.

2. Strang, G. and Fix, G., "An Analysis of the Finite Element Method," Prentice-Hall, New York (1973).

3. Girault, V. and Raviart, P.-A., "Finite element approximation of the Navier-Stokes equations," in Lecture Notes in Mathematics (A. Dold and B. Eckmann, eds.), Springer-Verlag, New York (1979).

4. Fix, G. J., Gunzburger, M. D., Nicolaides, R. A., and Peterson, J. S., in "AIAA 5th Computational Fluid Dynamics Conference Proceedings," Palo Alto, California, 265-271, June 22-23 (1981).

5. Gunzburger, M. D. and Peterson, J. S., "On conforming finite element methods for the inhomogeneous stationary Navier-Stokes equations," accepted for publication, Numerische Mathematik.

6. Fix, G. J., Gunzburger, M. D., and Nicolaides, R. A., "Theory and applications of mixed finite element methods," Proceedings of a symposium in honor of R. J. Duffin, 275-393, Academic Press (1979).

7. Fix, G. J., Gunzburger, M. D., and Nicolaides, R. A., Numerische Mathematik 38, 29-48 (1981).

8. Boland, James M., "Finite Elements and the Divergence Constraint for Viscous Flows," Ph.D. Thesis, Dept. of Math., Carnegie-Mellon University (1983).

9. Pitkaranta, J. and Stenberg, R., "Analysis of some mixed finite element methods for plane elasticity equations," Report HTKK-MAT-A194, Helsinki University of Technology (1982).

10. Sani, R., Gresho, P., Lee, R., and Griffiths, D., Int. J. Num. Meth. Fluids 1, Part 1, 17-44, Part 2, 171-204 (1981).

11. Fix, G. J., Lee, N. D., and Liang, G., Comp. & Maths. with Appls. 8, No. 5, 393-399 (1982).

12. Fix, G. J., Gunzburger, M. D., and Nicolaides, R. A., in "Computational Methods in Nonlinear Mechanics," (T. Oden, ed.), 245-260, North Holland (1980).

13. Gresho, P., private communication.

14. Fix- G. J., Gunzburger, M. D., and Nicolaides, R. A., "Least squares finite element methods," NASA-ICASE Report 77-18, revised version published in Math. and Comp. with Appls. 5, 87-98 (1979).

15. Fix, G. J. and Gunzburger, M. D., Int. J. for Numer. Mtd. in Eng. 12, 453-470 (1978).

16. Glowinski, R. and Pironneau, O., in "Convergence des solutions approchees," C.R. Acad. Sci. Paris, Série A, 286, 181-183 (1978).

17. Glowinski, R., this volume.

18. Fix, G. and Rose, M. E., "A comparative study of finite element and finite difference methods for Cauchy-Riemann type equations," SIAM J. of Numerical Analysis, to be submitted (1983).

19. Cox. C. L., Fix, G. J., and Gunzburger, "A least squares finite element scheme for transonic flow around harmonically oscillating wings," J. of Comp. Physics, to appear (1983).

20. Lee, Y.-K., "Shear Bands in Elasto-Perfectly Plastic Media," Ph.D. Thesis, Mech. Eng. Dept., Carnegie-Mellon University (1982).

21. Fix, G. J. and Stephan, E., "Finite element methods for the least square types for regions with corners," J. of Math. of Comp., (1981).

22. Fix, G. J. and Marin, S., J. of Comp. Phy. 28, 1-18 (1978).

23. Fix, G. J. and Nicolaides, R. A., SIAM J. Num. Anal. 17, No. 6, 779-786 (1980).

24. Fix, G. J., in "Nonlinear Partial Differential Equations in Engineering and Applied Science," Proceedings of a Conference Sponsored by Office of Naval Research, University of Rhode Island, (R. Sternberg, ed.), M. Dekker, Inc., 53-86 (1980).

25. Brekhovskikh, L. M., "Waves in Layered Media," Academic Press (1960).

26. Fix, G. J., SIAM J. on Appl. Math. 29, 371-387 (1975).

27. Aggarwal, S. K., Fix, G. J., Lee, D. N., and Sirignano, W., in "Advances in Computer Methods for Partial Differential Equations IV," (R. Vichnevetsky and R. Stepleman, eds.), Publ. IMACS, 317-323 (1981).

28. Aggarwal, S. L., Fix, G. J., Lee, D. N., and Sirignano, W., "Numerical optimization studies of axi-symmetric unsteady sprays," 1981 Aerospace Sciences Meeting, St. Louis, Missouri, January 12-15 (1981).

29. Aggarwal, S. L., Fix, G. J., Lee, D. N., and Sirignano, W., J. of Comp. Phys. 50, No. 1, 101-115 (1983).

SOLVING ELLIPTIC PROBLEMS BY DOMAIN DECOMPOSITION METHODS WTIH APPLICATIONS

Q.V. Dihn[1], R. Glowinski[2], and J. Périaux[1]

I. INTRODUCTION

Solving elliptic problems using *domain decomposition* is an idea which seems to originate with the classic Schwarz alternating method. Actually, many well-known methods for solving linear systems originating from the approximation of elliptic problems (e.g., block over-relaxation, block Gauss or Cholesky direct methods, etc.) take advantage of a subdomain decomposition.

The main goal of this paper is to discuss several variants of the Schwarz method. Used as preconditioners, they provide efficient tools for solving difficult nonlinear problems on complicated 2-D and 3-D geometries (cf: [1] (to which we also refer for more details on some of the methods discussed in this paper)).

II. FORMULATION OF THE MODEL PROBLEM. GENERALITIES

Let Ω be a bounded domain of $\mathbb{R}^N$ (N = 2,3 in practice) with a smooth boundary $\Gamma = \partial\Omega$. Consider the following problem

(2.1) $\quad -\Delta y = f \quad \text{in } \Omega, \quad y = g \quad \text{on} \quad \Gamma$

(where f and g are *given* functions). In the remainder of this paper we consider the following two types of decomposition of Ω by subdomains Ω_{ii}:

[1]AMD/BA - B.P. 300, 78 quai Carnot - 92214 ST CLOUD, France.

[2]Université P.et M. Curie - 4, place Jussieu - 75230 PARIS CEDEX 05 and INRIA.

ISBN 0-12-100560-7

(a) With

$$(2.2) \qquad \Omega = \overline{\overset{\circ}{\bigcup_i \Omega_{ii}}}$$

where $\overline{X}$ and $\overset{\circ}{X}$ denote, respectively, the *closure* and the *interior* of a set X, and also with

$$(2.3) \qquad \Omega_{ii} \cap \Omega_{jj} = \emptyset \quad \forall i,j,\ i \neq j,$$

we define γ_{ij} (possibly empty) by

$$\gamma_{ij} = \gamma_{ji} = \partial\Omega_{ii} \cap \partial\Omega_{jj} \quad \text{if } i \neq j .$$

Figure 1 illustrates this decomposition for the case i = 1,2 in (2.2).

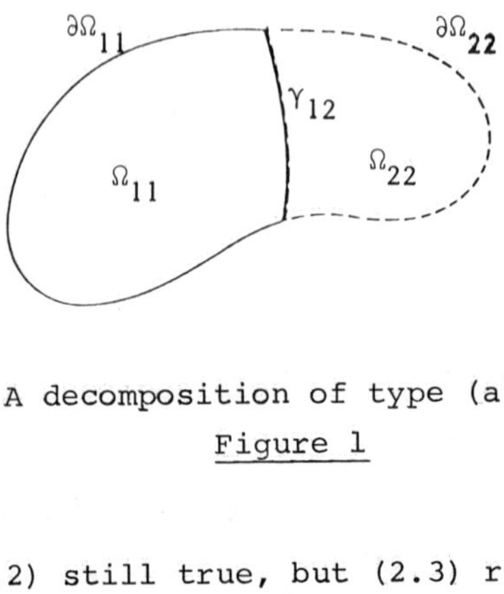

A decomposition of type (a)

Figure 1

(b) With (2.2) still true, but (2.3) replaced, for i ≠ j, by

$$(2.4) \qquad \begin{cases} \textit{either } \Omega_{ii} \cap \Omega_{jj} = \emptyset, \\ \\ \textit{or } \Omega_{ii} \cap \Omega_{jj} = \Omega_{ij} (= \Omega_{ji}) \textit{ with } \int_{\Omega_{ij}} dx > 0 , \end{cases}$$

(where $dx = dx_1 dx_2$). We also define $\gamma_i = \partial\Omega_{ii} - (\partial\Omega_{ii} \cap \Gamma)$. Figure 2 illustrates this situation (with i = 1,2,3); a more general decomposition of this type is shown on Figure 3.

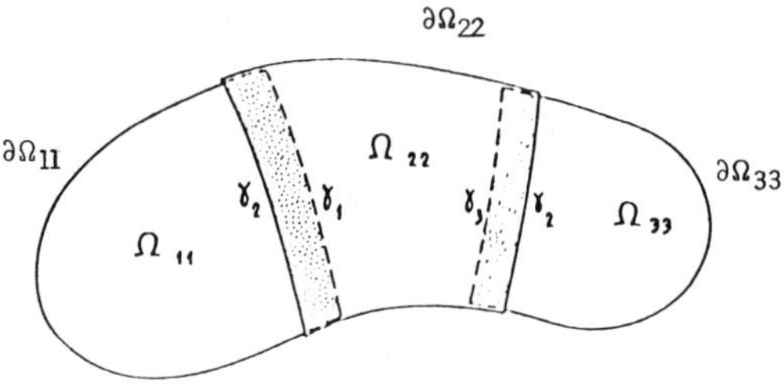

A decomposition of type (b)

Figure 2

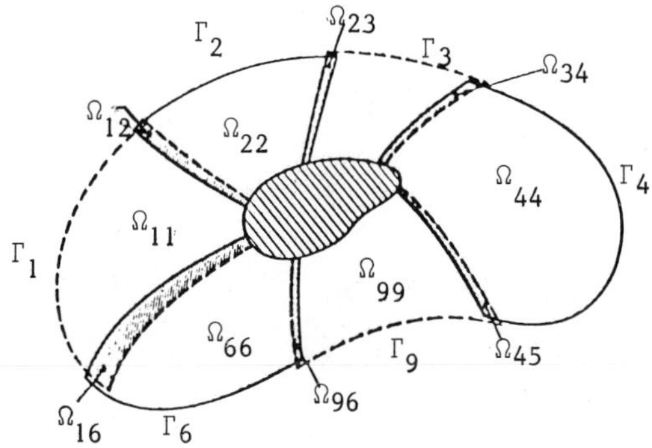

Figure 3

To conclude this section, we introduce some *Sobolev* functional spaces, which we use in later sections, namely

(2.5) $$H^1(\Omega) = \{v \mid v \in L^2(\Omega),\ \frac{\partial v}{\partial x_i} \in L^2(\Omega) \quad \forall i = 1,\dots,N\},$$

(2.6) $$H^1_o(\Omega) = \{v \mid v \in H^1(\Omega),\ v = 0 \quad \textit{on } \Gamma\}.$$

Note the space $H^1(\Omega)$ is a *Hilbert space* for the inner product

$$(2.7) \qquad (u,v)_{H^1(\Omega)} = \int_\Omega uv\,dx + \int_\Omega \nabla u \cdot \nabla v\,dx\ ;$$

moreover $H^1(\Omega)$ is a *closed* subspace of $H^1(\Omega)$, and if Ω is bounded (at least in one direction of $\mathbb{R}^N$), then $H_o^1(\Omega)$ is a Hilbert space if associated with the inner product

$$(2.8) \qquad (u,v)_{H_o^1(\Omega)} = \int_\Omega \nabla u \cdot \nabla v\,dx\ ,$$

and the corresponding norm

$$||v||_{H_o^1(\Omega)} = \left(\int_\Omega |\nabla v|^2 dx\right)^{1/2}$$

is equivalent to the norm induced by $H^1(\Omega)$. Suppose that $L^2(\Omega)$ has been identified to its dual space; then if $H^{-1}(\Omega)$ is the dual space of $H_o^1(\Omega)$ we have

$$(2.9) \qquad H_o^1(\Omega) \subset L^2(\Omega) \subset H^{-1}(\Omega).$$

Furthermore, operator $-\Delta$ is an *isomorphism* from $H_o^1(\Omega)$ onto $H^{-1}(\Omega)$; we shall denote by $\langle .,. \rangle$ the duality pairing between $H^{-1}(\Omega)$, such that

$$(2.10) \qquad \langle f,v \rangle = \int_\Omega fv\,dx \quad \forall f \in L^2(\Omega), \quad \forall v \in H_o^1(\Omega)\ .$$

If $f \in H^{-1}(\Omega)$, we have

$$(2.11) \qquad ||f||_{-1} = \sup_{v \in H_o^1(\Omega)-\{0\}} \frac{|\langle f,v \rangle|}{||v||_{H_o^1(\Omega)}}$$

For more properties and details on Sobolev spaces we refer to e.g., [2]-[4]. In the next sections, similar definitions will also be used for the subdomains Ω_{ii}, Ω_{ij}.

III. THE SCHWARZ ALTERNATING METHOD

The Schwarz alternating method was introduced around 1865 to solve Dirichlet problems by H.A. Schwarz, who proved its convergence using the *Maximum Principle*. It can be summarized as follows.

With Ω split as in Figure 4 into subdomains Ω_{11} and Ω_{22}, let f_i and g_i denote the restriction of f and g from (2.1) to Ω_{ii} and $\partial\Omega_{ii} \cap \Gamma$, respectively. Following the notation of Sec. II, the Schwarz algorithm is then defined by

Let η *be a function defined on* γ_1, *then for* $n \geq 0$ *construct a sequence* $\{y_1^n, y_2^n\}$ *as follows*

$$(3.1)\quad \begin{cases} -\Delta y_1^n = f_1 \text{ in } \Omega_{11}, \quad y_1^n = g_1 \text{ on } \partial\Omega_{11} \cap \Gamma , \\ \\ y_1^n|_{\gamma_1} = y_2^{n-1}|_{\gamma_1} \text{ if } n \geq 1, \quad y_1^o|_{\gamma_1} = \eta , \end{cases}$$

$$(3.2)\quad \begin{cases} -\Delta y_2^n = f_2 \text{ in } \Omega_{22}, \quad y_2^n = g_2 \text{ on } \partial\Omega_{22} \cap \Gamma , \\ y_2^n|_{\gamma_2} = y_1^n|_{\gamma_2} . \end{cases}$$

The mechanism of the above method is shown on Figure 4.

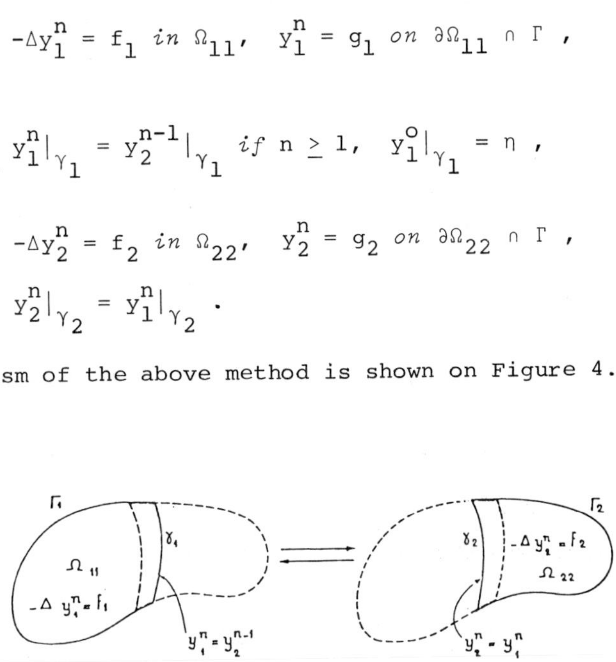

Mechanism of the Schwarz alternating method

Figure 4

For more properties of the Schwarz alternating method see [5].

IV. DOMAIN DECOMPOSITION METHODS WITH LAGRANGE MULTIPLIERS

IV.1. A first method

Problem (2.1) is equivalent to the following *minimization* problem

$$(4.1)\qquad \textit{Find } y \in V_g \textit{ such that } J(y) \leq J(z),\ \forall z \in V_g ,$$

where

$$(4.2)\qquad V_g = \{z \mid z \in H^1(\Omega),\ z = g \text{ on } \Gamma\},\ J(z) = \frac{1}{2}\int_\Omega |\nabla z|^2 dx - \int_\Omega fz dx.$$

We consider now a decomposition of Ω as done in case (a), Sec. II (we only consider a two-subdomain decomposition as in Figure 1); we associate to this decomposition the functionals $J_i\colon\ H^1(\Omega_{ii}) \to \mathbb{R}$ defined (with $f_i = f|_{\Omega_{ii}}$, $g_i = g|_{\partial\Omega_{ii}\cap\Gamma}$) by

$$(4.3)\qquad J_i(z_i) = \frac{1}{2}\int_{\Omega_{ii}} |\nabla z_i|^2 dx - \int_{\Omega_{ii}} f_i z_i dx,\quad i = 1,2\ .$$

We clearly have equivalence (in that $y_i = y|_{\Omega_{ii}}$, $\forall i = 1,2$) between (2.1), (4.1) and the following minimization problem

$$(4.4)\qquad \begin{cases} \textit{Find } \{y_1,y_2\} \in W \textit{ such that} \\[2ex] J_1(y_1) + J_2(y_2) \leq J_1(z_1) + J_2(z_2)\quad \forall\{z_1,z_2\} \in W \end{cases}$$

where $W = \{\{z_1,z_2\} \in H^1(\Omega_{11})\times H^1(\Omega_{22}),\ z_1 = z_2 \text{ on } \gamma_{12}\ ,$ $z_i|_{\partial\Omega_{ii}\cap\Gamma} = g_i,\ \forall i = 1,2\}$.

To overcome the main difficulty in (4.4), namely the linear relation

$$(4.5)\qquad z_1|_{\gamma_{12}} = z_2|_{\gamma_{12}}\ ,$$

we introduce a *Lagrange multiplier*, via the *lagrangian functional* L defined by

$$(4.6)\qquad L(z_1,z_2;\mu) = J_1(z_1) + J_2(z_2) + \int_{\gamma_{12}} \mu(z_2 - z_1)d\gamma\ .$$

We can prove the following

<u>Proposition 4.1.</u> *Define the spaces* V *and* Λ *by*

$$(4.7)\ \begin{cases} V = \{\{z_1,z_2\},\ \forall i = 1,2, z_i \in H^1(\Omega_{ii}),\ z_i|_{\partial\Omega_{ii}\cap\Gamma} = g_i\}, \\ \Lambda = L^2(\gamma_{12}) \end{cases}$$

and suppose that $\{y_1, y_2; \lambda\}$ *is a saddle-point of* L *over* $V\times\Lambda$ *i.e.* $\{y_1, y_2; \lambda\} \in V\times\Lambda$ *and*

$$(4.8)\quad L(y_1, y_2; \mu) \leq L(y_1, y_2; \lambda) \leq L(z_1, z_2; \lambda), \quad \forall\{z_1, z_2; \mu\} \in V\times\Lambda .$$

We have then, $\forall i = 1,2$,

$$y_i = y|_{\Omega_{ii}} \text{ and } \lambda = \frac{\partial y_1}{\partial n_1}\Big|_{\gamma_{12}} = -\frac{\partial y_2}{\partial n_2}\Big|_{\gamma_{12}}$$

where y *is the solution of* (2.1) *and* $\underset{\sim}{n}_i$ *the outward unit normal vector at* $\partial\Omega_{ii}$. *The reciprocal property holds if* y *is sufficiently smooth.*

From Proposition 4.1, we can replace the solution of (2.1), (4.1) by the solution of the saddle-point problem (4.8). It follows then from [6, Chapter 2 and Appendix 2] and [7] that (4.8) can be solved by a *saddle-point solver* like Uzawa's algorithm (and its conjugate gradient variants). The basic Uzawa algorithm corresponding to (4.8) is defined by

(4.9) *Let* $\lambda^0 \in \Lambda$ *be arbitrarily given,*

then for $n \geq 0$, *assuming* λ^n *known, we compute* y_1^n, y_2^n *and* λ^{n+1} *by*

$$(4.10)\quad \begin{cases} -\Delta y_1^n = f_1 \text{ in } \Omega_{11}, \\ y_1^n = g_1 \text{ on } \partial\Omega_{11}\cap\Gamma, \quad \dfrac{\partial y_1^n}{\partial n_1} = \lambda^n \text{ on } \gamma_{12}, \end{cases}$$

$$(4.11)\quad \begin{cases} -\Delta y_2^n = f_2 \text{ in } \Omega_{22}, \\ y_2^n = g_2 \text{ on } \partial\Omega_{22}\cap\Gamma, \quad \dfrac{\partial y_2^n}{\partial n_2} = -\lambda^n \text{ on } \gamma_{12}, \end{cases}$$

$$(4.12)\quad \lambda^{n+1} = \lambda^n + \rho(y_2^n - y_1^n)|_{\gamma_{12}}, \quad \rho > 0 .$$

The following can be shown about the convergence of algorithm (4.9)-(4.12).

Proposition 4.2.: *Suppose that the saddle-point problem* (4.8) *has a solution; then* $\forall \overset{\circ}{\lambda} \in \Lambda$, *we have, for* $i = 1,2$,

(4.13) $$\lim_{n \to +\infty} y_i^n = y_i = y|_{\Omega_{ii}} \quad \textit{strongly in } H^1(\Omega_{ii})$$

(*where* y *is the solution of* (2.1), (4.1)), *if*

(4.14) $$0 < \rho < 2/c^2$$

where, in (4.14), $c = \sup_{\{z_1,z_2\} \in Z - \{0\}} \dfrac{||z_2 - z_1||_{L^2(\gamma_{12})}}{||\{z_1,z_2\}||_Z}$, *and*

where

$$Z = \{\{z_1,z_2\} \in H^1(\Omega_{11}) \times H^1(\Omega_{22}), \quad z_i|_{\partial\Omega_{ii} \cap \Gamma} = 0 \quad \forall i = 1,2\},$$

$$||\{z_1,z_2\}||_Z = \left(\sum_{i=1}^{2} \int_{\Omega_{ii}} |\nabla z_i|^2 \, dx\right)^{1/2} .$$

Algorithm (4.9)-(4.12) is in fact a *gradient algorithm* with the *constant step* ρ, applied to the solution of the *dual problem* associated to (4.8), i.e.

(4.15) *Find* $\lambda \in \Lambda$ *such that* $J^*(\lambda) \leq J^*(\mu) \quad \forall \mu \in \Lambda$

where $J^*(\mu) = - \inf_{\{z_1,z_2\} \in V} L(z_1,z_2;\mu)$. It is easy to see that $J(\mu) = a(\mu,\mu)/2 - \ell(\mu)$, where $a(.,.)$ is *bilinear, continuous symmetric* over $\Lambda \times \Lambda$ (but *not* strongly-elliptic) and where $\ell(\cdot)$ is *linear continuous* over Λ. The above properties clearly suggest a *conjugate gradient* algorithm for solving (4.15), and therefore (2.1); to derive such an algorithm we define a linear *boundary* operator A as follows:

Let $\lambda \in \Lambda$; *to* λ *we associate, for* $i = 1,2$, $y_i(\lambda) = y_i$ *as the solution in* $H^1(\Omega_{ii})$ *of*

(4.16) $$\begin{cases} \Delta y_i = 0 \textit{ in } \Omega_{ii}, \quad y_i = 0 \textit{ on } \partial\Omega_{ii} \cap \Gamma, \quad \dfrac{\partial y_i}{\partial n_i} = (-1)^{i-1}\lambda \\ \textit{on } \gamma_{12}, \end{cases}$$

and then define A *by*

$$(4.17) \qquad A\lambda = (y_1(\lambda) - y_2(\lambda))|_{\gamma_{12}} .$$

Using *Green's formula*, we should prove that $\forall\lambda,\mu \in \Lambda$ we have

$$(4.18) \qquad \int_{\gamma_{12}} A\lambda\mu \, d\gamma = \sum_{i=1}^{2} \int_{\gamma_{12}} y_i(\lambda) \frac{\partial y_i}{\partial n_i}(\mu)\, d\gamma$$

$$= \sum_{i=1}^{2} \int_{\Omega_{ii}} \nabla y_i(\lambda) \cdot \nabla y_i(\mu)\, dx .$$

It follows from (4.18) that A is *symmetric* and *positive definite* over Λ; it is also *continuous* but *not* Λ-*elliptic*. We should observe that the bilinear form a(.,.) considered above satisfies

$$(4.19) \qquad a(\lambda,\mu) = \int_{\gamma_{12}} A\lambda\mu \, d\gamma, \quad \forall\lambda,\mu \in \Lambda .$$

The main result concerning the dual problem (4.15) is the following

Proposition 4.3.: *Let* λ *be the solution of the dual problem* (4.15); *then* λ *is also the solution of the following linear problem in* Λ:

$$(4.20) \qquad A\lambda = (y_{o2} - y_{o1})|_{\gamma_{12}}$$

where $\{y_{o1}, y_{o2}\} \in H^1(\Omega_{11}) \times H^1(\Omega_{22})$ *is defined (with* i = 1,2) *by*

$$(4.21) \qquad \begin{cases} -\nabla y_{oi} = f_i \text{ in } \Omega_{ii}, \quad y_{oi} = g_i \text{ on } \partial\Omega_{ii} \cap \Gamma, \\ \dfrac{\partial y_{oi}}{\partial n_i} = 0 \text{ on } \gamma_{12}. \end{cases}$$

From the above results we derive the following *conjugate gradient* variant of algorithm (4.9)-(4.12):

Step 0: *Initialization*

(4.22) $\lambda^o \in \Lambda$, *arbitrarily given,*

(4.23) $$g^o = A\lambda^o - (y_{o2} - y_{o1})|_{\gamma_{12}},$$

(4.24) $$w^o = g^o ;$$

then for $n \geq 0$, *assuming that* $\{\lambda^n, w^n, g^n\} \in \Lambda^3$ *are known, compute* λ^{n+1}, w^{n+1}, g^{n+1} *by*

Step 1: *Descent*

(4.25) $$\lambda^{n+1} = \lambda^n - \rho_n w^n \text{ with } \rho_n = \frac{\int_{\gamma_{12}} g^n w^n d\gamma}{\int_{\gamma_{12}} (Aw^n) w^n d\gamma} = \frac{\int_{\gamma_{12}} |g^n|^2 d\gamma}{\int_{\gamma_{12}} (A\,w^n) w^n d\gamma}$$

Step 2: *New descent direction*

(4.26) $$g^{n+1} = g^n - \rho_n Aw^n ,$$

(4.27) $$w^{n+1} = g^{n+1} + \gamma_n w^n, \text{ with } \gamma_n = \frac{||g^{n+1}||^2_\Lambda}{||g^n||^2_\Lambda}$$

Do $n = n+1$ *and go to* (4.25).

Despite its abstract formalism algorithm (4.22)-(4.27) is no more complicated to implement than algorithm (4.9)-(4.12) and like this later requires only the solution at each iteration of a Poisson problem on each Ω_{ii}; indeed g^o is obtained from λ^o (with $i = 1,2$) by

(4.28) $$\begin{cases} -\Delta y_i^o = f_i \text{ in } \Omega_{ii}, \quad y_i^o = g_i \text{ on } \partial\Omega_{ii} \cap \Gamma, \\ \dfrac{\partial y_i^o}{\partial n_i} = (-1)^{i-1}\lambda^o \text{ on } \gamma_{12}, \end{cases}$$

(4.29) $$g^o = (y_1^o - y_2^o)|_{\gamma_{12}} .$$

Concerning Steps 1 and 2, the key point is the calculation of Aw^n, which is done as follows:

We define $\{\chi_1^n, \chi_2^n\} \in H^1(\Omega_{11}) \times H^1(\Omega_{22})$ *by*

$$(4.30)\qquad \begin{cases} \Delta\chi_i^n = 0 \ \textit{in}\ \Omega_{ii}, \quad \chi_i^n = 0 \ \textit{on}\ \partial\Omega_{ii} \cap \Gamma, \\ \dfrac{\partial\chi_i^n}{\partial n_i} = (-1)^{i-1} w^n \ \textit{on}\ \gamma_{12}\,, \end{cases}$$

for $i = 1,2$, *and then*

$$(4.31)\qquad Aw^n = (\chi_1^n - \chi_2^n)|_{\gamma_{12}}\ ;$$

thus one iteration of (4.22)-(4.27) is no more costly than one iteration of (4.9)-(4.12). Another conjugate gradient variant of (4.9)-(4.12) is given in [8]; it is however *twice as costly per iteration* as (4.22)-(4.27).

IV.2. A second method

The methods discussed in Sec. IV.1 behave quite well as long as the Poisson problem to be solved has Dirichlet boundary conditions (or is at least Dirichlet boundary conditions dominated); actually methods with better convergence properties are obtained if we take as unknown the trace of the solution of (2.1) on γ_{12} and we adjust its values in order to satisfy $(\partial y_1/\partial n_1 + \partial y_2/\partial n_2 = 0)$ which appears then as a *linear constraint*, the corresponding *Lagrange multiplier* being $\lambda = y|_{\gamma_{12}}$. The following algorithm is founded on the above principle (and can be viewed too as an *Uzawa's algorithm*):

$$(4.32)\qquad \textit{Given}\ \lambda^o \in \Lambda_g\,,$$

then for $n \geq 0$, *assuming* λ^n *known, we compute* y_1^n, y_2^n, *by*:

$$(4.33)\qquad \begin{cases} -\Delta y_i^n = f_i \ \textit{in}\ \Omega_{ii}, \quad y_i^n = g_i \ \textit{on}\ \partial\Omega_{ii} \cap \Gamma, \quad y_i^n = \lambda^n \\ \textit{on}\ \gamma_{12};\ i = 1,2 \end{cases}$$

to compute λ^{n+1} *from* y_1^n, y_2^n *we first introduce*

$\delta^{n+1/2} = \lambda^{n+1} - \lambda^n$ *and we solve on* Λ_o *the variational problem*

$$(4.34)\quad \begin{cases} \text{Find } \delta^{n+1/2} \in \Lambda_o, \text{ such that } \forall\mu \in \Lambda_o \\ \sum_{i=1}^{2} \int_{\Omega_{ii}} \underset{\sim}{\nabla}\tilde{\delta}_i^{n+1/2} \cdot \underset{\sim}{\nabla}\tilde{\mu}_i dx \\ \qquad = -\rho \sum_{i=1}^{2} \left(\int_{\Omega_{ii}} \underset{\sim}{\nabla} y_i^n \cdot \underset{\sim}{\nabla}\tilde{\mu}_i dx - \int_{\Omega_{ii}} f_i \tilde{\mu}_i dx \right) ; \end{cases}$$

in (4.32)-(4.34) we have

$$(4.35)\quad \Lambda_o = \{\mu | \mu \in L^2(\gamma_{12}),\ \mu = \tilde{\mu}|_{\gamma_{12}} \text{ with } \tilde{\mu} \in H_0^1(\Omega)\} ,$$

$$(4.36)\quad \begin{cases} \Lambda_g = \{\mu | \mu \in L^2(\gamma_{12}),\ \mu = \tilde{\mu}|_{\gamma_{12}} \text{ with } \tilde{\mu} \in H^1(\Omega), \\ \tilde{\mu} = g \text{ on } \Gamma\} \end{cases}$$

and, in (4.34), $\tilde{q}_i$ is defined, for i = 1,2, from $q \in \Lambda_o$ as follows:

$$(4.37)\quad \tilde{q}_i \in H^1(\Omega_{ii}),\ \ \tilde{q}_i = 0 \text{ on } \partial\Omega_{ii} - \gamma_{12},\ \ \tilde{q}_i = q \text{ on } \gamma_{12} .$$

In practice to obtain *unique* extensions q_1, q_2 of q we introduce, for i = 1,2, H_i $(\subset H^1(\Omega_{ii}))$ by

$$(4.38)\quad H_i = \{v_i | v_i \in H^1(\Omega_{ii}),\ \ v_i = 0 \text{ on } \partial\Omega_{ii} - \gamma_{12}\} ,$$

and then P_i such that

$$(4.39)\quad H_i = H_o^1(\Omega_{ii}) \oplus P_i ;$$

if $q \in \Lambda_o$ is given we shall take for $\tilde{q}_i$ the *unique element* of P_i such that $\tilde{q}_i|_{\gamma_{12}} = q$. We observe using Green's formula that we have for the right-hand side in (4.34)

$$(4.40) \qquad \sum_{i=1}^{2} \left(\int_{\Omega_{ii}} \nabla y_i^n \cdot \nabla \tilde{\mu}_i \, dx - \int_{\Omega_{ii}} f_i \tilde{\mu}_i dx \right)$$

$$= \int_{\gamma_{12}} \left(\frac{\partial y_1^n}{\partial n_1} + \frac{\partial y_2^n}{\partial n_2} \right) \mu d\gamma \quad \forall \mu \in \Lambda_o \; ;$$

in (4.40) the left-hand side is easier to implement (in combination with the usual *finite element approximations*) than the right-hand side, since it does not require the explicit knowledge of the normal derivatives $\frac{\partial y_1^n}{\partial n_1}$, $\frac{\partial y_2^n}{\partial n_2}$ along γ_{12}.

To conclude Sec. IV.2 we shall describe a *conjugate gradient* variant of (4.32)-(4.34) defined as follows:

<u>Step 0</u>: *Initialization*

(4.41) $\lambda^o \in \Lambda_g$, *arbitrarily given,*

solve then for $i = 1,2$

$$(4.42) \quad \begin{cases} -\Delta y_i^o = f_i \text{ in } \Omega_{ii}, \quad y_i^o = g_i \text{ on } \partial\Omega_{ii} - \gamma_{12} , \\ y_i^o = \lambda^o \text{ on } \gamma_{12}, \end{cases}$$

and then

$$(4.43) \quad \begin{cases} \text{Find } g^o \in \Lambda_o \text{ such that } \forall \mu \in \Lambda_o \\ \displaystyle\sum_{i=1}^{2} \int_{\Omega_{ii}} \nabla \tilde{g}_i^o \cdot \nabla \tilde{\mu}_i \, dx = \sum_{i=1}^{2} \left(\int_{\Omega_{ii}} \nabla y_i^o \cdot \nabla \tilde{\mu}_i \, dx - \int_{\Omega_{ii}} f_i \tilde{\mu}_i dx \right) \end{cases}$$

and set

$$(4.44) \qquad w^o = g^n . \qquad \square$$

For $n \geq 0$, suppose that λ^n, g^n, w^n are known and compute λ^{n+1}, g^{n+1}, w^{n+1} as follows:

Step 1: *Descent*

Define $\{\chi_i^n\}_{i=1}^2 \in H^1(\Omega_{11}) \times H^1(\Omega_{22})$ *by*

$$\Delta\chi_i^n = 0 \text{ in } \Omega_{ii}, \quad \chi_i^n = 0 \text{ on } \partial\Omega_{ii} - \gamma_{12}, \quad \chi_i^n = w^n \text{ on } \gamma_{12}, \tag{4.45}$$

and then

$$\lambda^{n+1} = \lambda^n - \rho_n w^n, \text{ with } \rho_n = \frac{\sum_{i=1}^{2} \int_\Omega |\nabla \tilde{g}_i^n|^2 \, dx}{\sum_{i=1}^{2} \int_\Omega \nabla \chi_i^n \cdot \nabla \tilde{w}_i^n \, dx} \tag{4.46}$$

Step 2: *Descent direction*

Solve

$$\begin{cases} \text{Find } g^{n+1} \in \Lambda_0 \text{ such that } \forall \mu \in \Lambda_0 \\ \sum_{i=1}^{2} \int_\Omega \nabla \tilde{g}_i^{n+1} \cdot \nabla \tilde{\mu}_i \, dx = \sum_{i=1}^{2} \left(\int_\Omega \nabla \tilde{g}_i^n \cdot \nabla \tilde{\mu}_i \, dx - \rho_n \int_\Omega \nabla \chi_i^n \cdot \nabla \tilde{\mu}_i \, dx \right), \end{cases} \tag{4.47}$$

and set

$$w^{n+1} = g^{n+1} + \gamma_n w^n, \text{ with } \gamma_n = \frac{\sum_{i=1}^{2} \int_{\Omega_{ii}} |\nabla \tilde{g}^{n+1}|^2 \, dx}{\sum_{i=1}^{2} \int_{\Omega_{ii}} |\nabla \tilde{g}^n|^2 \, dx}. \tag{4.48}$$

Do $n = n+1$ *and go to* (4.45).

The notation used in (4.41)-(4.48) are those used previously in (4.32)-(4.34). Algorithm (4.41)-(4.48) (in fact *finite element* variants of it) has proved being very efficient for solving elliptic problems using domain decomposition. A generalization of the method of Sec. IV.2 will be discussed in the sequel; it concerns the solution of the full potential equation for transonic flows.

V. A NEW DOMAIN DECOMPOSITION METHOD

V.1. Motivation. An equivalence results

The two main goals of the method which follows are:

(i) Find a *conjugate gradient* variant of the *Schwarz's algorithm* (3.1), (3.2); and

(ii) Develop some experience of matching methods using *least squares*.

We suppose that Ω has been split in several subdomains according to situation (b) of Sec. II (whose notation is kept) and we suppose for simplicity that i,j = 1,2 in (2.2), (2.4). We introduce, for i = 1,2, the following spaces

$$(5.1)\quad V_i = \{v_i \in L^2(\gamma_i),\ \ v_i = \tilde{v}_i|_{\gamma_i}\ \ \textit{with}\ \tilde{v}_i \in H^1(\Omega_{ii}),$$

$$\tilde{v}_i = g_i \ \textit{on}\ \partial\Omega_{ii} \cap \Gamma\}\ ;$$

we have then the following (obvious):

Proposition 5.1: *The minimization problem*

$$\textit{Find}\ \{u_1,u_2\} \in V_1 \times V_2 \quad \textit{such that}$$

$$(5.2)\qquad J(u_1,u_2) \le J(v_1,v_2),\ \ \forall\ \{v_1,v_2\} \in V_1 \times V_2\ ,$$

where

$$(5.3)\qquad J(v_1,v_2) = \frac{1}{2}\int_{\Omega_{12}} \{|\underset{\sim}{\nabla}(y_2 - y_1)|^2 + (y_2 - y_1)^2\}\ dx$$

with y_i *a solution in* $H^1(\Omega_{ii})$ *(for* i = 1,2*) of*

$$(5.4)\quad -\Delta y_i = f_i\ \textit{in}\ \Omega_{ii},\ \ y_i = g_i\ \textit{on}\ \partial\Omega_{ii} \cap \Gamma,\ \ y_i = v_i\ \textit{on}\ \gamma_i,$$

has a unique solution such that

$$(5.5)\qquad u_i = y_i,\ \ \forall i = 1,2,\ \textit{where}\ y\ \textit{is the solution of}\ (2.1).$$

Problem (5.2) clearly has the structure of an *optimal control problem* (see [9]) with: v_1, v_2 the *control variables*; y_1, y_2 the state variables; (5.4) the *state equations*; $J(.,.)$ the *cost function*. Actually other cost functions than (5.3) may be used, such as

$$\int_{\Omega_{12}} |y_2 - y_1|^2 \, dx \; ;$$

it seems, however, that $J(.,.)$ defined by (5.3) is optimally suited for solving (2.1), via (5.2).

V.2. Conjugate gradient solution of (5.2). Further comments.

Any conjugate gradient algorithm for solving (5.2) is actually very close to the algorithms discussed in Sec. IV; it is clear that the natural metrics to use on the V_i's are those induced by the $H^1(\Omega_{ii})$'s; using these metrics is rather technical, therefore we shall consider only the case where the V_i's are equipped with the $L^2(\gamma_i)$-norms (*it is nonsense for the continuous problems, but not for the discrete ones*); we refer to [1] for a discussion of the case where the norm on V_i is the one induced by $H^1(\Omega_{ii})$.

Description of a conjugate gradient algorithm: For $i = 1,2$, define V_{oi} by

$$(5.6)\quad \begin{cases} V_{oi} = \{v_i \in L^2(\gamma_i), \quad v_i = \tilde{v}_i|_{\gamma_i} \text{ with } \tilde{v}_i \in H^1(\Omega_{ii}), \\ \tilde{v}_i = 0 \text{ on } \partial\Omega_{ii} \cap \Gamma\} \; ; \end{cases}$$

the conjugate gradient algorithm is then defined as follows

Step 0: *Initialization*

$$(5.7)\qquad u^o = \{u_1^o, u_2^o\} \in V_1 \times V_2 \text{ is arbitrarily given}$$

define then $g^o = \{g_i^o\}_{i=1}^2 \in V_{01} \times V_{02}$ *such that*

$$(5.8)\quad \sum_{i=1}^{2} \int_{\gamma_i} g_i^o z_i \, d\gamma = (J'(u^o), z), \quad \forall z = \{z_i\}_{i=1}^2 \in V_{01} \times V_{02} \; ,$$

where J' *denotes the differential of* J; *set then*

$$w^o = g^o \ . \tag{5.9}$$

Assuming that $u^n = \{u_i^n\}_{i=1}^2$, $g^n = \{g_i^n\}_{i=1}^2$, $w^n = \{w_i^n\}_{i=1}^2$ *are known, compute* u^{n+1}, g^{n+1}, w^{n+1} *by*

Step 1: *Steepest descent*

$$u^{n+1} = u^n - \rho_n w^n \text{ with } \rho_n = \operatorname*{Arg\,Min}_{\rho\in\mathbb{R}} J(u^n - \rho w^n) \ . \tag{5.10}$$

Step 2: *New descent direction*

Compute $g^{n+1} = \{g_i^{n+1}\}_{i=1}^2 \in V_{01} \times V_{02}$ *such that*

$$\sum_{i=1}^{2} \int_{\gamma_i} g_i^{n+1} z_i \, d\gamma = (J'(u^{n+1}), z) \ \forall z = \{z_i\}_{i=1}^2 \in V_{01} \times V_{02}, \tag{5.11}$$

and

$$w^{n+1} = g^{n+1} + \lambda_n w^n \text{ with } \lambda_n = \frac{\sum_{i=1}^{2} \int_{\gamma_i} |g_i^{n+1}|^2 \, d\gamma}{\sum_{i=1}^{2} \int_{\gamma_i} |g_i^n|^2 \, d\gamma} \tag{5.12}$$

Do n = n+1, *go to* (5.10).

Some remarks concerning algorithm (5.7)-(5.12): It appears clearly from the above description that the two non-trivial steps in algorithm (5.7)-(5.12) are finding ρ_n and the calculation of $J'(u^{n+1})$ in (5.11). Fortunately, since the cost function is a quadratic functional of $\{v_1, v_2\}$ we can compute ρ_n exactly; similarly using a recurrent relation satisfied by the g^n, we can reduce to 2 the number of Poisson's problems like (5.4) which must be solved on each Ω_{ii} at each iteration (instead of *one* for the Schwarz's method). The calculation of ρ_n and $J'(u^n)$ is discussed with many details in [1]. We shall find also in [1] a rather complete discussion of the solution of a *finite element approximation* of (2.1) by discrete variants of algorithm (5.7)-(5.12) and by other conjugate gradient algorithms. One may find also in [1] a discussion of the *direct* solution of (2.1) (in fact its *discrete* variants) via the least squares formulation (5.2); this solution technique involves matrices which approximate some *boundary operators* and is quite close to methods based on the concept of *capacitance matrices* (see [10] for more details and further references on capacitance matrix methods). The implementation of those direct methods on an array processor system is also discussed in [1] (with an evaluation of the computing performances).

VI. NUMERICAL EXPERIMENTS

We consider in this paper one family of numerical experiments, only; more results are discussed in [1], and particularly the solution of *nonlinear problems* in Fluid Mechanics by iterative methods using the above domain decomposition methods as *preconditioners*. The test problem is the following mixed linear elliptic boundary value problem

$$(6.1) \qquad \Delta\phi = 0 \text{ in } \Omega, \quad \phi = g_o \text{ on } \Gamma_o, \quad \frac{\partial\phi}{\partial n} = g_1 \text{ on } \Gamma_1$$

where Ω is the nozzle of Fig. 5, on which are also shown Γ_0, Γ_1 and a finite element triangulation. The calculations have been done using a 3 subdomain decomposition with overlapping as shown on Fig. 6. We have compared the performances of the

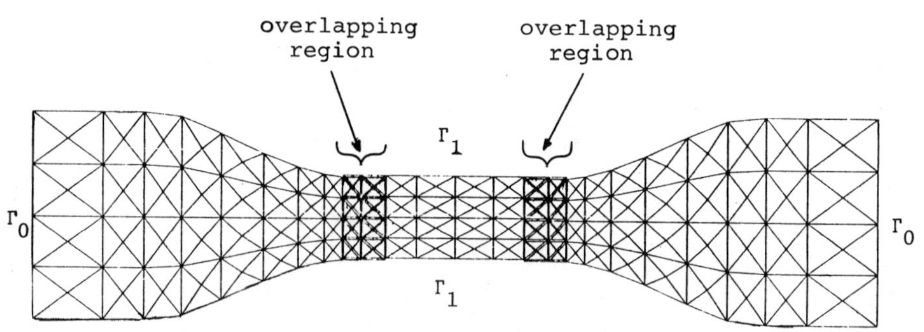

Figure 5

A convergent-divergent nozzle and its finite element triangulation

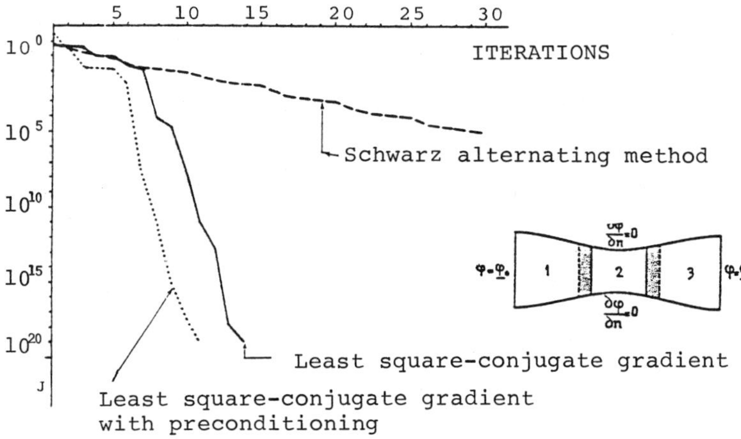

Figure 6

Schwarz method, of a discrete variant of algorithm (5.7)-(5.12) and of a preconditioned discrete variant of (5.7)-(5.12) (described in details in [1]). The results shown on Figure 6 clearly show the better performances of the method based on a least squares coupling on the overlapping region, combined with a conjugate gradient algorithm.

VII. APPLICATION TO THE NUMERICAL SIMULATION OF TRANSONIC FLOWS ON LARGE COMPUTATIONAL DOMAINS

VII.1. Synopsis

The numerical treatment of *nonlinear boundary value problems* involving *unbounded* regions of $\mathbb{R}^N$ is a nontrivial task. In the particular case of *two-dimensional transonic potential flows* around an airfoil we shall discuss a method combining a standard formulation close to the airfoil and an *exponential stretching transformation* to take the far field flow into account. The coupling between the corresponding local solutions is achieved through the method discussed in Sec. IV.2.

VII.2. Mathematical Formulation

The *potential transonic flows of compressible inviscid fluids* are governed by

(7.1) $$\nabla \cdot \rho u = 0 \quad \textit{in } \Omega ,$$

where Ω is the *flow region*, and where

(7.2) $$\rho = \rho_o \left(1 - \frac{\gamma-1}{\gamma+1} \frac{|u|^2}{c_*^2}\right)^{1/(\gamma-1)}, \quad u = \nabla\phi$$

where ϕ is the *velocity potential*; ρ is the *density* of the fluid; γ is the *ratio specific heats* ($\gamma = 1.4$ in air); c_* is the *critical velocity*. For an airfoil B (see Figure 7) the flow is assumed *uniform* at infinity (visualized by Γ_∞) and tangential at Γ_B.

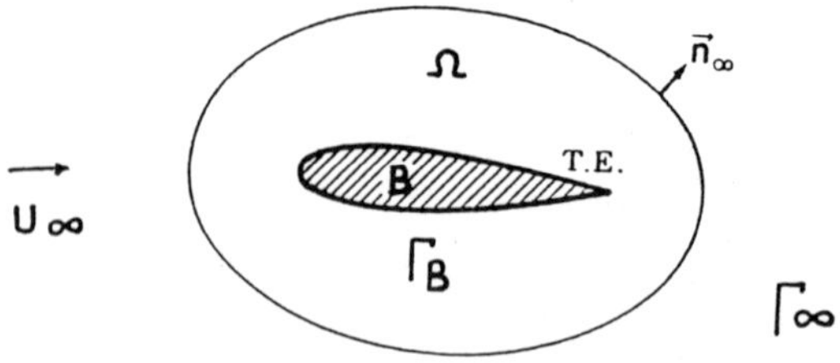

Figure 7

The corresponding boundary conditions are

(7.3) $$\frac{\partial\phi}{\partial n} = u_\infty \cdot n_\infty \text{ on } \Gamma_\infty, \quad \frac{\partial\phi}{\partial n} = 0 \text{ on } \Gamma_B .$$

Since Neumann boundary conditions are involved, the potential is determined only to within an arbitrary constant. To remedy this we can prescribe the value of ϕ at some point within $\Omega \cup \Gamma$. For example we may use

(7.4) $\phi = 0$ *at the trailing edge* T.E. *of* B.

In addition to the above boundary conditions, the *circulation* of $u = \nabla\phi$ has to be adjusted in order to have u satisfying the well-known *Kutta-Joukowsky condition*. In fact, this condition is not specific of transonic flows since it occurs also for other types of potential flows; we refer to e.g. [11], [12] for the numerical treatment of the Kutta-Joukowsky condition for 2-D and 3-D flows. Transonic flows of compressible inviscid fluids contain shocks; these shocks have to satisfy various conditions such that *Raukine-Hugoniot* and *entropy conditions* for which we refer to [13], [14] (and to [15], [16] for their numerical implementation).

VII.3. Least squares formulation and conjugate gradient solution of the continuous problem.

We do not consider the practical implementation of the Kutta-Joukowsky and entropy conditions, concentrating only on a variational formulation of (7.1)-(7.3) and of an associate nonlinear least squares formulation.

Consider the situation shown in Fig. 7. Under the assumptions that the airfoil and flow are symmetric and that the flow is subsonic at infinity; the Kutta-Joukowsky condition is then *automatically* satisfied. For practical reasons the airfoil is imbedded in a "large" bounded domain, we obtain then as flow formulation

$$(7.5) \quad \nabla\cdot\rho(\phi)\nabla\phi = 0 \textit{ in } \Omega, \quad \textit{with } \rho(\phi) = \rho_o\left(1 - \frac{\gamma-1}{\gamma+1}\frac{|\nabla\phi|^2}{c_*^2}\right)^{1/(\gamma-1)}$$

and

$$(7.6) \quad \rho\,\frac{\partial\phi}{\partial n} = 0 \textit{ on } \Gamma_B, \quad \rho\,\frac{\partial\phi}{\partial n} = \rho_\infty \mathbf{u}_\infty\cdot\mathbf{n}_\infty \textit{ on } \Gamma_\infty ;$$

on Γ $(= \Gamma_B \cup \Gamma_\infty)$ we define g by $g = 0$ *on* Γ_B, $g = \rho_\infty \mathbf{u}_\infty\cdot n$ *on* Γ_∞; we clearly have $\rho\,\frac{\partial\phi}{\partial n} = g$ on Γ with $\int_\Gamma g\,d\gamma = 0$. An equivalent formulation of (7.5), (7.6) is

$$(7.7) \quad \int_\Omega \rho(\phi)\nabla\phi\cdot\nabla v\,dx = \int_\Gamma gv\,d\gamma \quad \forall v \in H^1(\Omega),\ \phi \in W^{1,\infty}(\Omega)/\mathbb{R},$$

where, for $p \geq 1$, $W^{1,p}(\Omega) = \{v | v \in L^p(\Omega),\ \frac{\partial v}{\partial x_i} \in L^p(\Omega)\ \forall i\}$ (with $W^{1,2}(\Omega) = H^1(\Omega)$); ϕ is determined only to within an arbitrary constant. The space $W^{1,\infty}(\Omega)$ is a natural choice for ϕ since physical flows require (among other properties) a *positive density* ρ; therefore from (7.5), ϕ has to satisfy

$$|\nabla\phi| \leq \delta < \left(\frac{\gamma+1}{\gamma-1}\right)^{1/2} c_* \text{ a.e. on } \Omega.$$

Least squares formulation of (7.7): Since, for a genuine transonic flow, (7.7) is not equivalent to a standard problem of the Calculus of Variations (as it is the case for subsonic flows) we remedy this situation by introducing the following nonlinear least squares formulation:

(7.8) $\min_{\xi \in X} J(\xi)$ (X: *a set of feasible transonic flow solutions*)

and

(7.9) $J(\xi) = \frac{1}{2} \int_\Omega |\nabla y(\xi)|^2 \, dx$,

where $y(\xi)$ (= y) is the solution of the state equation

$$(7.10) \begin{cases} \int_\Omega \nabla y \cdot \nabla v \, dx = \int_\Omega \rho(\xi) \nabla\xi \cdot \nabla v \, dx - \int_\Gamma gv \, d\gamma \quad \forall v \in H^1(\Omega), \\ y \in H^1(\Omega)/\mathbb{R} \end{cases}$$

Conjugate gradient solution of the least squares problem (7.8), (7.9): We follow [1], [15] [16]; a preconditioned conjugate gradient algorithm for solving (7.8), (7.9) - with $X = H^1(\Omega)$ - is

Step 0: *Initialization*

(7.11) $\phi^o \in H^1(\Omega)$, *given*

then, compute g^o *by*

(7.12) $\int_\Omega \nabla g^o \cdot \nabla v \, dx = \langle J'(\phi^o), v\rangle \quad \forall v \in H^1(\Omega), \quad g^o \in H^1(\Omega)$,

and set

(7.13) $z^o = g^o$. □

Then for $n \geq 0$, *assuming* ϕ^n, g^n, z^n *known, compute* ϕ^{n+1}, g^{n+1}, z^{n+1} *by*

Step 1: *Descent*

(7.14) $$\phi^{n+1} = \phi^n - \lambda_n z^n \text{ with } \lambda_n = \operatorname*{Arg\,Min}_{\lambda \in \mathbb{R}} J(\phi^n - \lambda z^n) \ .$$

Step 2: *New descent direction. Define* g^{n+1} *by*

(7.15) $$\begin{cases} g^{n+1} \in H^1(\Omega), & \int_\Omega \nabla g^{n+1} \cdot \nabla v \, dx = \langle J'(\phi^{n+1}), v \rangle \\ \forall v \in H^1(\Omega) \ , \end{cases}$$

compute

(7.16) $$\gamma_n = \int_\Omega \nabla g^{n+1} \cdot \nabla (g^{n+1} - g^n) \, dx \Big/ \int_\Omega |\nabla g^n|^2 \, dx \ ,$$

(7.17) $$z^{n+1} = g^{n+1} + \gamma_n z^n \ . \quad \square$$

(7.18) *Do* $n = n+1$, *go to* (7.14).

We should prove that $J'(\xi)$ can be identified with the linear functional

(7.19) $$\eta \to \int_\Omega \rho(\xi) \nabla y \cdot \nabla \eta \, dx - 2K\alpha \int_\Omega (\rho(\xi))^{2-\gamma} \nabla \xi \cdot \nabla y \nabla \xi \cdot \nabla \eta \, dx$$

where y is defined from ξ by (7.10) (we have used the notation

$$K = \frac{1}{c_*^2} \frac{\gamma-1}{\gamma+1}, \quad \alpha = 1/(\gamma-1) \text{ and supposed that } \rho_0 = 1).$$

VII.4. Transonic flow simulations on large two-dimensional bounded computational domains by analytical transformation and decomposition methods.

Motivation: A genuine transonic flow is partly supersonic. If the supersonic region is large compared to the size of the airfoil, the computational domain itself has to be large compared to the size of the supersonic region, implying that standard finite element methods become inadequate. In the case of 2-D flows, we discuss below a method in which the far field is treated by an *exponential stretching* method which allows the use of very large computational domains.

Transformation of problem (7.5), (7.6): We consider again the situation depicted on Fig. 7. Let 0 be the origin of coordinates; it is then reasonable to take for Γ_∞ the circle $\{x | x \in \mathbb{R}^2,\ \Omega = R_\infty\}$, with $r = (x_1^2 + x_2^2)^{1/2}$. Suppose now that the disk of center 0 and radius R_o is sufficiently large to contain the airfoil B in its interior; we introduce then

$$\Omega_1 = \{x | x \in \mathbb{R}^2,\quad 0 \leq r < R_o,\quad x \notin B\},$$

$$\Omega_2 = \{x | x \in \mathbb{R}^2,\quad R_o < r < R_\infty\}$$

With $\{r,\theta\}$ a standard *polar coordinates* system associated to 0, we define the following new variables

$$(7.20)\qquad \begin{cases} \xi_1 = x_1, \quad \xi_2 = x_2 \quad \forall x = \{x_1,x_2\} \in \Omega_1 , \\ \\ \xi_1 = R_o(1 + \text{Log}\, \frac{r}{R_o}), \quad \xi_2 = \theta \quad \forall \{r,\theta\} \in \Omega_2 ; \end{cases}$$

we use the notation $\xi = \{\xi_1,\xi_2\}$. We introduce now $\tilde{\Omega}_1 = \Omega_1$ and

$$\tilde{\Omega}_2 = \{\xi | \xi \in \mathbb{R}^2,\quad R_o < \xi_1 < R_o(1 + \text{Log}\, \frac{R_\infty}{R_o}),\quad 0 < \xi_2 < 2\pi\} ;$$

we define then

$$\tilde{H}^1 = \{\tilde{v} = \{\tilde{v}_1,\tilde{v}_2\} \in H^1(\tilde{\Omega}_1) \times H^1(\tilde{\Omega}_2),\quad \tilde{v}_1(R_o \cos \xi_2,\ R_o \sin \xi_2)$$
$$= \tilde{v}_2(R_o,\xi_2)\ \forall\ \xi_2 \in]0,2\pi[\ ,\ \tilde{v}_2(\xi_1,0) = \tilde{v}_2(\xi_1,2\pi)$$
$$\forall\ \xi_1 \in]R_o,R_o(1 + \text{Log}\, \frac{R_\infty}{R_o})[\} .$$

The space $\tilde{H}^1$ is clearly *isomorphic* to $H^1(\Omega)$ via the transformation $\tilde{v} \to v$: $\tilde{H}^1 \to H^1(\Omega)$ defined by

$$(7.21)\quad v(x) = \begin{cases} \tilde{v}_1(x_1,x_2) \ \textit{if}\ x = \{x_1,x_2\} \in \Omega_1 , \\ \tilde{v}_2(R_o(1 + \text{Log}\, \frac{r}{R_o})\ \theta\ )\ \textit{if}\ x = \{r,\theta\} \in \Omega_2 . \end{cases}$$

Using the above properties, the continuity equation (7.10) is now formulated by

$$(7.22)\quad \begin{cases} \textit{Find } \tilde{\phi} \in \tilde{H}^1 \textit{ such that} \\ \int_{\tilde{\Omega}} \rho(\tilde{\phi}_1) \underset{\sim}{\nabla}\tilde{\phi}_1 \cdot \underset{\sim}{\nabla}\tilde{v}_1 d\xi + \int_{\tilde{\Omega}} \tilde{\rho}(\tilde{\phi}_2)(R_o \frac{\partial\tilde{\phi}_2}{\partial\xi_1}\frac{\partial\tilde{v}_2}{\partial\xi_1} \\ \quad + \frac{1}{R_o}\frac{\partial\tilde{\phi}_2}{\partial\xi_2}\frac{\partial\tilde{v}_2}{\partial\xi_2})d\xi \\ = R_\infty \rho_\infty \int_0^{2\pi} (u_{1\infty}\cos\xi_2 + u_{2\infty}\sin\xi_2)\tilde{v}_2(R_o(1+\mathrm{Log}\frac{R_\infty}{R_o}), \\ \quad \xi_2)d\xi_2 \quad \forall \tilde{v} \in \tilde{H}^1 , \end{cases}$$

with, in (7.22), $\{u_{1\infty}, u_{2\infty}\} = \underset{\sim}{u}_\infty$, and

$$(7.23)\quad \tilde{\rho}(\tilde{\phi}_2) = \rho_o[1 - \frac{\gamma-1}{\gamma+1}\frac{1}{c_*^2}(|\frac{\partial\tilde{\phi}}{\partial\xi_1}|^2 + R_o^{-2}|\frac{\partial\tilde{\phi}}{\partial\xi_2}|^2) \times e^{(2[R_o-\xi_1))/R_o}]^{1/(\gamma-1)} .$$

The discretization of (7.22), (7.23) can be done as follows:

In $\tilde{\Omega}_1$ (which is part of the *physical* space) use a standard finite element approximation taking into account the possible complexities of the geometry; on the other hand since $\tilde{\Omega}_2$ is a *rectangle* one may use on it a *finite difference* discretization which can be obtained in fact via a finite element approximation on a *uniform* triangular or rectangular grid.

From the above decomposition of the computational domain (in $\tilde{\Omega}_1$ and $\tilde{\Omega}_2$) it is natural to solve the approximate problems by iterative methods taking this decomposition into account, and also the fact that finite differences can be used in $\tilde{\Omega}_2$ (since finite differences allow special solvers on $\tilde{\Omega}_2$). Such a method is discussed below.

<u>On the solution of (7.22), (7.23) by a decomposition method</u>

Solving (7.22), (7.23) by a nonlinear least squares conjugate gradient method similar to the one discussed in Sec. VII.3 is quite easy in principle. The least squares formulation would be

$$(7.24)\qquad \operatorname*{Min}_{\tilde{\eta}\in\tilde{H}^1} J(\tilde{\eta}) \ ,$$

with

$$(7.25)\qquad J(\tilde{\eta}) = \frac{1}{2}\int_{\tilde{\Omega}_1} |\nabla \tilde{y}_1|^2 \, d\xi + \frac{1}{2}\int_{\tilde{\Omega}_2} \Big(R_o\Big|\frac{\partial \tilde{y}_2}{\partial \xi_1}\Big|^2 + R_o^{-1}\Big|\frac{\partial \tilde{y}_2}{\partial \xi_2}\Big|^2\Big) \, d\xi$$

where, in (7.25), y is obtained from $\tilde{\eta}$ via the solution of the state equation

$$(7.26)\qquad \begin{cases} \textit{Find } \tilde{y}\in\tilde{H}^1 \textit{ such that} \\ \displaystyle\int_{\tilde{\Omega}_1} \nabla \tilde{y}_1\cdot\nabla \tilde{v}_1 d\xi + \int_{\tilde{\Omega}_2} \Big(R_o \frac{\partial \tilde{y}_2}{\partial \xi_1}\frac{\partial \tilde{v}_2}{\partial \xi_1} + R_o^{-1}\frac{\partial \tilde{y}_2}{\partial \xi_2}\frac{\partial \tilde{v}_2}{\partial \xi_2}\Big) d\xi \\ \displaystyle = \int_{\tilde{\Omega}_1} \rho(\tilde{\eta}_1)\nabla \tilde{\eta}_1\cdot\nabla \tilde{v}_1 d\xi + \int_{\tilde{\Omega}_2} \tilde{\rho}(\tilde{\eta}_2)\Big(R_o \frac{\partial \tilde{y}_2}{\partial \xi_1}\frac{\partial \tilde{v}_2}{\partial \xi_1} \\ \displaystyle + R_o^{-1}\frac{\partial \tilde{\eta}_2}{\partial \xi_2}\frac{\partial \tilde{v}_2}{\partial \xi_2}\Big) d\xi - R_\infty \rho_\infty \int_0^{2\pi} (u_{1\infty}\cos\xi_2 \\ \displaystyle + u_{2\infty}\sin\xi_2)\tilde{v}_2\Big(R_o\Big(1+\operatorname{Log}\frac{R_\infty}{R_o}\Big),\xi_2\Big) d\xi_2 \quad \forall \tilde{v}\in\tilde{H}^1 , \end{cases}$$

where $\tilde{\rho}$ is still defined by (7.23). Solving (7.24)-(7.26) by a conjugate gradient algorithm using as preconditioning operator the elliptic operator associated to the bilinear form occurring in the left hand side of (7.26) is quite easy, and the crucial point is then to have efficient solvers for problems like

$$(7.27)\qquad \begin{cases} \textit{Find } \tilde{u}\in\tilde{H}^1 \textit{ such that} \\ \displaystyle\int_{\tilde{\Omega}_1} \nabla \tilde{u}_1\cdot\nabla \tilde{v}_1 \, d\xi + \int_{\tilde{\Omega}_2} \Big(R_o \frac{\partial \tilde{u}_2}{\partial \xi_1}\frac{\partial \tilde{v}_2}{\partial \xi_1} + R_o^{-1}\frac{\partial \tilde{u}_2}{\partial \xi_2}\frac{\partial \tilde{v}_2}{\partial \xi_2}\Big) d\xi \end{cases}$$

$$= L(\tilde{v}) \quad \forall \tilde{v}\in\tilde{H}^1$$

where L is linear and continuous over $\tilde{H}^1$; we can always suppose that $L(\tilde{v}) = L_1(\tilde{v}_1) + L_2(\tilde{v}_2)$ where L_i is for i = 1,2 linear and continuous over $H^1(\tilde{\Omega}_i)$. Solving linear problems like (7.27) (in fact discrete variants of (7.27)) by methods taking into account their special structure can be done by either direct or iterative techniques. Among these methods we have chosen to describe one which is a generalization of the conjugate gradient algorithm (4.41)-(4.48).

We denote by γ_{12} the circle of radius R_o and centre 0, and by Λ the space defined by

$$(7.28)\qquad \Lambda = \{\mu|\mu \in L^2(\gamma_{12}),\quad \mu = \tilde{\mu}_1|_{\gamma_{12}}\ \textit{where}\ \tilde{\mu}_1 \in H^1(\tilde{\Omega}_1)\}\ ;$$

we observe that if $\mu \in \Lambda$, then μ is a *periodic* function of the *polar angle* θ. We define also the following spaces

$$(7.29)\qquad \tilde{V}_{01} = \{\tilde{v}_1|\tilde{v}_1 \in H^1(\tilde{\Omega}_1),\quad \tilde{v}_1 = 0\ \textit{on}\ \gamma_{12}\}\ ,$$

$$(7.30)\qquad \begin{cases} \tilde{V}_2 = \{\tilde{v}_2|\tilde{v}_2 \in H^1(\tilde{\Omega}_2),\quad \tilde{v}_2(\xi_1,0) = \tilde{v}_2(\xi_1,2\pi) \\ \forall \xi_1 \in]R_o,\quad R_o(1 + \operatorname{Log}\frac{R_\infty}{R_o})[\}\ , \end{cases}$$

$$(7.31)\qquad \tilde{V}_{02} = \{\tilde{v}_2|\tilde{v}_2 \in \tilde{V}_2,\quad \tilde{v}_2(R_o,\xi_2) = 0\quad \forall \xi_2 \in]0,2\pi[\}\ .$$

It is also convenient to define the bilinear forms

$$a_i:\ H^1(\tilde{\Omega}_i) \times H^1(\tilde{\Omega}_i) \to \mathbb{R},\ i = 1,2 \text{ by}$$

$$(7.32)\qquad a_1(v_1,w_1) = \int_{\tilde{\Omega}_1} \tilde{\nabla} v_1 \cdot \tilde{\nabla} w_1\, d\xi\quad \forall v_1,w_1 \in H^1(\tilde{\Omega}_1)\ ,$$

$$(7.33)\qquad a_2(v_2,w_2) = \int_{\tilde{\Omega}_2} (R_o \frac{\partial v_2}{\partial \xi_1}\frac{\partial w_2}{\partial \xi_1} + R_o^{-1}\frac{\partial v_2}{\partial \xi_2}\frac{\partial w_2}{\partial \xi_2})d\xi$$

$$\forall v_2,w_2 \in H^1(\tilde{\Omega}_2)$$

respectively. The solution of the linear variational problem (7.27) by a domain decomposition method can be obtained through the following conjugate gradient algorithm which is also a generalization of algorithm (4.41)-(4.48).

Step 0: *Initialization*

(7.34) $\lambda^o \in \Lambda$, *arbitrarily given,*

solve then the linear variational problems below

(7.35)
$$\begin{cases} \text{Find } y_1^o \in H^1(\tilde{\Omega}_1),\ y_1^o = \lambda^o \text{ on } \gamma_{12}, \text{ such that} \\ a_1(y_1^o, z_1) = L_1(z_1) \quad \forall z_1 \in \tilde{V}_{01}, \end{cases}$$

(7.36)
$$\begin{cases} \text{Find } y_2^o \in \tilde{V}_2,\ y_2^o(R_o, \xi_2) = \lambda^o(R_o \cos \xi_2, R_o \sin \xi_2) \\ \qquad \forall \xi_2 \in \left]0, 2\pi\right[, \\ a_2(y_2^o, z_2) = L_2(z_2) \quad \forall z_2 \in \tilde{V}_{02}, \end{cases}$$

and then

(7.37)
$$\begin{cases} \text{Find } g^o \in \Lambda \text{ such that} \\ \sum_{i=1}^{2} a_i(\tilde{g}_i^o, \tilde{\mu}_i) = \sum_{i=1}^{2} [a_i(y_i^o, \tilde{\mu}_i) - L_i(\tilde{\mu}_i)] \quad \forall \mu \in \Lambda, \end{cases}$$

and set

(7.38) $w^o = g^o$. □

For $n \geq 0$, *suppose that* λ^n, g^n, w^n *are known and compute* λ^{n+1}, g^{n+1}, w^{n+1} *as follows*

Step 1: *Descent*

Define $\{\chi_i^n\}_{i=1}^2$ *by*

$$(7.39)\quad \begin{cases} \chi_1^n \in H^1(\tilde{\Omega}_1), \quad \chi_1^n = w^n \ on \ \gamma_{12}, \\ a_1(\chi_1^n, z_1) = 0 \quad \forall z_1 \in \tilde{V}_{01}, \end{cases}$$

$$(7.40)\quad \begin{cases} \chi_2^n \in \tilde{V}_2, \quad \chi_2^n(R_o, \xi_2) = w^n(R_o \cos \xi_2, R_o \sin \xi_2) \\ \qquad \forall \xi_2 \in]0, 2\pi[, \\ a_2(\chi_2^n, z_2) = 0 \quad \forall z_2 \in \tilde{V}_{02}, \end{cases}$$

then ρ_n *and* λ^{n+1} *by*

$$(7.41)\quad \lambda^{n+1} = \lambda^n - \rho_n w^n \ with \ \rho_n = \sum_{i=1}^{2} a_i(\tilde{g}_i^n, \tilde{g}_i^n) / \sum_{i=1}^{2} a_i(\chi_i^n, \tilde{w}_i^n)$$

respectively.

Step 2: *Calculation of the new descent direction*

Solve

$$(7.42)\quad \begin{cases} Find \ g^{n+1} \in \Lambda \ such \ that \ \forall \mu \in \Lambda \\ \sum_{i=1}^{2} a_i(\tilde{g}_i^{n+1}, \tilde{\mu}_i) = \sum_{i=1}^{2} a_i(\tilde{g}_i^n, \tilde{\mu}_i) - \rho_n \sum_{i=1}^{2} a_i(\chi_i^n, \tilde{\mu}_i), \end{cases}$$

compute

$$(7.43)\quad \gamma_n = \frac{\sum_{i=1}^{2} a_i(\tilde{g}_i^{n+1}, \tilde{g}_i^{n+1})}{\sum_{i=1}^{2} a_i(\tilde{g}_i^n, \tilde{g}_i^n)},$$

set finally

$$(7.44)\quad w^{n+1} = g^{n+1} + \gamma_n w^n . \quad \square$$

Do n = n+1 *and go to* (7.39).

The above algorithm involves functions extending on $\tilde{\Omega}_1$ and $\tilde{\Omega}_2$ the trace functions μ, g^n, w^n; more generally to define $\tilde{q}_1$,

$\tilde{q}_2$ from $q \in \Lambda$, we proceed as follows:

We introduce first $\tilde{\Lambda}_1$, $\tilde{\Lambda}_2$ such as

$$\tilde{\Lambda}_1 \subset H^1(\tilde{\Omega}_1), \quad H^1(\tilde{\Omega}_1) = \tilde{V}_{01} \oplus \tilde{\Lambda}_1 , \tag{7.45}$$

$$\tilde{\Lambda}_2 \subset \tilde{V}_2, \quad \tilde{V}_2 = \tilde{V}_{02} \oplus \tilde{\Lambda}_2 ; \tag{7.46}$$

there is an infinity of such $\tilde{\Lambda}_1$, $\tilde{\Lambda}_2$ satisfying (7.45), (7.46); in practice the discrete variants of the space $\tilde{\Lambda}_1$ (resp. $\tilde{\Lambda}_2$) will consist of functions vanishing outside a narrow neighborhood of γ_{12} in $\tilde{\Omega}_1$ (resp. of $\{\xi = \{\xi_1, \xi_2\} \mid \xi_1 = R_o, \quad \xi_2 \in]0, 2\pi[$ in $\tilde{\Omega}_2$); closely related spaces are described in [1].

Once $\tilde{\Lambda}_1, \tilde{\Lambda}_2$ have been defined, we define in turn $\tilde{q}_1, \tilde{q}_2$ as the unique elements of $\tilde{\Lambda}_1, \tilde{\Lambda}_2$ such as

$$\tilde{q}_1|_{\gamma_{12}} = q , \tag{7.47}$$

$$\tilde{q}_2(R_o, \xi_2) = q(R_o \cos \xi_2, R_o \sin \xi_2) \quad \forall \xi_2 \in]0, 2\pi[. \tag{7.48}$$

Numerical results obtained using the above decomposition method will be presented elsewhere.

VIII. FURTHER COMMENTS. CONCLUSION

The methods described in this paper are, obviously, closely related to the *Schwarz alternating method*; they are also closely related to those methods using the concepts of *low rank correction*, *capacitance matrix*, etc., and also to boundary integral methods (see [10], [17]; actually Reference [10] is also concerned with a family of domain decomposition methods).

The methods discussed in this paper have been restricted to the solution of *Poisson problems*; in fact they have been successfully used by the present authors for solving the *Stokes problem* for incompressible viscous fluids, using either finite element or spectral methods of approximation.

As a final comment we would like to point out that the comparison done in [10] between the methods discussed there and the methods in this paper is not very relevant since, as mentioned in our introduction, the least squares-overlapping approach was designed primarily as a technique to match, possibly different approximations, different mathematical modellings, etc., and not only to be an elliptic solver.

REFERENCES

[1] GLOWINSKI, R., PERIAUX, J., DINH Q.V., Domain decomposition methods for nonlinear problems in fluid dynamics, *Comp. Meth. Appl. Mech. Eng.* (to appear).

[2] NECAS, J., *Les Méthodes Directes en Théorie des Equations Elliptiques*, Masson, Paris, 1967.

[3] ADAMS, R.A., *Sobolev Spaces*, Academic Press, New York, 1975.

[4] ODEN, J.T., REDDY, J.N., *An Introduction to the Mathematical Theory of Finite Elements*, John Wiley and Sons, New York, 1976.

[5] LIONS, P.L., Interprétation stochastique de la méthode alternée de Schwarz, *Comptes Rendus Acad. Sc., Paris, T. 286-A*, (1978), pp. 325-328.

[6] GLOWINSKI, R., LIONS, J.L., TREMOLIERES, R., *Numerical Analysis of Variational Inequalities*, North Holland, Amsterdam, 1981.

[7] BENSOUSSAN, A., LIONS, J.L., TEMAM, R., Sur les méthodes de décomposition, de décentralisation, de coordination et applications, in *Sur les Méthodes Numériques en Sciences Physiques et Economiques*, J.L. Lions, G.I. Marchouk, eds., Dunod-Bordas, Paris, 1974, pp. 133-257.

[8] LEMONNIER, P., Résolution numérique d'équations aux derivées partielles, in *Sur les Methodes Numeriques en Sciences Physiques et Economiques*, J.L. Lions, G.I. Marchouk, eds., Dunod-Bordas, Paris, 1974, pp. 133-257.

[9] LIONS, J.L., *Contrôle Optimal des Systèmes Gouvernes par des Equations aux Dérivées Partielles*, Dunod-Gauthier Villars, Paris, 1968.

[10] BJORSTAD, P.E., WIDLUND, O.B., Solving elliptic problems on regions partitioned into substructures (these proceedings).

[11] BRISTEAU, M.O., GLOWINSKI, R., PERIAUX, J., PERRIER, P., PIRONNEAU, O., POIRIER, G., Application of Optimal Control and Finite Element Methods to the Calculation of Transonic Flows and Incompressible Viscous Flows, in *Numerical Methods in Applied Fluid Dynamics*, B. Hunt ed., Academic Press, London, 1980, pp. 203-312.

[12] BRISTEAU, M.O., GLOWINSKI, R., PERIAUX, J., PERRIER, P., PIRONNEAU, O., On the numerical solution of nonlinear problems in fluid dynamics by least squares and finite element methods (I) Least squares formulations and conjugate gradient solution of the continuous problems, *Computer Methods Appl. Mech. Eng.* 17/18, (1979), pp. 619-657.

[13] LAUDAU, L., LIFCHITZ, E., *Mécanique des Fluides*, Mir, Moscow, 1953.

[14] COURANT, R., FRIEDRICHS, K.O., *Supersonic flows and shock waves*, Springer-Verlag, New York, 1976.

[15] BRISTEAU, M.O., GLOWINSKI, R., PERIAUX, J., PERRIER, P., PIRONNEAU, O., POIRIER, G., Transonic flow simulations by finite elements and least squares methods, in *Finite Elements in Fluids*, R.H. Gallagher, D.H. Norrie, J.T. Oden, O.C. Zienkiewicz, eds., J. Wiley and Sons, Chichester, 1982, pp. 453-482.

[16] GLOWINSKI, R., *Numerical Methods for Nonlinear Variational Problems*, Springer-Verlag, New York, 1984.

[17] CONCUS, P., GOLUB, G.H., O'LEARY, D.P., A generalized conjugate method for the numerical solution of elliptic partial differential equations, in *Sparse Matrix Computations*, J.R. Bunch, D.J. Rose eds., Academic Press, New York, 1976, pp. 309-322.

COMPUTING ADDED MASS COEFFICIENTS

G. Birkhoff*, John F. Brophy**, R.E. Lynch**

I. PRELIMINARY REMARKS

The ultimate objective of packages of 'elliptic problem solvers' such as ELLPACK and the library of the Club Modulef is, of course, to obtain the accurate solution of _general_ elliptic problems by pushing the right button! So far, this goal has only come close to being reached in reasonably general domains for _linear_, _second-order_ differential equations (DE's) in _two_ space dimensions. Moreover even for the simplest elliptic DE, the Laplace equation, Dirichlet-type boundary conditions are often the only ones that are handled routinely, in theory and practice.[1]

This DE lies at the heart of the theory of _potential flows_, including especially that around a solid body moving through an 'ideal' (incompressible, inviscid) fluid. The theory of such flows, as developed by Kelvin and Kirchhoff, is by far the most satisfactory part of fluid dynamics from a mathematical standpoint. The resulting _forces_ and _moments_ on a body of general shape are all determined by a few added mass coefficients m_{jk}. In turn, these are Dirichlet inner products. Writing $\phi = \sum_{j=1}^{6} \dot{q}_j \Phi_j(\underline{q})$, where the q_j are the coordinates of the six degrees of freedom of motion, we have

*Harvard University and Naval Postgraduate School

**Mathematics Department, Purdue University

[1]For technical reasons, similar to those which led Gilbarg and Trudinger to treat only Dirichlet boundary conditions in their book "Elliptic Partial Differential Equations of Second Order", Springer, 1977.

ISBN 0-12-100560-7

$$(1.1) \qquad m_{jk} = \rho \iint_{\Omega} (\nabla \Phi_j \cdot \nabla \Phi_k) dR \,,$$

where the Φ_j are solutions of exterior Neumann problems of the form

$$(1.2) \qquad \nabla^2 \Phi_j = 0 \quad \text{in} \quad \Omega, \quad \partial \Phi_j / \partial n = f_j(\underline{x}) \quad \text{on} \quad \Gamma = \partial\Omega$$

Moreover all instantaneous velocity fields $\underline{u}(\underline{x};t) = \nabla\phi(x;t)$ are also constructable from the Φ_j by linear combination ($\phi = \sum c_k \Phi_k$) and Euclidean transformations $\underline{x} \to A\underline{x} + \underline{b}$, with A orthogonal.

Kelvin and Kirchhoff developed their general theories of added mass about 50 years after Green determined analytically (around 1835) the first potential flows around ellipsoids, and for another 50 years great analytical ingenuity was applied to determine the added masses of bodies of various special shapes.

In two or three dimensions one can use the method of inversion to replace exterior (Neumann or Dirichlet) problems by interior ones.

Expecially tractable is the case of 'two-dimensional' flows around cylinders of general cross-section, moving broadside, because in that case one can not only apply conformal mapping techniques, but one can convert the Neumann boundary conditions (1.2) to Dirichlet boundary conditions of the form

$$(1.3) \qquad \partial \Psi_j / \partial n = g_j(x) \quad \text{on} \quad \Gamma = \partial\Omega \,.$$

In (1.3), the stream function $\Psi_j(x,y)$ is the harmonic conjugate of the velocity potential $\Phi_j(x,y)$, and the sum $W_j = \Phi_j + i\Psi_j$ is an analytic complex function of $z = x + iy$.

II. HIGHER ANALYSIS

The pre-1955 literature on added mass was carefully reviewed by one of us in [1, Chap. VI]. The most thorough studies of added mass from the standpoint of classical analysis had been made a few years earlier by Schiffer and Szegö [7] and Pólya and Szegö [6]. Numerical values of added mass mass coefficients of various bodies (mostly axially symmetric cylinders and solids in translation) were specified (some very accurately) in these publications.

A. Analytical Methods

As libraries of transportable subroutines improve, it is natural to ask how much the special formulas and techniques of classical Analysis will continue to help applied mathematicians. One aim of this paper is to shed light on this question, through a case study of the added mass problem, about which so much is known analytically.

Three years ago, two of us reported in [2] on some numerical experiments on the added mass of various cylinders, for which Dirichlet boundary conditions can be used, as in (1.3). These experiments combined ELLPACK with some post-processors, designed to evaluate accurately Dirichlet inner products (1.1) and Dirichlet integrals (the case $j = k$ of (1.1)).[2] The latter are minimized by the exact solution, with respect to all other $\Psi(x,y)$ satisfying (1.3) that are 'regular at infinity'--i.e., in the Sobolev space $W_1^2(\Omega)$.

Since then, one of us has been writing a Ph.D. Thesis [4], applying formulas of Higher Analysis to achieve more accuracy with less effort. One such formula applicable to cylinders, is the following. Let[3]

$$f(t) = c_{-1}t^{-1} + c_0 + c_1 t + c_2 t^2 + \dots, \quad c_k = a_k + ib_k, \tag{2.1}$$

map the interior $|t| \leqq 1$ of the unit disk in the t-plane onto the exterior of a simple closed curve Γ in the z-plane. Then

$$m_{11} = 2\pi[a_{-1}^2 + b_{-1}^2 - a_{-1}a_1 + b_{-1}b_1] - \text{Area} \tag{2.2}$$

From (2.2), an elementary analysis of the effects of rotations $c_k \to e^{i\gamma}c_k$ of the dependent variable z gives m_{12} and m_{22}, and shows that every cylinder has the same added mass coefficients for translation as an equivalent elliptic cylinder.--The effect of expansion $c_k \to \beta c_k$ is to multiply m_{11} by β^2, as physical intuition and dimensional analysis

---

[2] With respect to suitable 'principal axes' for translation, $j \neq k$ implies $m_{jk} = 0$, and so only Dirichlet integrals need to be calculated if these are known. Any axis of symmetry is a principal axis.

[3] See [6, p. 244; 1, p. 160]; their t is the reciprocal of ours.

suggest, while translation $c_0 \to c_0 + \text{const.}$, and rotations $t \to e^{i\theta}t$ of the parameter t are without effect.

B. *Rotational Symmetry*

For cylinders having k-fold rotational symmetry $(k > 2)$, the expansion (2.1) is 'lacunary' in the elementary sense that many terms are missing:

$$z = f(t) = c_{-1}t^{-1} + c_{k-1}t^{k-1} + c_{2k-1}t^{2k-1} + \dots \quad (2.3)$$

It follows that $a_1 = b_1 = 0$, and so (2.2) simplifies to $m_{11} = 2\pi(a_{-1}^2 + b_{-1}^2)$ - Area. That is, the added mass matrix "tensor" for translation is a scalar matrix $m_{11}I$, having the same added mass coefficients $m_{11} = m_{22}$, $m_{12} = 0$, as some circular cylinder.

III. APPLYING ELLPACK 1978 TO CYLINDERS

Ideally, a package of subroutines for solving elliptic problems would enable the user to compute all added mass coefficients for a solid (or cylinder) of general shape, by simply specifying its shape (and size).[4] With this in mind, two of us used ELLPACK 1978, three years ago, to compute the added mass of cylinders of various shapes. We will next summarize and up-date the results, which were reported in [2].

By the preliminary steps described at the end of §1 (using the stream function as dependent variable and inversion), the computation of any velocity potential can be reduced to the solution of three Dirichlet problems for functions $\Psi_j(x,y)$ harmonic in some bounded domain Ω^*, and assuming known values on its boundary Γ^*:

$$\Psi_1(x,y) = \frac{x}{x^2+y^2}, \qquad \Psi_2(x,y) = \frac{y}{x^2+y^2}, \quad (3.1)$$

for translation, and

$$\Psi_3(x,y) = \frac{1}{x^2+y^2} \qquad \text{for rotation.} \quad (3.2)$$

[4]For bodies of the same shape, the added mass coefficients are proportional to the mass M (ρV or ρA) of fluid displaced; hence it is convenient to record the ratios m_{jk}/M.

From these, since two-dimensional Dirichlet inner products are invariant under conformal transformation, we can compute added mass coefficients, computed from the formula

$$m_{jk} = \iint_{\Omega^*} \nabla\Psi_j \cdot \nabla\Psi_k \, dx\, dy \,, \tag{3.3}$$

by numerical quadrature.

When $j \neq k$, the exact value of m_{jk} can only be inferred approximately from the internal consistency of the results. But when $j = k$, comparisons can also be based on the Thomson-Dirichlet Principle, which asserts that each $m_{jj} = D\langle\Psi_j,\Psi_j\rangle$ (with $k = j$) minimizes (3.3) with respect to the Dirichlet integral $D\langle h_j,h_j\rangle$, in the class of all continuous piecewise differentiable functions h_j in Ω^* assuming the same boundary values.

The accuracy of the m_{jk} computed in this way depends, obviously, on the accuracies of the algorithms used:

(a) to approximate the Laplace equation, in Ω^* and near Γ^*,

(b) to interpolate between the values of Ψ at mesh-points, and

(c) to calculate (3.3) by numerical quadrature.

As regards (a), we used the standard 5-point difference formula of [8, (6.38)] at interior points, for simplicity. We then turned our attention to (b), trying out each of three interpolation schemes, described in [2], as post-processes to ELLPACK. These all reduced the errors to about 40% of those associated with piecewise bilinear interpolation. Over interior mesh squares, the partial integrals of (3.3) over rectangles can be evaluated in closed form for each interpolation scheme. The accurate evaluation of Dirichlet integrals over curvilinear boundary regions is more troublesome.

IV. NETWORK ANALOGY

Also annoying was the fact that, at mesh points adjacent to curvilinear boundaries, the coefficient matrix given by [8, (6.38)] is not symmetric. IF we were to do these calculations again, we would utilize the following network analogy.[5]

[5] Contrast with the approximation recommended on p. 179 of C.-X. Chow, "Computational Fluid Dynamics", Wiley, 1979.

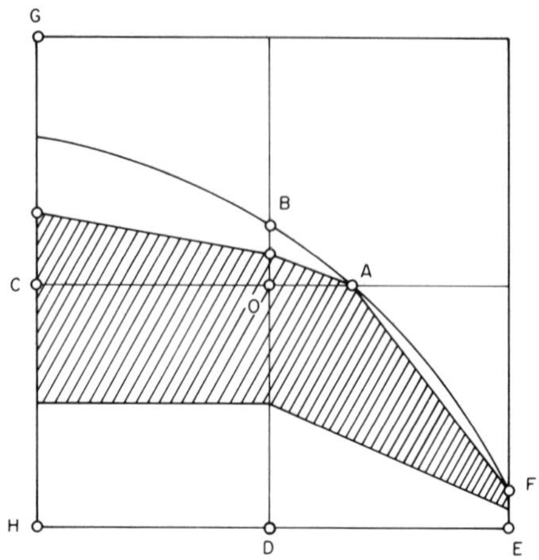

Resolve the current into its horizontal and vertical components, and imagine these as flowing along horizontal and vertical mesh segments, respectively, considered as conducting wires. Assume the conductance κ of each horizontal mesh segment to be proportional to its mean vertical width (cross section), and inversely proportional to its (horizontal) length. Assign an analogous conductance to each vertical segment. Thus, in the domain drawn above, the conductances of the "wires" $\overline{OC}$ and $\overline{OA}$ are assumed to be

$$(\overline{OB} + \overline{CG} + 2\overline{OD})/4\overline{OC} \quad \text{and} \quad (\overline{OB} + 2\overline{OD})/4\overline{OA} ,$$

respectively. A variational principle due to J.C. Maxwell states that, in a D.C. network, the current minimizes the rate of heating (conversion of electrical to thermal energy):

$$\text{(4.1)} \qquad \sum \kappa_\ell \nabla u_\ell^2 \quad \text{is a minimum.}$$

This is, for the configuration shown, equivalent to the following weighted mean value formula

$$\text{(4.2)} \qquad \kappa_a \Psi_a + \kappa_b \Psi_b + \kappa_c \Psi_c + \kappa_d \Psi_d = (\kappa_a + \kappa_b + \kappa_c + \kappa_d) \Psi_0 .$$

Actually, we would replace the usual 5-point by the 9-point formula having $O(h^6)$ accuracy.[6] This corresponds to a slightly more complicated network of wires, which would also have to be modified near the boundary.--Alternatively, we could use bicubic Hermite or other finite element approximations in the interior, suitably modified in the curvilinear boundary quadrilaterals.

V. DIPOLE MOMENT METHOD

As part of his Thesis [4], one of us has computed the translational added masses of a square and a cube using formulas from classical analysis. Thus, after inversion, the dipole moment of the potential flow around a cylinder turns out to be simply the derivative $\phi_x(0)$. Using this fact, and the greater capabilities of ELLPACK 83, it is possible to compute the translational added mass of a square by the following 25 line ELLPACK Program:

```
OPTION. CLOCKWISE $ MAX X-POINTS = 25 $ MAX Y-POINTS = 25
EQUATION.      UXX + UYY = 0.
BOUNDARY.      U = X/(X**2+Y**2)
 ON X =  1.+SIN(T), Y =      COS(T) FOR T =  0.0     TO      PI
 ON X =     SIN(T), Y = -1.+COS(T) FOR T =  0.5*PI TO 1.5*PI
 ON X = -1.+SIN(T), Y =      COS(T) FOR T = -    PI TO 0.0
 ON X =     SIN(T), Y =  1.+COS(T) FOR T = -0.5*PI TO 0.5*PI
FORTRAN.
      PRINT 10
   10 FORMAT(4X,
     A   '1/H  ADDED MASS   PERCENT ERROR   ERROR/H**2')
      DO 30 NGRID=9,25,4
GRID.              NGRID X-POINTS, -2.0 TO 2.0
                   NGRID Y-POINTS, -2.0 TO 2.0
DISCRETIZATION. COLLOCATION
SOLUTION.          BAND GE
FORTRAN.
         DIPOLE = UX(0.,0.)
         ADMS   = 2.*PI*DIPOLE-1.
         ADMSEX = 1.1884396152
         ADMSER = 100.*ABS(ADMS-ADMSEX)/ADMSEX
         PRINT 20, NGRID-1, ADMS, ADMSER, ADMSER*(NGRID-1)**2
   20    FORMAT(I6,F13.7,F13.5,F14.2)
   30    CONTINUE
END.
```

The output from this program is displayed below:

1/H	ADDED MASS	PERCENT ERROR	ERROR/H**2
8	1.2394228	4.28993	274.55
12	1.2137711	2.13149	306.94
16	1.2015295	1.10144	281.97
20	1.1970284	.72269	289.08
24	1.1945212	.51173	294.76

---

[6] L.V. Kantorovich and V.I. Krylov, "Approximate Methods of Higher Analysis", Interscience, 1964, p. 185.

It is easy to explain this program. Line 1 states that at most a 25×25 grid will be used. Lines 4-7 define the domain Ω^*: the 'four-leaf clover' obtained by inverting the exterior of a square. Lines 9-11 specify the output; lines 12-14 describe the five grids used: $N = 9$ to 25, step 4. (The discretization is by collocation [3, Chap. 6, §6], and the resulting linear algebraic equations are solved by (Gaussian) band elimination.) Lines 19-24 instruct the computer to compute and print the dipole moment, the approximate and exact added mass, and their difference (the error)."

VI. ADDED MASS OF SOLIDS

The added mass coefficient for translation of an axially symmetric solid along its axis can be computed using ELLPACK, by methods similar to those used for cylinders. Again, using the stream function $\Psi(x)$ as dependent variable, we can reduce the added mass problem from a Neumann to a Dirichlet problem. One can also reduce the exterior to an interior Dirichlet problem by inversion. The singularity in the differential equation along the axis of symmetry causes no difficulty if one approximates using reflection symmetry.

More challenging, naturally, is the problem of computing the added mass coefficients of a general solid in $\mathbb{R}^3$. Those for ellipsoids are given by analytical formulas due to Green, Stokes, and others. Standard difference and finite element methods are expensive; moreover ELLPACK's three-dimensional capabilities are limited to box-shaped domains. Therefore, more innovative methods seem called for. In [4], the added mass of a cube was computed, using integral equation methods and a systematic analysis of translational added (alias "virtual") mass and polarization by Schiffer and Szegö [7], summarized in a later monograph by Pólya and Szegö [6]. This contains many new inequalities as well as some new exact analytical solutions.

Whereas the velocity potential ϕ induced by translation equals the change in the potential $\underline{h} \cdot \underline{x}$ of a uniform field $\underline{h}$ 'induced' by postulating the Neumann condition $\partial\phi/\partial n = 0$ on the boundary Γ of a solid, the 'polarization' potential is the harmonic function ψ expressing the change in $\underline{h} \cdot \underline{x}$ 'induced' by a conductor, whose surface Γ must be at a constant potential, i.e., must satisfy $\psi \equiv c$ for some constant c. For a body that is symmetric about $\underline{x} = \underline{0}$, like $\underline{h}\cdot\underline{x}$, we will have $c = 0$. But in general, the value of c must be determined by the condition $\iint \frac{\partial\psi}{\partial n}\, dS = 0$ of zero net charge on Γ.

The dipole moments D and E of the induced charges, giving the asymptotic character of the 'far fields' associated with these potentials, are positive definite quadratic forms related to the added mass tensor $W = ||m_{ij}||$ and the polarization tensor $p = ||p_{ij}||$ by the identities (cf. [6, p. 241])

$$W = 4\pi D - V, \quad P \equiv 4\pi E - V , \tag{6.1}$$

where V is the geometric volume.

It is observed in [6, p. 24] that "the quantities D and E show a more regular behavior than the forms W and P." In particular (as was first observed by G.I. Taylor) they depend monotonically on Ω: if $\Omega \subset \Omega_1$, then $D(\Omega) \leqq D(\Omega_1)$ and $E(\Omega) \leqq E(\Omega_1)$.

Added mass of cube. Using (6.1), so as to take advantage of the smoother domain dependence of D, one of us [4] has recently computed the (translational) added mass of a cube very accurately, by the following variant of the 'boundary element' method. For Neumann problems, this method approximates the given normal derivative $\partial\phi/\partial n$ by a single layer of charge,[7] of suitable surface density $\sigma(\underline{y})$ distributed over the boundary Γ of the domain Ω. The function $\sigma(\underline{y})$ then satisfies (on Γ) a Fredholm integral equation of the second kind

$$f(\underline{y}) = \sigma(\underline{y}) + \lambda \int K(\underline{y};\underline{\eta})\, \sigma(\underline{\eta})\, dS , \tag{6.2}$$

whose kernel $K(\underline{y};\eta)$ is the normal derivative of the Green's function.

By applying collocation to piecewise constant charge densities $\sigma(\underline{y})$ on subdivisions of Γ into $6N^2$ rectangles

[7]As in [5, p. 287]. The 'boundary element method' essentially implements, using piecewise constant or polynomial densities, the classic integral equation method of potential theory.

with $N = 4, 8, 16$, and extrapolating by Aitken's δ-process, the approximate virtual mass of 0.638 for the cube was calculated by one of us. We believe that this value is correct to within ±0.5%. We also note that, because of its symmetry, the translational added mass tensor of a cube is a scalar matrix mI: $m_{11} = m_{22} = m_{33}$ and $m_{ij} = 0$ if $i \neq j$.

Although the boundary element method with discontinuous, piecewise constant charge densities gives a very accurate determination of this numerical 'domain constant', the associated velocity field is not at all accurate. This illustrates an important general principle: 'boundary element methods' are most effective for estimating domain constants; they are less effective for determining the velocity distributions.

Presumably, a considerably more accurate approximation to the flow field could be obtained by using continuous, piecewise bilinear charge densities, and solving (6.2) at the vertices instead of the midpoints of the rectangles into which the faces are subdivided. However, the application of extrapolation methods to approximate flow fields is evidently much more complicated than using them to determine domain constants, while (6.1) sheds no light on the velocity field. Moreover, in any case, the singularities in the velocity field near the corners of a cube seem to be unknown, and are very hard to describe analytically, so to obtain an accurate approximation to the velocity field for flow around a cube is orders of magnitude more difficult than to approximate the added mass accurately.

REFERENCES

[1] G. Birkhoff, "Hydrodynamics: A Study in Logic, Fact, and Similitude", 2n ed., Princeton Univ. Press, 1960.

[2] G. Birkhoff and R.E. Lynch, "Added mass of cylinders", Math. and Computers in Simulation XXII (1980), 291-7.

[3] G. Birkhoff and R.E. Lynch, "Numerical Solution of Elliptic Problems", SIAM Publications, 1983.

[4] John Brophy, "Computation of capacity, polarization, and added mass", Ph.D. Thesis, Purdue University, 1983.

[5] O.D. Kellogg, "Potential Theory", Springer, 1929.

[6] G. Pólya and G. Szegö, "Isoperimetric Inequalities in Mathematical Physics", Princeton Univ. Press, 1951.

[7] M. Schiffer and G. Szegö, "Polarization and virtual mass", Trans. Am. Math. Soc. 67 (194-), 130-205.

[8] R.S. Varga, "Matrix Iterative Analysis", Prentice-Hall, 1962.

A PRIORI LOCAL GRID REFINEMENT IN THE MULTIGRID METHOD

Joel E. Dendy, Jr.*

In this work we study the possibility of a priori local grid refinement in the multigrid method for the single group neutron diffusion equation

$$\begin{aligned} -\nabla\cdot(D(x,y)\ \nabla U(x,y)) + \sigma(x,y)\ U(x,y) &= f(x,y),(x,y)\ \varepsilon\ \Omega \\ \nu(x,y)\cdot D(x,y)\ \nabla U(x,y) + \gamma(x,y)\ U(x,y) &= 0,\ (x,y)\ \varepsilon\ \partial\Omega\ . \end{aligned} \tag{1}$$

Here Ω is a bounded region in R^2 with boundary $\partial\Omega$, ν is the outward normal to $\partial\Omega$, D is positive, and σ and γ are nonnegative. D, σ, and f are piecewise constant with jumps across internal boundaries Γ of Ω, where U and $\mu\cdot(D\nabla U)$ are continuous at almost every $(x,y)\ \varepsilon\ \Gamma$ for a smooth normal vector μ to Γ. The emphasis is on jumps of orders of magnitude.

For simplicity we assume Ω to be a rectangle $(0,X) \times (0,Y)$ and Γ to be composed of horizontal and vertical line segments. We also assume that the grid G^M, on which (1) is approximated, has a uniform mesh size h_M and that Γ is a subset of the grid; relaxation of these

* Theoretical Division, Los Alamos National Laboratory, Los Alamos, NM 87545

ISBN 0-12-100560-7

assumptions is discussed below. Let us write the difference approximation to (1) on a grid G^M as

$$L^M U^M = F^M \quad ; \tag{2}$$

details of the difference approximation (2) may be found in (1,9).

In the simplest form of the multigrid method, one constructs a sequence of grids $G^1, \ldots, G^M$ with corresponding mesh sizes $h_1, \ldots, h_m$, where $h_{i-1} = 2h_i$. One does a fixed number, IM, of relaxation sweeps (Gauss-Seidel, for example) on (2) and then drops down to grid G^{M-1} and the equation

$$L^{M-1} V^{M-1} = f^{M-1} \equiv I_M^{M-1}(F^M - L^M v^M) \quad , \tag{3}$$

where V^{M-1} is to be the coarse grid approximation to $V^M \equiv U^M - u^M$, where $v^M = u^M$ is the last iterate on grid G^M, and where I_M^{M-1} is an interpolation operator from G^M to G^{M-1}. To solve equation (3) approximately, one resorts to recursion, taking ID relaxation sweeps on grid G^k before dropping down to grid G^{k-1}, $M-1 \geq k \geq 2$ and the equation

$$L^{k-1} V^{k-1} = f^{k-1} \equiv I_k^{k-1}(f^k - L^k v^k) \quad . \tag{4}$$

When grid G^1 is reached, the equation $L^1 V^1 = f^1$ can be solved directly and $v^2 \leftarrow v^2 + I_1^2 v^1$ performed. Then one does IU relaxation sweeps on grid G^{k-1} before forming $v^k \leftarrow v^k + I_{k-1}^k v^{k-1}, 3 \leq k \leq M$.

For equations with well behaved coefficients, it suffices to take I_{k-1}^{k} to be bilinear interpolation and I_{k}^{k-1} to be $(I_{k-1}^{k})^{T}$ or just injection. However, because of the jump discontinuities in D, a more exotic choice of these operators is required (1,5): We take $I_{k-1}^{k} = J_{k-1}^{k}$ and $I_{k}^{k-1} = (J_{k-1}^{k})^{T}$, where J_{k-1}^{k} is as follows. Suppose that at $(IF+1,JF) \varepsilon G^{k}$, L^{k} has the pointwise template

$$\begin{bmatrix} -T_{IF+1,JF+1}^{k} & -W_{IF+1,JF+1}^{k} & -R_{IF+2,JF+1}^{k} \\ -Q_{IF+1,JF}^{k} & S_{IF+1,JF}^{k} & -Q_{IF+2,JF}^{k} \\ -R_{IF+1,JF}^{k} & -W_{IF+1,JF}^{k} & -T_{IF+2,JF}^{k} \end{bmatrix} \tag{5}$$

and that $(IC,JC) \varepsilon G^{k-1}$ is the same point as $(IF,JF) \varepsilon G^{k}$. At (IF,JF), J_{k-1}^{k} is just given by replacement:

$$(J_{k-1}^{k} v^{k-1})_{IF,JF} = v_{IC,JC}^{k-1} \quad .$$

At (IF+1,JF), we have

$$(J_{k-1}^{k} v^{k-1})_{IF+1,JF} = \frac{\tilde{Q}_{IF+1,JF} \, v_{IC,JC}^{k-1} + \bar{Q}_{IF+2,JF} \, v_{IC+1,JC}^{k-1}}{\tilde{S}_{IF+1,JF}^{k}} \quad ,$$

where

$$\tilde{Q}_{IF+1,JF}^{k} = T_{IF+1,JF+1}^{k} + Q_{IF+1,JF}^{k} + R_{IF+1,JF}^{k} \quad ,$$

$$\tilde{S}_{IF+1,JF}^{k} = W_{IF+1,JF}^{k} - S_{IF+1,JF}^{k} - W_{IF+1,JF+1}^{k} \quad ,$$

$$\bar{Q}^k_{IF+1,JF} = T^k_{IF+2,JF} + Q^k_{IF+2,JF} + R^k_{IF+2,JF+1} .$$

(We have just summed (5) vertically to average out its y-dependence.) A similar formula is used for points like (IF,JF+1). In each case, one performs $v^k \leftarrow v^k + J^k_{k-1} v^{k-1}$. Enough information is now present to use the difference equation at points like (IF+1, JF+1) to solve for $v^k_{IF+1,JF+1}$ in terms of its eight neighbors. Finally, we need to specify L^k, $k < M$, and this is done by taking $L^k = (J^{k+1}_k)^T L^{k+1} J^{k+1}_k$.

Let us now consider an example. Take $\Omega = (0.,48.) \times (0.,48.)$, $\Omega' = (23.,25.) \times (23.,25.)$,

$$D(x,y) = \begin{cases} 1., \text{ if } (x,y) \in \Omega' \\ 1000., \text{ otherwise} \end{cases} ,$$

$$\sigma(x,y) = \begin{cases} 1/3, \text{ if } (x,y) \in \Omega' \\ 0., \text{ otherwise} \end{cases} , \qquad (6)$$

$$f(x,y) = \begin{cases} 1., \text{ if } (x,y) \in \Omega' \\ 0., \text{ otherwise} \end{cases} .$$

The boundary conditions are $\frac{\partial U}{\partial \nu} = 0$ along y = 0. and along x = 0. and $\frac{\partial U}{\partial \nu} = -\frac{1}{2D}U$ along x = 48. and along y = 48. Results of solving this problem with a uniform grid are described below.

In (9), Wachspress derived an accuracy criterion for problems like the one above. This criterion gives the mesh size h necessary for accuracy in terms of σ and D; it is $\sqrt{\frac{\sigma}{D}}\, h \ll 1$. A theoretical explanation for this criterion was given only for one space dimension, but its utility has been verified many times for multi-dimensional problems. For the problem above, it suggests that the mesh size in Ω' should be less than $\sqrt{3}$ but that in $\Omega\backslash\Omega'$ the mesh size can be much larger. Such problems have been solved traditionally by using Cartesian products of nonuniform meshes; for many practical problems, however, the number and placement of regions needing small mesh sizes is such that this approach leads to a globally small mesh size. Thus, although the Cartesian product approach is quite useful in the above problem, we will retain this problem as a model problem for an illustration of how multigrid and local grid refinement can be successfully combined.

The multigrid local grid refinement we use is the composite grid approach (2,3). In this approach, only the coarsest grid, G^1, needs to cover all of Ω; see, for example Figure 1. As in (2) we use an adaptation of the full approximation storage (FAS) version of multigrid instead of the correction version described above.

Let G^k_m denote the set of points of G^k which are inner points of a finer level G^m (that is, points where the G^m difference equations are defined); then the difference equations on G^k are

$$L^k\ U^k = \bar{F}^k \quad ,$$

where

$$\bar{F}^k = F^k \text{ in } G^k\backslash G^k_{k+1} \quad , \tag{7a}$$

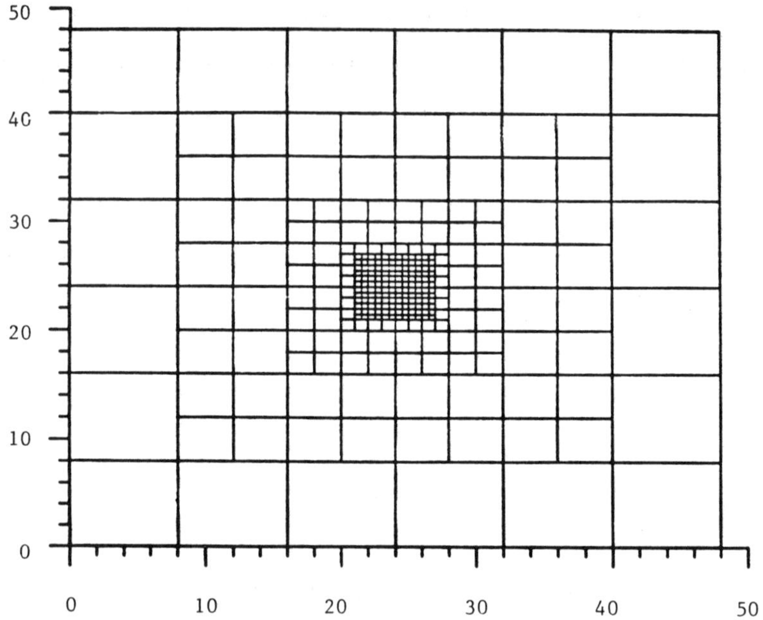

Figure 1a

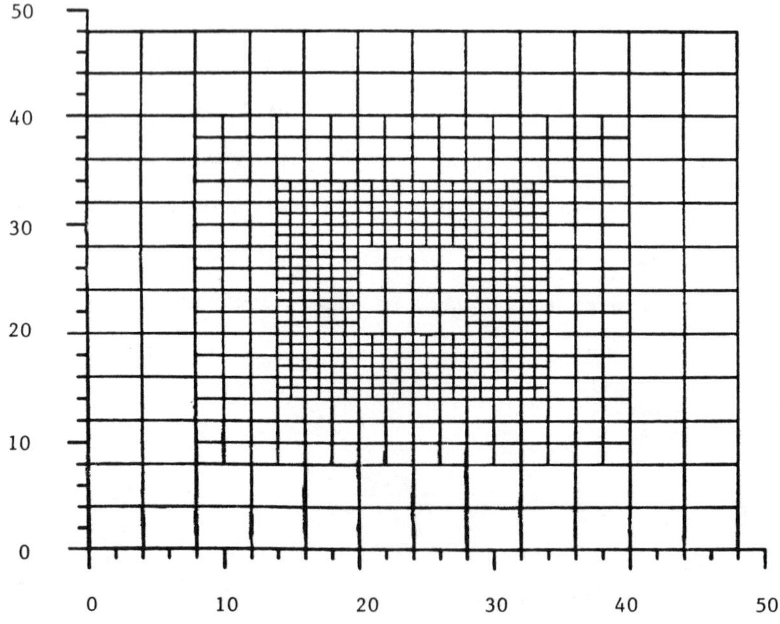

Figure 1b

$$\bar{F}^k = F^k_{k+1} = I^k_{k+1}(\bar{F}^{k+1} - L^{k+1}\, u^{k+1}) + L^k(\tilde{I}^k_{k+1} u^{k+1}) \quad , \text{ in } G^k_{k+1} \qquad (7b)$$

where $\tilde{I}^k_{k+1} u^{k+1}$ is injection from G^{k+1} to G^k. Thus on the subdomain where G^{k+1} is not defined, G^k plays the role of the finest grid, and on the subdomain where finer subgrids are present, G^k serves for calculating the coarse grid correction. In interpolation, $u^{k+1} \leftarrow u^{k+1} + I^{k+1}_k(u^k - \tilde{I}^k_{k+1}\, u^{k+1})$ is used in G^k_{k+1} and $u^{k+1} \leftarrow u^{k+1} + I^{k+1}_k\, u^k$ is used on ∂G^{k+1}. Thus the approximate solution u^{k+1} of (7a) is interpolated to provide boundary data for the difference equations on G^{k+1}.

An additional specification is needed in the above description, which did not arise in the global grid case of reference (5) and that is for the definition of L^k on ∂G^k_{k+1}. L^k can be defined directly in $G^k \setminus G^k_{k+1}$ and can be defined as $(I^{k+1}_k)^T\, L^{k+1}\, I^{k+1}_k$ in G^k_{k+1}. If it were not for requirements of symmetry and energy conservation, it could also be defined directly on ∂G^k_{k+1}. Instead we use a hybrid definition, which can be illustrated by considering a point P of a vertical part of ∂G^k_{k+1} which is not a corner point. Assume also that G^k_{k+1} lies to the east of ∂G^k_{k+1} and that part of $G^k \setminus G^k_{k+1}$ lies to the west of ∂G^k_{k+1}. Then at P we define the north, northwest, west, southwest, and south coefficients directly; the northeast, east, and southeast coefficients are defined from $(I^{k+1}_k)^T\, L^{k+1}\, I^{k+1}_k$; the central coefficient is defined by requiring consistency.

It is clear from the above description that there is a limit to the width of refinement near an interface. If

G^{k+1}, for example, ends too close to an interface, then the boundary data from G^k will be inaccurate since it will be playing the role of the finest grid and using a mesh size that is too large. The minimum width to avoid such a situation is $2h_{k+1}$, and in the situation depicted in Fig. 1a, $3h_{k+1}$ is necessary in order that G^{k+1} terminates on a grid line of G^k. In the example (6) note that the fine grids do not always need to cover Ω'. If Ω' were (17., 31.) x (17., 31.) instead, then refining in the neighborhood of $\partial\Omega'$ might give a picture like Fig. 1b.

We now give a comparison of uniform grid multigrid and *a priori* local grid refinement multigrid for problem 6. In the uniform case we did problem 6 three times, with h = 1., ½, and ¼ respectively. We used Richardson extrapolation on h = ½ and h = ¼ to obtain an answer at (24., 24.) of .346818, which we accept as the true answer for purposes of comparison. In each case we solved the problem until the discrete L^2 norm of the residual was less than 10^{-6}. We also used the capability of multigrid to compare residuals on grid G^M and G^{M-1} to solve just to truncation error (2,5). In this case we list the number of times, NCYC, that the finest grid was visited.

In the local grid refinement case, we worked the problem with M = 5 and 6 respectively. The grids used were:

G^1: uniform with h_1 = 16.,

G^2: uniform with h_2 = 8.,

G^3: (8., 40.) x (8., 40.) with h_3 = 4.,

G^4: (12., 32.) x (12., 32) with h_4 = 2.,

G^5: (20., 28.) x (20., 28.) with h_5 = 1.,

G^6: (21., 27) x (21., 27) with h_6 = ½;

the M = 6 case is shown in Fig. 1a. The comparisons are given in Table 1; T.E. stands for "truncation error". Both methods start on the coarsest grid and "bootstrap" themselves up to the finest grid before beginning simple cycling (5).

TABLE 1.

Case	Answer, Uniform grids	Error, Uniform grids	Answer Local grids	Error, Local grids
M = 5, TOL=10^{-6}	.306501	.040	.307590	.039
M = 6, TOL=10^{-6}	.334839	.012	.334776	.012
M = 7, TOL=10^{-6}	.343824	.003	---	---
M = 5, T.E.	.306501,NCYC=3	.040	.307590, NCYC=3	.039
M = 6, T.E.	.334858,NCYC=3	.012	.334486, NCYC=3	.012
M = 7, T.E.	.343841,NCYC=3	.003	---	---

The comparison between answers for the local grid method versus the uniform grid method is rather good, and a comparison between the number of unknowns suggests that the local grid method should be about eight and twenty times faster than the uniform grid method in the M = 5 and 6 case, respectively. As currently implemented, however, the local grid method code runs four and ten times faster than the uniform grid method code in the M = 5 and 6 case, respectively. Since both codes take roughly the same amount of time on a uniform grid problem, this suggests that the startup time for the relatively shorter loops (see comments on data structure below) is preventing the local grid code from achieving its full potential. Since the local grid method is still being investigated, we have not yet tried to minimize this startup time.

What about the restriction that Γ lie on grid lines of the finest grid? Certainly there is no difficulty about using the above approach with nonuniform grids adjusted so as to satisfy this constraint, but we claim that such an approach is not necessary. Another common approach to interfaces with uniform grids is to assume Γ lies halfway between grid lines; in this case the difference equation coefficients are formed using harmonic averages of values of D instead of arithmetic averages. A comparison of the two approaches in (5) suggested equivalent accuracy. It is easy to devise schemes--and we have--which use a combination of arithmetic and harmonic averaging, reducing to the first if Γ lies on grid lines and to the second if Γ lies halfway between grid lines. Thus one need only come within $h/2$ of Γ to assure accuracy and can thereby use only locally uniform grids.

What about the restriction that Γ consist of horizontal and vertical segments? In this application, this restriction is no undue burden, but in other applications--such as front tracking in time dependent codes--being able to refine locally only horizontally and vertically has been shown to be inefficient.(8) One solution is to allow grids with local coordinates, one of which is nearly parallel to the front.(4) Given the formidable logical difficulties of communicating between such grids and the global uniform grids, it seems more attractive to us to abandon uniform grids in such cases but to retain the logically rectangular structure of the grids; this leads to difference schemes on quadrilaterals whose vertices can be moved with the front. This technique leads to unacceptable tangling of the mesh in many cases, but recent ideas by J. M. Hyman (6) may solve this difficulty, in which case the above multigrid local mesh refinement techniques--requiring a logically rectangular structure--may be used.

We have called the local mesh refinement in this paper *a priori* refinement since one knows beforehand where to place the refinement. This is in contrast to adaptive refinement, where decisions of where to refine are made in the course of computation. We have examined such procedures (2,3) in the case of (1), and it is reassuring that the refinements they would dictate are not far from the refinements that we choose in an *ad hoc* manner. It is clear, therefore, that adaptive refinements are inefficient in this case since they expend unnecessary work to obtain refinements that one could have chosen *a priori* anyway.

In this regard, we comment briefly on the data structure we use for local grid refinement. It is basically the same linked list structure described in (3,7). Each grid has a pointer to the beginning of storage for that grid, and each grid is organized as a collection of strings. There is one row string which gives the beginning and end J values for the grid. For each J there is a beginning and end I value and a pointer to the next I string. There may be more than one I string for each J, as in Fig. 1b. In the adaptive mode, when additional points need to be added to an I string, the pointer to that I string is simply changed to point to the next available storage location and the augmented I string is stored beginning there. The pointer to the next I string is unchanged. This system is efficient for serial machines but not apparently so for vector machines like the Cray 1 (which demands vectors stored in contiguous storage locations or, at best, a constant offset) since relaxation and other procedures require a given I string to communicate with I strings above and below it. The finest grid in Fig. 1b could be organized into the bottom six rows, then an intermediate row, then two of seven rows, then an intermediate row, then the top six rows. Four of these six sets would vectorize well. Thus, in the *a priori* mode, it appears to be possible to take advantage of vector efficiency. In the adaptive mode however, such a conclusion is not clear. Suppose additional points must be added to an I string. Then it appears necessary to move all storage locations after this I string to make room for it, in order to keep storage contiguous. Using vector moves will amortize part of the expense of this operation, but in the worst case the expense could be proportional to the number of grid points. This matter needs further investigation, and in any case the development of an adaptive procedure for the above data structure still awaits the development of suitable software, such as that being developed by Ophir (7).

In conclusion, the promising comparison shown in Table 1 and other comparisons that we have performed indicate that this type of local grid multigrid method is worthy of further study.

References

1. Alcouffe, R. E., Brandt, A., Dendy, J. E., Jr., and Painter, J. W., (1981). SIAM J. Sci. Stat. Comput., 2, 430.

2. Brandt, A., (1977). Math. Comp. 31, 333.

3. Brandt, A., (1977). In "Mathematical Software III" (John R. Rice, ed.), 273. Academic Press, New York.

4. Brandt, A., (1981). In "Multigrid Methods," (W. Hackbush, V. Trottenberg, eds.), Springer-Verlag.

5. Dendy, J. E., Jr., (1982). J. Comp. Phys. 48, 366.

6. Hyman, J. M., Self-adjusting grid methods for one-dimensional hyperbolic conservation laws, contributed talk at SIAM meeting, Palo Alto, CA., June, 1982.

7. Ophir, D., (1979). Language for processes of numerical solutions to differential equations, Ph.D. thesis, The Weizmann Institute of Science, Rehovat, Israel to appear.

8. Swartz, B., Courant-like conditions limit reasonable mesh refinement to order h^2, to appear in SIAM J. Numer. Anal.

9. Wachspress, E. L., (1966). "Iterative solution of elliptic systems and applications to the neutron diffusion equations of reactor physics". Prentice-Hall, Englewood Cliffs, N.J.

ABSTRACT MULTI-GRID WITH APPLICATIONS TO ELLIPTIC BOUNDARY-VALUE PROBLEMS

Craig C. Douglas[1]

I. GENERAL THEORY

In this section, we discuss the approximate solution of an abstract elliptic variational problem. While this theory is applied to solving the large sparse linear systems which arise in connection with finite element or finite difference procedures for solving self adjoint elliptic boundary value problems in the next section, it can also be applied to problems which do not arise from partial differential equations. Our scheme involves a sequence of finite dimensional spaces $\mathcal{M}_j$, j = 1,2,...,k. We approximate the solution using the largest space. Under certain circumstances, the solution to a problem can be approximated well using smaller spaces. Since we require the smaller spaces to have geometrically fewer unknowns than the largest one, the savings in computation can be substantial. In fact, we prove that such procedures are of optimal order under appropriate conditions.

Assume we are given a triple,

$$\{H,\ a(u,v),\ f(v)\},$$

where H is a Hilbert space with norm $\|\cdot\|$, a(u,v) is a continuous symmetric real valued bilinear form on $H \otimes H$, and $f(v): H \to R$ is a continuous real valued linear functional. Furthermore, we assume that there exists a constant $a_0 > 0$ such that

$$a(v,v) \ \geq\ a_0\|v\|^2, \qquad \text{for all } v \in H.$$

The bilinear form $a(\cdot,\cdot)$ induces the *energy norm* $|||u||| \equiv [a(u,u)]^{1/2}$. We seek an approximation to the unique solution of

Problem 1: Given {H, a(u,v), f(v)}, find $u \in H$ such that $a(u,v) = f(v)$ for all $v \in H$.

---

[1]Department of Computer Science, Yale University, New Haven, Connecticut

ISBN 0-12-100560-7

We now consider the finite dimensional approximation of Problem 1. Let $\mathcal{M}_j$, $j \geq 1$, be a sequence of N_j-dimensional spaces. Associated with each space $\mathcal{M}_j$ is a continuous, symmetric, positive-definite bilinear form $a_j(u,v)$ and a continuous, bounded linear form $f_j(v)$. We require that

$$N_j \sim \sigma N_{j-1}, \qquad \text{for some } \sigma \geq 2. \tag{1}$$

We will see that σ is important: when $\sigma > 2$, we can always construct algorithms of optimal order to approximate the solution to Problem 1. For $j \geq 1$, the finite dimensional approximations of Problem 1 are

Problem 2: Given $\{\mathcal{M}_j, a_j(u,v), f_j(v)\}$, find $u_j \in \mathcal{M}_j$ such that $a_j(u_j,v) = f_j(v)$ for all $v \in \mathcal{M}_j$.

We assume that linear operators exist which project H onto $\mathcal{M}_j$, inject $\mathcal{M}_j$ into H, and interpolate between adjacent solution spaces:

$$p_j\colon H \xrightarrow[\text{onto}]{} \mathcal{M}_j \quad \text{and} \quad i_j\colon \mathcal{M}_j \xrightarrow[1-1]{} H \tag{2}$$

$$E_j\colon \mathcal{M}_{j-1} \to \mathcal{M}_j \quad \text{and} \quad R_j\colon \mathcal{M}_j \xrightarrow[\text{onto}]{} \mathcal{M}_{j-1}$$

We do not require commutativity, i.e., $i_{j-1}R_jp_jv$ does not have to equal $i_jE_jp_{j-1}v$, $v \in H$.

Associated with each space $\mathcal{M}_j$ are eigenvalues $\lambda_i^{(j)}$ and eigenfunctions (eigenvectors) $\psi_i^{(j)}$, $1 \leq i \leq N_j$, satisfying

$$a_j(v,\psi_i^{(j)}) = \lambda_i^{(j)} (v,\psi_i^{(j)})_j \qquad \text{for all } v \in \mathcal{M}_j,$$

where $(\cdot,\cdot)_j$ denotes the inner product in $\mathcal{M}_j$. Let δ_{ik} be the Kronecker delta. Without loss of generality,

$$\begin{gathered} 0 < \lambda_1^{(j)} \leq \lambda_2^{(j)} \leq \dots \leq \lambda_{N_j}^{(j)} \equiv \Lambda_j, \\ (\psi_i^{(j)},\psi_k^{(j)})_j = \delta_{ik} \text{ and } a_j(\psi_i^{(j)},\psi_k^{(j)}) = \lambda_i^{(j)}\delta_{ik}, 1 \leq i,k \leq N_j. \end{gathered} \tag{3}$$

With each space $\mathcal{M}_j$, we define discrete norms

$$|||v|||_s^2 \equiv \sum_{i=1}^{N} c_i^2(\lambda_i^{(j)})^s, \quad \text{for } v \equiv \sum_{i=1}^{N} c_i\, \psi_i^{(j)},$$

where we have suppressed the j subscript on the norm and $-2 \leq s \leq 2$. Note that $|||v|||_1 = |||v|||$ is the usual energy norm on level j. Hereafter, we drop both the superscripts from the eigenvalues and eigenfunctions (eigenvectors) and the subscript from the dimension of the spaces.

We require the bilinear forms on adjacent spaces to have a consistency relationship. As a consequence, the energy norms on two adjacent spaces are uniformly consistent.

Hypothesis 1 (Energy Norm Consistency): For all $j > 1$ there exists a positive constant C_1, independent of j, such that

$$a_j(E_jv,E_jw) = C_1\, a_{j-1}(v,w), \qquad \text{for all } v,w \in \mathcal{M}_{j-1}.$$

Since $|||v|||^2 = a_j(v,v)$. This hypothesis implies that for any $v \in \mathcal{M}_{j-1}$, the energy norm of v on level $j-1$ is equal to a constant times the energy norm of E_jv on level j.

We now define and analyze a k-level iterative procedure for solving Problem 2. The process involves solving problems like Problem 2 sequentially for $j=1, 2,\dots, k$. The *k-level scheme* has three parameters: m and n, which determine the number of smoothing iterations used; and p, which is used in a recursion iteration. Note that computation begins with the largest space and uses the smaller ones only to solve correction problems.

Algorithm MGI(k, m, n, p): Given an integer $k > 0$ and $\{\mathcal{M}_j, a_j(\cdot,\cdot), f_j(\cdot)\}_{j=1}^{k}$, we want to approximate $u_k \in \mathcal{M}_k$, where $a_k(u_k,v) = f_k(v)$ for all $v \in \mathcal{M}_k$.

(a) If $k = 1$, then solve directly.

(b) If $k > 1$, then one iteration of the k-level scheme takes an initial guess $z_0 \in \mathcal{M}_k$ to a final approximation $z_{m+n+1} \in \mathcal{M}_k$ in three steps:
(i) if $n > 0$, define z_i, $1 \le i \le n$, by

$$(z_i - z_{i-1}, v)_k = \Lambda_k^{-1}[f_k(v) - a_k(z_{i-1},v)], \qquad \text{for all } v \in \mathcal{M}_k. \tag{4}$$

(ii) Let $q \in \mathcal{M}_{k-1}$ be the approximation of $\bar{q} \in \mathcal{M}_{k-1}$ obtained by applying p iterations of the $(k-1)$-level scheme to the residual equation

$$a_{k-1}(\bar{q},v) = C_1^{-1}\{f_k(E_kv) - a_k(z_n,E_kv)\} \equiv \bar{f}_{k-1}(v), \quad \text{for all } v \in \mathcal{M}_{k-1}, \tag{5}$$

starting from an initial guess zero. Then set $z_{n+1} = z_n + E_kq$.
(iii) If $m > 0$, then define z_i, $n+2 \le i \le m+n+1$, by (4).

In the *correction recursion iteration* (step (ii)), we approximately compute the elliptic projection of the error in $\mathcal{M}_{k-1}$ using p iterations of the $(k-1)$-level scheme applied to a problem of Problem 2's form with $j = k - 1$. In the *smoothing iterations* (steps (i) and (iii)), error components whose oscillation are "large" are damped. A simultaneous displacement procedure is used in this step.

Later in this section we will see that Λ_k^{-1} can be replaced by a particular type of bound (see Hypothesis 2). We will also see that (4) can be replaced by other iterations which are computationally more attractive, but do not affect the character of our convergence results. The use of iteration (4) simplifies the initial analysis of the convergence. In Bank and Douglas [4] general smoothers are allowed, including Symmetric Gauss-Seidel, Chebyshev, Conjugate Gradients, and Conjugate Residuals.

There are three cases of note in Algorithm MGI: (a) when $n > 0$ and $m = 0$, (b) when $n = 0$ and $m > 0$, and (c) when $n > 0$ and $m > 0$. Case (a) is the scheme analyzed by Astrakhantsev [1], Bank and Dupont [5], Hackbusch [11], Nicolaides [13] for finite element discretizations of various elliptic boundary value problems. Federenko [8, 9] and Bakhvalov [2] analyzed this case for finite difference discretizations. Each of these authors assumed in their convergence proofs that $p > 1$. Brandt [6] gave a heuristic analysis of (b) and (c) for finite difference discretizations using local mode analysis. Hackbusch [12] analyzed case (c) for $p \geq 1$. The motivation for studying cases (b) and (c) comes from trying to understand the behavior of a large finite difference program which is described in Douglas [7]. It was observed empirically that case (b) sometimes required fewer correction recursions to achieve the same accuracy as case (a).

Before stating a convergence theorem for Algorithm MGI, we need to state one definition and two more hypotheses. The first hypothesis is a bound for the largest eigenvalues and the other is an error estimate for the correction produced by the (k−1)-level iteration. Finally, we state a lemma describing the smoothing iteration's effect on the error. The error on level k at the ith stage of Algorithm MGI will be denoted by

$$e_i \equiv u_k - z_i.$$

Hypothesis 2 (Maximum Eigenvalue): There exist positive constants δ and C_2, each independent of j, such that

$$\Lambda_j \leq C_2 N_j^{2\delta}, \qquad \text{for all } j \geq 1.$$

The use of Λ_k in the smoothing iteration (4) may be replaced by any upper bound satisfying Hypothesis 2. The last hypothesis is a norm estimate:

Hypothesis 3 (Approximating Error Estimate): For some α with $0 < \alpha \leq 1$, there exists a positive constant C_3 such that

$$|||E_k\bar{q}-e_n|||_{1-\alpha} \leq C_3^{\alpha} N_k^{-2\alpha\delta} |||e_n|||_{1+\alpha}.$$

For problems derived from elliptic boundary value problems, the value of α depends on the spatial domain.

The following lemma is used in the convergence proof to analyze the effect of the smoothing on elements in the solution spaces.

Lemma 1: Let $n > 0$ be any integer and $z_0 \in \mathcal{M}_k$. Then the smoothing iteration (4) is a contraction operator: $|||e_n||| \leq |||e_0|||$. Further, for every fixed $0 \leq \omega \leq 2$ and $0 < \alpha \leq 1$, $|||e_n|||_\omega \leq C_2^{\alpha/2} N_k^{\alpha\delta} (2n + \alpha)^{-\alpha/2} |||e_0|||_{\omega-\alpha}$.

The convergence of Algorithm MGI is stated in Theorem 1. In essence, this result says that the error on level k can be reduced by any positive constant less than one provided the correction recursion problem on level k−1 can be solved sufficiently accurately.

Theorem 1 (Convergence of Algorithm MGI): Assume that Hypotheses 1, 2, and 3 hold. Let $p > 1$ be any fixed integer. For any constant $0 < \gamma < 1$ there exist a nonnegative integer I which depends only on p and γ, such that

$$|||e_{m+n+1}||| \leq \gamma |||e_0|||, \quad \text{for all } m+n \geq I. \tag{6}$$

The proof ends with

$$|||e_{m+n+1}||| \leq [C_4(2m + \alpha)^{-\alpha/2}(2n + \alpha)^{-\alpha/2} + \gamma^p] \, |||e_0|||, \tag{7}$$

where $C_4 \equiv C_2^\alpha C_3^\alpha$. Finally, choose I such that

$$C_4(2m + \alpha)^{-\alpha/2}(2n + \alpha)^{-\alpha/2} \leq \gamma - \gamma^p, \quad \text{for all } m+n \geq I. \tag{8}$$

When $p = 1$, the proof of Theorem 1 breaks down because the right hand side of (8) is zero. If we assume that k is a small natural number, then we can define a sequence $\{\gamma_j\}_{j=1}^k$ such that

$$\gamma_j - \gamma_{j-1}^p = C_5 > 0, 2 \leq j \leq k \text{ and } p > 0,$$

$$\gamma_1 > 0, \text{ and } \gamma_k = \gamma < 1.$$

Corresponding to Theorem 1 is

Theorem 2: Assume that Hypotheses 1, 2, and 3. Let $p > 0$ be any fixed integer. For any constant $0 < \gamma < 1$ there exist a nonnegative integer I which depends only on p, γ, and C_5, such that $|||e_{m+n+1}||| \leq \gamma |||e_0|||$, for all $m+n \geq I$.

A different approach to the $p = 1$ case is presented by Hackbusch [12]. His method of proof shows that

$$|||e_{m+n+1}||| \leq C\cdot(C+m+n)^{-1}\cdot|||e_0|||,$$

where C does not depend on the number of levels used. However, the analysis requires that $\alpha = 1$. For $0 < \alpha < 1$, Bank and Douglas [4] show that this method of proof leads only to the less than useful bound $|||e_{m+n+1}||| \leq |||e_0|||$, which is sharp.

We now analyze the cost of one iteration of Algorithm MGI. Let F(N) be the cost of reducing the error by a factor of γ for a problem with N unknowns. We assume that the cost of the smoothing iterations (4) (or an iteration with similar properties) on the finest level, level k, can be bounded by $C_6(m+n)N_k \equiv C_7N_k$, where C_6 is independent of k. The cost of the correction recursion (5) is $pF(N_{k-1})$. Thus,

$$F(N_k) \sim pF(N_{k-1}) + C_7N_k. \tag{9}$$

Since $N_k \sim \sigma N_{k-1}$ (see (1)), the asymptotic (in k) solution of (9) is

$$F(N) \leq \begin{cases} C_8N, & \text{if } 1 \leq p < \sigma \\ C_8N\log N, & \text{if } p = \sigma \\ C_8N^{\log p}, & \text{if } p > \sigma, \end{cases}$$

where the logarithms are base σ. One of the surprises of multigrid is that implementations (e.g., Bank [3] or Douglas [7]) exhibit the asymptotic convergence rates for very small k's.

Choosing $1 \leq p < \sigma$ leads to an algorithm of optimal order in the sense that the error can be reduce by a fixed factor of γ each iteration with work proportional to the number of unknowns. When k is finite, the cost for the direct solution of the one level problem may not be majorized by C_7N_1. In this case, the algorithm is not of optimal order.

We may want to reduce the initial error by a factor of $N^{-q\delta}$ for some fixed q. The obvious implementation would then require F(N)logN operations. We assume the solutions u_j of Problem 2 satisfy

$$|||p_ju-u_j||| \leq KN_j^{-q\delta}, \quad j \geq 1,$$

where K is a constant independent of N_j. Denote by $\widetilde{u}_j$ the computed solution of the j-level scheme (Algorithm MGI). To avoid the extra logN factor, we use $E_j\widetilde{u}_{j-1}$ as the initial guess to $\widetilde{u}_j$, $j > 1$, and prove that the initial error is small.

Approximate solutions $\widetilde{u}_j$ of finite dimensional Problem 2 are generated using the following:

Algorithm MGO(j, r, m, n, p): Given an integer $j > 0$ and $\{\mathcal{M}_i, a_i(\cdot,\cdot), f_i(\cdot)\}_{i=1}^{j}$, we want to approximate $u_j \in \mathcal{M}_j$, where $a_j(u_j,v) = f_j(v)$ for all $v \in \mathcal{M}_j$.

(a) If $j = 1$, then solve Problem 2 directly.

(b) If $j > 1$, then starting from an initial guess $z_0 \equiv E_j\widetilde{u}_{j-1}$, apply r iterations of MGI(j, m, n, p) to Problem 2 to obtain $\widetilde{u}_j$.

This algorithm actually has four parameters. The parameter r determines the number of inner iterations of MGI(j,m,n,p) to use. Recall that m and n are the number of smoothing iterations and that p is the number of correction recursion iterations in Algorithm MGI. Unlike Algorithm MGI, computation begins with the smallest space and winds its way "down" to the largest space.

The convergence properties of Algorithm MGO are stated in the Theorem 3.

Theorem 3: Assume that Hypotheses 1, 2, and 3 hold. Let $r \geq 1$ be any fixed integer. Suppose

$$|||p_ju-u_j||| \leq KN_j^{-q\delta}, \quad j \geq 1, \qquad |||p_ju-E_jp_{j-1}u||| \leq KN_j^{-q\delta}, \quad j \geq 2,$$

$$\text{and} \qquad |||u_1-\widetilde{u}_1||| \leq KN_1^{-q\delta}.$$

Then (a) For any integer $p \geq 1$, there exists a nonnegative integer I such that for $j \geq 1$,

$$|||u_j-\widetilde{u}_j||| \leq KN_j^{-q\delta}, \qquad \text{for all } m+n \geq I.$$

(b) $|||p_ju-\widetilde{u}_j||| \leq 2KN_j^{-q\delta}$

(c) the cost of computing $\widetilde{u}_j$ is bounded by $C_9F(N_j)$, where $C_9 \equiv \sigma r/(\sigma-1)$ is independent of j.

It is worth pointing out that $r \geq 1$ is independent of j. This tells us that Algorithm MGO is of optimal order whenever $1 \leq p < \sigma$.

II. ELLIPTIC BOUNDARY-VALUE PROBLEMS EXAMPLES

In this section, we summarize the application of the algorithms of Section I to the solution of large sparse linear systems which arise in connection with finite element and finite difference procedures for solving self adjoint elliptic boundary value problems. The actual proofs can be found in Douglas [7].

Using C^0-piecewise linear triangular elements, we can apply the general theory of the previous section to the model problem

$$-\nabla\cdot(P\nabla u) + Su = f \text{ in } \Omega; \quad u = 0 \quad \text{or} \quad u_n = 0 \text{ on } \partial\Omega,$$

where Ω is a polygonal domain in R^2. We assume that $P \in C^1(\bar{\Omega})$, $S \in C(\bar{\Omega})$, and that there exist positive constants $\bar{p}$, $\tilde{p}$, $\bar{s}$, and $\tilde{s}$ such that

$$0 < \bar{p} \leq P(x) \leq \tilde{p} \text{ and } 0 < \bar{s} \leq S(x) \leq \tilde{s}, \qquad \text{for all } x \in \bar{\Omega}. \tag{10}$$

The proofs use either the energy norm or the L^2 norm. When using the latter, Ω must be convex. The energy norm proofs allow for the elements to have one curved edge, which allows us to solve problems in quite general domains. Since none of the constants can be computed easily and the proofs are standard, we omit the details.

We also can apply this general theory on uniform meshes to finite difference approximations to the model problems

$$-(Pu_x)_x + Su = f \text{ in } \Omega \equiv (0,1); \quad u(0) = u(1) = 0,$$

$$-(Pu_x)_x - (Pu_y)_y + Su = f \text{ in } \Omega \equiv (0,1)x(0,1); \quad u = 0 \text{ on } \partial\Omega, \text{ and} \tag{11}$$

$$-\nabla\cdot(P\nabla u) + Su = f \text{ in } \Omega; \quad u = 0 \text{ on } \partial\Omega.$$

We assume that $f \in L^2(\Omega)$ and that P and S are constrained as in (10).

A. One Dimensional Definitions

Let $k > 1$ and $N_0 > 0$ be fixed integers. We define $N_j \equiv 2N_{j-1} + 1$, $j \geq 1$. Following (1), we know that $\sigma = 2$. We define k uniform grids Ω^j, $1 \leq j \leq k$, by

$$h_k \equiv (N_k + 1)^{-1}, \quad h_j \equiv 2^{k-j}h_k, \text{ and } \Omega^j \equiv \{ ih_j \mid 1 \leq i \leq N_j \}.$$

For the moment, we assume that $P \equiv 1$ and $S \equiv 0$ in (11). The general case will be analyzed at the end of the section. We discretize (11) by approximating the second derivatives by central differences to obtain a system of linear equations

$$A_k u_k = h_k f_k,$$

where u_k and f_k approximate the solution u and the right-hand side f on Ω^k. The N_k x N_k tridiagonal matrix A_k is given by

$$A_k \equiv h_k^{-1} [-1, 2, -1]. \tag{12}$$

Factoring and solving this algebraic system directly requires only $5N_k$ multiplications and $3N_k$ additions [10]. We will show that the multigrid algorithms of Section I require $O(N_k)$ operations to solve this problem to accuracy comparable to the discretization error, i.e., $O(N_k^{-2})$.

We associate a *solution space* $\mathcal{M}_j$, $1 \leq j \leq k$, with each grid. Let $N \equiv N_k$. Then

$$\begin{aligned} \mathcal{M}_k &\equiv \{ v \mid v \in R^N \} \quad \text{and} \\ \mathcal{M}_{k-1} &\equiv \{ v \mid v \in \mathcal{M}_k \text{ and } v = < 0, v_1, 0, v_2, 0, \ldots, 0, v_{N_{k-1}}, 0 > \}. \end{aligned} \tag{13}$$

The spaces $\mathcal{M}_{k-2}, \ldots, \mathcal{M}_1$ are defined recursively.

We define *piecewise-linear interpolation* operators between adjacent solution spaces

$$E_2\colon \mathcal{M}_{j-1} \rightarrow \mathcal{M}_j \quad \text{and} \quad R\colon \mathcal{M}_j \rightarrow \mathcal{M}_{j-1}$$

by means of the matrices

$$E_2 \equiv \begin{bmatrix} 0 & .5 & 0 & 0 & & & & & & & \\ 0 & 1 & & & & & & & & & \\ 0 & .5 & 0 & .5 & 0 & 0 & & & & & \\ \cdot & & & 1 & & & & & & & \\ \cdot & 0 & 0 & .5 & 0 & .5 & 0 & 0 & & & \\ \cdot & & & & & 1 & & & & & \\ & & & 0 & 0 & .5 & 0 & .5 & 0 & 0 & \\ & & & & & & \cdot & & & & \\ & & & & & & & \cdot & & & \cdot \\ & & & & & & & & \cdot & & \cdot \\ & & & & & & & & 1 & & \cdot \\ & & & & & 0 & 0 & .5 & 0 & .5 & 0 \\ & & & & & & & & & 1 & 0 \\ & & & & & & & 0 & 0 & .5 & 0 \end{bmatrix}$$

and $R \equiv E_2^T$.

Problems on coarser grids are defined using the interpolation matrices. Linear systems on coarser grids, A_j, $1 \leq j < k$, are defined recursively:

$$A_{k-j} \equiv E_2^T A_{k-j+1} E_2.$$

Let $r \in R^N$ be a right-hand side for level k. Then the *(k−1)-level problem* is defined by

$$A_{k-1}\,\bar{q} = E_2^T r \equiv f_{k-1}, \bar{q} \in \mathcal{M}_{k-1}.$$

Half the rows and columns of A_{k-1} are zero. We can re-order the matrix so that the nonzero rows and columns are the first N_{k-1} rows and columns. Then A_{k-1} has a submatrix whose form is identical to A_k. The re-ordered solution space $\mathcal{M}_{k-1}$ has the form $\{ v \in \mathcal{M}_k \mid v = < v_1, v_2, \dots, v_{N_{k-1}}, 0, \dots, 0 > \}$. The (k−i)-level problems are defined similarly.

Before the theory of Section I can be applied, we must complete the definitions required for the triples used by Problems 1 and 2. For $j \geq 1$, a bilinear form $a_j(\cdot,\cdot)$, a linear functional $f_j(v)$, and an inner product $(\cdot,\cdot)_j$ are defined for all $u,v \in \mathcal{M}_j$ by

$$a_j(u,v) \equiv u^T A_j\, v, \quad f_j(v) \equiv h_j\, f_j^T v, \quad \text{and} \quad (u,v)_j \equiv h_j\, u^T v.$$

The bilinear form $a_j(\cdot,\cdot)$ induces the energy norm $|||u|||^2 \equiv a_j(u,u)$ for all $u \in \mathcal{M}_j$. Associated with each $a_j(\cdot,\cdot)$ are $N \equiv N_j$ nonzero eigenvalues $\lambda_i^{(j)}$ and eigenvectors $\psi_i^{(j)}$ satisfying (3). Let Λ_j be the largest eigenvalue. For $-2 \leq s \leq 2$, discrete norms are defined by

$$|||v|||_s^2 \equiv \sum_{i=1}^{N} \beta_i^2 (\lambda_i^{(j)})^s \quad \text{for } v \equiv \sum_{i=1}^{N} \beta_i\, \psi_i^{(j)}, \tag{14}$$

where we have suppressed the j subscript on the norm. Note that $|||v|||_1 = |||v|||$ is the usual energy norm on level j. Hereafter, we drop the superscripts from the eigenvalues and eigenvectors.

B. Two Dimensional Definitions

Let $k > 1$ and $\bar{N}_0 > 0$ be fixed integers. We define $\bar{N}_j \equiv 2\bar{N}_{j-1} + 1$, $j \geq 1$. We define k uniform grids Ω_2^j, $1 \leq j \leq k$, as products of the one dimensional domains Ω^j:

$$h_k \equiv (\bar{N}_k + 1)^{-1}, \quad h_j \equiv 2^{k-j} h_k, \quad \text{and} \quad \Omega_2^j \equiv \Omega^j \otimes \Omega^j.$$

Each grid Ω_2^j covers the interior of Ω with $N_j \equiv \bar{N}_j^2$ points. Following (1), we know that $\sigma \equiv 4$.

For the moment, we assume that $P \equiv 1$ and $S \equiv 0$ in (11). The general case will be analyzed in Section II.C. We discretize (11) by central differences to obtain a system of linear equations

$$B_k u_k = h_k^2 f_k,$$

where u_k and f_k approximate the solution u and the right-hand side f on Ω_2^k. The $N_j \times N_j$ matrix B_k is given by

$$B_k \equiv h_k \{A_k \otimes I_N + I_N \otimes A_k\},$$

where I_N is the $N_k x N_k$ identity matrix and A_k is defined in (12).

We define solution spaces $\mathcal{M}_j$, $1 \leq j \leq k$, as the tensor-product of the one dimensional solution spaces defined in (13). Following the techniques of Section II.A, we define piecewise-bilinear interpolation operators between adjacent solution spaces

$$E_b \equiv E_2 \otimes E_2: \mathcal{M}_{j-1} \to \mathcal{M}_j.$$

Problems on coarser grids are defined using the interpolation matrices. Linear systems on coarser grids, B_j, $1 \leq j < k$, are defined recursively:

$$B_{k-j} \equiv E_b^T B_{k-j+1} E_b.$$

Let $N \equiv N_k$ and $r \in R^N \otimes R^N$ be a right-hand side for level k. Then the *(k−1)-level problem* is defined by

$$B_{k-1} \bar{q} = E_b^T r \equiv f_{k-1}, \quad \bar{q} \in \mathcal{M}_{k-1}.$$

Similar to the one dimensional case, we define for all $u,v \in \mathcal{M}_j$,

$$a_j(u,v) \equiv u^T B_j v, \ f_j(v) \equiv h_j^2 f_j^T v, \text{ and } (u,v)_j \equiv h_j^2 u^T v.$$

The bilinear form $a_j(\cdot,\cdot)$ induces the energy norm $|||u|||^2 \equiv a_j(u,u)$ for all $u \in \mathcal{M}_j$. Associated with each $a_j(\cdot,\cdot)$ are N_j nonzero eigenvalues $\lambda_{\rho\omega}$ and eigenvectors $\psi_{\rho\omega}$ satisfying (3). Let Λ_j be the largest eigenvalue. For $-2 \leq s \leq s$, discrete norms $|||v|||_s$ are defined similarly to (14).

C. Convergence

In order to establish the convergence of Algorithms MGI and MGO for the finite difference cases of this section, we must verify Hypotheses 1, 2, and 3. In this section, E will refer to either E_2 or E_b. To show that (the Energy Norm Consistency) Hypothesis 1 holds is done by using the definition of the linear systems A_{j-1} and B_{j-1}, $j > 1$.

Lemma 2: Let $j > 1$ be an integer. Then

$$a_j(Ev,Ew) = a_{j-1}(v,w), \qquad \text{for all } v,w \in \mathcal{M}_{j-1}.$$

That (the Maximum Eigenvalue) Hypothesis 2 holds is a simple consequence of the explicit formula for the eigenvalues of A_j and B_j.

Lemma 3: $\Lambda_j \leq 4dh_j^{-2}$, where d is the dimension of the problem.

The proof that Hypothesis 3 holds requires Lemma 4 and is contained in Douglas [7].

Lemma 4: Let $k > 1$ be an integer and d the dimension of the probelm. Let $C(s,1) = 4^{(s-1)/2}$ and $C(s,2) = 8^{s/2}$. For $s = 0$, 1, or 2 and $u \in \mathcal{M}_k$,

$$\min_{v \in \mathcal{M}_{k-1}} |||u-Ev|||_s \leq C(s,d)\, h_k^{2-s}\, |||u|||_2. \tag{15}$$

This proves that Algorithm MGI converges at the rates specified by Theorems 1 and 2 for the finite difference cases of this section. As a consequence of Lemma 4 we can estimate the constant C_4 in (7), which governs the number of smoothing steps required in Algorithm MGI. For Poisson's equation in one dimension $C_4 = 4$ and in two dimensions $C_4 = 16 \cdot 2^{1/2}$. For the parameter choice of $m + n = 2$ and $p = 2$, the number of multiplies is approximately $22N_k$.

Before we can show that the j-level scheme (Algorithm MGO) holds, we need to define projection and injection operators between the spaces H and $\mathcal{M}_j$, $1 \leq j \leq k$ (see (2)):

p_j: H $\xrightarrow[\text{onto}]{}$ $\mathcal{M}_j$ as evaluation of $u \in H$ at the grid points of Ω^j and

i_j: $\mathcal{M}_j$ $\xrightarrow[\text{1-1}]{}$ H as piecewise-linear interpolation.

These operators have properties which are worth pointing out, namely,

$$p_j i_j = \text{Identity on } \Omega^j \text{ and } i_j E p_{j-1} = i_{j-1} p_{j-1} \text{ on } \Omega^{j-1}.$$

We can verify that the assumptions of Theorem 3 hold using these properties and simple known facts about the model problems. Details are contained in Douglas [7]. This shows that Algorithm MGO converges at the rate specified in Theorem 3.

We conclude this section by considering (11) when the problems are no longer restricted to $P \equiv 1$ and $S \equiv 0$. We discretize (11) by central differences to get

$N_k \times N_k$ linear system of equations. As before, we define coarse grid approximations of the problems using the interpolation matrices. We define inner products $\tilde{a}_j(u,v)$ using the linear systems of equations. Energy norms are defined by $||||u||||^2 \equiv \tilde{a}_j(u,v)$. The discrete norms $||||u||||_s$, $0 \leq s \leq 2$ and $u \in \mathcal{M}_j$, are defined as usual. We prove the following theorem in Douglas [7]:

Theorem 4: Let $j \geq 1$ be a fixed integer and d the dimension of the problem. Then the following holds:
(a) the linear systems are symmetric and have N_j positive eigenvalues which are bounded by Ch_j^{-2}, where C is independent of j.
(b) For $\beta = 0$, 1, or 2, there exist positive constants $C_{1,\beta}$ and $C_{2,\beta}$ such that

$$C_{1,\beta}|||u|||_\beta \leq ||||u||||_\beta \leq C_{2,\beta}|||u|||_\beta \quad \text{for all } u \in \mathcal{M}_j.$$

(c) $||||E_b\bar{q}-e_n||||_0 \leq K(d)\, C_{2,1}\, h_k^2\, ||||e_n||||_2$, where $K(1) = 1$ and $K(2) = 8^{1/2}$.

Theorems 1, 2, and 3 can be reproven using the $||||\cdot||||$ norm instead of the $|||\cdot|||$ norm. This proves that the variable coefficient Dirichlet problems (11) can be solved using the theory of Section I in $O(N_k)$ operations.

ACKNOWLEDGEMENTS

This research was supported in part by ONR Grant N00014-82-K-0184, FSU-ONR Grant F1.N00014-80-C-0076, and NSF MCS-8106181. Martin Schultz and Randy Bank deserve special mention for their advice during the research reported in this paper.

REFERENCES

[1] G. P. Astrakhantsev, An iterative method of solving elliptic net problems, *Z. Vycisl. Mat. i. Mat. Fiz.*, 11 (1971), pp. 439-448.

[2] N. S. Bakhvalov, On the convergence of a relaxation method under natural constraints on an elliptic operator, *Z. Vycisl. Mat. i. Mat. Fiz.*, 6 (1966), pp. 861-883.

[3] R. E. Bank, *PLTMG User's Guide*, Technical Report, Univeristy of California at San Diego, 1981.

[4] R. E. Bank and C. Douglas, *Sharp Estimates for Multigrid Rates of Convergence with General Smoothing and Acceleration*, Technical Report 277, Department of Computer Science, Yale University, 1983.

[5] R. E. Bank and T. Dupont, An optimal order process for solving elliptic finite element equations, *Math. Comp.*, 36 (1981), pp. 35-51.

[6] A. Brandt, Multi-level adaptive solution to boundary-value problems, *Math. Comp.*, 31 (1977), pp. 333-390.

[7] C. Douglas, *Multi-Grid Algorithms for Elliptic Boundary-Value Problems*, PhD Thesis, Department of Computer Science, Yale University, 1982. Also, Computer Science Department Technical Report 223.

[8] R. P. Federenko, A relaxation method for solving elliptic difference equations, *Z. Vycisl. Mat. i. Mat. Fiz.*, 1 (1961), pp. 922-927.

[9] R. P. Federenko, The speed of convergence of one iteration process, *Z. Vycisl. Mat. i. Mat. Fiz.*, 4 (1964), pp. 559-563.

[10] G. Forsythe and C. Moler, *Computer Solution of Linear Algebraic Systems*, Prentice-Hall, Englewood Cliffs, 1967.

[11] W. Hackbusch, Convergence of multi-grid iterations applied to difference equations, *Math. Comp.*, 34 (1980), pp. 425-440.

[12] W. Hackbusch, Multi-grid convergence theory, W. Hackbusch and U. Trottenberg, ed., *Multigrid Methods*, Springer-Verlag, New York, 1982, pp. 177-219.

[13] R. A. Nicolaides, On the l^2 convergence of an algorithm for solving finite element equations, *Math. Comp.*, 31 (1977), pp. 892-906.

A NEW ORDERING SCHEME FOR THE HERMITE BICUBIC COLLOCATION EQUATIONS

Wayne R. Dyksen[1,2]
John R. Rice[1,3]

I. INTRODUCTION

We consider a second order, linear elliptic problem on a rectangle R

$$\begin{aligned} Lu &= au_{xx} + bu_{xy} + cu_{yy} + du_x + eu_y + fu = g \quad (x,y) \in R \\ Mu &= \alpha u + \beta u_x + \gamma u_y = \delta \quad (x,y) \in \partial R. \end{aligned}$$

Here, a, b, c, d, e, f, g, α, β, γ and δ are functions of x and y. A Hermite cubic is a C^1 piecewise cubic polynomial and a Hermite bicubic is the tensor product of Hermite cubics in x and in y. We consider a collocation method using a rectangular grid on R and Hermite bicubic basis functions with grid lines as breakpoints to approximate $u(x,y)$ by

$$U(x,y) = \sum a_{ij}\, h_i(x)\, h_j(y) \approx u(x,y).$$

---

1. This work supported on part by Department of Energy contract DE-AC02-81ER10997

2. Department of Mathematics and Computer Science, Calvin College, Grand Rapids, Michigan, 49506.
3. Division of Mathematical Sciences, Purdue University, West Lafayette, Indiana 47907.

ISBN 0-12-100560-7

The coefficients a_{ij} of U are determined by the *collocation equations* at four interior points of each grid square plus at points along the boundary:

$$LU(x_k, y_k) = g(x_k, y_k) \quad k = 1,\ldots,N$$
$$MU(x_k, y_k) = \delta(x_k, y_k) \quad k = N+1,\ldots,K$$

where $N = 4n^2$, $K-N = 4(2n+1)$ for a grid of n^2 rectangles (Rice, 1983).

The unknowns and equations of the collocation equation are usually numbered so as to give a coefficient matrix of the linear system which is somewhat bi-diagonal (i.e., with the finite element ordering). With $n=3$ and a Poisson problem with Dirichlet boundary conditions, the pattern of non-zero elements is as shown in Figure 1.

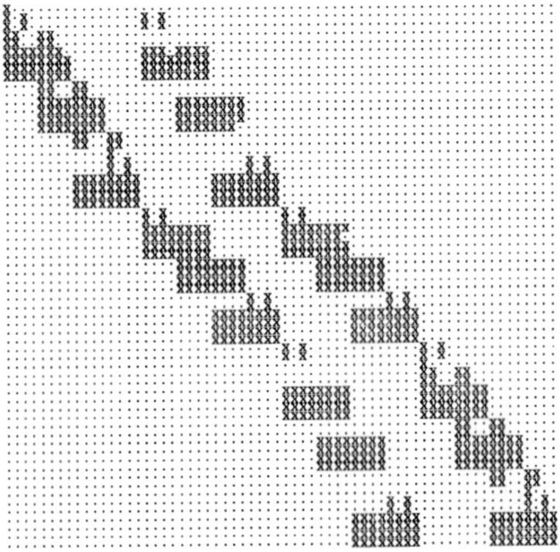

FIGURE 1. The pattern of non-zero elements in the collocation equations with the usual orderings of the equations and unknowns. This is for a 3×3 grid and a Poisson problem with Dirichlet boundary conditions.

The pattern seen in Figure 1 is not the expected somewhat tri-diagonal pattern of non-zeros and, potentially more serious, one cannot use any of the standard iteration methods on the equations because of the large number of zero elements on the diagonal. Thus we were led to search for a new ordering which has non-zeros on the diagonals and which is more "natural". Such an ordering, called collorder, is presented in the next section. The tensor product ordering is also such an ordering.

Unfortunately, standard iteration methods do not converge for these equations in the collorder ordering (see (Balart, 1982), (Dyksen, 1982) for iteration methods that do converge). We thus turned our attention to the behavior of Gauss elimination applied with this ordering. We discovered that the standard LINPACK routines (Dongarra, 1979) give unacceptable results because of scaling problems. The scaling of these equations is analyzed in (Dyksen and Rice, 1983); we merely report here that *a priori* scaling is difficult (at best) and that one should use scaled partial pivoting which we call BAND GE here. Once we switched to BAND GE, we found round-off error effects to be remarkably low. Even more surprising, we found that pivoting is **not** needed with the collorder ordering, and that the collorder ordering is essentially the ordering one obtains when applying scaled partial pivoting to the equations in the ordering of in Figure 1.

In Section III we present the results of an experimental study which strongly supports the hypothesis: **Gauss elimination without pivoting is safe for the collorder ordering of the collocation equations.**

Even more surprising to us is the fact: **not pivoting in the Gauss elimination does not reduce the computational time in solving these equations.** The usual operations count approach suggests that the arithmetic should be halved; this does not happen. There is about a 40% savings in the total memory required for the computation.

II. THE COLLORDER ORDERING

Once the grid is placed over the rectangle R, the grid intersection points are numbered in a natural way from west to east, south to north. The collocation points (equations) are then associated with the nearest grid point and are numbered in groups of four in the order of their corresponding grid point. For example, with a 3×3 grid, there are 64 collocation points whose ordering is given in Figure 2.

The important idea in the above ordering is that the grid points are first numbered in a natural way and then the collocation points are numbered in the order of their corresponding grid point. We could number the grid points south to north, west to east as long as we suitably modify the rest of the ordering scheme described in the following sections. Similarly, the four collocation points associated with a grid point may be locally ordered in any way.

51	52	55	56	59	60	63	64
49	50	53	54	57	58	61	62
35	36	39	40	43	44	47	48
33	34	37	38	41	42	45	46
19	20	23	24	27	28	31	32
17	18	21	22	25	26	29	30
3	4	7	8	11	12	15	16
1	2	5	6	9	10	13	14

FIGURE 2. The collorder ordering of the collocation points for a 3×3 grid with 64 unknowns.

To the $i^{\text{th}}[j^{\text{th}}]$ $x[y]$ grid coordinate there are associated two Hermite cubic functions: $\varphi_i^x[\varphi_j^y]$ which takes on function values and $\psi_i^x[\psi_j^y]$ which takes on derivative values. The Hermite bicubic basis functions (unknowns) are then ordered corresponding to their support in a natural way from west to east, south to north.

This ordering of the collocation points and the basis functions yields an intermediate collorder matrix with a nice structure. It is a band matrix with band width $4n+11$. The matrix is block symmetric in that it consists of 4×4 blocks which if $B_{ij} \neq 0$ then $B_{ji} \neq 0$. There are at most 16 non-zero entries per row occurring in 4 blocks. All of the symmetric pairs of off-diagonal blocks can be stored in their natural order within one 4×4 block. However, the intermediate collorder matrix still has some zero diagonal elements corresponding to boundary conditions.

To further reorder this matrix to remove all zeros from the diagonal we identify eight types of boundary points labeled in an obvious way by Northwest, North, Northeast, East, Southeast, South, Southwest and West.

For Dirichlet boundary conditions the 4×4 block in the upper left is associated with the southwest grid point and has zeros on the diagonal in rows 2 and 3 as shown below on the left. If we permute the second and third basis functions (columns 2 and 3), we obtain a non-zero diagonal in this block; that is,

$$\begin{matrix} X & 0 & 0 & 0 \\ X & 0 & X & 0 \\ X & X & 0 & 0 \\ X & X & X & X \end{matrix} \quad \rightarrow \quad \begin{matrix} X & 0 & 0 & 0 \\ X & X & 0 & 0 \\ X & 0 & X & 0 \\ X & X & X & X \end{matrix}$$

By considering all nine possible boundary conditions at the southwest corner, one can always reorder the columns to obtain a non-zero diagonal in this 4×4 diagonal block. Table 1 gives the possible boundary conditions at the southwest corner with the corresponding permutations.

TABLE 1. Column permutations at the southwest corner.

Southwest corner

Boundary condition		Block before	Column permutation	Block after
West	South			
Dirichlet	Dirichlet	XOOO XOXO XXOO XXXX	2 → 3 3 → 2	XOOO XXOO XOXO XXXX
Neumann	Dirichlet	XOOO XOXO OOXX XXXX	2 → 4 3 → 2 4 → 3	XOOO XXOO OXXO XXXX
Mixed	Dirichlet	XOOO XOXO XXXX XXXX	2 → 3 3 → 2	XOOO XXOO XXXX XXXX
Dirichlet	Neumann	OXOO OXOX XXOO XXXX	1 → 3 2 → 1 3 → 4 4 → 2	XOOO XXOO XOXO XXXX
Neumann	Neumann	OXOO OXOX OOXX XXXX	1 → 4 2 → 1 4 → 2	XOOO XXOO OXXO XXXX
Mixed	Neumann	OXOO OXOX XXXX XXXX	1 → 4 2 → 1 4 → 2	XOOO XXOO XXXX XXXX
Dirichlet	Mixed	XXOO XXXX XXOO XXXX	2 → 3 3 → 2	XOXO XXXX XOXO XXXX
Neumann	Mixed	XXOO XXXX OOXX XXXX	2 → 4 4 → 2	XOOX XXXX OXXO XXXX
Mixed	Mixed	XXOO XXXX XXXX XXXX	2 → 3 3 → 2	XOXO XXXX XXXX XXXX

Similar reorderings can be made for every type of boundary grid point to obtain a non-zero diagonal in the corresponding diagonal block. A complete set of tables describing the reordering of the boundary basis functions is given in (Dyksen and Rice, 1982). The result for the Poisson equation with Dirichlet boundary conditions is shown in Figure 4. Appendix B of (Dyksen, 1981) shows a complete set of example matrices with the collorder ordering.

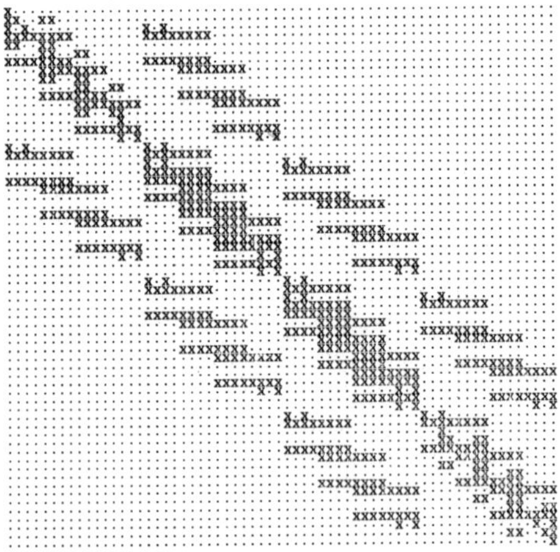

FIGURE 4. The collorder matrix for a 3×3 *grid with* 64 *unknowns.*

This ordering is implemented in the ELLPACK system as the indexing module HERMITE COLLORDER.

III. PIVOTING IN GAUSS ELIMINATION WITH THE COLLORDER ORDERING

We consider the following hypothesis: **Gauss elimination without pivoting is safe for the collorder ordering of the collocation equations.** We would, of course, like to establish this by analytical means but have been unable to do so. We do not believe this phenomena is restricted to collocation; it is a more general phenomena in the numerical solution of elliptic problems.

We test this hypothesis in the ELLPACK environment by considering the following modules for solving elliptic problems: HERMITE COLLOCATION, P3C1 COLLOCATION, HERMITE COLLORDER, BAND GE, and BAND GE NO PIVOTING. The modules HERMITE COLLOCATION and P3C1 COLLOCATION generate the same equations, except that HERMITE COLLOCATION scales the equations (as best we know how). BAND GE uses the LINPACK routines SGBFA and SGBSL modified to do scaled partial pivoting, while BAND GE NO PIVOTING is the same module without pivoting. HERMITE COLLORDER applies the collorder ordering to the equations.

A method is a combination of these modules and they are denoted using the following scheme:

S means use scaling = HERMITE COLLOCATION

C means use collorder ordering = HERMITE COLLORDER

P means use pivoting = BAND GE

NP means use no pivoting = BAND GE NO PIVOTING

If P3C1 COLLOCATION is used, then no letter is included in the scheme. Thus, methods SP and CNP are HERMITE COLLOCATION + BAND GE and P3C1 COLLOCATION + HERMITE COLLORDER + BAND GE NO PIVOTING.

These methods were applied to the following 20 elliptic problems from the population of (Rice et al, 1981): 2-1, 3-1, 5-1, 6-1, 9-2, 10-2, 10-3, 12-3, 17-2, 20-2, 22-1, 23-6, 33-1, 35-3, 38-1, 40-1, 50-1, 53-3, 54-2, 59-1 There is a variety of operators in this set; 15 of these problems have Dirichlet boundary conditions (5 are homogeneous).

We state below more specific, testable hypotheses for accuracy in terms of "method A" = "method B" along with an interpretation (pivoting always means scaled partial pivoting).

H1: SCNP = SP: For scaled equations, collorder and no pivoting is the same as pivoting

H2: CNP = P: For unscaled equations, collorder and no pivoting is the same as pivoting

H3: SCNP = SCP: For scaled equations, pivoting is not needed for the collorder ordering

H4: SCP = SP: For scaled equations, collorder has no effect if pivoting is used.

These four hypotheses are tested with the sample described above and accuracy measured as follows:

(a) subjective evaluation of accuracy from plots for each PDE,

(b) grid size required to achieve 5%, 0.5% and 0.05% accuracy.

The second set of measures is complicated by the fact that the accuracy is dominated by round-off effects for fine grids. However, we are fortunate in that all four measures give consistent results. We give data for measure (a) in Table 2 since it is simpler.

TABLE 2. Accuracy results in testing the four hypotheses with 20 PDE problems.

Hypothesis	No difference	First method better	Second method better
H1: SCNP = SP	13	5	2
H2: CNP = P	13	5	2
H3: SCNP = SCP	13	5	2
H4: SCP = SP	19	1	0

None of the differences observed are statistically significant using the Friedman, Kendall and Babington-Smith test (Hollander and Wolfe, 1973).

We also constructed a set of eight PDE problems whose exact solutions are bicubic polynomials. Every method under consideration here gives the exact solution except for round-off. For this population we studied the growth of round-off effects for solving the collocation equations in the collorder ordering using pivoting. We varied n from 2 to 28 (36 to 3364 unknowns) and computed the ratios of the maximum to minimum round-off effects; these results are summarized below.

PDE	$\frac{\text{Maximum round-off}}{\text{Minimum round-off}}$
1	27
2	17
3	13
4	10
5	30
6	17
7	12
8	13

For 4 of the problems there is a definite linear trend in the log-log plot of error versus n; the maximum slope is 0.75 suggesting that round-off grows at worst like $n^{0.75}$ with the collorder ordering and no pivoting. In 2 of the problems there is a hint of a linear trend and in the other 2 the round-off effect is constant over this range of problem sizes.

For comparison, we solved these same eight PDE problems with the unscaled equations using the LINPACK software; the round-off growth ratios observed range from 500 to 15,000 (average of 7500). Even these growth ratios are small compared to what one would expect since the condition numbers of the linear systems range from 10^7 to 10^{11} when $n=28$. Data for three other problems is given in Table 3; these numbers differ from those given above where round-off growth is measured with respect to the minimum observed round-off and not in units in the last place.

TABLE 3. Round-off effects (in units in the last place) observed using LINPACK on the collorder ordering on the unscaled collocation equations.

Problem	Number of equations	Condition number	Round-off No pivoting	Round-off Pivoting
2-1	1156	$1.8*10^6$	50	400
	3364	$9.8*10^6$	500,000	2,000
22-1	1156	$2.2*10^{10}$	<20	500,000
	3364	$2.4*10^{11}$	80	10,000,000
59-1	1156	$1.0*10^8$	8	40,000
	3364	$9.1*10^8$	50	300,000

We now summarize other relevant data and work. We computed the growth factor and maximum multiplier for Gauss elimination on the scaled equations with the collorder ordering. Typical results indicate that the equations do not become more difficult as the size of the system grows.

	Number of equations 84	1228	4492
Growth factor with pivoting	1.21	1.21	-
Growth factor without pivoting	2.04	1.56	-
Maximum multiplier without pivoting	22.4	22.5	22.5

In (Trottenberg and Witsch, 1981) it is shown that, for a fixed continuous elliptic problem, the condition of the corresponding discretized problem is independent of h, the fineness of the discretization. Note that the condition of a numerical computation may be much worse than that of the problem itself. This suggests that there are numerical methods for elliptic problems which are numerically stable even for discretizations with a very large number of equations. One data point to support this conclusion is seen in Figure 5(c) of (Rice, 1981) where full machine accuracy is obtained by a high order finite difference plus FFT method involving over 65,000 unknowns. Another data point is Example 7.6.2 in (Dyksen, 1982) where no round-off effect is observed for a tensor product method with 82,368 unknowns.

The study of (Hammarling and Wilkinson, 1980) includes the usual finite difference systems arising from the Laplace operator. They test whether the relative error grows like the condition number (about n^2 or $1/h^2$) of the linear system. The experimental data stop at $n=30$ or 35 and do not show conclusive trends. A similar analysis and study appears in (Birkhoff and Lynch, 1983) and they also suggest that the error due to solving the finite difference equations grows like n^2. Implicit in both these analyses is the use of ordinary Gauss elimination; the FFT example cited above shows that the error growth can be much less.

The error in solving the collocation equations is much less correlated with the condition number of the linear system as is seen from the data in Table 3. For the collorder ordering the round-off grows at worse like $n^{0.75}$ while the condition number grows like n^4. We used the Galerkin method instead of collocation on Problem 5-1 and solved the linear system by the LINPACK software for symmetric positive definite band systems. The condition number of the linear system is $2.3*10^9$ but only 1600 units in the last place of accuracy are lost.

IV. WHY PIVOTING IS NOT NEEDED WITH THE COLLORDER ORDERING

The reason that pivoting is not needed with the collorder ordering is that this ordering is essentially the order obtained when one does scaled partial pivoting of the scaled equations. Recall that the role of scaling is to rationalize partial pivoting, so if one knows in advance what the partial pivoting order is, then one does not need to do scaling. We have tried without success to prove that the collorder ordering is as we state, so our statement is a conjecture rather than a fact. We have checked this conjecture on a large number of PDEs and found it to hold.

A typical example is shown in Figure 5 for $n=3$. At the top we show the collocation points (equations) ordered two ways. At the bottom we show corresponding elimination orderings using scaled partial pivoting. We see that the elimination orderings are almost identical to each other and to the collorder ordering. The differences in the elimination and collorder ordering is almost always restricted to permutation of the order of the four points associated with a grid point. The subordering of collorder used at a grid point is somewhat arbitrary and we believe it to be somewhat irrelevant.

FINITE ELEMENT

43	44	45	50	51	56	57	58
42	47	48	53	54	62	63	60
41	46	49	52	55	61	64	59
26	28	29	32	33	38	39	36
25	27	30	31	34	37	40	35
3	7	8	13	14	22	23	20
2	6	9	12	15	21	24	19
1	4	5	10	11	16	17	18

HERMITE COLLORDER

51	52	55	56	59	60	63	64
49	50	53	54	57	58	61	62
35	36	39	40	43	44	47	48
33	34	37	38	41	42	45	46
19	20	23	24	27	28	31	32
17	18	21	22	25	26	29	30
3	4	7	8	11	12	15	16
1	2	5	6	9	10	13	14

PARTIAL PIVOTING

50	51	56	53	60	57	63	62
52	49	54	55	58	59	61	64
34	35	38	37	42	41	46	45
36	33	39	40	44	43	47	48
18	19	22	21	26	25	30	29
20	17	24	23	28	27	31	32
4	2	5	6	9	10	13	15
1	3	8	7	12	11	16	14

PARTIAL PIVOTING

52	49	55	54	59	58	64	61
50	51	53	56	57	60	62	63
34	35	38	37	42	41	45	46
36	33	39	40	44	43	48	47
18	19	22	21	26	25	29	30
20	17	24	23	28	27	32	31
3	2	5	6	9	10	15	13
1	4	8	7	12	11	14	16

FIGURE 5. (Top) Two orderings of the scaled collocation equations. (Bottom) The order of elimination of the collocation equations using scaled partial pivoting.

V. COMPUTATIONAL EFFICIENCY

For a grid of n^2 rectangles we have a linear system with $4(n+1)^2$ unknowns and band width $4n+11$. Standard operation counts for memory and arithmetic are (neglecting lower order terms).

	Multiplies	Memory
Elimination with pivoting	$64\,n^4$	$48\,n^4$
Elimination without pivoting	$32\,n^4$	$32\,n^4$

Naive Gauss elimination with pivoting would reflect these counts directly. We found that BAND GE NO PIVOTING actually requires **more** computing time than BAND GE for these equations. The reason is that no pivoting leads to much more fill-in in the band as seen from the number of calls to the BLAS routine SAXPY in the case of $n=28$ (3364 equations).

Module used	Calls on SAXPY	Zero returns	Difference	Total seconds to run
BAND GE	410,866	104,979	305,887	767
BAND GE NO PIVOTING	406,146	40,785	365,361	846

Recall that SAXPY computes the vector sum $a\mathbf{x}+\mathbf{y}$ and is in the inner loop of Gauss elimination so the total computation time should be (and is) proportional to the number of its executions. A zero return from SAXPY does not involve any significant computation.

For a sample of 45 cases, we observed that no pivoting requires an average of 8% more execution time than pivoting with a range of 4 to 14% more. Pivoting requires about 40% more total memory than not pivoting.

VI. CONCLUSIONS AND CONJECTURES

The collorder ordering has several nice properties: (1) the matrix has a non-zero diagonal and is somewhat symmetric, (2) the linear system can be solved by elimination without pivoting or scaling, and the computation is exceptionally stable, (3) there is a savings in the elimination of about 40% in memory at the cost of 5–10% longer execution time.

We conjecture that: **it is completely safe to use a "near collorder" ordering where the equations are ordered according to the grid points together with local partial pivoting used on the 4 by 4 blocks associated with the grid points.** This would greatly simplify the algorithm to obtain the collorder ordering by eliminating the reordering associated with boundary conditions. It is not likely to increase the speed and would increase the memory requirement by a factor of $1+O(h)$.

We make a second and somewhat more amorphous conjecture: **it is essential for numerical stability that the elimination order follow a continuous geometric progression through the domain of the problem.** We believe this to be a general property of discretizations of elliptic problems. We have experimented with both finite elements and finite differences and observed that "disrupting" the geometric pattern during elimination usually produces numerical instability, sometimes of enormous magnitude. It is as though the round-off error satisfies an initial-value problem which is parabolic and smoothing if the geometry is followed properly and is hyperbolic and unstable if the geometry is not followed.

REFERENCES

Balart, R., Houstis, E. N. and Papatheodorou, T. S. (1982), On the iteration solution of collocation method equations, *10th IMACS Congress*, 98-100.

Birkhoff, G. and Lynch, R. E. (1983), "Numerical Solution of Elliptic Partial Differential Equations", SIAM Publications.

Dongarra, J. J. et al (1979), "LINPACK User's Guide", SIAM Publications.

Dyksen, W. R. (1981), A new ordering scheme for the Hermite bicubic collocation equations, CSD-TR 364, CS Department, Purdue University.

Dyksen, W. R. (1982), Tensor product generalized alternating direction implicit methods for solving separable second order linear elliptic partial differential equations, Ph. D. Thesis, Purdue University.

Dyksen, W. R. and Rice, J. R. (1982), A new ordering scheme for the Hermite bicubic collocation equations, CSD-TR 425, CS Department, Purdue.

Dyksen, W. R. and Rice, J. R. (1983), The importance of scaling in the collocation method, to appear.

Hammarling, S. and Wilkinson, J. H. (1980), On linear systems arising from finite difference approximations to elliptic differential equations, Rpt. DNACS 34/80, National Physical Laboratory, England.

Hollander, M. and Wolfe, D. A. (1973), "Non parametric Statistical Methods", John Wiley.

Rice, J. R. (1981), ELLPACK: Progress and plans, in *Elliptic Problem Solvers* (M. Schultz, ed.) Academic Press, 135-162.

Rice, J. R., Houstis, E. N. and Dyksen, W. R. (1981), A population of linear, second order elliptic partial differential equations on rectangular domains, *Math. Comp.*, **36**, 475-484.

Rice, J. R. (1983), "Numerical Methods, Software and Analysis", McGraw-Hill.

Trottenberg, U. and Witsch, K. (1981), Zur Kondition diskreter elliptischer Randwertaufgaben, Preprint No. 457, GMD-Bonn.

SOLUTION METHODS FOR THE FINITE ELEMENT EQUATIONS IN NONLINEAR SHELL ANALYSIS[1]

Kelton Flinn and Lois Mansfield[2]

We compare the use of an iterative method with the use of direct methods for solving the linear systems arising when Newton's method along with continuation is used to solve the nonlinear equations associated with the large displacement analysis of shells or plates using the finite element method. Representative program execution times will be given which show that the conjugate gradient method is considerably more efficient than Gauss elimination for solving these linear systems.

The general form of the algebraic problem to be solved is

$$F_h(x^h)-\lambda f = 0 \tag{1}$$

where F is a nonlinear algebraic operator, f is a vector representing the applied force and λ is the load parameter. One wants to trace the load–displacement curve which will involve finding x^h for a sequence of values of λ, starting with λ = 0. The continuation method for (1) we consider is

initialize $n := 0,\ \lambda_n := 0,\ x_0^h := 0$

<u>repeat</u>

(i) set $\lambda_{n+1} = \lambda_n + \Delta\lambda$

(ii) guess $x_{n+1}^h = x_n^h + \Delta\lambda \dfrac{\partial x_n^h}{\partial \lambda}$

---

1. Research supported by NSF Grant MCS80–16532.
2. Department of Applied Mathematics and Computer Science, University of Virginia, Charlottesville, Virginia.

ISBN 0-12-100560-7

(iii) use Newton's method to find x^h_{n+1} to the desired accuracy

<u>until</u> λ_n is sufficiently large.

We shall assume that we remain on that part of the load displacement curve where λ is an appropriate parameter, i.e. before any bifurcation point or limit point is reached. The motivation for investigating iterative methods to solve the linear systems arising in (iii) is that in calculating the next Newton iterate one has a good initial guess in the previous iterate.

The usual method of obtaining the nonlinear system of partial differential equations which is solved in the large displacement analysis of plates and shells is by minimizing the potential energy functional. We have found, however, in [3], [4] that the so-called mixed formulation which adds the stresses and bending moments to the set of unknowns is considerably more efficient, even though there are many more equations to be solved. This greater efficiency comes from the fact that for the mixed formulation the equations are quadratic in the unknowns whereas for the potential energy formulation the equations are cubic in the unknowns. In addition, we have found that the mixed equations are less stiff and so that much large load increments may be taken.

In the conventional shell theory the transverse displacement w occurs with second partial derivatives. This means that the finite element approximation w^h to w should be in C^1. The simplest conforming finite element spaces over a triangulation of the domain seem to be those derived from Clough-Tocher cubics. To achieve optimal convergence and to satisfy the compatibility requirements between subspaces which all mixed methods require (see [2]), one should then choose u^h and v^h, the approximations to the tangential displacements u and v, to be piecewise C^0 quadratics while the stresses N^h and M^h should be piecewise linear polynomials with possible jumps across edges of triangles. All of this means that the number of unknowns on each triangle of the triangulation is really quite large.

The second partial derivatives of w arise from assuming Kirchhoff's hypothesis. To try to reduce the complexity of the problem we were led to consider the Reissner formulation which does not assume Kirchhoff's hypothesis although it does require the introduction of more variables, the rotations ϕ_1, ϕ_2 and the twisting moments Q_1, Q_2. Let the shell surface be defined parametrically by the position vector $r(x,y)$ which is a vector function of two curvilinear surface coordinates x and y. This new problem is to find a stationary value of

$$\frac{A}{2}\int_\Omega \left[N_{11}^2 - 2\nu N_{11}N_{22} + 2(1+\nu)N_{12}^2 + N_{22}^2\right]\alpha_1\alpha_2 dxdy \tag{2}$$

$$+ \frac{D}{2}\int_\Omega \left[M_{11}^2 - 2\nu M_{11}M_{22} + 2(1+\nu)M_{12}^2 + M_{22}^2\right]\alpha_1\alpha_2 dxdy$$

$$+ \frac{A_Q}{2}\int_\Omega \left[Q_1^2 + Q_2^2\right]\alpha_1\alpha_2 dxdy$$

$$+ \int_\Omega \left[\frac{1}{\alpha_1}u_x + k_{11}w + 1/2\left[\frac{1}{\alpha_1}w_x\right]^2 - A\left[N_{11} - \nu N_{22}\right]\right]N_{11}\alpha_1\alpha_2 dxdy$$

$$+ \int_\Omega \left[\frac{1}{\alpha_2}v_y + k_{22}w + 1/2\left[\frac{1}{\alpha_2}w_y\right]^2 - A\left[N_{22} - \nu N_{11}\right]\right]N_{22}\alpha_1\alpha_2 dxdy$$

$$+ \int_\Omega \left[\frac{1}{\alpha_2}u_y + \frac{1}{\alpha_1}v_x + k_{12}w + \frac{1}{\alpha_1\alpha_2}w_x w_y - A\left[1+\nu\right]N_{12}\right]N_{12}\alpha_1\alpha_2 dxdy$$

$$+ \int_\Omega \left[\frac{1}{\alpha_1}\frac{\partial\phi_1}{\partial x} - D\left[M_{11} - \nu M_{22}\right]\right]M_{11}\alpha_1\alpha_2 dxdy$$

$$+ \int_\Omega \left[\frac{1}{\alpha_2}\frac{\partial\phi_2}{\partial y} - D\left[M_{22} - \nu M_{11}\right]\right]M_{22}\alpha_1\alpha_2 dxdy$$

$$+ \int_\Omega \left[\frac{1}{\alpha_1}\frac{\partial\phi_2}{\partial x} + \frac{1}{\alpha_2}\frac{\partial\phi_1}{\partial y} - D\left[1+\nu\right]M_{12}\right]M_{12}\alpha_1\alpha_2 dxdy$$

$$+ \int_\Omega \left[\phi_1 + \frac{1}{\alpha_1}w_x - A_Q Q_1\right]Q_1 + \left[\phi_2 + \frac{1}{\alpha_2}w_y - A_Q Q_2\right]Q_2\alpha_1\alpha_2 dxdy$$

$$- \lambda\int_\Omega f w \alpha_1\alpha_2 dxdy$$

where Ω is the shell domain,

$$\alpha_1^2 = \sum_{j=1}^{3}\left[\frac{\partial r_j}{\partial x}\right]^2 \quad ,\alpha_2^2 = \sum_{j=1}^{3}\left[\frac{\partial r_j}{\partial y}\right]^2 \quad .$$

A, D, and A_Q are coefficients, and the k_{ij} denote the curvature of the shell.

The first three terms along with the last term in (2) represent the strain energy. The other terms are constraints which are introduced using Lagrange multipliers. These Lagrange multipliers can be shown to actually be the stresses and bending moments which appear in (2). If Q_1 and Q_2 are assumed to be zero, then

$$\phi_1 = \frac{-1}{\alpha_1}w_x \quad , \quad \phi_2 = \frac{-1}{\alpha_2}w_y,$$

which can then be substituted into the seventh, eighth, and ninth terms of (2) giving second partials of w.

In our work we have assumed that the shell is clamped and so took the boundary conditions to be

$$u = v = w = w_n = 0 \ on \ \partial\Omega.$$

In our numerical experiments reported on here we have taken the shell domain to be that part of the sphere of radius 10 centered at (1/2, 1/2, 0) which is contained within the parallelepiped defined by $0 \leq x \leq 1$, $0 \leq y \leq 1$, $0 \leq z < \infty$. This makes Ω the unit square on which we imposed a uniform triangular grid obtained from a rectangular grid by drawing in the diagonal from the upper left corner to the lower right corner on each rectangle. We took the force f to be a uniform pressure. Of course, one expects that singularities will occur near the corners so that a nonuniform grid should be more accurate. In this study, however, our objective was to evaluate computation times and so we weren't particularly concerned with accuracy.

In the tables below we compare the number of operations per element (i.e. triangle) required to set up the Jacobian and system itself for each iteration of Newton's method, and also the amount of storage required for the conventional Kirchhoff formulation with Clough–Tocher cubic finite elements used to approximate w as opposed to the Reissner formulation using either piecewise linear polynomials or piecewise quadratic polynomials to approximate w (Reissner1 or Reissner2). Although no convergence results have been obtained for the Reissner formulation it seems one should then use piecewise linear or piecewise quadratic polynomials to approximate u, v, ϕ_1, ϕ_2. For the Kirchhoff formulation, it was shown in [2] that compatibility of the finite element subspaces is achieved in the mixed method if the linearized constraints between displacements and stresses and bending moments are satisfied exactly neglecting curvature. This suggests one should use piecewise constants to approximate $\underline{N}$ and $\underline{M}$ and piecewise linear polynomials with possible jumps across edges of the triangles to approximate Q_1 and Q_2 or piecewise linear polynomials to approximate $\underline{N}$ and $\underline{M}$ and piecewise quadratic polynomials to approximate Q_1 and Q_2, all with possible jumps across edges of the triangles. Since the Reissner formulation with piecewise linear polynomials for w was so much simpler we used it exclusively in our numerical experiments.

Table 1

Operation Counts		
	Jacobian	Systems
Kirchhoff	7,153	12,084
Reissner 1	456	1,202
Reissner 2	2,180	6,714

Table 2

Storage Requirements		
	per element	total for 10x10 grid
Kirchhoff	2,916	2,996,000
Reissner 1	841	870,000
Reissner 2	6,400	6,560,000

For mixed methods the matrix associated with the system of linear equations which must be solved in each iteration of Newton's method is not positive definite. However, the system can be written in the form

$$\begin{bmatrix} M & B \\ B^T & C \end{bmatrix} \begin{bmatrix} \sigma \\ u \end{bmatrix} = \begin{bmatrix} f \\ g \end{bmatrix},$$

where the vector σ represents all of the unknowns associated with the stresses and bending moments and u represents the unknowns associated with the displacements and rotations. Because we allow the finite elements used to approximate the stresses and bending moments to have jumps across edges of the triangles the submatrix M is block diagonal. Our choice of finite element spaces gave 3x3 blocks. In addition M is independent of u and σ and so only needs to be set up once. The submatrix C is 0 when $\lambda = 0$ and represents a totally nonlinear contribution to the problem. This suggests the iterative algorithm repeat

(i) solve $M\sigma = f - Bu$ for a new σ

(ii) use $B^T\sigma + cu = g$ to update u

until convergence.

The system (i) is easy to solve. For (ii) we chose to use the conjugate gradient method. Note that if we substitute

$$\sigma = M^{-1}f - M^{-1}Bu$$

into

$$B^T\sigma + Cu = g,$$

we get

$$(-B^T M^{-1} B + C)u = g - B^T M^{-1} f$$

or

$$Au = q'.$$

The matrix A is positive definite when $\lambda=0$, since it is then the matrix for the linearized potential energy formulation of the problem. Presumably A remains positive definite until the first bifurcation point or limit point is encountered. Of course we never form the matrix $B^T M^{-1} B$ explicitly.

The algorithm we used was

input u^0

solve $M\sigma^0 = f - Bu^0$

$r^0 := g - B^T\sigma^0 - Cu^0$; $p^0 := r^0$; $n := 0$

repeat

$n := n+1$

solve $My = Bp^{n-1}$

$z := Cp^{n-1} - B^T y$

$$\gamma := \frac{(r^{n-1}, p^{n-1})}{(z, p^{n-1})}$$

$u^n := u^{n-1} + \gamma p^{n-1}$

solve $M\sigma^n = f - Bu^n$

$r^n := g - B^T\sigma^n - Cu^n$

$\alpha := \frac{(z, r^n)}{(z, p^{n-1})}$

$p^n := r^n - \alpha p^{n-1}$

<u>until</u> $\| r^n \|$ is sufficiently small.

The direct method we used for comparison was Gauss elimination with the linear system stored as a banded system. We used the natural ordering for the nodes with all unknowns associated with a given node ordered consecutively.

In Tables 3–6 below we give program execution times in seconds on a VAX 11/780 to solve the shell equations for a single continuation step using both the conjugate gradient algorithm just described and Gauss elimination. In Tables 3 and 4 we took $\Delta\lambda = 10$. Table 3 gives the times to obtain a solution for the first continuation step where the initial guess was taken to be zero while Table 4 gives the times further along the curve where the initial guess was obtained as indicated in step (ii) of our continuation algorithm. Tables 5 and 6 give corresponding results where the much larger step size $\Delta\lambda = 1000$ was used.

In all cases the conjugate gradient method proved to be considerably more efficient than Gauss elimination, with the greatest advantage coming where the initial guess would be expected to be best. In cases where a fairly large number of conjugate gradient iterations were required, further efficiencies may be expected if an appropriate preconditioning were developed and used. This is a subject for further research.

Gauss elimination does have one advantage over iterative methods in tracing the load curve in that it is easy to monitor the determinant and thus detect the presence of a bifurcation point by the fact that the determinant of the Jacobian has changed sign. In experimental results for the circular arch in [1] we have discovered that at a bifurcation point the smallest eigenvalue of the Jacobian changes sign. This can easily be monitored using a few iterations of the inverse power method and seems worthy of additional study.

Table 3

Start up with $\Delta\lambda$ = 10					
unknowns	Newton iterations	conj. grad. iterations	conj.grad. time	Gauss el. time	conj. grad/ Gauss el.
509	27	393	4,826	10,089	.478
780	29	607	15,538	34,600	.449
1109	36	987	48,376	110,620	.437

Table 4

Along curve with $\Delta\lambda$ = 10					
unknowns	Newton iterations	conj. grad. iterations	conj. grad. time	Gauss el. time	conj. grad./ Gauss el.
509	1	4	73.3	382.5	.19
780	1	4	122.3	1197	.102
1109	1	4	404.0	3074	.13

Table 5

Start up with $\Delta\lambda$ = 1000					
unknows	Newton iterations	conj. grad. iterations	conj. grad. time	Gauss el. time	conj. grad./ Gauss el.
296	27	337	1658	2531	.655

Table 6

Along curve with $\Delta\lambda$ = 1000					
unknowns	Newton iterations	conj. grad. iterations	conj.grad. time	Gauss el. time	conj. grad./ Gauss el.
296	5	34	189.9	468.7	.405

References

1. Flinn, K., "Buckling of shallow circular arches," M.S. Thesis, University of Virginia, January, 1982.

2. Mansfield, L., "Finite element methods for nonlinear shell analysis," Numer. Math., 37 (1981), 121-131.

3. Mansfield, L., "Comparison of methods of solution of the finite element equations for the large displacement analysis of arches," in Elliptic Problem Solvers, ed., M. H. Schultz, Academic Press, New York, 1981, 361–367.

4. Mansfield, L., "On the solution of the finite element equations for nonlinear shell analysis," SIAM J. Sci. Stat. Comput., 3 (1982), 447–459.

V

Advances in Modeling and Physical Applications

DISCRETIZATION AND MULTILEVEL SOLUTION TECHNIQUES FOR NONLINEAR ELLIPTIC SYSTEMS

*Randolph E. Bank**
*Donald J. Rose***

1. INTRODUCTION

This paper identifies some numerical techniques that have proved successful in semiconductor device simulation (see [BR81], [BR82], [BRF83]) and suggests that these techniques may be applied more generally to solve other systems of nonlinear elliptic partial differential equation. Although we do not dwell on the appropriate definition of "nonlinear elliptic system," our discussion operationally would suggest a system in which the diagonal operators in the matrix of linear operators corresponding to the system operator Jacobian are elliptic. Indeed we focus on systems in which these diagonal operators are essentially self-adjoint and uniformly elliptic (positive definite).

To motivate our ideas, we consider a single nonlinear PDE in sections 2 and 3. We discuss scaling the equation and discretization procedures such that the linearized PDE is self-adjoint and uniformly elliptic. We consider approximate Newton techniques as in [BR81], [BR82] applied to the nonlinear operator and then think of discretizing the linear operator. We show that the "box method" of discretization leads to both finite difference and finite element equations with desirable properties.

We illustrate how to solve elliptic systems in section 4, which focuses on the semiconductor equations and considers nonlinear iteration at the operator level. To solve a PDE system, $g(z) = 0$, we discuss both block nonlinear iteration (a nonlinear Gauss-Seidel-like operator iteration) and a Newton-like iteration. Both approaches involve linearization and both require the solution of linear self-adjoint PDEs as discussed above. This allows techniques and software for single (linear) PDEs to be applied to the nonlinear PDE system in a natural manner. Section 5 shows how various multilevel techniques can be extended to such systems of PDEs.

* University of California, San Diego, California 92115
** Bell Laboratories, Murray Hill, New Jersey 07974

ISBN 0-12-100560-7

2. SCALING

In this section we consider the solution of nonlinear elliptic equations of the form

$$-\nabla(a(x,y,u)\nabla u) - f(x,y) = 0 \quad \text{in } \Omega\subset R^2 , \tag{2.1}$$

$$u = 0 \quad \text{on } \partial\Omega$$

by finite difference and finite element techniques. We assume Ω is a closed, bounded region of R^2, that (2.1) has a unique solution and that $0 < \underline{a} \leqslant a \leqslant \overline{a}$ for the relevant ranges of (x,y,u). Allowing f to depend on u with $\frac{\partial f}{\partial u} \leqslant 0$ or allowing more general boundary conditions does not alter our basic results but does complicate the exposition. We shall consider cases where $a(x,y,u)$ is of the following forms:

$$a(x,y,u) = a(u) \; ; \tag{2.2a}$$

$$a(x,y,u) = c(x,y)b(u), \, c > 0, \, b > 0 \, . \tag{2.2b}$$

First, we consider the case (2.2a). If $a(u)$ is constant, then (2.1) is a linear self adjoint uniformly elliptic (positive definite) equation, and its solution is relatively easy by a variety of techniques. For $a(u)$ nonlinear we will examine iterations with linear differential operators which maintain these properties.

One obvious approach is to use a Picard iteration of the form

$$-\nabla(a(u_k)\nabla u_{k+1}) - f(x,y) = 0 \quad \text{in } \Omega \, , \tag{2.3}$$

$$u_{k+1} = 0 \quad \text{on } \partial\Omega \, .$$

The usual discretizations of (2.3) will then lead to systems of equations with sparse, symmetric, positive definite matrices. If one writes (2.3) as

$$-\nabla(a(u_k)\nabla(u_{k+1}-u_k)) = f + \nabla(a(u_k)\nabla u_k) \equiv -r_k \quad \text{in } \Omega \, , \tag{2.4}$$

$$u_{k+1} - u_k = 0 \quad \text{on } \partial\Omega \, ,$$

then the iteration can be viewed as an approximate Newton method in which the Jacobian has been approximated by the operator $-\nabla(a(u_k)\nabla *)$ ($*$ is a "place holder" function). The rate of convergence of such a scheme will be linear in general with the rate depending on the size of $\frac{\partial a}{\partial u}$. If $a(u)$ is a highly nonlinear function, then this rate of convergence may be unacceptably slow. On the other hand, if one were to use Newton's method, one could achieve in theory quadratic convergence. However, at each step one must solve a linear elliptic equation of the form

$$-\nabla(a(u_k)\nabla\delta_k) - \left\{\nabla\left[\frac{\partial a}{\partial u}(u_k)\nabla u_k\right]\right\}\delta_k - \frac{\partial a}{\partial u}(u_k)\nabla u_k\cdot\nabla\delta_k = -r_k \text{ in } \Omega,$$

$$\delta_k = 0 \quad \text{on } \partial\Omega \tag{2.5}$$

where $\delta_k = u_{k+1} - u_k$ and r_k is as in (2.4).

If $\frac{\partial a}{\partial u}$ is large, straightforward discretizations of (2.5) may lead to highly nonsymmetric and possibly indefinite sets of linear equations. On the other hand, one can manipulate (2.5) into the form

$$-\Delta(a(u_k)\delta_k) = -r_k, \tag{2.6}$$

a simple elliptic equation for $a(u_k)\delta_k$. In this case we have apparently recovered the symmetry of the original problem. However, care must be taken in the discretization procedure so that the system of nonlinear equations to be solved will reflect these properties in a computationally attractive way. In particular, if (2.5) is regarded as a linear elliptic equation for δ_k, then (2.6) may be regarded as a linear elliptic equation for the scaled variable $\bar{\delta}_k = a(u_k)\delta_k$. The discretization process should be such that a simple scaling, (i.e., by a diagonal matrix) can be used to make the analogous discrete change of variables.

The situation is nearly as nice when $a(x,y,u) = c(x,y)b(u)$ as in (2.2b). Then the analogue of (2.6) is

$$-\nabla\{c(x,y)\nabla(b(u_k)\delta_k)\} = -r_k. \tag{2.7}$$

When $a(x,y,u)$ is not separable, there appears to be no simple scaling which leads directly to self adjoint, positive definite elliptic equations like (2.7) to be solved when Newton's method is applied. However, one can still use an approximate Newton method. Consider

$$a(x,y,u) = \frac{a(x,y,u)}{b(u)}\, b(u) \equiv c(x,y,u)b(u)$$

where $b(u) > 0$ is chosen to (hopefully) make $|\frac{\partial c}{\partial u}|$ small. Then the analogue of (2.7) (with $c(x,y)$ replaced by $c(x,y,u)$) can be regarded as an approximate Newton method. The true Newton iteration in this case is

$$-\nabla\{c(x,y,u_k)\nabla(b(u_k)\delta_k)\} - \nabla\left\{\frac{\partial c}{\partial u}(x,y,u_k)b(u_k)\delta_k\nabla u_k\right\} = -r_k \tag{2.8}$$

The approximate Newton strategy can be analyzed by the techniques given in

[BR81]. Those results show that the rate of convergence will be governed by the quantity

$$\alpha_k = ||\nabla(\frac{\partial c}{\partial u}(x,y,u_k)b(u_k)\delta_k \nabla u_k)||/||r_k|| \tag{2.9}$$

where $||\cdot||$ is an appropriate norm. In particular, $\alpha_k \leqslant \alpha < 1$ will guarantee linear convergence with rate α for a damped version of (2.7).

3. DISCRETIZATION

In this section we consider finite difference and finite element discretization procedures for equations of the form (2.7). We assume that an appropriate (possibly nonuniform) rectangular grid or a triangular finite element mesh is placed on the region Ω. The size of the mesh is characterized by a small parameter h, and the discrete solution will be denoted by u_k.

We begin by observing that the accuracy of the discrete solution u_k depends *only* on the discretization of the residual r_k, and in particular, is independent of the discretization of the Jacobian. While it is a common practice to discretize the residual and then compute the Jacobian matrix of the discrete set of nonlinear equations, this may not always be the best procedure.

Our main goal in discretizing r_k is to obtain an accurate solution u_k, while our main goal in discretizing the Jacobian is to obtain a set of linear equations which can be easily solved. Thus we suggest it may be advantageous to use slightly different discretizations for residual and Jacobian even though it will sacrifice quadratic convergence of Newton's method. If we formally have an $O(h^P)$ discretization of all terms, then the discrete Jacobian will agree with the Jacobian of the discrete residual to terms of order h^P. The resulting scheme can be regarded as an approximate Newton method with a linear rate of convergence which is asymptotically $O(h^P)$. Since we expect $||u - u_n|| = O(h^P)$ this rate is almost as good as quadratic convergence in terms of generating an approximation to the solution of the continuous problem.

We now consider a finite difference discretization of the Jacobian of (2.7) using the box method [V62]. Consider the grid point x_i,y_j in a possibly nonuniform rectangular grid.

We construct a box B around the grid point using the perpendicular bisectors of the edges connecting it to adjacent grid points, as illustrated in Fig. 3.1. Then

$$\iint_B -\nabla\{c(x,y,u)\nabla(b(u)\delta)\}dxdy = -\int_{\partial B} c(x,y,u)\nabla(b(u)\delta)\cdot n ds \tag{3.1}$$

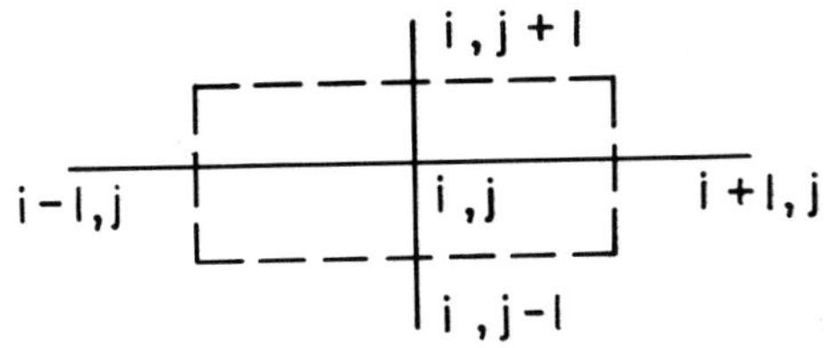

Fig. 3.1

where n is the outward pointing unit normal. We approximate the right hand side of (3.1) by

$$-\int_{\partial B} c(x,y,u)\nabla(b(u)\delta)\cdot n ds \approx \tag{3.2}$$

$$\frac{(k_{j-1}+k_j)}{2}\{(c_{i+1/2,j}\,(b_{ij}\delta_{ij}-b_{i+i,j}\delta_{i+1,j})/h_i)+(c_{i-1/2,j}\,(b_{ij}\delta_{ij}-b_{i-1,j}\delta_{i-1,j})/h_{i-1})\}$$

$$+\ \frac{(h_{i-1}+h_i)}{2}\{\left[c_{i,j+1/2}(b_{ij}\delta_{ij}-b_{i,j+1}\delta_{i,j+1})/k_j)+(c_{i,j-1/2}(b_{ij}\delta_{ij}-b_{i,j-1}\delta_{1,j-1})/k_{j-1})\}$$

where $h_i = x_{i+1} - x_i$, $k_j = y_{j+1} - y_j$, and

$$b_{ij} = b(u(x_i,y_j)) \ , \tag{3.3}$$

$$\delta_{ij} = \delta(x_i,y_i) \ .$$

There are several possible choices for terms like $c_{i\pm 1/2,j}$ and $c_{i,j\pm 1/2}$, for example,

$$c_{i+1/2,j} = c\left(\frac{x_i+x_{i+1}}{2},\ y_j,\ \frac{u(x_i,y_j)+u(x_{i+1},u_j)}{2}\right) ; \tag{3.4a}$$

or

$$c_{i+1/2,j} = (c(x_i y_j u_{ij}) + c(x_{i+1} y_j u_{i+1,j}))/2 \tag{3.4b}$$

Typically one expects (3.4a) to have a smaller truncation error and (3.4b) to require fewer function evaluations. In specific cases where c has several terms or is of special form, there may be a more appropriate alternative, perhaps involving a combination of (3.4a,b).

Grid points on or near the boundary are treated in an analogous fashion, although details vary according to the boundary conditions. In any event, one obtains a five point finite difference discretization of the partial differential equation. The coefficient matrix M is of the form

$$M = AD \tag{3.5}$$

where A is symmetric and positive definite (a Stieltjes matrix, in fact) and $D = \operatorname{diag}[b_{ij}]$ is diagonal.

There are many interesting possibilities for solving linear systems of the form $MX = B$. For example, one can set $Y = DX$ and solve

$$AY = B \tag{3.6a}$$

$$DX = Y \tag{3.6b}$$

The explicit scaling $DX = Y$ allows one to use algorithms and software for sparse symmetric positive definite matrices in (3.6a). However, it is also possible to treat the scaling implicitly and still accrue many of the benefits. For example, since A can be factored as

$$A = L\bar{D}L^T \tag{3.7}$$

where L is lower triangular and $\bar{D}$ diagonal, we may factor

$$M = LU \; ; \; U = \bar{D}L^T D \tag{3.8}$$

without pivoting. Also, if $A = \bar{A} - (\bar{A} - A)$ is splitting of A with iteration matrix

$$G = I - \bar{A}^{-1}A \, , \tag{3.9}$$

then the splitting $M = \bar{A}D - (\bar{A} - A)D$ has the iteration matrix

$$I - (\bar{A}D)^{-1}AD = D^{-1}GD \, . \tag{3.10}$$

Thus any convergent iterative method for A (including preconditioned conjugate gradient methods in which D will enter in the inner product computation) will have a correspondingly convergent iterative method for M.

We now consider a finite element discretization of (2.7) using continuous piecewise linear finite elements on a triangular mesh. In this case we must compute matrix elements by approximating integrals of the form,

$$\int_\Omega c(x,y)\nabla(b(u_k)\theta_j)\cdot\nabla\theta_i d\Omega \approx m_{ij} \, , \tag{3.11}$$

where $\{\theta_k\}$ are the usual modal basis functions in the finite element subspace.

Our goal is to find a quadrature rule of sufficient accuracy which will lead to a stiffness matrix M of the form (3.5). Many of the usual quadrature rules associated with the finite element method do not have this property. However, using the box method as a guide, we shall derive one appropriate rule.

Suppose that all interior angles of the triangular elements are smaller than $\pi/2$. Then about each vertex in the mesh we may construct a polygonal box B using the perpendicular bisectors of the edges incident to that vertex; see Fig.3.2.

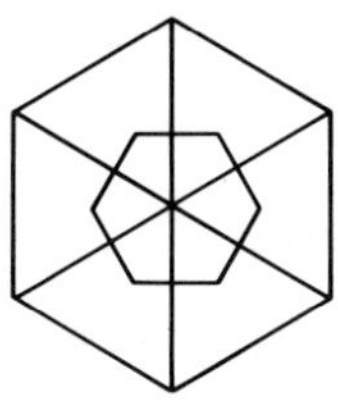

Fig. 3.2

The box method can now be applied to these polygons in a fashion analogous to the case of rectangles. However, since the finite element assembly process is not commonly oriented to equation by equation assembly but rather to element assembly, we may reformulate the box method as follows.

For each element, we connect the perpendicular bisectors of the three edges as in Fig. 3.3. The element is thus partioned into portions of three boxes and the corresponding integrals contribute to equations for unknowns associated with the vertices of the element. By performing these partial integrals, we can assemble a 3×3 element stiffness matrix, as in the normal finite element procedure.

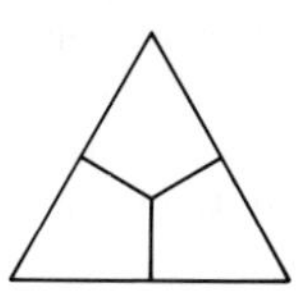

Fig. 3.3

For the operator $\mathscr{L}(u) = -\Delta u$, one can easily verify that the element stiffness matrices obtained from the box method are identical to those obtained by exact integration of (3.1) with $b = c = 1$. In the general case, one can interpret the matrix obtained as having resulted from the application of a specific quadrature rule to the integrals required in the usual assembly process.

Once the quadrature rule has been determined, one can disregard their derivation and simply use them in the usual elementwise assembly process. Even the requirement that all interior angles of the element be bounded by $\pi/2$ can be relaxed. However, in this more general setting a careful selection of the quadrature points for c is important to insure that the global stiffness matrix remains positive definite.

4. SEMICONDUCTOR DEVICE MODELS

A simplified model of an MOS semiconductor is given by the set of partial differential equations

$$-\Delta u + e^{u-v} - e^{w-u} - k_1(x,y) = 0\,, \tag{4.1}$$

$$-\nabla(\mu_n e^{u-v}\nabla v) - k_2(u,v,w) = 0\,,$$

$$-\nabla(\mu_p e^{w-u}\nabla w) + k_2(u,v,w) = 0\,.$$

Here u is the electrostatic potential, and v and w are the quasi-fermi levels for electrons and holes, respectively; k_1 is the doping profile for the device. For this discussion we take the mobilities, normally functions of ∇u, as given by $\mu_n = \mu_p = 1$ and the generation-recombination term $k_2 = 0$. The domain is typically rectangular, or the union of rectangles, and the boundary conditions are a combination of homogeneous Neumann and non-homogeneous Dirichlet.

More complete descriptions and derivations of these model equations can be found in [BFR83, FRB83]. Our concern here is not with the details of this model per se, but rather as a representative example of an important system of elliptic equations.

We shall symbolically denote the set of equations (4.1) as

$$g(z) = \begin{bmatrix} g_1(z) \\ g_2(z) \\ g_3(z) \end{bmatrix} = 0\,; \quad z = \begin{pmatrix} u \\ v \\ w \end{pmatrix}\,. \tag{4.2}$$

If (4.2) is linearized about a given point $(u,v,w)^T$, the resulting 3×3 operator Jacobian has the form

$$g' = \begin{bmatrix} -\Delta * + (e^{u-v} + e^{w-u})* & -e^{u-v}* & -e^{w-u}* \\ -\nabla(*e^{u-v}\nabla u) & -\nabla(e^{u}\nabla(e^{-v}*)) & 0 \\ \nabla(*e^{w-u}\nabla w) & 0 & -\nabla(e^{-u}\nabla(e^{w}*)) \end{bmatrix} \tag{4.3}$$

In the "matrix vector product" $g'\delta z$, $\delta z^T = (\delta u, \delta v, \delta w)$, the symbol $*$ in each entry shows where the corresponding entry of δz should be inserted.

There are two fundamental strategies for solving (4.2) which we will consider. The first, which we will call block nonlinear iteration (BNL) consists of using the equation for g_1 to update u, that for g_2 to update v and g_3 to update w. The second is to update all three variables using some Newton-like scheme (FN) as in (4.4)

$$(a) \quad g_k' \delta z_k = - g_k \tag{4.4}$$

$$(b) \quad z_{k+1} = z_k + t_k \delta z_k \ .$$

Here $g_k = g(z_k)$ $g_k' = g'(z_k)$, and t_k is a scalar chosen to make the method globally convergent [BR81].

In either case, the discretized versions of the diagonal blocks of (4.3) will play a major role. If the BNL scheme is used, and Newton's method used for each block, then the required Jacobian matrices correspond exactly to the diagonal of (4.3). If the full Newton method is used, the full Jacobian is required, but the linear systems corresponding to (4.4a) could be solved by a block iterative method.

If (4.1a) is regarded as an equation for u with v and w fixed, the Jacobian (g_{11}' of (4.3)) is self adjoint and positive definite. Likewise, if the second and third equations are regarded as equations for v and w, respectively, with u fixed, the Jacobian entries g_{22}' and g_{33}' of (4.3) are both of the form described in Section 2. If we discretize these operators using the methods of Section 3, we will be able to exploit the symmetric positive definite matrices which will result in the solution process.

It is not known (theoretically) whether the set of equations (4.1) satisfies assumptions necessary to guarantee the convergence of either the BNL or FN schemes. Some related theoretical studies on existence are surveyed by Jerome [J83] in this volume. Empirically, of course, both methods are known to converge and are widely used in device modeling. In this regard, it is clear that the size of the off-diagonal entries in (4.3) will be the primary factor effecting the rate of convergence of the BNL scheme, since essentially it is the coupling between the equations which is being iterated in this scheme. Similarly, in the FN approach, if block iteration, (say block Gauss Seidel) is used to solve the large linear system, then the rate of convergence of that method will also depend on the size of the off diagonal blocks. This will be an important consideration in the use of system multi-grid techniques described below.

5. MULTI-LEVEL TECHNIQUES

The ideas of multi-level iterative techniques can be employed in (at least) three distinct ways in the solution of nonlinear systems of the type described in Section 4. Let $\mathcal{D}_1, \mathcal{D}_2, \ldots, \mathcal{D}_j$ be a sequence of finite difference or finite element discretizations of (4.1), corresponding to systems of nonlinear equations of dimension $N_1, N_2, \ldots, N_j$. Let $\mathcal{M}_j$ denote the N_j-dimensional solution space corresponding to $\mathcal{D}_j$.

In the present case, such a sequence of discretizations can be generated inductively from an initial coarse discretization $\mathcal{D}_1$. To compute $\mathcal{D}_j$ for $j > 1$, we start from $\mathcal{D}_{j-1}$ and refine some or all of the elements or grid blocks. We assume that all discretizations produce qualitatively (though not quantitatively) accurate approximations $(u_j, \mathrm{v}_j, w_j)^T$ to (u, v, w) of (4.1), and that the dimensions of the problems increase rapidly, say

$$N_j \geqslant \beta N_{j-1}, \quad \beta > 2. \tag{5.1}$$

We also assume $\mathcal{M}_j \subset \mathcal{M}_{j+1}$ for $j > 1$. In the finite element context, where the $\mathcal{M}_j$ are spaces of piecewise polynomials, this is a natural assumption. In the finite difference context, where the $\mathcal{M}_j$ correspond to point values of the function, such an inclusion must be defined, and this is usually done in terms of natural interpolation operators. We are, of course, interested in obtaining the solution for the finest of the discretizations.

The first way in which the sequence of discretizations can be employed is through a procedure known variously as grid continuation, mesh refinement, one way multi-grid, full multi-grid, and nested iteration. In this scheme, a sequence of approximate solutions $\tilde{z}_j^T = (\tilde{u}_j, \tilde{\mathrm{v}}_j, \tilde{w}_j)$ to $z_j^T = (u_j, \mathrm{v}_j, w_j)$ are computed as follows:

Algorithm 5.1.

(1) Solve (4.1) using the discretization $\mathcal{D}_1$, with some iterative or direct method and an appropriate initial guess to obtain $\tilde{z}_1$.

(2) For $j > 1$, approximately solve (4.1) using the discretization $\mathscr{D}_j$, and some iterative method starting from an initial guess $\tilde{z}_{j-1} \in \mathscr{M}_{j-1} \subset \mathscr{M}_j$ obtaining $\tilde{z}_j \in \mathscr{M}_j$.

Such schemes, when applied to nonlinear equations, are analyzed in detail in [BR82]. We will not repeat that analysis here but merely state the result pertinent to the present discussion. In particular, we have shown:

Theorem. Let N_j satisfy (5.1) and let

$$0 \leqslant \delta\beta^q < 1, \epsilon = \delta(1+\beta^q)(1-\delta\beta^q)^{-1}, q > 0 .$$

Assume

$$||z_j - z|| \leqslant c_0 N_j^{-q} , \tag{5.2}$$

$$||\tilde{z}_1 - z_1|| \leqslant \epsilon c_0 N_j^{-q} , \tag{5.3}$$

and

$$||\tilde{z}_j - z_j|| \leqslant \delta \, ||\tilde{z}_{j-1} - z_j|| . \tag{5.4}$$

Then

$$||\tilde{z}_j - z|| \leqslant (1+\epsilon) c_0 N_j^{-q} , \tag{5.5}$$

where $||\cdot||$ is an appropriate norm.

This theorem says that if there is a notion of convergence (5.2), if the first problem is solved accurately (5.3), and if in the remaining problems, the initial error is reduced by a modest amount δ as in (5.4), then the computed solutions $\tilde{z}_j$ converge to z at the same rate as the exact discrete solutions z_j.

The principle advantage of this scheme is that, irrespective of whether the BNL or FN iterations of Section 4 are used, most of the iterations will be done on the first problem where computational costs are relatively low. Relatively few iterations will be required on the subsequent problems, where the computational costs are higher. In fact, if the full Newton iteration is used, the results in [BR82] show that, under reasonable assumptions, only one approximate Newton step per level will be required for j sufficiently large.

The second way in which multi-level techniques can be used is in the solution of systems of equations corresponding to the diagonal blocks of g' in (4.3). In both the BNL and FN methods, these equations correspond to scalar linear elliptic boundary value problems. In solving such a problem on level j as a part of Algorithm 1, one may employ the j-level iterative method, using discretizations $\mathcal{D}_1$ to $\mathcal{D}_j$. Again the details are not important in the present context. What is important is that such methods are empirically known to be extremely efficient, and in many cases can be proved to be optimal order (i.e., the work required to reduce the error by any fixed amount is proportional to the number of unknowns).

Finally, we consider the application of multi-level techniques directly to the BNL or FN iterations as a whole on level j. Symbolically, we represent the system of equations on level j as

$$h(x) = 0 . \tag{5.6}$$

For the BNL scheme, the equations are nonlinear and

$$h(x) = \begin{bmatrix} g_{1j}(u_j,v_j,w_j) \\ g_{2j}(u_j,v_j,w_j) \\ g_{3j}(u_j,v_j,w_j) \end{bmatrix} = 0; \quad x = \begin{bmatrix} u_j \\ v_j \\ w_j \end{bmatrix} . \tag{5.7}$$

For the FN scheme, the equations are linear and of the form

$$h(x) = g'x + g = 0 \; ; \; x = \begin{bmatrix} \delta u_j \\ \delta v_j \\ \delta w_j \end{bmatrix} . \tag{5.8}$$

as in (4.3). Given some initial iterate x_0, we can define a single j-level iteration for computing an updated iterate x_{m+1}. This is done inductively as follows:

for $1 \leqslant k \leqslant m$, compute $x_k \epsilon \mathcal{M}_j$ using

$$x_k = S(x_{k-1}) . \tag{5.9}$$

Then (approximately) solve the system of equations

$$h(x_m + \delta) = 0 \tag{5.10}$$

for $\delta \epsilon \mathcal{M}_{j-1}$ using, say, two iterations of the $j-1$ level scheme (except for $j = 1$) and initial guess zero. From this step one obtains an approximation $\tilde{\delta} \epsilon \mathcal{M}_{j-1}$ to $\delta \epsilon \mathcal{M}_{j-1}$. Then we set

$$x_{m+1} = x_m + \tilde{\delta} . \tag{5.11}$$

Equation (5.9) is called a smoothing iteration. When applied to the BNL scheme, it is the block nonlinear iteration itself. When applied to the FN scheme, it can represent (perhaps) a block Gauss-Seidel iteration applied to the system of linear equations.

Equation (5.10) is called a coarse grid correction. After the error $x - x_k$ has been smoothed in (5.9), one anticipates that it can be well approximated in the smaller dimensional space $\mathscr{M}_{j-1}$. Thus (5.10) represents a set of N_{j-1} nonlinear (BNL) or linear (FN) equations to be solved. These equations can be defined by the restriction of the solution space $\mathscr{M}_j$ of the original problem to the subspace $\mathscr{M}_{j-1}$ of the coarse grid correction. This problem is of the same general form as the original problem (5.6) and its solution is approximated by the (recursive) application of the $j - 1$ level scheme. The 1-level scheme consists of solving the problem exactly, either by iteration or direct methods.

The goal of this use of multi-grid techniques is to weaken the dependence of the rate of convergence of the BNL or FN schemes on the size of the off diagonal entries of the Jacobian g' of (4.3). In particular, when the coupling between the equations is strong (corresponding to the physical situation of high currents), the size of these entries can be quite large. This use of multi-grid techniques essentially allows the "smooth" part of the coupling to be resolved on the coarser grids through the recursive use of (5.10). This should make the overall iteration, as viewed from the final grid, much less sensitive to the strength of the coupling and thereby increase the overall efficiency of the solution process.

REFERENCES

[BR81] R. E. Bank and D. J. Rose, Global approximate Newton methods. Numer Math, 37 (1981), 279-295.

[BR82] R. E. Bank and D. J. Rose, Analysis of a multilevel iterative method for nonlinear finite element equations. Math of Comp, 160 (1982), 453-465.

[BRF83] R. E. Bank, D. J. Rose, and W. Fichtner, Numerical methods for semiconductor device modeling. SIAM J Sci Stat Comp, to appear.

[J83] J. W. Jerome, Analysis of nonlinear elliptic systems arising in reaction-diffusion modeling. This volume.

[FRB83] W. Fichtner, D. J. Rose, R. E. Bank, Semiconductor device modeling. SIAM J Sci Stat Comp, to appear.

[V62] R. S. Varga, Matrix Iterative Analysis, Prentice-Hall, Englewood Cliffs, NJ, 1962.

ANALYSIS OF NONLINEAR ELLIPTIC SYSTEMS ARISING IN REACTION/DIFFUSION MODELING

Joseph W. Jerome[1]

I. INTRODUCTION

We shall discuss in this paper the analytical foundations associated with certain nonlinear elliptic systems of partial differential equations. These systems will be components of boundary value problems, defining stationary solutions of parabolic evolutions, or will appear recursively in semidiscrete approximations to solutions of the initial/boundary-value problem. Although the mathematical formulations of these models are effectively of reaction-diffusion type, the physical modeling does cover reaction/convection/diffusion Brownian motion models for which the Einstein relations, linking mobility and diffusion, are valid. In fact, semiconductor modeling, discussed in detail in this paper, is of this type. This model and a model from chemical catalysis are developed in Section 2; they will serve as

1. Mathematics Dept., Northwestern University, Evanston, Illinois.

ISBN 0-12-100560-7

illustrations, respectively, of two independent approaches used in verifying aspects of pointwise stability. The attendant ideas are developed in Sections 3 and 4; the latter approach is known in the literature as the method of invariant rectangles.

Numerical procedures, particularly variationally based ones, require a well-defined operator map as a necessary preliminary to effective approximation. We shall use the pointwise stability estimates to provide well-defined maps of appropriate Hilbert space subsets into themselves, or into natural dual spaces. The two cases just cited, viz., invariant sets and dual sets, provide the settings for two distinct frameworks within which we examine existence for the semiconductor model: the Schauder theory and the pseudomonotone theory. These are developed in Sections 5 and 6, respectively. The latter theory is quasi-constructive, and is based upon (weak) convergence of solutions of finite-dimensional variational inequalities to a solution of an elliptic variational inequality, equivalent to the elliptic system. This, in turn, raises interesting connections between nonlinear elliptic systems, satisfying generalized maximum principles, and penalty methods, defined by such principles, as a class of solution methods for approximate solution of the variational inequalities.

The systems to be studied are not gradient systems, hence are not amenable to the direct methods of convex analysis. The individual equations may be of gradient type when uncoupled, however, or trivial perturbations of such, and this partly accounts for the success of solution maps defined via uncoupling. The iteration convergence implied in such solution maps, however, is currently at best formal from the standpoint of numerical computation, though observed in the physical regime.

The uncoupling is of enormous theoretical import as revealed in the sequel, however. Space limitations prevent a discussion here of numerical estimation or linearization algorithms. However, it is worth mentioning that fully discrete versions of the stability estimates of Sections 3 and 4 are possible. The reader should consult the paper of (Fichtner, Rose, and Bank, 1983) for numerical results and (Bank, Rose and Fichtner, 1983) for algorithm discussion.

II. MODEL DEVELOPMENT

A. Semiconductor Simulation: A Brownian Motion Model

According to the band theory of solids, the outer electrons of the atoms of a crystalline lattice move in an electron gas in an electric field due to the periodic array of atomic cores and the other electrons. The allowed energy bands are stratified. Without the introduction of impurities, the electron/hole generation is strictly thermal. Such semiconductors are called intrinsic. Doping, however, establishes donor/acceptor energy levels in the forbidden zone between conduction and valence bands, thus facilitating pair generation by gap narrowing.

The legitimacy of macroscopic modeling rests upon quantum mechanical foundations (cf. Shockley, 1950) permitting the realization of wave packets as particles. One introduces the dependent variables u, n, and p for the electrostatic potential, the free electron concentration, and the free hole concentration, respectively. The carrier motion is due to drift, induced by the ambient electric field $E = -\nabla u$, and to diffusion, induced by the (negative) concentration gradients, $-\nabla n$ and $-\nabla p$. If q denotes the charge modulus, if μ_n and μ_p denote field-dependent carrier mobilities, and if D_n and D_p denote carrier diffusion coefficients, then the carrier current components are represented by (cf. Sze, 1981)

$$J_n = -q(-\mu_n nE - D_n \nabla n) , \tag{2.1a}$$

$$J_p = q(\mu_p pE - D_p \nabla p) . \tag{2.1b}$$

Note that the mobility-field products in (2.1) yield velocity, and that the sign convention in (2.1) reflects electron movement against the field and hole movement with the field. The total current $J = J_n + J_p$ is the primary quantity of interest to the physical scientist.

The evolution system is defined by the conjunction of Gauss' law, linking E with charge density, and the continuity/mass balance equations for the carriers. If N denotes the ionized doping, then the net charge density is given as, $-q(n-p-N)$. Choosing units so that $q = 1$, and denoting the dielectric function by ϵ, we have

$$-\nabla \cdot (\epsilon \nabla u) + n - p = N , \tag{2.2a}$$

for the Gauss law and

$$\frac{\partial n}{\partial t} = \nabla \cdot J_n - P , \tag{2.2b}$$

$$\frac{\partial p}{\partial t} = -\nabla \cdot J_p - P , \tag{2.2c}$$

for the continuity equations. Here P is the net recombination given by

$$P(n,p) = Q(n,p)(np - n_i^2) , \tag{2.3}$$

according to standard mass action laws (cf. Evans, 1969). Q is a positive rational function of n and p, and $n_i^2 = 1$ is an average equilibrium product concentration.

A composite macroscopic/statistical model is derived from (2.1)-(2.2) if the motion is assumed Brownian, with negligible fluctuations. In this case the Einstein relations

$$\mu_n = D_n \ , \quad \mu_p = D_p \ , \tag{2.4}$$

hold (cf. Nelson, 1979) if the carrier probability densities satisfy the Fokker-Plank equation; here we emphasize the assumption of thermodynamic equilibrium, and units have been chosen so that the product of Boltzmann's constant and temperature is unity. If the formal change of variable,

$$n = \exp(u - \mathrm{v}) \ , \quad p = \exp(w - u) \ , \tag{2.5}$$

is introduced for the new dependent variables v and w, then (2.1)-(2.2) assume a form in steady-state given by

$$-\nabla\cdot(\epsilon\nabla u) + e^{u-\mathrm{v}} - e^{w-u} = N \ , \tag{2.6a}$$

$$-\nabla\cdot(\mu_n(\nabla u)e^{u-\mathrm{v}}\nabla \mathrm{v}) - Q(u,\mathrm{v},w)(e^{w-\mathrm{v}} - n_i^2) = 0 \ , \tag{2.6b}$$

$$-\nabla\cdot(\mu_p(\nabla u)e^{w-u}\nabla w) + Q(u,\mathrm{v},w)(e^{w-\mathrm{v}} - n_i^2) = 0 \ . \tag{2.6c}$$

The transformation (2.5) is valid whenever the Dirichlet boundary values for n and p are bounded above zero on that portion of the semiconductor on which biasing occurs; the remainder of the device is assumed insulated with respect to current flow. In the special case when the electrons are assumed to satisfy a Fermi-Dirac statistical density, for the (energy) state occupancy probability for particles with half-integer spin, then v and w may be interpreted as quasi-Fermi levels, if energy localization in the Boltzmann tail is assumed. In this case, with appropriate normalization, the transformation (2.5) results from integrating particle probability densities over $[\mathrm{v},\infty)$ and

$(-\infty, w]$, respectively (cf. Evans, 1969).

B. Chemical Catalysis

In catalytic chemistry, porous pellets serve as carrier structures for the adsorption of chemical reactants, at sites on the pellet surface, with spontaneous reaction, and subsequent interior diffusion and further reaction. The concentrations of the chemical reactants are modeled by continuity equations similar to (2.2b, c). If the reactions are not isothermal, an enthalpy balance equation must be adjoined to the continuity equations. For example, for a single reaction in a non-isothermal pellet, under an assumption of constant diffusion and negligible convection, we have the system

$$\frac{\partial u}{\partial t} - \nabla^2 u + P(u, \mathrm{v}) = 0 , \qquad (2.7a)$$

$$\frac{\partial \mathrm{v}}{\partial t} - \nabla^2 \mathrm{v} + R(u, \mathrm{v}) = 0 , \qquad (2.7b)$$

for the temperature u and chemical concentration v; the system is coupled through the reaction terms P and R. The Robin boundary conditions are suitable for this problem (cf. Aris, 1975).

III. MAXIMUM PRINCIPLES FOR THE SEMICONDUCTOR SYSTEM

We shall indicate in this section the fundamental role played by the maximum principles in defining a solution map T for the boundary-value problem consisting of (2.6) and the associated boundary values; the latter are specified by $u = \bar{u}$, $\mathrm{v} = \bar{\mathrm{v}}$, $w = \bar{w}$ on $\Sigma_D \subset \partial\Omega$, with vanishing normal derivatives of u, v and w on the essential complement Σ_N of Σ_D. We proceed heuristically as follows to uncouple the system. Begin with an arbitrary pair $[\tilde{\mathrm{v}}, \tilde{w}]$ of essentially bounded functions, and solve the weak formulation for $\tilde{u}$ corresponding to

$$-\nabla\cdot(\epsilon\nabla\tilde{u}) + e^{\tilde{u}-\tilde{v}} - e^{\tilde{w}-\tilde{u}} = N\ , \tag{3.1a}$$

$$\tilde{u}|_{\Sigma_0} = \bar{u}|_{\Sigma_0}\ , \quad \frac{\partial\tilde{u}}{\partial\nu}|_{\Sigma_N} = 0\ . \tag{3.1b}$$

This fractional step yields

$$[\tilde{v},\tilde{w}] \to \tilde{u}(\tilde{v},\tilde{w})\ . \tag{3.2}$$

Note that (3.1) is of gradient form. The image

$$[v,w] = T[\tilde{v},\tilde{w}] \tag{3.3}$$

is now derived by solving separately for v and w the gradient equations

$$-\nabla\cdot(\mu_n(\nabla\tilde{u})e^{\tilde{u}-\tilde{v}}\nabla v) - P(\tilde{u},v,\tilde{w}) = 0\ , \tag{3.4a}$$

$$v|_{\Sigma_0} = \bar{v}|_{\Sigma_0}\ , \quad \frac{\partial v}{\partial\nu}|_{\Sigma_N} = 0\ , \tag{3.4b}$$

and

$$-\nabla\cdot(\mu_p(\nabla\tilde{u})e^{\tilde{w}-\tilde{u}}\nabla w) + P(\tilde{u},\tilde{v},w) = 0\ , \tag{3.5a}$$

$$w|_{\Sigma_0} = \bar{w}|_{\Sigma_0}\ , \quad \frac{\partial w}{\partial\nu}|_{\Sigma_N} = 0\ . \tag{3.5b}$$

T is called a solution map, since $(\tilde{u}(v,w),v,w)$ solves the boundary-value problem for (2.6) if $[v,w]$ is a fixed point for T. The choice of definition of domain K for T is motivated not only by the intent that (3.1), (3.4) and (3.5) be well-defined, but that K be invariant under the action of T. Suppose that there exist constants $\alpha \leqslant \beta$, independent of $\tilde{u}$, $\tilde{v}$ and $\tilde{w}$, such that the solutions v and w of (3.4) and (3.5) are confined to $[\alpha,\beta]$. Then we may define K to be the closed, convex subset of

$$H = L^2(\Omega) \times L^2(\Omega)\ , \tag{3.6}$$

given by

$$K = \{[f,g] \in H : \alpha \leqslant f \leqslant \beta, \alpha \leqslant g \leqslant \beta\}. \tag{3.7}$$

For the function P given by (2.3), the system coupling is such that α and β may be defined by

$$\alpha = \min(\inf_{\Sigma_D} \bar{v}, \inf_{\Sigma_D} \bar{w}), \tag{3.8a}$$

$$\beta = \max(\sup_{\Sigma_D} \bar{v}, \sup_{\Sigma_D} \bar{w}). \tag{3.8b}$$

In addition, $\tilde{u}$ satisfies the bounds $\gamma \leqslant \tilde{u} \leqslant \delta$, where

$$\gamma = \min(\inf_{\Sigma_D} \bar{u}, \gamma'), \tag{3.8c}$$

$$\delta = \max(\sup_{\Sigma_D} \bar{u}, \delta'), \tag{3.8d}$$

where γ' and δ' are uniquely defined by

$$e^{\gamma' - \alpha} - e^{\alpha - \gamma'} - \inf_{\Omega} N = 0, \tag{3.8e}$$

and

$$e^{\delta' - \beta} - e^{\beta - \delta'} - \sup_{\Omega} N = 0. \tag{3.8f}$$

Proofs of these facts may be found in (Jerome, 1983a).

IV. INVARIANT RECTANGLES FOR CHEMICAL SYSTEMS

For simplicity, we shall confine our remarks to the system (2.7); a more general setting, allowing for degeneracies due to negligible diffusion, and including a partitioning of the variables into those of concentration and potential type, may be found in (Jerome, 1983b). Literature references may also be found in this source.

Definition 4.1. Let $Q = [a,b] \times [c,d]$ be a given rectangle, with $a \leqslant 0 \leqslant b$ and $c \leqslant 0 \leqslant d$, such that $b - a > 0$ and $d - c > 0$. Define the faces

$$Q_a = \{(u, \mathrm{v}) \in Q : u = a\}, \; Q_c = \{(u, \mathrm{v}) \in Q : \mathrm{v} = c\}, \tag{4.1}$$

with analogous definitions for Q_b and Q_d, wherein u and v assume the values b and d, respectively. The vector field $(-P, -R)$ of (2.7) is said to be *inward pointing* if

$$P(u, \mathrm{v}) \leqslant 0 \,(\text{resp.} \geqslant 0), \; (u, \mathrm{v}) \in Q_a \,(\text{resp.} \,(u, \mathrm{v}) \in Q_b), \tag{4.2a}$$

$$R(u, \mathrm{v}) \leqslant 0 \,(\text{resp.} \geqslant 0), \; (u, \mathrm{v}) \in Q_c \,(\text{resp.} \,(u, \mathrm{v}) \in Q_d). \tag{4.2b}$$

P and R are said to satisfy an *ordering of sign regions* property if, on Q,

$$P(u', \mathrm{v}) \leqslant 0, \; P(u'', \mathrm{v}) \geqslant 0 \Rightarrow u' \leqslant u'', \tag{4.3a}$$

for each fixed v,

$$R(u, \mathrm{v}') \leqslant 0, \; R(u, \mathrm{v}'') \geqslant 0 \Rightarrow \mathrm{v}' \leqslant \mathrm{v}'', \tag{4.3b}$$

for each fixed u.

If Ω denotes the physical pellet, then the boundary conditions are said to be of Robin type for (u, v) if, for specified nonnegative functions η and θ, $\int_{\partial\Omega} \eta > 0$, $\int_{\partial\Omega} \theta > 0$, u and v are required to satisfy

$$\frac{\partial u}{\partial \nu} + \eta u = 0, \text{ on } \partial\Omega, \tag{4.4a}$$

$$\frac{\partial \mathrm{v}}{\partial \nu} + \theta \mathrm{v} = 0, \text{ on } \partial\Omega, \tag{4.4b}$$

where ν denotes the outward normal. Finally, Q is said to be an invariant rectangle for (2.7)/(4.4) if, whenever the initial datum pair (u_0, v_0) has range in Q, the (unique) solution of the initial/boundary-value problem has range in Q.

Remark 4.1. It can be shown that Q is an invariant rectangle under the inward pointing and ordering of sign region hypotheses on P and R. However, it is also true, and demonstrated in (Jerome, 1983b), that Q is invariant for the recursively generated fully implicit semidiscretization of (2.7), if P and R are represented as $P(u_k, v_{k-1})$, $R(u_{k-1}, v_k)$, thereby uncoupling the system. Thus, we have an alternative to standard maximum principles for obtaining pointwise stability estimates for elliptic reaction-diffusion systems arising from an application of the backward Euler method.

V. COMPACTNESS AND CONTINUITY OF THE SCHAUDER MAP: EXISTENCE

The existence of solutions for the semiconductor system is based upon the Schauder fixed point theorem, a version of which we state as follows (cf. Collatz, 1966).

Lemma 5.1. Suppose T is a continuous mapping of a closed convex subset K of a Banach space X into itself such that T is compact, i.e., the range of T has compact closure in K. Then T has a fixed point in K.

Remark 5.1. We have retained the notation of T and K given by (3.3) and (3.7), respectively; also, the identification $X = H$, where H is given by (3.6), is made. The maximum principles discussed in Section 3 show that K is invariant under T. K is closed since L^2 sequential convergence implies pointwise subsequential convergence almost everywhere. The verification of the continuity of T is somewhat technical; T is actually continuous into $H^1(\Omega) \times H^1(\Omega)$.

The compactness of T is immediate from its range boundedness in $H^1(\Omega) \times H^1(\Omega)$. Details and precise hypotheses are given in (Jerome, 1983a). Among the more salient hypotheses are the monotone decrease and increase of P (cf. (2.3)) in the arguments v and w, respectively; the ellipticity hypothesis that ϵ, μ_n and μ_p are bounded above zero; and certain boundedness and continuity assumptions associated with these functions, as well as certain natural hypotheses on the boundary data and the domain Ω. As noted in Section 3, a fixed point [v,w] of T yields a solution for the boundary-value problem via $(\tilde{u}(\mathrm{v},w),\mathrm{v},w)$.

Remark 5.2. Versions of this existence theorem for the semiconductor system have been obtained by (Mock, 1972) and (Seidman, 1980). Neither of these authors, however, uses methods capable of treating a variant of the semiconductor problem, dictated by the MOS-FET technology, in which the domain for n and p is a subset of the domain Ω for u, thus entailing interface conditions. This problem is discussed in (Jerome, 1983a), where it is noted that the interface boundary conditions are natural for v and w.

VI. A QUASI-CONSTRUCTIVE FRAMEWORK; PSEUDOMONOTONE OPERATORS

The Schauder fixed point theorem is a nonconstructive method for proving existence of fixed points, or zeros of mappings. In an effort to provide a quasi-constructive framework, we note that existence for the semiconductor model can be demonstrated by use of pseudomonotone operators. If

$$Y = H^1(\Omega) \times H^1(\Omega) \times H^1(\Omega) , \tag{6.1a}$$

$$Y_0 = \{(u,\mathrm{v},w) \in Y : u|_{\Sigma_*} = \mathrm{v}|_{\Sigma_*} = w|_{\Sigma_*} = 0\} , \tag{6.1b}$$

then the pair A, K, such that $A: K \subset Y \to Y^*$ is a pseudomonotone mapping associated with (2.6), is defined as follows (cf. (3.8)):

$$K = \{(u,\mathrm{v},w) : (u,\mathrm{v},w) \in (\bar{u},\bar{\mathrm{v}},\bar{w}) + Y_0 \text{ and} \tag{6.2}$$

$$\gamma \leqslant u \leqslant \delta \ , \ \alpha \leqslant \mathrm{v} \leqslant \beta \ , \ \alpha \leqslant w \leqslant \beta\} \ ,$$

and A is defined in the obvious way by the variational formulation corresponding to the left hand side of (2.6), with test functions in Y. A is pseudomonotone, i.e., A is bounded, and satisfies a type of weak lower semicontinuity property; and A is coercive, i.e., A maps unbounded sets onto unbounded sets in a precise manner. It follows that there is a solution of the variational inequality

$$\langle A(u,\mathrm{v},w) - (N,0,0) \ , \ (U,V,W) - (u,\mathrm{v},w)\rangle \geqslant 0 \ , \tag{6.3}$$

for all $(U,V,W) \in K$. Under reasonable regularity conditions, (6.3) is equivalent to (2.6) with boundary conditions, i.e.,

$$A(u,\mathrm{v},w) = (N,0,0) \ . \tag{6.4}$$

The pseudomonotone framework guarantees that the finite dimensional approximations defined by (6.3) are subsequentially convergent to a solution (u,v,w). The concomitant inequalities in finite dimensions may be loosely interpreted as inducing potential penalization for violation of the pointwise maximum principles (cf. (Hess, 1978 and Jerome, 1983a) for approaches of this type).

ACKNOWLEDGEMENT

The author thanks the Bell Laboratories for supporting this research, and Donald Rose for his stimulating professional advice.

REFERENCES

Aris, R. (1975). The Mathematical Theory of Diffusion and Reaction in Permeable Catalysts (Vol. 1), Oxford Univ. Press (Clarendon), London and New York.

Bank, R., Rose, D. J. and Fichtner, W. (1983). Numerical methods for semiconductor simulation, to be published jointly in SIAM J. Stat. Sci. Comp. and IEEE J. Electron Devices.

Collatz, L. (1966). Functional Analysis and Numerical Mathematics, Academic Press, New York.

Evans, D. A. (1969). Statistics of electrons and applications to devices, in, Solid State Theory (P. T. Landsberg, Ed.), Wiley-Interscience, London, pp. 259-324.

Fichtner, W., Rose, D. J. and Bank, R. (1983). Semiconductor device simulation, to be published jointly in SIAM J. Stat. Sci. Comp. and IEEE J. Electron Devices.

Hess, P. (1978). On a second order nonlinear elliptic boundary value problem, in, Nonlinear Analysis (L. Cesari, R. Kannan, and H. Weinberger, Eds.), Academic Press, New York, pp. 99-107.

Jerome, J. (1983a). Consistency of semiconductor modeling: An existence analysis for the stationary Van Roosbroeck system, to be submitted.

Jerome, J. (1983b). Approximation of Nonlinear Evolution Systems, Academic Press, New York.

Mock, M. (1972). On equations describing steady-state carrier distributions in a semiconductor device, Comm. Pure Appl. Math. 25, 781-792.

Nelson, E. (1979). Connection between Brownian motion and quantum mechanics, Lecture Notes in

Physics 100 (H. Nelkowski et al., Eds.), Springer-Verlag, Berlin, pp. 168-179.

Seidman, T. (1980). Steady state solutions of diffusion reaction systems with electrostatic convection, Nonlinear Anal. 4, 623-637.

Shockley, W. (1950). Electrons and Holes in Semiconductors, Van Nostrand, New York.

Sze, S. M. (1981). Physics of Semiconductor Devices (2nd ed.), Wiley-Interscience, New York.

NUMERICAL METHODS FOR SOLVING COUPLED SEMICONDUCTOR EQUATIONS ON A MINICOMPUTER

James L. Blue* and Charles L. Wilson*

I. INTRODUCTION

A general mathematical model for analyzing currents and fields in semiconductor devices requires three coupled nonlinear elliptic partial differential equations in two dimensions.

A general-purpose solver for systems of coupled nonlinear elliptic PDEs is used. The numerical framework is linear elements on triangles, with nonuniform triangulations. The nonlinear finite element equations are solved by approximate Newton methods; the linearized equations are solved by sparse Gaussian elimination and by multi-level iteration. For accurate solutions, triangle refinemenets are generated adaptively.

Interactive operation and graphics are essential, both understanding the solutions and for developing robust solution strategies.

We have developed a package which is capable of solving a wide variety of semiconductor device problems. The package analyzes coupled systems of semiconductor equations in two dimensions. The package is hierarchical, consisting of a technology-dependent top level, or driver, which calls a

*National Bureau of Standards, Washington, D.C.

ISBN 0-12-100560-7

general-purpose elliptic solver (B2DE). With this organization, we can change the physical model of the system easily, and we can test the package on problems with known solutions.

II. THE MATHEMATICAL MODEL

The model is formulated as three coupled elliptic partial differential equations for u (the electrostatic potential), v (the electron quasi-Fermi potential), and w (the hole quasi-Fermi potential). In normalized units, the equations are:

$$\nabla \cdot (\nabla u) + N - n + p = 0$$

$$\nabla \cdot (\mu_n \, n \, \nabla v) - R = 0$$

$$\nabla \cdot (\mu_p \, p \, \nabla w) + R = 0$$

Charge is carried by electrons and holes; n and p are the electron and hole densities; R is the recombination/generation rate; μ_n and μ_p are electron and hole mobilities. The doping density, N, is a given spatial function, independent of u, v, and w. The other quantities depend on u, v, and w; the exact form of the dependence varies with the physical model chosen.

When n and p are not too large, Boltzmann statistics may be used. The carrier densities are

$$n = \exp(u-v), \quad p = \exp(w-u).$$

With this simple form, and especially when μ_n and μ_p are constants, the three PDEs appear mathematically tractable, and useful transformations may be made (Bank and Rose, 1983). However, in many practical devices n and p get too large for the Boltzmann approximation to be valid. The next simpler approximation is Fermi-Dirac statistics,

$$n = N_c\, F(u - v - u_1), \quad p = N_v\, F(w - u - u_1),$$

where N_c, N_v, and u_1 are constants, and F is proportional to the Fermi-Dirac integral of order 1/2,

$$F(t) = \int_0^\infty s^{1/2}\, ds \,/\, [1 + \exp(s-t)].$$

This is not mathematically tractable.

At low values of the electric field, $-\nabla u$, μ_n and μ_p are independent of u, v, and w, although they have a weak dependence on N. At high electric field, the mobilities are inversely proportional to the magnitude of the electric field. A typical form for R is

$$R = \frac{pn - 1}{\tau_1 (n + n_1) + \tau_2 (p + p_1)},$$

where τ_1, τ_2, n, and p_1 are constants. Other forms of R are sometimes important.

General-purpose PDE software is desirable so that we can change the model easily. Our philosophy from the beginning has been to be willing to sacrifice speed for flexibility, since we want to be able to model many kinds of devices in many modes of operation.

III. SPECIAL FEATURES OF THE SEMICONDUCTOR DEVICE EQUATIONS

In this section, we discuss the special nature of the semiconductor equations, and how it affects our methods. There are several interrelated topics.

First and most obviously, there are three coupled PDEs. Individually, they are in conservation form, and in simple cases standard numerical methods will generate symmetric, positive-definite matrices. As a coupled set, all matrices will be nonsymmetric.

The equations contain exponential nonlinearities. At room temperature, the normalization is one volt equals 40 units in u, v, or w. Since typical logic levels in circuits are 0 and 5 volts, we must be prepared for a dynamic range of 200 in u; some kinds of devices require a much larger dynamic range. On the VAX 11/780, for example, overflow occurs just past exp(88) and underflow just past exp(-88). We need to be very careful; often we can evaluate exp(u-v) at a solution, but not exp(u) and exp(-v).

The exponentials also insure that systems of nonlinear equations arising from the PDEs will be difficult.

Semiconductor devices typically contain steep fronts whose locations cannot be predicted before the device is modeled. The steep fronts arise because of the form of the equations and because of steep fronts in the doping profile, N(x,y). As a result of the steep fronts, a good initial guess for the solution is not known, nor is a good triangulation. Uniform meshes are wasteful, since the solutions are slowly-varying over much of the region. Because the steep fronts generally occur on curved boundaries, even tensor product meshes are wasteful, leading us to finite elements on triangle rather than finite differences. Automatic adaptive mesh generation adaptive mesh generation is needed.

Over much of the region, semiconductor devices typically

have exp(v-u) << 1 and |R| << 1, and similarly for exp(u-w). Practically speaking, any Jacobian matrices will be highly rank-deficient. Some form of regularization is essential.

When we began this project, R. E. Bank and A. H. Sherman had just finished the first version of PLTMG (Bank and Sherman, 1979). PLTMG had many of the features we required, but could solve only one linear elliptic PDE.

We began with PLTMG and developed a general-purpose program that could also solve semiconductor device equations. B2DE solves systems of m coupled elliptic PDEs, each in conservation form.

$$\nabla \cdot (a_i \nabla u_i) = f_i, \; i = 1, 2, \ldots, m$$

The functions a_i and f_i may depend on x, y, all the u_j, and all the ∇u_j. The region is a two-dimensional union of triangles. On each boundary edge, each PDE has a nonlinear boundary condition, either of Dirichlet type or of mixed type.

$$g_i = 0 \quad \text{or} \quad \partial u_i / \partial \nu + g_i = 0$$

The functions g_i may depend on x, y, and all the u_j.

The formulation must be at least this general to model known physical effects in semiconductor devices. For example, μ_n and μ_p may behave like

$$\mu \approx \mu_0 / \sqrt{1 + C\,|\nabla u|^2}\,.$$

However, we make no claim that B2DE can solve all the systems of PDEs that the formulation allows, since the formulation allows ridiculous systems of PDEs.

Only in simple, fairly uninteresting cases, such as linear systems, is it easy to solve the systems of PDEs. In general, we are forced to rely on the physical insight, the intelligence, and the insight of the user. B2DE has to make it easy for the user to gain an understanding of the iteration process and to generate a solution strategy.

We convert the PDEs to a system of nonlinear equations by the standard finite-element method (Strang and Fix, 1973). Given a triangulation of the region, we have the standard linear finite elements Φ_j; Φ_j is 1 at the jth vertex, 0 at all other vertices, and is linear on each triangle. Multiplying the PDEs by Φ_j, integrating over the entire region, and then integrating by parts, we have

$$\int\int (a_i \nabla\Phi_j \cdot \nabla u_i + \Phi_j f_i) = \int \Phi_j a_i g_i$$

The right-hand side integral is only over the regions of the boundary with the mixed boundary condition. For (i,j) corresponding to a Dirichlet-type boundary condition, we have the nonlinear boundary condition instead of the above. After substituting

$$u_i = \sum_j \alpha_{ij}\Phi_j,$$

a large set of nonlinear equations for $\{\alpha_{ij}\}$ has replaced the PDEs.

We know no algorithm guaranteed to solve such a set of nonlinear equations, and would be suspicious of any claims for one. Instead, we provide some standard methods for the user, and make it easy for the user to try alternative solution strategies. This is best done interactively.

The basic method for nonlinear equations is a damped Newton's method. Linearizing the nonlinear finite element equations and the nonlinear Dirichlet boundary conditions gives us a set of linear equations for the correction to the alphas. These linear equations are like those which would arise from using Galerkin's method on a set of linear PDEs. We use the full correction only if the integral L_2 norm of the residuals decreases; otherwise we use only a part of the correction.

The user of B2DE specifies the PDEs by providing Fortran subroutines to calculate the functions a_i, f_i, and g_i, and their first partial derivatives. The user also provides an initial triangulation of the region, T_1, and an initial guess $u^{(0)}$.

The basic cycle includes generating a suitable triangulation adaptively.

repeat until done
{ use $u^{(m-1)}$ as a guess
solve the nonlinear Galerkin equations on T_m
estimate the error in the solution, $u^{(m)}$
refine the worst triangles to get T_{m+1}
increase m by 1
}

At each step of the solution process, the user has many possible choices. We list some of the opportunities. Space prohibits a full discussion; selected items will be discussed briefly.

Method for linear equation solving
Accuracy for linear equation solving
Strategy for nonlinear equation solving
Accuracy for nonlinear equation solving
How to estimate error in solution
How to refine the mesh
How to regularize
How to scale the individual PDEs
Iterate on all PDEs jointly?
On individual PDEs?

Since we have no assurance that there is a globally-appropriate algorithm, we have provided an interactive driver program as part of B2DE. The user selects the next action from menus; selecting an action from one menu may elicit another menu. Heavy use is made of interactive graphics on the user's video terminal.

For a particular class of problems, an expert user can use B2DE interactively to discover appropriate robust solution strategies and initial guesses. With this knowledge, the expert can implement a "black box" program for the non-expert to use.

IV. SELECTED DISCUSSION OF CHOICES

A. Linear Equation Solving

On the initial triangulation, T_1, we use sparse Gaussian elimination (Sherman, 1978). The Yale Sparse Matrix Package storage scheme is used. There are usually 100 to 300 vertices.

Usually we keep all triangulations, and use multi-level iteration to solve on the finer triangulations. In our experience, Gauss-Seidel smoothing usually works, although there is no reason to expect that it should, since the matrices are generally not diagonally-dominant.

The time to solve the linear equations is much less than the time to generate them, at least when there are only 3 PDEs and a few hundred vertices in T_1.

B. Nonlinear Equation Solving

As discussed previously, we use Newton's method with stepsize control; with Fermi-Dirac statistics, we almost always are able to use a full step. Generally, we accept a solution (and refine the triangulation) when the maximum Newton correction to any α_{ij} is less than 0.1. Because of the exponentials, the iteration process is seldom in the super-linear convergence regime of Newton's method.

A typical strategy is

(1) First iterate on u, holding v and w fixed
(2) Then iterate on u and v, holding w fixed
(3) Then iterate on all three

C. Regularization

Many possibilities exist. We allow the user to specify (small) parameters s_i. In the Jacobian calculation, but not in calculating the residuals, a_i is replaced by $a_i + s_i$.

D. Refinement of Triangulations

Given an approximate solution $u^{(k)}$ on a triangulation T_k, B2DE estimates the error on the t-th triangle, ε_t, by a local method. The user specifies the relative weights assigned to each equation, and a threshold parameter, p, $0 \leq p < 1$. The user may choose to refine the triangles with $\varepsilon_t > p\ \varepsilon_{max}$ or to refine a fraction p of the triangles with the highest ε_t.

If the errors are localized, and few triangles are refined, the user may decide to merge the new triangulation with the previous one. In many problems, this is done several times in the initial stages of a solution; the triangulation is being adapted to fit the solution. Later in the solution process, this is not necessary; the refinement merely reduces the truncation error in the solution.

V. GRAPHICS

The user needs graphics to understand the solution process. After a solution is found, the user needs graphics to understand the solution. Samples of some of the available graphics may be found in (Blue and Wilson, 1983). All of these are available interactively, on several kinds of video terminals; the plots may also be written to data files for post-porcessors.

VI. REFERENCES

Bank, R.E., and Rose, D.J., "Multi-level Solution of (Semiconductor-like) Systems of Nonlinear Partial Differential Equations," this volume (1983).

Bank, R.E., and Sherman, A.H., "PLTMG User's Guide," CNA 152, Center for Numerical Analysis, University of Texas at Austin (1979).

Blue, J.L., and Wilson, C.L., "Two-Dimensional Analysis of Semiconductor Devices using General-Purpose Interactive PDE Software," to be published jointly in IEEE Trans. Elec. Dev. and SIAM J. Sci. Stat. Comp. (1983).

Sherman, A.H., ACM Trans. Math. Soft., 4 391, (1978).

Strang, G., and Fix, G., *An Analysis of the Finite Element Method*, Prentice-Hall, Englewood Cliffs, N.J., (1973).

Numerical Solution of a Model Equation Near the Onset of the Rayleigh-Bénard Instability

P. E. Bjørstad[1]
W. M. Coughran, Jr.[2]
H. S. Greenside[2]
D. J. Rose[2]
N. L. Schryer[2]

1. Introduction

The Oberbeck-Boussinesq approximation to the general fluid equations is derived by assuming that all parameters with the exception of the fluid density are independent of temperature and is given by

$$DU = \sigma \Delta U - \nabla p + \begin{pmatrix} 0 \\ 0 \\ \sigma\Theta \end{pmatrix}$$

$$D\Theta = Rw + \Delta\Theta$$

$$\nabla \cdot U = 0$$

where

$u(x,y,z,t)$, $v(x,y,z,t)$, and $w(x,y,z,t)$ are the x, y, and z components of the velocity respectively,

$U(x,y,z,t) \equiv (u,v,w)^T$,

$\Theta(x,y,z,t)$ is the deviation from the linear temperature profile of the conducting state,

$p(x,y,z,t)$ is the pressure, which can be eliminated using the continuity equation,

R is the Rayleigh number,

σ is the Prandtl number,

$D \equiv \partial_t + U \cdot \nabla$.

This nonlinear system describes buoyancy-induced convection in a three-

1 *Veritas, FDIV, Hovik, Norway*
2 *Bell Laboratories, Murray Hill, New Jersey*

ISBN 0-12-100560-7

dimensional cell [5]. Figure 1 shows the cell geometry we are interested in here.

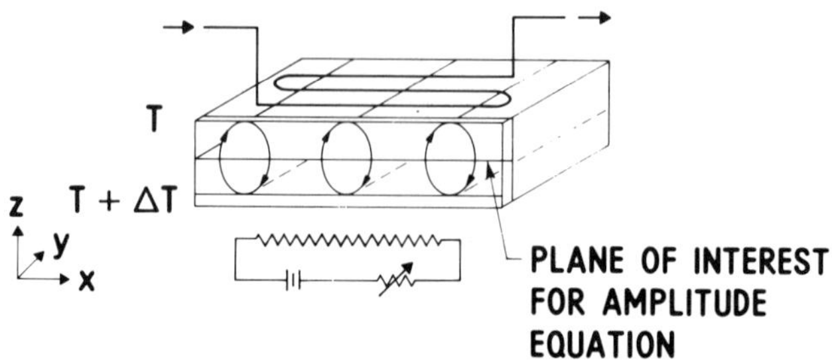

Figure 1. Schematic diagram of a Rayleigh-Bénard convection cell.

As the Rayleigh number (equivalently, the temperature difference between the horizontal plates) is increased, there is a transition, known as the Rayleigh-Bénard instability, where conduction gives way to convection. Rayleigh-Bénard convection has been studied both theoretically and experimentally since it is a simple system and the flow complexity increases slowly as R is increased. For large aspect ratios (ratio of the largest horizontal dimension to the depth of the fluid) and small Prandtl numbers ($\sigma \leqslant 1$), the fluid becomes turbulent as soon as it changes from conduction to convection; the nature of this turbulence is under intense experimental and theoretical investigation.

It is known that sufficiently close to the onset of convection and in a sufficiently wide cell, the fluid motion is weakly three-dimensional in that only slow spatial and temporal modulations of a locally two-dimensional roll pattern can occur. The five three-dimensional Boussinesq equations can then be replaced by a single two-dimensional equation that describes the evolution of a complex amplitude or envelope function whose phase and magnitude slowly modulate the sinusoidal oscillations of the rolls [10,13]. Amplitude equations have been helpful in understanding which of the various, highly degenerate, solutions to the linearized Boussinesq equations persist and are stable when lateral boundaries and nonlinear effects are included [10,16,14].

In this paper, we describe some efficacious numerical techniques for solving a two-dimensional model equation that reduces, in perturbation theory, to the same amplitude equation as that derived from the Boussinesq system. We also discuss the solution of an augmented system of two-dimensional equations that includes the generation of vertical vorticity by roll curvature.

2. A Two-Dimensional Equation for Three-Dimensional Convection

A lowest-order nontrivial amplitude equation is obtained by a multiple-scale analysis of the Oberbeck-Boussinesq equations near onset [10,13]. The perturbation expansion is made in terms of the small parameter

$$\epsilon = \frac{R - R_C}{R_C}$$

where R_C is the critical Rayleigh number at the Rayleigh-Bénard instability for a laterally infinite container. By including the rapid oscillations in a real amplitude field $\Psi(x,y,\tau)$, where $\tau = \epsilon^{-1} t$ is the slow-time variable, a rotationally invariant model equation can be derived [16,6]. This equation is given by

$$\Psi_\tau = \left[\epsilon - (\Delta + q_0)^2\right]\Psi - \Psi^3 \equiv L\Psi - \Psi^3,$$

where q_0 is the wavenumber minimizing R_C; in what follows, we normalize q_0 to be 1 so a roll diameter is 2π. Note that the Prandtl number enters only in higher-order perturbation terms.

This equation is one of the simplest nontrivial models describing the effects of boundaries and nonlinearities on pattern formation. Analysis of the linearized equation indicates that the trivial solution $\Psi = 0$, corresponding to conduction, is the only stable solution for ϵ below a critical value, depending on the aspect ratio. Above this critical value, nonzero solutions Ψ exist and evolve such that the rolls are everywhere perpendicular to the boundaries and the heat transport is maximized. For large aspect ratios, the fluid motion is dominated by the interaction of irregularities in roll patterns, which we call defects by analogy with similar imperfections in crystals [8].

The model equation can be variationally derived from a Lyapunov functional

$$F[\Psi(\tau)] = \frac{1}{2}\int \frac{\Psi^4(\tau)}{2} - \epsilon\Psi^2(\tau) + [(\Delta+1)\Psi(\tau)]^2 dx\, dy,$$

which decreases monotonically for all initial conditions; also, it can be shown that the model equation is purely relaxational [6]. F is an "energy" measure for solutions and provides a means of ordering possible roll patterns according to their stability [6].

The hydrodynamic variables can be recovered from the two-dimensional field Ψ as follows:

$$U(x,y,z,t) = \begin{pmatrix} -i\Psi_x(x,y,\tau)u_0(z) \\ -i\Psi_y(x,y,\tau)u_0(z) \\ \Psi(x,y,\tau)w_0(z) \end{pmatrix} + O(\epsilon)$$

$$\Theta(x,y,z,t) = \Psi(x,y,\tau)\Theta_0(z) + O(\epsilon),$$

where u_0, w_0, and Θ_0 are the lowest eigenvalues of the linearized Boussinesq system. Also, the heat transport is known since the Nusselt number can be computed from Ψ to within $O(\epsilon)$.

We are interested in solving the amplitude equation in a rectangular domain $\Omega = [0, L_x] \times [0, L_y]$ with boundary conditions

$$\Psi|_{\partial\Omega} = \Psi_n|_{\partial\Omega} = 0,$$

where the subscript n denotes differentiation with respect to the exterior unit normal. These boundary conditions are consistent with those imposed on the complex amplitude function to lowest order in ϵ [6].

We can apply a backward-Euler discretization in time to give a series of static problems of the form

$$\frac{\Psi(\tau) - \Psi(\tau - \Delta\tau)}{\Delta\tau} = L\Psi(\tau) - \Psi^3(\tau).$$

(We abuse notation by using Ψ as a grid function as well as the continuous variable.) The Laplacian and biharmonic operators in the semi-discrete problem can be spatially discretized on a grid of $n_x \times n_y$ points by the usual second-order five- and thirteen-point formulas respectively. The fully discretized problem requires solving a system of nonlinear equations at each time step; a Newton-like method is an obvious choice to solve the nonlinear system.

The time step $\Delta\tau$ can be regulated by taking a step with $\Delta\tau$, reintegrating with two steps $\Delta\tau/2$, and then estimating the local error while extrapolating [12]. A first-order method is more than adequate for most of our problems; in fact, the low-accuracy requirements and the time-step restrictions necessitated by the Newton-iterative methods described later make first-order methods a reasonable choice. An estimate of the local error at each time step $e(\tau)$ is made to satisfy the inequality

$$\|e(\tau)\|_\infty \leqslant \epsilon_{rel} \|\Psi(\tau)\|_\infty + \epsilon_{abs}.$$

There are efficient algorithms for solving the discrete-biharmonic equation with Dirichlet data

$$\left[\Delta^2 + a\Delta + b\right] u = f_1 \in \Omega$$

$$u|_{\partial\Omega} = f_2$$

$$u_n|_{\partial\Omega} = f_3$$

for fixed scalars a and b, where Δ^2 and Δ now represent the discretizations of the biharmonic and Laplacian operators respectively [3,4]. In particular, there is an implementation of an $O(n_x n_y \log \max(n_x, n_y))$ algorithm for an $n_x \times n_y$ grid on a rectangle. (The solution process makes extensive use of discrete-sine transforms, which can be optimized for a vector machine like the Cray-1 [3,15].)

Given a fast solver for biharmonic equations with constant coefficients, we employ a Newton-iterative method (for example, see [2]) to solve the nonlinear equations

$$g[\Psi(\tau)] \equiv \left[(\Delta\tau)^{-1} - L + \Psi^2(\tau)\right]\Psi(\tau) - (\Delta\tau)^{-1}\Psi(\tau-\Delta\tau) = 0,$$

where L is now the discretization of the continuous L. The Jacobian of this system is

$$J = M - N$$

where

$$M = (\Delta\tau)^{-1} - L,$$

$$N = -3\Psi^2(\tau).$$

Starting with an initial guess of $\Psi_0 = \Psi(\tau-\Delta\tau)$, we approximately solve the Newton equations

$$J(\Psi_j)x_j \equiv J_j x_j = -g(\Psi_j) \equiv -g_j$$

where

$$x_j = \Psi_{j+1} - \Psi_j.$$

We stop the Newton iteration when

$$\|x_{m-1}\|_\infty < \epsilon_{rel}\|\Psi_m\|_\infty + \epsilon_{abs}$$

and then set

$$\Psi(\tau) = \Psi_m.$$

We approximate x_j by repeatedly solving the inner-iteration equations

$$M_j(x_{jk} - x_{j,k-1}) = -(J_j x_{j,k-1} + g_j) \equiv -r_{jk},$$

with $x_{j0} = 0$, until

$$\frac{\|r_{jm_j}\|}{\|g_j\|} \leqslant \alpha\left(\frac{\|g_j\|}{\|g_0\|}\right)^\beta$$

and then setting

$$x_{j+1} = x_j + x_{jm_j}.$$

This stopping criterion for the inner iteration is known to give superlinear convergence (with Q-order $1+\beta$) under certain technical assumptions [2].

In practice, the Jacobian is written as

$$J = (M - c\|\Psi^2\|_\infty) - (N - c\|\Psi^2\|_\infty)$$

and the inner iteration is based on the implied splitting to enhance convergence. The form $c\|\Psi^2\|_\infty$ suggests itself because the linearized model equation supports a primarily sinusoidal variation having an average value of $3\|\Psi^2\|_\infty/2$. The choice of c can substantially alter the convergence rate of the inner iterations, although the iteration is guaranteed to converge for sufficiently small $\Delta\tau$. We have empirically found that $\Delta\tau > \Delta\tau_{max} \approx 1$ will often cause the splitting to diverge. As a result, we have forced $\Delta\tau < \Delta\tau_{max}$. ($\Delta\tau << 1$ during the initial transient but it is constrained by $\Delta\tau_{max}$ as the integration continues.)

We performed a series of computations to investigate the effects of the various parameters. All of the calculations used

$$\epsilon = 0.25$$

$$L_x = L_y = 5\pi$$

$$n_x = n_y = 33$$

$$\epsilon_{abs} = 10^{-4}$$

with the same random initial conditions, forced to obey the boundary conditions by the application of a cosine window and scaled so the spatial average of Ψ^2 was ϵ, in order to activate the nonlinear term. The following table summarizes the numerical tests:

Amplitude-equation results

ϵ_{rel}	c	α	β	Time Steps	Average Outers	Average Inners	$F[\Psi(\tau=30)]$
0.1	1.5	0.1	0.5	21	1.11	1.86	-0.3018
0.01	1.5	0.1	0.5	33	1.38	1.40	-0.3026
0.1	0.75	0.1	0.5	21	1.14	1.73	-0.3029
0.1	3	0.1	0.5	21	1.14	2.59	-0.2996
1	1.5	0.1	0.5	19	1.00	1.95	-0.3012
0.1	1.5	0.025	0.5	21	1.06	2.76	-0.3027
0.1	1.5	0.4	0.5	21	1.15	0.83	-0.2987
0.1	1.5	0.1	1	21	1.11	1.89	-0.3018
0.01	1.5	0.025	1	33	1.16	2.08	-0.3035

We note that, although the relative-error tolerance ϵ_{rel} was fairly large, the computed solutions have a relative error on the order of one to five per cent.

Taking $\beta=0.5$ is somewhat better than taking $\beta=1$; the Newton-iterative method typically provides good linear, rather than quadratic, convergence for $\beta=1$. We find that a moderate value of α between 0.1 and 0.3 provides good performance, with one or two Newton iterations required per time step. Smaller values of α lead to wasted inner iterations whereas larger values require additional work at the time-stepping or Newton-iteration levels. The influence of the splitting constant, c, is easily seen.

Figure 2 illustrates a particular evolution; figure 2b represents a metastable state, which persists for a long time until arithmetic noise disturbs it. They show how a set of rolls parallel to the longer side of the cell slowly evolve into rolls parallel to the shorter side, which have a lower-energy state. Figure 3 shows a plot of the spatial average of Ψ^2 as a function of time.

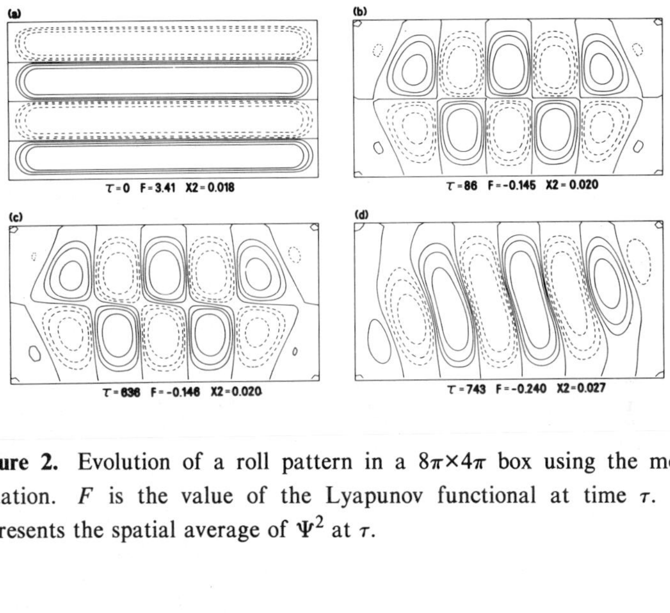

Figure 2. Evolution of a roll pattern in a $8\pi\times4\pi$ box using the model equation. F is the value of the Lyapunov functional at time τ. $X2$ represents the spatial average of Ψ^2 at τ.

Numerical experiments show that the model equation qualitatively reproduces many of the observed experimental features of pattern formation and also predicts some new features [8]. These results are significant for various reasons. First, it is possible to simulate accurately three-dimensional flow on short- and long-length scales via a two-dimensional equation, at least near the onset of convection. Second, the solutions give explicit examples of stationary patterns, an ordering of their relative stability, and the corresponding heat transport, thereby providing an important link between theory and experiment. Finally, the numerical methods presented here have explored a new domain of large cells and long-time scales (simulations have been carried out to 2.5×10^4 natural time units) that, to date, have been inaccessible with three-dimensional codes.

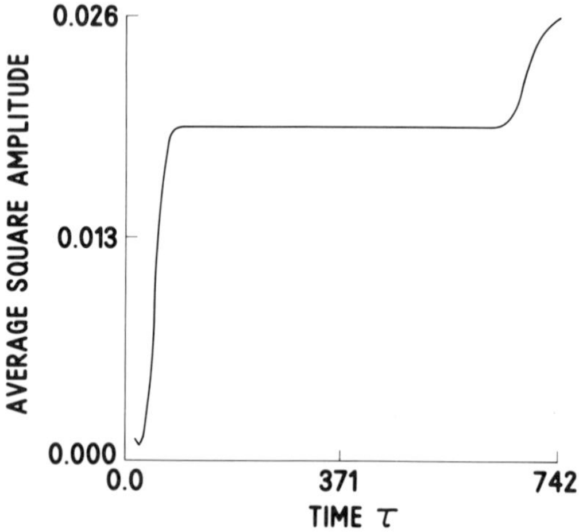

Figure 3. The spatial average of Ψ^2 as a function of time τ, corresponding to figure 2.

3. A Model System with Vorticity

The amplitude equation is purely relaxational and all initial conditions evolve to a time-independent equilibrium state. Terms occur in higher-order multiple-scale analysis that make the resulting equations nonrelaxational and allow the study of the onset of turbulence. A model system that includes the most-important new physical feature, the generation of vertical vorticity from roll curvature, can be derived [9]. This amplitude-vorticity system is given by

$$\Psi_\tau = \left[\epsilon - (\Delta+1)^2 - \sigma^{-2}(\delta_y \partial_x - \delta_x \partial_y)\right] \Psi + \Gamma(\Psi) \equiv L_1(\delta)\Psi + \Gamma(\Psi)$$

$$\Delta\delta = \Psi_y \Delta\Psi_x - \Psi_x \Delta\Psi_y \equiv L_2(\Psi)$$

with boundary conditions

$$\Psi|_{\partial\Omega} = \Psi_n|_{\partial\Omega} = \delta|_{\partial\Omega} = 0.$$

Here δ is a vorticity potential, which gives the vertical-vorticity component $\zeta = -\Delta\delta$. This system admits the long-wavelength skewed-varicose instability, which is suspected to be important in the onset of turbulence [5, 14].

The nonlinear term may be

$$\Gamma(\Psi) = -\Psi^3,$$

which is appropriate for perfectly conducting plates, or

$$\Gamma(\Psi) = \nabla \cdot (\|\nabla \Psi\|_2^2 \nabla \Psi),$$

which is appropriate for poorly conducting plates [11].

We can discretize with backward Euler in time and standard second-order spatial differencing. We can then use a nonlinear Gauss-Seidel iteration to solve the nonlinear system

$$\begin{pmatrix} g_1 \\ g_2 \end{pmatrix} [\Psi(\tau), \delta(\tau)] \equiv \begin{pmatrix} \{(\Delta\tau)^{-1} - L_1[\delta(\tau)]\}\Psi(\tau) - \Gamma[\Psi(\tau)] - (\Delta\tau)^{-1}\Psi(\tau-\Delta\tau) \\ \Delta\delta(\tau) - L_2[\Psi(\tau)] \end{pmatrix} = 0,$$

instead of a Newton-iterative method.

The nonlinear Gauss-Seidel iteration consists of solving for Ψ_{j+1} in

$$g_1(\Psi_{j+1}, \delta_j) = 0$$

given δ_j and then solving for δ_{j+1} in

$$g_2(\Psi_{j+1}, \delta_{j+1}) = 0.$$

(The actual implementation solves for the change in Ψ and δ.) The g_1 equation is very similar to the nonlinear system for the amplitude equation; in particular, the Newton-iterative method and termination criteria described in the previous section are directly applicable. The g_2 equation requires the solution of a Poisson equation with Dirichlet boundary conditions [1].

We performed a series of computations with $\Gamma = -\Psi^3$, $\epsilon = 0.25$, $L_x = L_y = 5\pi$, $n_x = n_y = 33$, $\epsilon_{abs} = 10^{-4}$, and the same random initial conditions, modified as discussed in the previous section. The computation for $0 \leqslant \tau \leqslant 1$ is done with $\sigma^{-2} = 0$ to anneal out deformations in the initial random pattern; for $\tau > 1$, $\sigma^{-2} = 50$ is used. The numerical experiments are summarized in the following table:

Amplitude-vorticity-system results						
ϵ_{rel}	α	β	Time Steps	Average Outers	Average Inners	$F[\Psi(\tau=10)]$
0.1	0.1	0.5	58	1.06	0.89	-0.1447
0.01	0.1	0.5	64	1.78	2.01	-0.1453
1	0.1	0.5	56	1.00	0.98	-0.1446
0.1	0.025	0.5	59	1.06	1.84	-0.1448
0.1	0.4	0.5	56	1.15	0.28	-0.1445
0.1	0.1	1	58	1.06	0.93	-0.1447
0.01	0.025	1	64	1.78	4.07	-0.1454

The computed solutions have a relative error on the order of one to five per cent.

Small values of the inner-iteration parameter α lead to substantially more work than moderate values. The influence of the exponent β seems to be minor. The time steps are much smaller than for the amplitude equation alone since the system exhibits smaller-time-scale behavior and the Gauss-Seidel iteration only converges for sufficiently small $\Delta\tau$.

Figure 4 illustrates a particular evolution in a small box. Defects do not anneal out in the same way as observed with the amplitude equation. A non-periodic "pinching" of the rolls is a predicted new feature and is similar to that found in recent experiments [7]. The spatial average of Ψ^2 as a function of time is shown in figure 5.

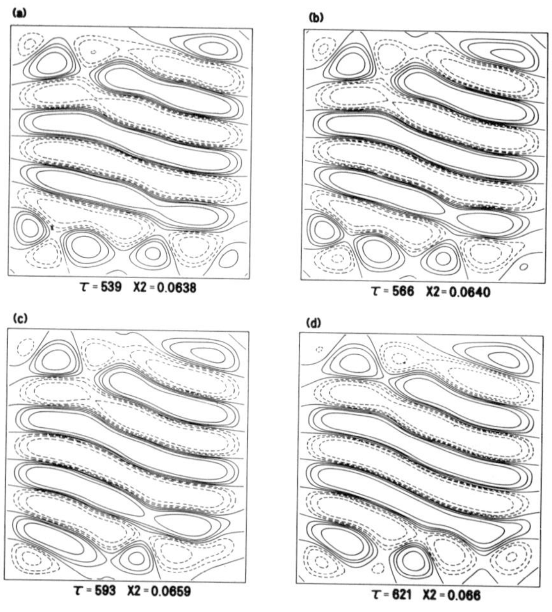

Figure 4. Evolution of a roll pattern in a $5\pi\times5\pi$ box using the augmented system with vorticity. $X2$ represents the spatial average of Ψ^2 at τ.

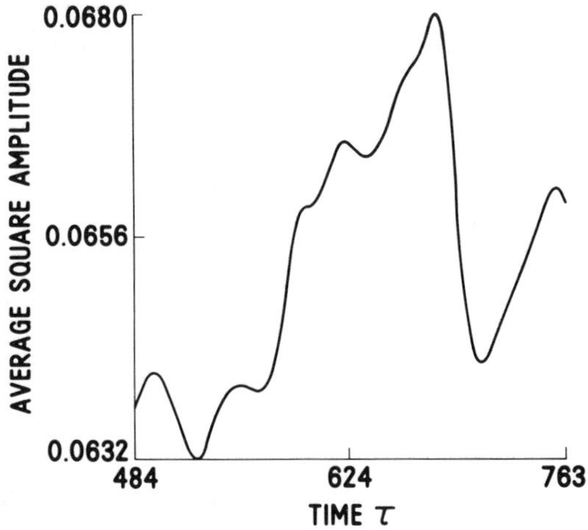

Figure 5. The spatial average of Ψ^2 as a function of time τ, corresponding to figure 4.

The amplitude-vorticity system should model the Boussinesq system more closely than the amplitude equation alone. However, the results of extensive simulations have not yet been analyzed. Our calculations to date show that a nonperiodic time dependence formed by roll pinching can occur for some initial conditions (but not all) for sufficiently large boxes. A linear-stability analysis of the amplitude-vorticity system with various Γ's shows that L_2, the vorticity operator, overemphasizes the cross-roll instability (which is discussed in [5,14]), making it the most dangerous; a suitable set of equations without this problem has not yet been found.

4. Conclusions

The model equation of section 2 approximates the Boussinesq system near the Rayleigh-Bénard instability and agrees quite well with experiment [7]. The amplitude-equation approach reduces the dimensionality of the underlying problem by one and reduces the number of equations from five to one. Although the resulting equation is relaxational, it does qualitatively reproduce the stationary flows found near the convective instability and predicts some new features [8]. The amplitude-vorticity system extends the amplitude equation to include the generation of vertical vorticity without leaving the two-dimensional domain. We are currently studying these equations to see whether this simple generalization can explain some of the properties of the turbulence observed at onset.

The combination of efficient control of the time step, approximate-Newton methods with appropriate criteria for terminating the iterations, and existing software for solving biharmonic and Poisson equations has proven effective for these problems. The application of these ideas has led to a better understanding of Rayleigh-Bénard convection and should be useful for related problems having a continuous instability at a finite wavelength such as dendritic-crystal growth, flame propagation, Taylor-vortex flow, and reaction-diffusion systems.

References

1. R. E. Bank, "Algorithm 527: A Fortran Implementation of the Generalized Marching Algorithm," *ACM Transactions on Mathematical Software* **4**, pp. 165-176 (1978).
2. R. E. Bank and D. J. Rose, "Global Approximate Newton Methods," *Numerische Mathematik* **37**, pp. 279-295 (1981).
3. P. E. Bjørstad, "Numerical Solution of the Biharmonic Equation," Stanford University Computer Science Department Report STAN-CS-80-834 (1980).
4. P. E. Bjørstad, "Fast Numerical Solution of the Biharmonic Dirichlet Problem on Rectangles," *SIAM Journal on Numerical Analysis* **20**, pp. 59-71 (1983).
5. F. H. Busse, "Non-linear Properties of Thermal Convection," *Reports on Progress in Physics* **41**, pp. 1929-1967 (1978).
6. M. C. Cross, "Ingredients of a Theory of Convective Textures Close to Onset," *Physical Review A* **25**, pp. 1065-1076 (1982).
7. J. Gollub, A. R. McCarriar, and J. F. Steinman, "Convective Pattern Evolution and Secondary Instabilities," *Journal of Fluid Mechanics* **125**, pp. 259-281 (1982).
8. H. S. Greenside, W. M. Coughran, Jr., and N. L. Schryer, "Nonlinear Pattern Formation near the Onset of Rayleigh-Bénard Convection," *Physical Review Letters* **49**, pp. 726-729 (1982).
9. P. Mannville, "A Two-Dimensional Model of Three-Dimensional Convection," *Journal de Physique* (in press).
10. A. C. Newell and J. A. Whitehead, "Finite Bandwidth, Finite Amplitude Convection," *Journal of Fluid Mechanics* **38**, pp. 279-303 (1969).
11. L. M. Pismen, *A Reduced Equation for Three-Dimensional Nonlinear Long-Wave Benard Convection*, preprint, 1982.
12. N. L. Schryer, *An Extrapolation Step-Size and Order Monitor for use in Solving Differential Equations*, Proceedings ACM National Meeting, San Diego, 1974.

13. L. A. Segel, "Distant Side-walls Cause Slow Amplitude Modulation of Cellular Convection," *Journal of Fluid Mechanics* **38**, pp. 203-224 (1969).

14. E. D. Siggia and A. Zippelius, "Pattern Selection in Rayleigh-Bénard Convection near Threshold," *Physical Review Letters* **47**, pp. 835-838 (1981).

15. P. N. Swartztrauber, "Vectorizing the FFTs," pp. 51-83 in *Parallel Computations*, ed. G. Rodrigue (1982).

16. J. Swift and P. C. Hohenberg, "Hydrodynamic Fluctuations at the Convective Instability," *Physical Review A* **15**, pp. 319-328 (1977).

APPROXIMATION OF A NONSYMMETRIC, SINGULARLY PERTURBED PROBLEM

E.C. Gartland, Jr.[1]

I. INTRODUCTION

Consider the following model <u>convection-diffusion</u> equation:

$$\begin{aligned} -\varepsilon\Delta\phi + \phi_x &= f(x,y), \quad (x,y) \in R \equiv (0,a) \times (0,b) \\ \phi(x,y) &= g(x,y), \quad (x,y) \in \partial R. \end{aligned} \tag{1}$$

This is a simple example of a type of problem encountered in a variety of applications, which include convective heat transport, vorticity transport, oil reservoir simulation, and the like. The parameter ε is constant, positive, and small compared to a; it measures the relative strength of dispersive to convective effects.

The solution ϕ of (1) is rather complicated but has been thoroughly analyzed. In general, it possesses an $O(\varepsilon)$ boundary layer at the "downstream" end $x = a$ and $O(\varepsilon^{1/2})$ shear layers along the lower and upper edges $y = 0$ and $y = b$ (cf. Eckhaus, 1979). The numerical approximation of this solution is difficult for two reasons: first, it is difficult to construct "good" discretizations of such a strongly nonsymmetric differential operator, and second, the rapid variation of the solution in the boundary layers degrades the performance of any discretization there.

---

[1]Department of Mathematics
Southern Methodist University
Dallas, Texas

ISBN 0-12-100560-7

The goal here is to achieve a uniform approximation to the solution over the entire rectangle. This will consist of a tabulation (produced by a finite-difference approximation) plus a method (local bilinear interpolation) of compatible accuracy for approximation of ϕ off the finite-difference grid. A five-point difference scheme, which combines the properties of "upwinding" and central differencing, is derived using the HODIE Method of Lynch and Rice (1977). This is combined with a decomposition of the solution into "inner" and "outer" components to make the downstream boundary layer more manageable.

II. A HODIE SCHEME

The HODIE Method is a systematic approach to deriving high-order difference approximations by using compact discretizations with several evaluations of the source term (Lynch and Rice, 1977). For the sake of illustration, consider the following one-dimensional model of our problem:

$$\begin{aligned} &-\varepsilon\phi_{xx} + \phi_x = f(x), \qquad 0 < x < a. \\ &\phi(0) = g_0 \quad \text{and} \quad \phi(a) = g_a . \end{aligned} \tag{2}$$

Given three successive mesh points, x_{j-1}, x_j, and x_{j+1} and auxiliary (HODIE) points $x_{j-1} \leq \tau_1 < \tau_2 < \dots < \tau_J \leq x_{j+1}$, one seeks a finite-difference approximation of the form

$$\alpha_{-1}\phi_{j-1} + \alpha_0\phi_j + \alpha_1\phi_{j+1} = w_1 f(\tau_1) + \dots + w_J f(\tau_J) \tag{3}$$

subject to some convenient normalization, like $w_1 + \dots + w_J = 1$. The α's, w's, and τ's (these may be fixed or variable) are chosen so that the scheme is exact on some chosen class of functions (typically polynomials of a certain order).

For example, in the case of a uniform mesh, the HODIE scheme of the form

$$\alpha_{-1}\phi_{j-1} + \alpha_0\phi_j + \alpha_1\phi_{j+1} = f_j$$

that is exact on $\{1,x,x^2\}$ is precisely the standard central difference approximation

$$\left(\frac{-\varepsilon}{h^2} - \frac{1}{2h}\right)\phi_{j-1} + \frac{2\varepsilon}{h^2}\phi_j + \left(\frac{-\varepsilon}{h^2} + \frac{1}{2h}\right)\phi_{j+1} = f_j .$$

The scheme of the form

$$\alpha_{-1}\phi_{j-1} + \alpha_0\phi_j + \alpha_1\phi_{j+1} = f_{j-\frac{1}{2}}$$

exact on the same class is given by

$$\left(\frac{-\varepsilon}{h^2} - \frac{1}{h}\right)\phi_{j-1} + \left(\frac{2\varepsilon}{h^2} + \frac{1}{h}\right)\phi_j + \left(\frac{-\varepsilon}{h^2}\right)\phi_{j+1} = f_{j-\frac{1}{2}} .$$

This three-point stencil is exactly that of <u>upwind differencing</u>. The central difference scheme has a truncation error that is $\mathcal{O}(h^2)$ but for $h > 2\varepsilon$ it is not of <u>monotone type</u>, in the sense of Collatz (1966). For this reason, upwinding (which is unconditionally monotone) is preferred in this range despite the fact that its formal accuracy is $\mathcal{O}(h)$--actually the truncation error of the upwind scheme above is $\mathcal{O}(\varepsilon h + h^2)$, which is "effectively" $\mathcal{O}(h^2)$ when $h > \varepsilon$.

An improved discretization can be derived in the following way. Seek a scheme of the form

$$\alpha_{-1}\phi_{j-1} + \alpha_0\phi_j + \alpha_1\phi_{j+1} = f(\tau_j)$$

with τ_j now <u>variable</u>. By requiring that this be exact for $\{1,x,x^2,x^3\}$, we get

$$\left(-\frac{\varepsilon}{h^2}\sigma - \frac{1}{2h}\right)\phi_{j-1} + \frac{2\varepsilon}{h^2}\sigma\phi_j + \left(-\frac{\varepsilon}{h^2}\sigma + \frac{1}{2h}\right)\phi_{j+1} = f(x_j - \varepsilon(\sigma - 1))$$
$$\sigma = \sqrt{1 + \frac{1}{3}\left(\frac{h}{\varepsilon}\right)^2} . \qquad (4)$$

This scheme is $O(h^2)$, uniformly in ε, and unconditionally monotone. The variable HODIE point, τ_j, lies upwind of x_j between $x_j - h/\sqrt{3}$ and x_j converging to the former point as $\varepsilon \to 0$ and to the latter as $h \to 0$.

This one-variable-HODIE-point approximation successfully blends together the desirable properties of upwind and central differences. The two-dimensional version of (4) was used in the numerical experiments reported in the sequel and was found to provide an improvement in accuracy of about 15% (on the average) over the two standard schemes. The form of the stencil and the location of the variable evaluation point are consistent with the "upwind nature" of the problem. Why this should be so can be seen from an examination of an exact local representation of the solution.

If we restrict our attention to the subinterval $[x_{j-1}, x_{j+1}]$ and treat the values of $\phi(x_{j-1})$ and $\phi(x_{j+1})$ as known, then we can regard the solution ϕ of (2) as the solution of the subproblem

$$-\varepsilon\phi_{xx} + \phi_x = f(x), \qquad x_{j-1} < x < x_{j+1},$$

$$\phi(x_{j-1}) = \phi_{j-1} \quad \text{and} \quad \phi(x_{j+1}) = \phi_{j+1}.$$

For $x_{j-1} \leq x \leq x_{j+1}$, this solution can be represented

$$\phi(x) = \phi_{j-1}s_j(x) + \phi_{j+1}t_j(x) + \int_{x_{j-1}}^{x_{j+1}} g_j(x,y)f(y)\,dy,$$

where $\{s_j, t_j\}$ is an appropriate basis for the null space of the differential operator and g_j is the local Green's function. Evaluation at $x = x_j$ produces the exact three-point scheme

$$-s_j(x_j)\phi_{j-1} + \phi_j - t_j(x_j)\phi_{j+1} = \int_{x_{j-1}}^{x_{j+1}} g_j(x_j,y)f(y)\,dy.$$

The expression $w_1 f(\tau_1) + \ldots + w_J f(\tau_J)$ in the general HODIE scheme (3) is an approximation to the quadrature above. The "optimal" HODIE points are the Gauss points with respect to this weighted quadrature, with weight function $g_j(x_j,y)$. If the differential operator were self-adjoint, then $g_j(x_j,\cdot)$ would be symmetric and would typically look like the figure below (Fig. 1), indicating that the optimal HODIE points would be symmetrically distributed throughout the interval. For our (non-self-adjoint) convection-diffusion operator, however, $g_j(x_j,\cdot)$ is strongly skewed in the upwind direction (Fig. 2), indicating that for $h >> \varepsilon$ the optimal points are located upwind. These ideas are exploited, for the case of one-dimensional singular-perturbation problems, in Gartland (1983).

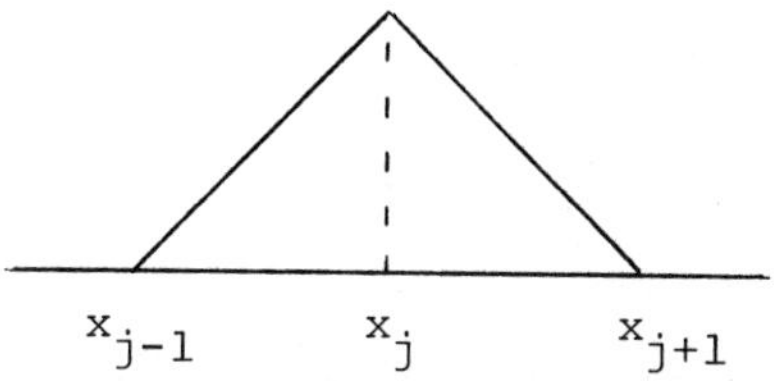

FIGURE 1
Symmetric Green's function

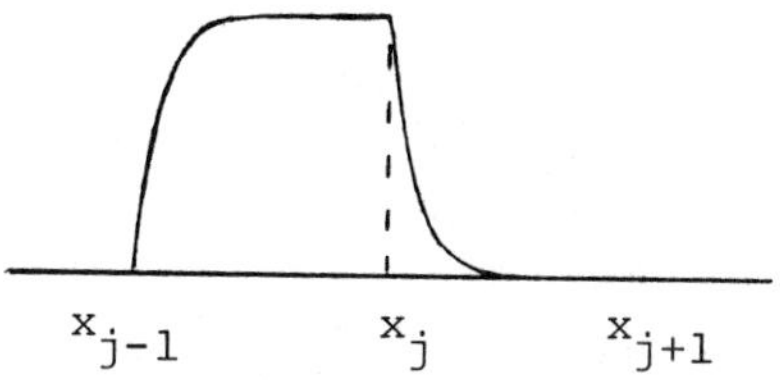

FIGURE 2
Convection-Diffusion Green's function

III. INNER/OUTER DECOMPOSITION

Because of the very different natures of the solution in the regions outside of and inside of the downstream boundary layer, it is to our advantage to somehow break up the problem accordingly; this will allow us to attack two separate problems each of a consistent character. This can be accomplished as follows.

A. The Decomposition

Let $\partial R = \Gamma_1 \cup \Gamma_2$ be a decomposition of the boundary of R into "upstream," $\Gamma_1 = \{(0,y) : 0 \le y \le b\} \cup \{(x,y) : y = 0 \text{ or } b, 0 \le x \le a\}$, and "downstream," $\Gamma_2 = \partial R \setminus \Gamma_1$, components. Consider the following equations satisfied by auxiliary functions u, v, and w:

$$\begin{aligned} u_x &= \varepsilon u_{yy} + f(x,y), && (x,y) \in R, \\ u(x,y) &= g(x,y), && (x,y) \in \Gamma_1 ; \end{aligned} \tag{5}$$

$$\begin{aligned} -\varepsilon \Delta v + v_x &= f(x,y), && (x,y) \in R, \\ v(x,y) &= g(x,y), && (x,y) \in \Gamma_1 , \\ v(x,y) &= u(x,y), && (x,y) \in \Gamma_2 ; \end{aligned} \tag{6}$$

$$\begin{aligned} -\varepsilon \Delta w + w_x &= 0 , && (x,y) \in R, \\ w(x,y) &= 0 , && (x,y) \in \Gamma_1 , \\ w(x,y) &= g(x,y) - u(x,y), && (x,y) \in \Gamma_2 . \end{aligned} \tag{7}$$

The solution to (1) is then given by $\phi = v + w$.

The "parabolic outer approximation" u has no x diffusion. It is used to provide a boundary-layer-free downstream boundary condition for v. The "outer solution" v, then, has no regular boundary layer at $x = a$. For this reason, it can be approximated to high accuracy much more easily than the original ϕ. The "inner solution" w adjusts to satisfy the true downstream boundary condition, and it dies off rapidly upstream.

This decomposition is very similar in spirit to but differs slightly from the approach of <u>singular perturbation theory</u>, where one would take as outer approximation the solution to the $\varepsilon = 0$ case, $u_x = f(x,y)$ in R. The parabolic approximation (5) has the advantage, from a computational point of view, of carrying with it the boundary data from the top and bottom edges.

B. The Parabolic Outer Approximation

The numerical approximation of the solution u of (5) can be efficiently accomplished by either explicit or implicit methods. The equation is of parabolic type with x the "time" variable. A natural (and satisfactory) choice of integrator is the Crank-Nicolson scheme. The presence of the small parameter ε makes it also possible to use explicit methods effectively. The stability condition for the simplest explicit one-step method is given by $\Delta x \leq \Delta y^2/(2\varepsilon)$. So we are allowed to take a much larger x spacing than such conditions normally allow. This is helpful because considerations of truncation error and the error in local bilinear interpolation (using asymptotic approximations for u and its derivatives) suggest that it is most efficient to use a mesh spacing where $\Delta y/\Delta x \propto \varepsilon^{\frac{1}{2}}$. In the numerical experiments conducted, the performances of the explicit and the Crank-Nicolson schemes were very nearly equal.

The shear layers, which can develop along the top and bottom edges, were not present in the model problem on which the numerical experiments were performed because for this problem (described later) the boundary term g was continuous at the upwind corners (0,0) and (0,b) and constant along the top and bottom edges. The additional level of complication provided by the presence of these layers can be handled, without much difficulty, by appropriately grading the y mesh in the vicinity of these edges.

C. *The Outer Solution v*

The problem (6) for the outer solution v is an elliptic, convection-dominated problem; it has the advantages that its solution has no downstream boundary layer and that the convection is parallel to an axis. In such circumstances, a five-point stencil with upwind differencing is satisfactory; however the one-variable-HODIE-point scheme performs better and was used below. The solution can be approximated on the same mesh as the u solution--it can use the same array since once the downstream u values have been obtained the interior values are no longer needed.

In the employment of iterative methods to solve the resulting algebraic equations, we have the advantage that a very good initial approximation to v is available from the approximate u solution. In fact, a simple application of the Maximum Principle gives us that

$$\| u - v \|_\infty \leq \varepsilon \, a \, \| u_{xx} \|_\infty \ .$$

The stencil is nonsymmetric with a strong upwind bias. It is of the usual diagonally dominant or monotone type, and very efficient methods (as discussed elsewhere in these proceedings) have been developed for such problems. The features of having a strong upwind bias and having few vertical mesh lines make a certain direct method, the marching algorithm of Bank and Rose (1977), a viable alternative solution method. This method is inherently unstable--exponential growth of roundoff errors--but in this situation can be used to solve modest sized problems (40 by 20 or 50 by 10, say, on a CDC 6600) (cf. Gartland, 1982).

D. The Inner Solution w

The solution w of (7) contains the boundary layer correction and decays rapidly upstream; it readily follows from the Maximum Principle that

$$|w(x,y)| \leq \| g(a,\cdot) - u(a,\cdot)\|_{\infty} \cdot e^{\frac{x-a}{\varepsilon}} .$$

Because of this, w can be approximated on a <u>reduced domain</u> $(a-\ell,a) \times (0,b)$, where the width ℓ of this subdomain must contain about $2\frac{1}{3}\ \varepsilon$ for each desired decimal digit of accuracy-- $\ell = 10\varepsilon$ is typical.

To accurately resolve this boundary solution with uniform x and y meshes would require an extremely fine x mesh (roughly $\Delta x/\Delta y \propto \varepsilon$) and a very large number of mesh points, even on this reduced domain. It is more efficient to use a graded mesh in the x direction that clusters the nodes near the $x = a$ edge. There seems to be little point in trying to construct an "optimal" mesh: it would have to be done adaptively, making it too expensive, and the error is not very sensitive to the particular mesh grading function. The <u>arctangent scheme</u> of Israeli (Orszag and Israeli, 1974) was found to perform satisfactorily.

IV. NUMERICAL RESULTS AND CONCLUSIONS

Since we have a good amount of information about the solution ϕ (and u,v, and w), it is not hard to generate "good numbers." As an example of attainable accuracy per amount of work, consider the following table for a model problem (with a downstream, regular boundary layer but no shear layers) with $\varepsilon = .01$, in which case the leading order asymptotic approximation gives an error of .89(-1).

The model problem is the following:

$$-\varepsilon\Delta\phi + \phi_x = 0, \quad 0 < x,y < 1$$

$$\phi(0,y) = 1 + \sin \pi y, \quad \phi(1,y) = 1 + 2 \sin \pi y, \quad 0 < y < 1$$

$$\phi(x,0) = \phi(x,1) = 1, \quad 0 < x < 1,$$

the solution of which is given by

$$\phi(x,y) = 1 + \exp\left(\frac{x-1}{\varepsilon}\right) [2 \sinh \tau x - \exp\left(\frac{1}{2\varepsilon}\right) \sinh \tau(x-1)] \times \sin \pi y/\sinh \tau.$$

Here $\tau = (1 + 4\pi^2\varepsilon^2)^{1/2}/2\varepsilon$.

TABLE 1. Maximum Error (Discretization plus Interpolation) for Model Problem with $\varepsilon = .01$ and $\ell = .1$

Outer Mesh ($N_y \times N_x$)	Inner Mesh ($N_y \times N_x$)	Maximum Error: Upwind Differences (outer) Central Differences (inner)	Maximum Error: HODIE
16 × 4	16 × 16	.44(-2)	.37(-2)
32 × 8	32 × 32	.11(-2)	.93(-3)
64 × 16	64 × 64	.27(-3)	.23(-3)
128 × 32	128 × 128	.69(-4)	.58(-4)

We can conclude that the simple HODIE scheme used is a promising step towards developing high-order discretizations for these types of problems. It outperforms, by a noticeable amount, the standard upwind scheme in the outer region and central differences in the inner region, successfully blending together the desirable properties of each. Decomposing the solution into inner and outer components can only be expected to improve the efficiency of solution--it can never do worse.

REFERENCES

Bank, R.E. and Rose, D.J. (1977), SIAM J. Numer. Anal. 14, 792.

Collatz, L. (1966). "The Numerical Treatment of Differential Equations." Springer-Verlag, Berlin.

Eckhaus, W. (1979). "Asymptotic Analysis of Singular Perturbations." North-Holland, Amsterdam.

Gartland, E.C. (1982). SIAM J. Sci. Stat. Comput. 3, 460.

Gartland, E.C. (1983). In preparation.

Lynch, R.E. and Rice, J.R. (1977). CSD-TR 223, Department of Computer Science, Purdue University.

NUMERICAL SOLUTION OF A CONFINED LAMINAR DIFFUSION FLAME*

Mitchell D. Smooke**, Reginald E. Mitchell** and Joseph F. Grcar**

I. INTRODUCTION

The flame type of most practical combustion devices is the diffusion flame. Although diffusion flames are important in combustion applications, they have received relatively little attention in theoretical flame studies. Part of this neglect is due to the two-dimensional nature of the problem coupled with the complexities associated with the combined effects of transport phenomena and chemical processes. Theoretical models are needed, however, to obtain an understanding of the factors that influence flame structure and chemistry, in particular, pollutant formation during combustion.

Many of the numerical studies that attempt to predict the concentration, thermal and aerodynamic distributions established in diffusion flames employ a time dependent relaxation approach (see e.g., Mitchell et al., 1980). In such methods,

---

*This work supported by DOE.

**Sandia National Laboratories, Livermore, California.

ISBN 0-12-100560-7

the governing equations are discretized using finite difference techniques and the solution is advanced in time until steady-state profiles are obtained. However, the slow relaxation of time-dependent methods to steady-state can make parameter studies employing complicated nonlinear transport properties and chemical kinetics extremely time consuming. Steady-state methods, on the other hand, have the potential of solving the diffusion flame equations in much less time. In this paper we develop a numerical method which enables us to integrate the two-dimensional, time independent equations directly. We apply a global finite difference method in which the nonlinear difference equations are solved by a damped-modified Newton method. Grid points are inserted adaptively in regions where the dependent solution components exhibit high spatial activity. This procedure is continued on successively finer grids until several termination criteria are satisfied. We apply the method to a methane-air flame in which we utilize the flame sheet assumption of the Burke and Schumann model (1928).

II. PROBLEM FORMULATION

Fundamentally, diffusion flames exist whenever fuel and oxidizer gases are separated prior to ignition. In Figure 1 we illustrate a laminar fuel jet discharging into a laminar air stream. The tubes through which the fuel and oxidizer flow are concentric and have radii R_I and R_O, respectively. The two gases make contact at the outlet of the inner tube; the resulting flame has a definite shape similar to that of a candle flame. Our goal in such problems is to solve numerically the appropriate conservation laws governing the transport of mass, energy and momentum in the system. As a result, we obtain theoretical predictions of the mass fractions, $Y_k, k=1,2,...,N$, the temperature T, the density ρ, and the radial and axial velocities v_r and v_z as functions of the independent coordinates r and z. The computational domain is the rectangular region enclosed by $0 \leq r \leq R_O$, $0 \leq z \leq L$, where L is chosen large enough so that we can realistically apply the boundary conditions in (2.10).

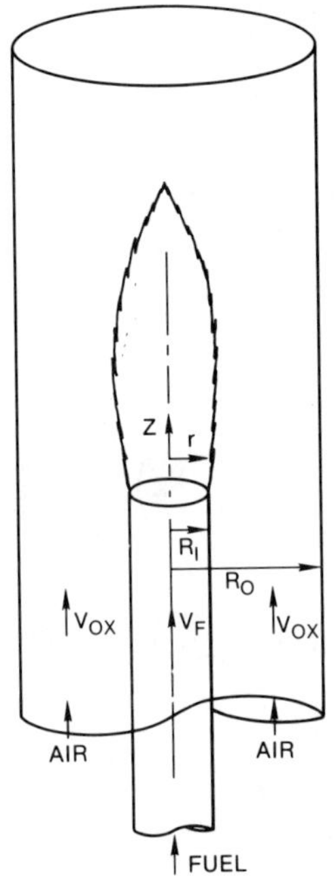

Figure 1. Physical model of a gaseous laminar jet diffusion flame in concentric coflowing annular airstream.

The complete set of governing equations is derived from well established conservation laws of mass, momentum, and energy transfer. We can reduce the size of the system to be solved if we introduce the vorticity and stream function (Gosman et al., 1969). In particular, formulation of the vorticity transport equation serves to eliminate the pressure as one of the dependent variables. The vorticity ω is a measure of the amount of counter clockwise rotation possessed by a fluid element and is defined as

$$\omega = \frac{\partial v_r}{\partial z} - \frac{\partial v_z}{\partial r} . \tag{2.1}$$

The stream function ψ is used to replace the radial and axial components of the velocity vector by a single function. It is defined in such a way that the continuity equation is identically satisfied. We have

$$\rho r v_r = - \frac{\partial \psi}{\partial z} ; \qquad \rho r v_z = \frac{\partial \psi}{\partial r} . \tag{2.2}$$

As a result, our governing equations become

Stream Function:

$$\frac{\partial}{\partial z}\left(\frac{1}{r\rho}\frac{\partial \psi}{\partial z}\right) + \frac{\partial}{\partial r}\left(\frac{1}{r\rho}\frac{\partial \psi}{\partial r}\right) = -\omega \ . \tag{2.3}$$

Species Conservation:

$$\begin{aligned} &-\frac{\partial}{\partial r}\left(\frac{\partial \psi}{\partial z} Y_k\right) + \frac{\partial}{\partial z}\left(\frac{\partial \psi}{\partial r} Y_k\right) - \frac{\partial}{\partial r}\left(r\rho D_{k,mix}\frac{\partial Y_k}{\partial r}\right) \\ &\quad - \frac{\partial}{\partial z}\left(r\rho D_{k,mix}\frac{\partial Y_k}{\partial z}\right) - r\dot{w}_k = 0, \quad k=1,2,\ldots,N-1, \end{aligned} \tag{2.4}$$

and $\qquad Y_N = 1 - \sum_{k=1}^{N-1} Y_k \ .$

Energy:

$$\begin{aligned} &- c_p \frac{\partial}{\partial r}\left(\frac{\partial \psi}{\partial z} T\right) + c_p \frac{\partial}{\partial z}\left(\frac{\partial \psi}{\partial r} T\right) - \frac{\partial}{\partial r}\left(r\lambda \frac{\partial T}{\partial r}\right) - \frac{\partial}{\partial z}\left(r\lambda \frac{\partial T}{\partial z}\right) \\ &- \sum_{k=1}^{N}\left\{r\rho D_{k,mix}\, c_{pk}\left(\frac{\partial T}{\partial r}\frac{\partial Y_k}{\partial r} + \frac{\partial T}{\partial z}\frac{\partial Y_k}{\partial z}\right) - rh_k\dot{w}_k\right\} = 0 \ . \end{aligned} \tag{2.5}$$

Vorticity:

$$\begin{aligned} &r\frac{\partial}{\partial z}\left(\omega \frac{\partial \psi}{\partial r}\right) - r\frac{\partial}{\partial r}\left(\omega \frac{\partial \psi}{\partial z}\right) - \frac{\partial}{\partial z}\left(r^3 \frac{\partial}{\partial z}\left(\frac{\mu}{r}\omega\right)\right) \\ &\quad - \frac{\partial}{\partial r}\left(r^3 \frac{\partial}{\partial r}\left(\frac{\mu}{r}\omega\right)\right) + \omega\frac{\partial \psi}{\partial z} - r\,S = 0 \ , \end{aligned} \tag{2.6}$$

where

$$S = - r\bar{g}\,\frac{\partial \rho}{\partial r} - 2\,\frac{\partial \mu}{\partial r}\,\frac{\partial}{\partial z}\left[\frac{1}{r}\,\frac{\partial}{\partial r}\,(rv_r) + \frac{\partial v_z}{\partial z}\right]$$

$$+ 2\,\frac{\partial \mu}{\partial z}\,\frac{\partial}{\partial r}\left[\frac{1}{r}\,\frac{\partial}{\partial r}\,(rv_r) + \frac{\partial v_z}{\partial z}\right] \tag{2.7}$$

$$+ 2\,\frac{\partial^2 \mu}{\partial z^2}\,\frac{\partial v_z}{\partial r} - 2\,\frac{\partial^2 \mu}{\partial r^2}\,\frac{\partial v_r}{\partial z} + 2\,\frac{\partial^2 \mu}{\partial z\,\partial r}\left(\frac{\partial v_r}{\partial r} - \frac{\partial v_z}{\partial z}\right)$$

$$- r\,\frac{\partial \rho}{\partial z}\left(v_r\,\frac{\partial v_r}{\partial r} + v_z\,\frac{\partial v_z}{\partial r}\right) + r\,\frac{\partial \rho}{\partial r}\left(v_r\,\frac{\partial v_r}{\partial z} + v_z\,\frac{\partial v_z}{\partial z}\right),$$

and where values for v_r and v_z can be determined from (2.2).

Equation of State:

$$\rho = \frac{P\bar{W}}{RT}\,. \tag{2.8}$$

The boundary conditions are given by

Axis of Symmetry (r=O):

$$\frac{\partial T}{\partial r} = \frac{\partial Y_k}{\partial r} = \psi = \omega = 0, \qquad k=1,2,\ldots,N-1. \tag{2.9}$$

Exit ($z\to\infty$):

$$\frac{\partial T}{\partial z} = \frac{\partial Y_k}{\partial z} = \frac{\partial \psi}{\partial z} = \frac{\partial \omega}{\partial z} = 0, \qquad k=1,2,\ldots,N-1. \tag{2.10}$$

Inlet (z=O):

$r < R_I$

$$T = T_F$$

$$Y_F = 1.0,\ \text{all other } Y_k = 0,$$

$$\psi = \rho_F\, v_F\, r^2/2, \tag{2.11}$$

$$\omega = 0\,,$$

$R_I < r < R_O$

$$T = T_{OX} ,$$
$$Y_F = 0 , \quad Y_{N_2} = .768, \quad Y_{O_2} = .232 ,$$
$$\psi = \frac{1}{2} \rho_F v_F R_I^2 + \frac{1}{2} \rho_{OX} v_{OX} (r^2 - R_I^2) , \tag{2.12}$$
$$\omega = 0 .$$

Wall ($r=R_O$):

$$T = T_{wall},$$
$$\frac{\partial Y_k}{\partial r} = 0, \qquad k=1,2,\ldots,N-1,$$
$$\psi = \frac{1}{2} \rho_F v_F R_I^2 + \frac{1}{2} \rho_{OX} v_{OX} (R_O^2 - R_I^2), \tag{2.13}$$
$$\omega = \frac{1}{3} (R_O - R_{NP}) \frac{\partial \omega}{\partial r} + \frac{2}{R_O \rho (R_O - R_{NP})} \frac{\partial \psi}{\partial r} ,$$

where R_{NP} is the grid point nearest the wall.

In addition to the variables already defined, $D_{k,mix}$ denotes the diffusion coefficient of species k with respect to the mixture; $\dot{w}_k$, the rate of production of the kth species by chemical reaction; c_p, the specific heat capacity of the mixture at constant pressure; λ, the thermal conductivity of the mixture; c_{pk}, the specific heat capacity of the kth species at constant pressure; h_k, the specific enthalpy of the kth species with respect to the mixture of remaining species; μ, the viscosity of the mixture; g the gravitational constant; $\overline{W}$, the mean molecular weight of the mixture; R, the universal gas constant; $Y_{F(OX)}$, $\rho_{F(OX)}$, $T_{F(OX)}$, $v_{F(OX)}$, the mass fraction of the fuel (oxidizer), the density of the fuel (oxidizer), the temperature of the fuel (oxidizer) and the incoming velocity of the fuel (oxidizer) respectively and T_{wall} the wall temperature.

III. METHOD OF SOLUTION

The set of equations in (2.3-2.13) forms a set of nonlinear elliptic boundary value problems. Our goal is to obtain a discrete solution of these equations on the mesh $\mathcal{M}$, the nodes of which are formed by intersection of the lines given by the mesh $\mathcal{M}_r$

$$\mathcal{M}_r = \left\{ 0 = r_0 < r_1 < \ldots < r_{M_r} = R_0 \right\}, \tag{3.1}$$

and the mesh $\mathcal{M}_z$

$$\mathcal{M}_z = \left\{ 0 = z_0 < z_1 < \ldots < z_{M_z} = L \right\}. \tag{3.2}$$

We let, $h_r = r_i - r_{i-1}$, $i=1,2,\ldots,M_r$, and $h_z = z_j - z_{j-1}$, $j=1,2,\ldots,M_z$.

A. *Newton's Method*

By replacing the differential operators in (2.3-2.13) by appropriate difference expressions, we convert the problem of finding an analytic solution to this set of equations to one of finding an approximation to it at each point of the mesh $\mathcal{M}$. We employ a nine point difference stencil and in particular we approximate first derivative terms using upwind difference expressions. If we denote the vector of dependent variables at each point (r_i, z_j) of the mesh $\mathcal{M}$ by $U = (U_{ij})$, then we seek the solution of the system of nonlinear difference equations

$$F(U) = 0, \tag{3.3}$$

where the $N(M_r+1)(M_z+1)$ vector F corresponds to the discretization of the governing equations and boundary conditions at each point of the mesh $\mathcal{M}$. The nonlinear equations can be solved by Newton's method. We write

$$J(U^k)\,(U^{k+1} - U^k) = -\,\lambda^k\, F(U^k), \qquad k=0,1,\ldots, \tag{3.4}$$

where U^k denotes the kth solution iterate, λ^k the kth damping parameter $(0 < \lambda^k \leq 1)$ and $J(U^k) = \partial F(U^k)/\partial U$ the Jacobian matrix. When the cost of forming the Jacobian matrix (either analytically or numerically) is a significant part of the cost of a full Newton step, we apply the modified Newton method in which the Jacobian matrix is only periodically re-evaluated. We employ an error estimator (Smooke, 1983) to determine whether the sequence of successive modified Newton iterates is converging at the appropriate rate. If the rate of convergence is too slow we revert back to a full Newton method and employ a damping strategy.

B. *Adaptive Gridding*

Problems similar in form to (2.3-2.13) often give rise to solutions which exhibit regions of high spatial activity. It is important for reasons of accuracy and computational efficiency that grid points be placed adaptively in these regions. In two-dimensional problems the use of an equispaced or mildly nonuniform grid can require an excessively large number of points to obtain a solution to a preset level of accuracy.

We determine the grid points of the mesh $\mathcal{M}$ by equidistributing positive weight functions over mesh intervals in both the r and z directions. Specifically, we attempt to equidistribute the mesh $\mathcal{M}_r$ with respect to the non-negative function W_r and constant C_r for each of the M_z+1 horizontal grid lines. We write

$$\int_{r_i}^{r_{i+1}} W_r \, dr \leq C_r, \qquad i=0,1,\ldots,M_r-1, \tag{3.5}$$

for $j=0,1,\ldots,M_z$. Similarly, we attempt to equidistribute the mesh $\mathcal{M}_z$ with respect to the non-negative function W_z and constant C_z for each of the M_r+1 vertical grid lines. We have

$$\int_{z_j}^{z_{j+1}} W_z \, dz < C_z, \qquad j=0,1,\ldots,M_z-1, \tag{3.6}$$

for $i=0,1,\ldots,M_r$. The choice of the functions W_r and W_z

is somewhat flexible. They should be chosen, however, such that grid points are placed in regions of high spatial activity with the goal of reducing the local discretization (and hopefully global) error. We equidistribute a weighted combination of the difference in the computed solution and its gradient between consecutive mesh intervals in both the r and z directions (W_r and W_z are weighted combinations of the first and second derivatives of the computed solutions). In implementing this adaptive mesh algorithm, we first solve the boundary value problem on an initial coarse grid. We then test the inequality in (3.5) one r subinterval at a time for all the "j" grid lines and all the dependent solution components. If the inequality is not satisfied a grid point is inserted at the midpoint of the r subinterval in question for $j=0,1,\ldots,M_z$. Once this procedure has been carried out in the r direction, we check the inequality in (3.6) one z subinterval at a time for all "i" grid lines and insert points at the midpoints of the z intervals in question. While there can be a loss of efficiency due to the introduction of "unwanted" nodes in this two-dimensional global grid refinement procedure, the Jacobian matrix formed in this type of algorithm can be evaluated efficiently on vector machines (see below). Once the inequalities in (3.5-3.6) are satisfied, the previously converged numerical solution is interpolated onto the new mesh and the result serves as an initial solution estimate for Newton's method on this finer grid. The process is continued until the inequalities are satisfied.

C. Computational Considerations

The Jacobian matrix in (3.4) can be written in block nine diagonal form. For problems involving complicated transport coefficients and complex chemistry, it is often more efficient to evaluate the Jacobian numerically as opposed to analytically. The numerical method we implement extends the procedure outlined by Curtis, Powell, and Reid (1974). The idea is to form several columns of the Jacobian simultaneously using vector function evaluations and the Jacobian's given sparsity structure. We point out that, while the Jacobian can be assembled one node at a time, such a procedure is not as efficient computationally when implemented on a vector machine as forming the Jacobian with vector function evaluations (the total number of function component evaluations being the same in both cases). If to each column of the Jacobian we associate the i and j values of the node corresponding to the column's diagonal block, then all columns of the Jacobian having the same value of the parameter

$$\alpha = (i+3j) \bmod 9, \tag{3.7}$$

can be evaluated simultaneously. Ideas along these lines have been explored in a more abstract setting by Newsam and Ramsdell (1981) and Coleman and More (1982). Once the Jacobian is formed we solve the Newton equations with a block line SOR method.

We terminate the Newton iteration when

$$||\Delta U^{k+1}||_2 = ||U^{k+1} - U^k||_2 \leq \text{TOL}, \qquad k=0,1,2,..., \tag{3.8}$$

where we typically take TOL $\leq 10^{-5}$.

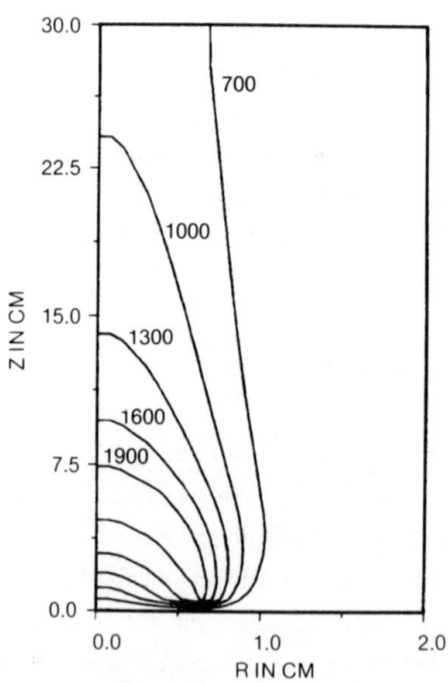

Figure 2. Temperature isotherms (K) for the methane-air diffusion flame.

IV. NUMERICAL RESULTS

In this section we apply the solution method to an atmospheric pressure methane-air flame (Mitchell et al., 1980). To reduce the computational complexities, the numerical study utilizes the major feature of the Burke and Schumann model (1928). The flame front is represented as a sheet at which fuel reacts instantaneously with a stoichiometric amount of oxygen. At the fuel-oxygen interface, chemical reaction is assumed to be infinitely fast and to be represented by a single, irreversible global reaction which accounts for the overall stoichiometry of the reactions considered. An assumed temperature profile (see Figure 2) was used to help generate a numerical solution for the remaining dependent variables. The calculation was performed on an initial 11x14 nonuniform grid, and after four grid refinements we obtained a solution on a 30x31 grid. The calculation took approximately 10 minutes of

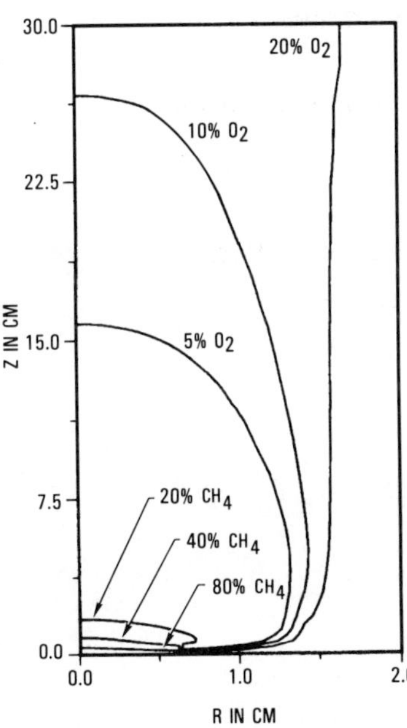

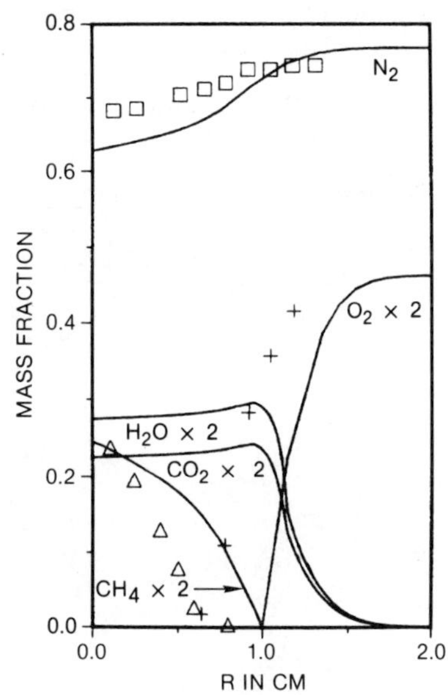

Figure 3. (left) Calculated methane-oxygen isopleths (mass fraction) for the methane-air diffusion flame.

Figure 4. (right) Calculated (solid lines) N_2, CH_4, O_2, H_2O, and CO_2 profiles and experimental data for N_2 (□), CH_4 (Δ) and O_2 (+) at a height 2.4 cm above the inlet.

CPU time on a CRAY-1S. In Figure 3 we illustrate the calculated methane-oxygen isopleths, and in Figure 4 we compare our calculations with the experimental measurements of Mitchell (1975) at a height of 2.4 cm above the inlet.

REFERENCES

Burke, S. P., and Schumann, T. E. W. Indust. Eng. Chem. 29 (1928), 998.

Coleman, T. F. and More, J. J. Cornell University Technical Report #TR-82-502, 1982.

Curtis, A. R., Powell, M. J. D. and Reid, J. K. J. Inst. Math. Appl. 13 (1974), 117.

Gosman, A. D., Pun, W. M., Runchal, A. K. and Spalding, D. B. "Heat and Mass Transfer in Recirculating Flows," (1969), Academic Press, London.

Mitchell, R. E. Sc.D. Thesis (1975), MIT.

Mitchell, R. E., Sarofim, A. E. and Clomburg, L. A. Comb. and Flame 37 (1980), 227.

Newsam, G. N. and Ramsdell, J. D. Harvard University Technical Report (1981) #&R-17-81.

Smooke, M. D. J. Comp. Phys. 48 (1982), 72.

Smooke, M. D. J. Opt. Theory and Applications, in press, (1983).

Index